HANDBOOK OF MATHEMATICAL CONCEPTS AND FORMULAS

FOR STUDENTS IN SCIENCE AND ENGINEERING

HANDBOOK OF MATHEMATICAL CONCEPTS AND FORMULAS FOR STUDENTS IN SCIENCE AND ENGINEERING

Mohammad Asadzadeh
Reimond Emanuelsson
Chalmers University of Technology, Sweden

World Scientific

NEW JERSEY • LONDON • SINGAPORE • BEIJING • SHANGHAI • HONG KONG • TAIPEI • CHENNAI • TOKYO

Published by

World Scientific Publishing Europe Ltd.
57 Shelton Street, Covent Garden, London WC2H 9HE
Head office: 5 Toh Tuck Link, Singapore 596224
USA office: 27 Warren Street, Suite 401-402, Hackensack, NJ 07601

Library of Congress Cataloging-in-Publication Data
Names: Asadzadeh, Mohammad, author. | Emanuelsson, Reimond, author.
Title: Handbook of mathematical concepts and formulas for students in science and engineering / Mohammad Asadzadeh, Reimond Emanuelsson, Chalmers University of Technology, Sweden.
Description: New Jersey : World Scientific, [2024] | Includes bibliographical references and index.
Identifiers: LCCN 2022045046 | ISBN 9781800613317 (hardcover) | ISBN 9781800613324 (ebook) | ISBN 9781800613331 (ebook other)
Subjects: LCSH: Mathematics. | Science--Mathematics. | Engineering mathematics.
Classification: LCC QA37.3 .A83 2024 | DDC 510--dc23/eng20221230
LC record available at https://lccn.loc.gov/2022045046

British Library Cataloguing-in-Publication Data
A catalogue record for this book is available from the British Library.

For any available supplementary material, please visit
https://www.worldscientific.com/worldscibooks/10.1142/Q0393#t=suppl

Desk Editors: Soundararajan Raghuraman/Adam Binnie/Shi Ying Koe

Typeset by Stallion Press
Email: enquiries@stallionpress.com

Preface

In view of the challenges in efficient use of mathematical concepts/formulas for university students, a somewhat comprehensive text in this area would serve as a useful tool. In this regard, we find that including some proofs/definitions and examples in a handbook of formulas would help for a better understanding of the context of the introduced concepts. To this approach we have designed a layout emphasizing the following aspects as

1. Augmented coverage of the topics of mathematical concepts/formulas:
   a. We include material beyond the elementary concepts in undergraduate math.
   b. We present proofs for important theorems having key roles in the subject.
2. For more enlightenment, we work out a number of examples, and discuss applications.

The current text represents the authors' efforts to introduce these aspects in a classical handbook. We feel that an adequate text should be sufficiently complete and have enough scope to warrant a place on a personal bookshelf of the student after leaving school and anyone with interest in mathematics and its applications. More specifically, this handbook presents mathematical definitions, formulas, and theorems in a comprehensive way so that, in addition to access to basic formulas and concepts, the reader is guided through concise mathematical reasonings with less detailed or too sketchy arguments. The intention is to go beyond just listing mathematical relations/formulas, and including insights to the introduced concepts and their interactions, if any.

Our plan is, through PDF file/QR-code, to provide supplementary material available for the users (see page 625).

The book covers material of interest from upper high school- to upper undergraduate-level university students, beginning graduates, and instructors, in the natural science and engineering disciplines, as well as industrial applicants. Our hope is that the challenges in concise proofs and arguments will have tempting effects so that the users are encouraged to try their own way of ending some reasonings: *we did circumvent most of the tedious details.*

The book is organized into 22 chapters and 6 appendices, starting with elementary set theory, algebra, and geometry/trigonometry, and continuing with rather augmented concepts such as vector- and linear-algebra, algebraic structures, logic, and number theory. Next come single variable calculus, derivative, and integral. So far the contents are considered to be suitable for first year undergraduates. Slightly advanced chapters concern the following: differential equations, numerical analysis, differential geometry, series, and sequences. The next level contains transform theory, complex analysis, calculus of several variables, vector analysis, topology, integration theory, and functional analysis. These, rather advanced, chapters can be of interest for the advanced undergraduates as well as starting graduates.

In the concluding chapter, we introduce some basic concepts of Mathematical Statistics. The appendices concern Mathematica and MATLAB programming, a short introduction to Mechanics, some tables, and key concepts.

For easy access to literature: in the bibliography, the main subject of each item appears in **bold face**.

We hope to receive your suggestions, corrections, and constructive criticism that would improve the quality of the material in the book and the presentation. Finally, we hope the users will find the book helpful in finding answers to their math questions, and enjoy consulting it for some overviews.

About the Authors

Mohammad Asadzadeh, PhD, is a Professor of Applied Mathematics at the Department of Mathematics, Chalmers University of Technology and the University of Gothenburg in Sweden. His primary research interest includes the numerical analysis of hyperbolic PDEs, as well as convection-diffusion and integro-differential equations. His work is mainly focused on the analysis of the finite element methods for the above PDE types. He is the author of several textbooks and compendiums in, e.g., analysis, linear algebra, finite element methods for PDEs, Fourier analysis, and wavelets. Professor Asadzadeh has half a century of teaching experience, both in undergraduate and graduate levels, from Iran, Sweden, and the USA.

Reimond Emanuelsson is a Lecturer of Mathematics at the Department of Mathematics, Chalmers University of Technology and the University of Gothenburg in Sweden. His primary research interest is in singular differential operators. He is the author of several textbooks and compendiums in, e.g., linear algebra and calculus. He has over three decades of teaching experience, mainly in undergraduate mathematics and mathematical statistics.

Acknowledgments

We are grateful for the many useful discussions and the most valuable help that we got from our colleague and friend Sakib Sistek. We also wish to thank our wonderful families for their moral support and patience.

Contents

Part I

Elementary Set Theory, Algebra, and Geometry

Chapter 1

Set Theory

1.1 Basic Concepts

(i) A set, initially denoted by M, is a collection of *elements* (objects), e.g., $M = \{-1, 3, 6, 3, a, b, b\}$. In this case, the elements are -1, 3, 6, a, and b.

(ii) The parenthesis "{" and "}" are used to start and end a presentation of a variety of elements.

(iii) The mutual order of the elements or their repetition do not matter for a set.
For example, $M = \{-1, 3, 6, 3, a, b, b\} = \{-1, 3, 6, b, a\} = \{a, b, 3, -1, 6\}$.

(iv) That 3 is an element of $M = \{a, b, -1, 3, 6\}$ is written as $3 \in M$. This reads "3 belongs to M". That 2 does not belong to $M = \{3, -1, 6, a, b\}$ is written as $2 \notin M$.

(v) A set that contains no elements is called an *empty set* denoted by $\emptyset$.

(vi) Two sets A and B are equal-if they contain the same elements.

(vii) A subset A of a set B is a set such that all elements of A can be found in B. For instance $\{-1, 3\}$ is a subset of $\{3, -1, 6\}$. $\{-1, 3\}$ is a *proper* subset of $\{3, -1, 6\}$ since $\{-1, 3\} \neq \{3, -1, 6\}$.

(viii) That A is a subset/proper subset of B is written as

$$A \subseteq B \text{ or } B \supseteq A, \quad \text{and} \quad A \subset B \text{ or } B \supset A. \tag{1.1}$$

(ix) By a *universal set* Ω (or sometimes X) is meant a set which contains all considered elements. The designation X is used on the following pages.

Definition 1.1 (Operations between sets). The *union* of two sets A and B is the set that consists of all elements in A and B, and is denoted by

$$A \cup B. \tag{1.2}$$

The *intersection* between two sets A and B means the set consisting of all common elements (i.e., which are in both sets), and is denoted by

$$A \cap B. \tag{1.3}$$

The *difference* between two sets A and B is the set of elements in A-that are not in B, and is written as

$$A \setminus B. \tag{1.4}$$

The *complement* of a set $A \subseteq X$ is the set $X \setminus A$. The complement is also denoted A^c.

The *symmetric difference* of A and B is the set

$$A \Delta B := (A \cup B) \setminus (A \cap B) = \{\text{can also be written as}\} = (A \setminus B) \cup (B \setminus A).$$

$A \Delta B$ consists of the elements that are in A or B but not in both. It is illustrated on page 6.

Following yield for different collections of sets:

- For finite class of sets $\{A_i,\ i = 1, 2, \ldots, n\}$, the following applies:

$$\begin{aligned} \cup_{i=1}^{n} A_i &= A_1 \cup A_2 \cup \cdots \cup A_n \\ &= \{x; \quad x \in A_i \text{ for some } i = 1, 2, \ldots, n\}. \\ \cap_{i=1}^{n} A_i &= A_1 \cap A_2 \cap \cdots \cap A_n \\ &= \{x; \quad x \in A_i \text{ for all } i = 1, 2, \ldots, n\}. \end{aligned} \tag{1.5}$$

- For a countable class of sets $\{A_i,\ i = 1, 2, \ldots\}$, the following applies:

$$\begin{aligned} \cup_{i=1}^{\infty} A_i &= A_1 \cup A_2 \cup \cdots \\ &= \{x; \quad x \in A_i \text{ for some } i = 1, 2, \ldots\}. \\ \cap_{i=1}^{\infty} A_i &= A_1 \cap A_2 \cap \cdots \\ &= \{x; \quad x \in A_i \text{ for all } i = 1, 2, \ldots\}. \end{aligned} \tag{1.6}$$

• For a class of sets $\{A_i, \quad i \in I\}$, the following applies:

$$\begin{aligned} \bigcup_{i\in I} A_i &= \{x;\ x \in A_i \text{ for some } i \in I\}. \\ \bigcap_{i\in I} A_i &= \{x;\ x \in A_i \text{ for all } i \in I\}. \end{aligned} \tag{1.7}$$

Remark.

(1.7) coincides with (1.5) in the case when $I = \{1, 2, \ldots, n\}$ and (1.7) coincides with (1.6) in the case when $I = \{1, 2, \ldots\} = \mathbb{N}$.

Definition 1.2. $\{A_i \subset X,\ i \in I\}$ is called a *partition* of the set X if

(1) $\cup_{i\in I} A_i = X$ and (2) $A_i \cap A_j = \emptyset$, if $i \neq j$.

Equivalent logical notations

$$\begin{aligned} x \in A \wedge x \in B &\Longleftrightarrow \quad x \in A \cap B. \\ x \in A \vee x \in B &\Longleftrightarrow \quad x \in A \cup B. \end{aligned}$$

Here $\wedge$: means "logical and", $\vee$: means "logical or".
The following hold true:

$$\begin{aligned} &\emptyset \subseteq A \subseteq X, \quad A \subseteq A. \\ &(A \subset B) \wedge (B \subset C) \Longrightarrow A \subset C. \\ &A \subseteq B \Longrightarrow A \cup B = B, \text{ and } A \cap B = A. \end{aligned}$$

Example 1.1.

$$\begin{aligned} &A = \{1, 2, a\}, \quad B = \{a, b, c\} \quad \Longrightarrow \\ &A \cup B = \{1, 2, a, b, c\}, \quad A \cap B = \{a\}, \quad A \setminus B = \{1, 2\}. \end{aligned}$$

In Figure 1.1: The green marked surface in A is the set difference $A \setminus B$.

Likewise, the olive colored surface in B is the set difference $B \setminus A$.

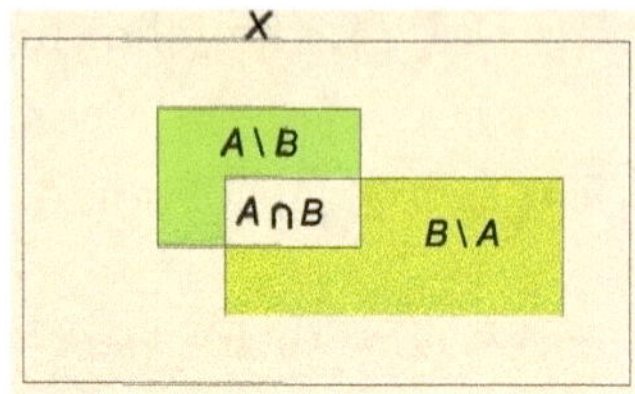

Figure 1.1: A universal set X and two sets A and B. The middle rectangle part illustrates the intersection $A \cap B$ and the union of the two colored sets, the symmetric difference $A \Delta B$.

Further set theoretical identities

$$
\begin{array}{c|c}
A \cup A = A \cap A = A & A \cap X = A \cup \emptyset = A \\
A \cup X = X & A \cap \emptyset = \emptyset \\
A \setminus B = A \cap B^c & \emptyset^c = X \\
X^c = \emptyset & A \cup A^c = X \\
A \cap A^c = \emptyset & (A^c)^c = A \\
A = (A \setminus B) \cup (A \cap B) & (A \cup B) = (A \setminus B) \cup (A \cap B) \cup (B \setminus A)
\end{array}
\tag{1.8}
$$

The penultimate identity is the partition of A with respect to B. The last identity refers to Figure 1.1, and is a partition of $A \cup B$.

Theorem 1.1. *The following yields for inclusion and equality between two sets:*

- *An equality between two sets, $A = B$, is the same as both $A \subseteq B$ and $A \supseteq B$. Formally:*

$$A = B \Longleftrightarrow (A \subseteq B \text{ and } B \subseteq A).$$

- *$A \subseteq B$ is equivalent to $A^c \supseteq B^c$.*
- *$A \subseteq B$ is also equivalent to $x \in A \Longrightarrow x \in B$, that is, all x in A are also in B.*
- *$x \notin B \Longrightarrow x \notin A$, according to the previous two points, is equivalent to $A \subseteq B$.*

With the index set $I = \emptyset$ (the empty set) and universal set X, we have that

$$\bigcup_{i\in\emptyset} A_i = \emptyset, \qquad \bigcap_{i\in\emptyset} A_i = X. \tag{1.9}$$

Standard identities

$$A \cap B = B \cap A,\ A \cup B = B \cup A,\ A\Delta B = B\Delta A \quad \text{(Commutative laws)}. \tag{1.10}$$

$$A \cap (B \cap C) = (A \cap B) \cap C, A \cup (B \cup C) = (A \cup B) \cup C, \quad A\Delta(B\Delta C) = (A\Delta B)\Delta C \quad \text{(Associative laws)}. \tag{1.11}$$

$$\emptyset \cap A = \emptyset, \emptyset \cup A = A\Delta\emptyset = A, A\Delta A = \emptyset. \tag{1.12}$$

$$\begin{cases} A^c \cap B^c = (A \cup B)^c \\ A^c \cup B^c = (A \cap B)^c \end{cases} \quad \text{(De Morgan's laws)}. \tag{1.13}$$

De Morgan's laws have the general forms

$$\cap_i A_i^c = (\cup_i A_i)^c \quad \text{and} \quad \cup_i A_i^c = (\cap_i A_i)^c, \text{ respectively.} \tag{1.14}$$

$$\begin{cases} A \cap (B \cup C) = (A \cap B) \cup (A \cap C) \\ A \cup (B \cap C) = (A \cup B) \cap (A \cup C) \end{cases} \quad \text{(Distributive laws)}. \tag{1.15}$$

Definition 1.3. For a sequence of sets A_m, m, $n = 1, 2, \ldots$, form the sets $B_n = \cup_{m=n}^{\infty} A_m$ and $C_n = \cap_{m=n}^{\infty} A_m$. Then

$$\begin{aligned} &(1) \quad \bigcap_{n=1}^{\infty} B_n = \lim_{n\to\infty} B_n =: \limsup_{n\to\infty} A_n. \\ &\quad\ \text{and} \\ &(2) \quad \bigcup_{n=1}^{\infty} C_n = \lim_{n\to\infty} C_n =: \liminf_{n\to\infty} A_n. \end{aligned} \tag{1.16}$$

If (1) and (2) are equal, their common value is written as

$$\lim_{m\to\infty} A_m.$$

Theorem 1.2. *The formulas in* (1.16) *can be rewritten as*

$$(1)\ \limsup_{n\to\infty} A_n = \bigcap_{n=1}^{\infty}\bigcup_{m\geq n} A_m \quad \textit{and} \quad (2)\ \liminf_{n\to\infty} A_n = \bigcup_{n=1}^{\infty}\bigcap_{m\geq n} A_m.$$

Let $\{A_m, m = 1, 2, \ldots\}$ *be a class of sets, and* B_n *and* C_n *be as in* (1.16), *then*

$$\begin{aligned}\limsup\nolimits_{n\to\infty} A_n &= \{x : x \in A_m \textit{ for infinitely many } m\}.\\ \liminf\nolimits_{n\to\infty} A_n &= \{x : x \in A_m \textit{ for all } m, \textit{ except finitely many}\}.\end{aligned} \tag{1.17}$$

1.1.1 *Product set*

Definition 1.4. *The product* of two sets A and B is the set

$$A \times B = \{(x, y) : x \in A, \quad y \in B\}. \tag{1.18}$$

If A_k, $k = 1, 2, \ldots, n$ is a countable finite class of sets or an infinitely countable class, A_k, $k = 1, 2, \ldots$, then the product sets are written as

$$\prod_{k=1}^{n} A_k = A_1 \times A_2 \times \cdots \times A_n = \{(x_1, x_2, \ldots, x_n) : x_k \in A_k\}$$

and

$$\prod_{k=1}^{\infty} A_k = A_1 \times A_2 \times \cdots = \{(x_1, x_2, \ldots) : x_k \in A_k\}, \tag{1.19}$$

respectively.

Each A_k is a *factor set.*

In particular, if the product consists of finite number of equal factor sets, that is $A_k = A$, $k = 1, 2, \ldots, n$, then the product is written as follows:

$$A^n := \underbrace{A \times A \times \cdots \times A}_{n \text{ sets}}. \tag{1.20}$$

Distributive relations involving union, intersection, and product

$$\begin{aligned}
(A \cup B) \times C &= (A \times C) \cup (B \times C),\\
(A \cap B) \times C &= (A \times C) \cap (B \times C),\\
A \times (B \cup C) &= (A \times B) \cup (A \times C),
\end{aligned}$$

more generally,

$$\begin{aligned}
A \times \bigcup_{j\in I} B_j &= \bigcup_{j\in I}(A \times B_j),\\
A \times (B \cap C) &= (A \times B) \cap (A \times C),
\end{aligned}$$

and

$$A \times \bigcap_{j\in I} B_j = \bigcap_{j\in I}(A \times B_j).$$

1.2 Sets of Numbers

Definition 1.5.

Designation description	The set of...
$\mathbb{N} = Z_+ = \{1, 2, 3, \ldots\}$	natural numbers.
$\mathbb{Z} = \{0, \pm 1, \pm 2, \ldots\}$	integers.
$\mathbb{Q} = \{p/q : p,\, q \in \mathbb{Z},\, q \neq 0\}$	rational numbers.
$\mathbb{Q}_+ = \{x \in \mathbb{Q};\, x > 0\}$	positive rational numbers.
$\mathbb{R} = (-\infty, \infty)$	real numbers.
$\mathbb{R}_+ = \{x \in \mathbb{R};\, x > 0\}$	positive real numbers.
$\mathbb{C} = \{z = x + i\,y,\, x, y \in \mathbb{R}\}$	complex numbers.

(i) A positive integer ≥ 2, which has only 1 and itself as divisor, is called a *prime number*.

(ii) A number α, which is a zero to a polynomial with only integer coefficients, is called *algebraic*. Here, the set of algebraic numbers is denoted $\mathbb{A}$. It yields that $\mathbb{Q} \subset \mathbb{A}$.

(iii) The set of irrational numbers is $\mathbb{R} \setminus \mathbb{Q}$.

(iv) The set of transcendental number is denoted $\mathbb{T} = \mathbb{R} \setminus \mathbb{A}$.

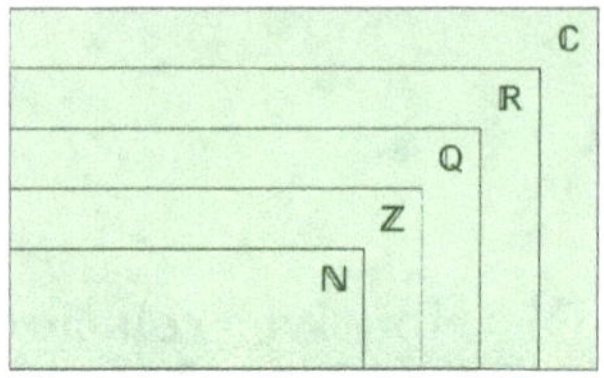

Illustration of the relations of some number sets: $\mathbb{N} \subset \mathbb{Z} \subset \mathbb{Q} \subset \mathbb{R} \subset \mathbb{C}$.

Also, the inclusions $\mathbb{Z}_+ \subset \mathbb{Q}_+ \subset \mathbb{R}_+$ *hold.*

Decimal expansion

(i) Decimal expansion of a real number $x \geq 0$ is given by

$$x = a_0 + a_1 \cdot 10^{-1} + a_2 \cdot 10^{-2} + \cdots, \tag{1.21}$$

where $a_i \geq 0$, $i = 0, 1, \ldots$, are integers such that $0 \leq a_i \leq 9$ for $i = 1, 2, \ldots$

The real number x in (1.21) is written in the position system as

$$x = a_0 \cdot a_1 \cdot a_2 \cdots \tag{1.22}$$

(ii) A real number, on the form (1.22), has a periodic decimal expansion if its decimal expansion contains a finite (repeated) sequence $a_{k+1}\,a_{k+2}\,\cdots\,a_n$, of length $n-k$, for which

$$\begin{aligned} a_{n+(n-k)j+1} &= a_{k+1},\ a_{n+(n-k)j+2} \\ &= a_{k+2}, \ldots, a_{2n+(n-k)j-k} = a_n \end{aligned} \tag{1.23}$$

for $j = 0, 1, \ldots$

(iii) Generally, for any positive integer $b > 1$, $b-$expansion of x is given by

$$x = a_0 + a_1 \cdot b^{-1} + a_2 \cdot b^{-2} + \cdots, \tag{1.24}$$

where $a_i \geq 0$, $i = 0, 1, \ldots$, are integers $0 \leq a_i \leq b-1$ for $i = 1, 2, \ldots$ In the position system

$$x = a_0 \cdot a_1 \cdot a_2 \cdots$$

Theorem 1.3.

$$x \text{ is a rational number}$$
$$\Longleftrightarrow$$
$$x \text{ has after a finite number of positions, a periodic expansion.} \tag{1.25}$$

A real number has a unique decimal expansion.

Remarks. For example, the number $x = \frac{1441733}{33330}$ in decimal form is

$$x = 43.2\,5631\,5631\,5631\,\underbrace{5631}_{\text{period}}\,\ldots$$

Testing of (1.23): Using the notation in (1.23), the digit $a_2 = 5$, that is $k+1 = 2$ and the period is $n - k = 4$, so $n = 5$. By (1.23) a_2 must equal

$$a_{5+4\cdot j+1} = a_{6+4j} \text{ for all } j = 0, 1, \ldots$$

We may take $j = 2$ to get $a_{14} = 5$, as desired.

A binary expansion has the base $b = 2$ and uses only the digits 0 and 1. For example, $x = 10111_2$, that is written in base $b = 2$, has the decimal expansion (in base 10)

$$\begin{aligned} 10111_2 &= 1\cdot 2^4 + 0\cdot 2^3 + 1\cdot 2^2 + 1\cdot 2^1 + 1\cdot 2^0 \\ &= 16 + 4 + 2 + 1 = 23. \end{aligned}$$

The binary base is crucial in computer operations.

Hexadecimal expansion uses base $b = 16$ and thus needs 16 digits, where the last six are symbolized as

$$10 = A,\ 11 = B,\ 12 = C,\ 13 = D,\ 14 = E \text{ and } 15 = F.$$

For example, $44 \equiv 44_{10} = 2C_{16}$ is seen by rewriting

$$2C_{16} = 2\cdot 16^1 + C\cdot 16^0 = 32 + 12 = 44.$$

The decimal expansion is unique as it ends with an infinite sequence of only 9s like $x = 1.99999\ldots$, which is assumed to be $2.0000\ldots$, etc.

An irrational number is characterized by the property that its decimal expansion is not periodic.

The set of real numbers (denoted by $\mathbb{R}$) is the disjoint union of rational and irrational numbers.

A number $x \in \mathbb{N}$ can uniquely be written as

$$x = \sum_{k=0}^{n} a_k \cdot 10^k = \{\text{or expressed in the position system}\}$$

$$= a_n\, a_{n-1} \ldots a_1\, a_0, \tag{1.26}$$

where $a_k = 0, 1, 2, \ldots, 9$.

For a positive integer x written as in (1.26), the following two rules hold true:

$$9|x \Longleftrightarrow 9|a_0 + a_1 + \cdots + a_n \text{ (The rule of 9)},$$

and

$$3|x \Longleftrightarrow 3|a_0 + a_1 + \cdots + a_n \text{ (The rule of 3)}.$$

1.3 Cardinality

Definition 1.6.

- (i) (a) For a set A containing only a finite number of elements, $|A|$ means the number of elements in A.
  - (b) For $\mathbb{Z}_+$, the number of elements is $|Z_+| = \aleph_0$ ("aleph zero").
  - (c) For $\mathbb{R}_+$, the number of elements is $|\mathbb{R}_+| = c$.
  - (d) The number $|A|$ is called *the cardinality* for the set A.
- (ii) Two sets A and B have the same cardinality if there is a bijective mapping $f : A \to B$. The inequality $|A| < |B|$ applies if there is an injection $f : A \to B$ but no injection in the other direction.
- (iii) The *power set* $\mathcal{P}(A)$ means the set of all subsets of A.
- (iv) A set of cardinality $\aleph_0$ is infinitely countable.
- (v) A set with infinite cardinality $> \aleph_0$ is called uncountable.

Theorem 1.4.

- (i) $\aleph_0 = |\mathbb{Z}_+| = |\mathbb{Z}| = |\mathbb{Q}| = |\mathbb{A}|$.
- (ii) $c = |\mathbb{R}| = |\mathbb{R}^n| = |\mathbb{C}|$.
- (iii) $c = 2^{\aleph_0} = |\mathcal{P}(\mathbb{Z}_+)|$.
- (iv) *(Schröder–Bernstein's theorem) If there are injective (or surjective) mappings $f : A \to B$ and $g : B \to A$, then $|A| = |B|$.*

(v) $\aleph_0$ *is the smallest infinity.*

(vi) *If* $|A| = n < \infty$ $|\mathcal{P}(A)| = 2^n$, *and the number of inclusions* $C \subseteq B \subseteq A$ *is* 3^n.

(vii) $|A| < |\mathcal{P}(A)|$.

(viii) $|A \times B| = |A| \cdot |B|$, *if both sets have finite cardinality.*

Theorem 1.5. *Arithmetic for some cardinal numbers.*

$$\begin{array}{|ll|}\hline \aleph_0 = \aleph_0 + \aleph_0 = n \cdot \aleph_0 = \aleph_0 + \aleph_0 + \ldots, & n = 1, 2, \ldots \\ \hline c = 2^{\aleph_0} = n^{\aleph_0} = \aleph_0^{\aleph_0} = c^n = c^{\aleph_0}, & n = 2, 3, \ldots \\ \hline 2^c = n^c = (\aleph_0)^c = c^2 = c^c, & n = 3, 4, \ldots \\ \hline \end{array} \tag{1.27}$$

Remark. It is uncertain whether there exists a cardinal number x between $\aleph_0$ and c or not, i.e., $\aleph_0 < x < c$. (By Cantor's continuum hypothesis there is no such x.)

Chapter 2

Elementary Algebra

2.1 Basic Concepts

A mathematical *expression* is written with numbers or variables, and with operations on or between them. The most usual operations are, e.g., $+, -, \cdot, \times, \div, \sqrt{}$ and so on, for instance $\dfrac{\sqrt{3x}}{2^n - 1}$.

In an *equality* $a = b$, as written, a is called the left-hand side (LHS) and b, the right-hand side (RHS).

An equation is an equality (=) between two different expressions.

An identity is an equality between two expressions that are valid for all values of their variables. The equality sign "=" in an identity is sometimes denoted by "≡".

For the equality sign, the following apply:

$$
\begin{gathered}
a = a, \\
a = b \Longleftrightarrow b = a, \\
(a = b \text{ and } b = c) \Longrightarrow a = c.
\end{gathered} \tag{2.1}
$$

2.2 Rules of Arithmetics

Axiom. For real/complex numbers a, b, and c, the following are commutative and associated laws for addition:

$$a + b = b + a \quad a + (b + c) = (a + b) + c. \tag{2.2}$$

Corresponding laws for multiplication are

$$a \cdot b = b \cdot a \quad \text{and} \quad a \cdot (b \cdot c) = (a \cdot b) \cdot c, \quad \text{respectively.} \tag{2.3}$$

Further, the *distributive law* reads as

$$a(b + c) = ab + ac. \tag{2.4}$$

Distributive rule:
The area of the left and right rectangles are $a \cdot b$ and $a \cdot c$, respectively. The sum of their areas is thus $a \cdot b + a \cdot c$ as well as $a \cdot (b + c)$.

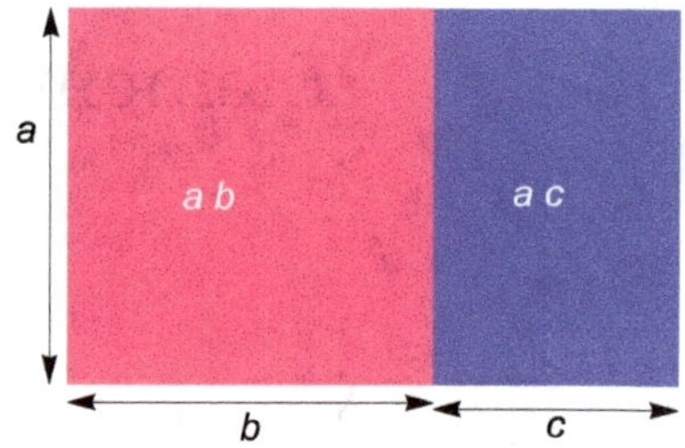

Remarks. When one of the factors is symbolized by a letter, the multiplication operator "$\cdot$" is generally not written out. Equation (2.3) can thus be written $ab = ba$ and $a(bc) = (ab)c$. For example, $\pi \cdot 2$, is written as 2π.

Note that this is not suitable for numbers: $2 \cdot 3 \neq 23$.

$$(a + b)(c + d) \underset{\longleftarrow \text{factorization}}{\overset{\longrightarrow \text{expansion}}{=}} ac + ad + bc + bd.$$

This is called expansion (of parentheses) and factorization (in parentheses), respectively.

2.2.1 *Fundamental algebraic rules*

Theorem 2.1.

(a) $a^2 - b^2 = (a - b)(a + b)$, (b) $(a + b)^2 = a^2 + 2ab + b^2$,
(c) $(a - b)^2 = a^2 - 2ab + b^2$, (d) $a^3 - b^3 = (a - b)(a^2 + ab + b^2)$,
(e) $a^3 + b^3 = (a + b)(a^2 - ab + b^2)$, (2.5)
(f) $(a + b)^3 = a^3 + 3a^2b + 3ab^2 + b^3$,
(g) $(a - b)^3 = a^3 - 3a^2b + 3ab^2 - b^3$.

$$(a+b+c)^2 = a^2 + b^2 + c^2 + 2(ab + bc + ca). \tag{2.6}$$

Reduction of double fractions:

$$\text{main fraction bar} \rightarrow \frac{\frac{a}{b}}{\frac{c}{d}} = \frac{ad}{bc}. \tag{2.7}$$

(The main fraction bar stands at the same height as the equal sign.)

2.2.2 *The binomial theorem*

Definition 2.1.

(i) $n!$ reads "n-factorial" and is defined as $0! = 1$ and $n! = 1 \cdot 2 \cdot \ldots \cdot n$.
(ii) A binomial coefficient is given by

$$\binom{n}{k} := \frac{n!}{k!(n-k)!} = \frac{n \cdot (n-1) \cdot \ldots \cdot (n-k+1)}{k!}. \tag{2.8}$$

Theorem 2.2 (The Binomial theorem).

$$(a+b)^n = \sum_{k=0}^{n} \binom{n}{k} a^k b^{n-k}, \quad n = 0, 1, 2, \ldots \tag{2.9}$$

The identity (2.9) is called *binomial expansion* of $(a+b)^n$. The coefficients in the binomial expansion of $(a+b)^n$ for $n = 0, 1, 2, \ldots$ can be recursively obtained using Pascal's triangle (compare with identity (7.7) page 133).

$n=0$						1					
$n=1$					1		1				
$n=2$				1		2		1			
$n=3$			1		3		3		1		
$n=4$		1		4		6		4		1	
$n=5$	1		5		10		10		5		1

Some common identities containing binomial terms

$$
\begin{aligned}
&(n+1)! = n!\cdot(n+1), \quad \binom{n}{n-k} = \binom{n}{k},\\
&\binom{n}{k} + \binom{n}{k+1} = \binom{n+1}{k+1},\\
&\binom{n}{0} + \binom{n+1}{1} + \binom{n+2}{2} + \cdots + \binom{n+k}{k} = \binom{n+k+1}{k},\\
&\binom{n}{0} + \binom{n}{1} + \cdots + \binom{n}{n} = 2^n.
\end{aligned}
\tag{2.10}
$$

Definition 2.2.

(i) Let $k_1, k_2, \dots, k_r$ be integers ≥ 0 such that $k_1 + k_2 + \cdots + k_r = n$. A *multinomial coefficient* is defined as

$$\begin{pmatrix} & & n & \\ k_1 & k_2 & \dots & k_r \end{pmatrix} := \frac{n!}{k_1!k_2!\cdot\ldots\cdot k_r!}. \tag{2.11}$$

Theorem 2.3. *A multinomial coefficient can be presented by binomial coefficients as follows:*

$$
\begin{aligned}
&\begin{pmatrix} & & n & \\ k_1 & k_2 & \dots & k_r \end{pmatrix}\\
&\quad = \binom{n}{k_1}\binom{n-k_1}{k_2}\cdot\ldots\cdot\binom{n-k_1-k_2\ldots-k_{r-1}}{k_r}.
\end{aligned}
\tag{2.12}
$$

The multinomial theorem

The multinomial expansion of $(a_1 + a_2 + \cdots + a_r)^n$ is

$$(a_1 + a_2 + \cdots + a_r)^n = \sum_{|k|=n} \begin{pmatrix} & & n & \\ k_1 & k_2 & \cdots & k_r \end{pmatrix} a_1^{k_1} a_2^{k_2} \cdot\ldots\cdot a_r^{k_r}, \tag{2.13}$$

where $n = |k| = k_1 + k_2 + \cdots + k_r$ and $k_i \geq 0$ are integers, $i = 1, 2, \dots, r$.

Semi factorial identities

$$1\cdot 3\cdot 5\ldots(2n-1) = (2n-1)!!, \quad 1\cdot 2\cdot 4\ldots 2n = (2n)!!$$

Stirling's formula

$$n! = \sqrt{2\pi}n^{n+1/2}e^{-n}(1+\varepsilon_n), \quad \text{where} \quad \varepsilon_n \to 0 \quad \text{as } n \to 0.$$

2.3 Polynomials in One Variable

Definition 2.3.

(i) A monomial in x is

$$x^n = \underbrace{x \cdot x \cdot \ldots \cdot x}_{n\,\text{factors}},\ n = 1, 2, \ldots \quad \text{and} \quad x^0 = 1. \tag{2.14}$$

(ii) Polynomials of degree 1 and 2 are (in the variable x) given by

$$ax + b \quad \text{and} \quad ax^2 + bx + c, \text{ respectively } (a \neq 0). \tag{2.15}$$

(iii) A polynomial of degree n, $n = 0, 1, 2, \ldots$ in the variable x is given by

$$\begin{aligned} f(x) &= a_n x^n + a_{n-1}x^{n-1} + \cdots + a_1 x + a_0 \\ &= \sum_{k=0}^{n} a_k x^k, \quad a_n \neq 0. \end{aligned} \tag{2.16}$$

The numbers $a_1, a_2, \ldots, a_n$ are called *coefficients.*

(iv) An equation of the first and second degree is an equation which can be written as

$$ax+b = 0 \quad \text{and} \quad ax^2+bx+c = 0, \text{ respectively}, \quad a \neq 0. \tag{2.17}$$

An equation (or polynomial equation) of degree n can be written as

$$a_n x^n + a_{n-1}x^{n-1} + \cdots + a_1 x + a_0 = 0, \quad a_n \neq 0. \tag{2.18}$$

(v) A concise form for a polynomial of degree n in *two variables* x and y is given by

$$\sum_{m=0}^{n} \sum_{k+l=m,} a_{k,l} x^k y^l, \quad (k, l = 0, 1, 2, \ldots, m), \tag{2.19}$$

where $a_{k,n-k} \neq 0$ for some k.

Definition 2.4 (The nth root of a real number). Assume that n is a positive integer. For a real number a, its nth root is defined by

$$\sqrt[n]{a} = a^{1/n} \qquad \text{in two cases:} \tag{2.20}$$

(i) $a \geq 0$. $b := \sqrt[n]{a} \geq 0$ is the non-negative number that satisfies $b^n = a$.
(ii) $a < 0$ and n odd. $b := \sqrt[n]{a} < 0$ is the real number that satisfies $b^n = a$.

$\sqrt{a} = \sqrt[2]{a}$ reads "square root of a" and $\sqrt[3]{a}$ reads the "cubic root of a". Generally, $\sqrt[n]{a} := a^{1/n}$ is the $n:th$ root of a. For rational exponent m/n, where m, n are relatively prime, positive integers (i.e., m and n have only 1 as common factor), one has the equality

$$b^{m/n} = \sqrt[n]{b^m}.$$

Rules of roots

$$\begin{aligned} (\sqrt[n]{a})^n &= a, \quad \begin{cases} a \geq 0, & n \text{ even}, \\ a \in \mathbb{R}, & n \text{ odd}, \end{cases} \\ \sqrt[n]{a^n} &= a, \quad a \in \mathbb{R}, \quad n \text{ odd}, \\ \sqrt[n]{a^n} &= |a|, \quad a \in \mathbb{R}, \quad n \text{ even}. \end{aligned} \tag{2.21}$$

$$\begin{aligned} \sqrt[n]{a} \cdot \sqrt[n]{b} &= \sqrt[n]{a \cdot b}, \\ \frac{\sqrt[n]{a}}{\sqrt[n]{b}} &= \sqrt[n]{\frac{a}{b}}, \\ \sqrt[m]{\sqrt[n]{a}} &= \sqrt[m \cdot n]{a} = \sqrt[n]{\sqrt[m]{a}}. \end{aligned} \tag{2.22}$$

The identities of (2.22) apply to a and b as long as the roots are well-defined. Powers and rules for powers can be found on page 34.

Theorem 2.4.

(i) *Solution of second-degree equation (p and q are real numbers)*

$$x^2 + px + q = 0$$

$$\Longleftrightarrow \begin{cases} x = -\frac{p}{2} \pm \sqrt{\left(\frac{p}{2}\right)^2 - q}, & \left(\frac{p}{2}\right)^2 - q \geq 0, \\ x = -\frac{p}{2} \pm i\sqrt{q - \left(\frac{p}{2}\right)^2}, & \left(\frac{p}{2}\right)^2 - q < 0, \end{cases} \tag{2.23}$$

where i is the imaginary unit ($i^2 = -1$).

(ii) *Solution of second-degree equation* (a, b, *and* c *reals and* $a \neq 0$)

$$ax^2 + bx + c = 0$$
$$\iff \begin{cases} x = \dfrac{-b \pm \sqrt{b^2 - 4ac}}{2a}, & b^2 - 4ac \geq 0, \\ x = \dfrac{-b \pm i\sqrt{4ac - b^2}}{2a}, & b^2 - 4ac < 0. \end{cases} \tag{2.24}$$

(iii) $\Delta := b^2 - 4ac$ *is called discriminant.*
(iv) *If* x_1 *and* x_2 *are the two roots, then*

$$x_1 + x_2 = -p \quad \text{and} \quad x_1 x_2 = q \text{ in (2.23).}$$
$$x_1 + x_2 = -b/a \quad \text{and} \quad x_1 x_2 = c/a \text{ in (2.24).}$$

Theorem 2.5. *Solution of equation of third degree*

(i) *Any polynomial equation of third degree (after division by the coefficient of highest degree) can be written as*

$$x^3 + \alpha x^2 + \beta x + \gamma = 0. \tag{2.25}$$

(ii) *"Eliminating" the term of second degree by using the substitution* $x - \alpha/3 = t$*-yields*

$$t^3 + \frac{3\beta - \alpha^2}{3} t + \frac{2\alpha^3}{27} - \frac{\alpha\beta}{3} + \gamma = 0.$$

(iii) *With the new coefficients denoted by* a *and* b*, respectively, it ends up with*

$$t^3 + at + b = 0.$$

(iv) *This equation has the following solution formula for one of its roots:*

$$t = t_1 = \sqrt[3]{-\frac{b}{2} + \sqrt{\left(\frac{a}{3}\right)^3 + \left(\frac{b}{2}\right)^2}} - \sqrt[3]{\frac{b}{2} + \sqrt{\left(\frac{a}{3}\right)^3 + \left(\frac{b}{2}\right)^2}}. \tag{2.26}$$

(v) *Then, with* $x_1 = t_1 + \alpha/3$ *and dividing the LHS in* (2.25) *by* $x - x_1$, *an equation of second degree is obtained, with zeros given by* (2.23).

(vi) *The absolute value of a root* $x = x^*$ *of a polynomial is limited by the coefficients of the polynomial,* *see page* 268.

Remarks. Basically, equation of fourth degree is solvable by a formula containing similar root expressions like the $p-q$ formula. There is no "algebraic" expression for the roots to equations of degree five and higher: a result proved (independently) by Abel, Galois, and Ruffini.

> Omar Khayyam (Persian Poet and Mathematician 1048–1131) discovered a geometrical method of solving cubic equations by intersecting a parabola with a circle.
> Each equation, in variable x, can be written on the form $f(x) = 0$. With a "root" x of an equation $f(x) = 0$ means a number x-that satisfies the equation. Then, x is called a *zero* of the function $f(x)$.

Theorem 2.6 (The factor theorem). *The following equivalent relations generally hold for a polynomial* $f(x)$ *of degree* $n = 1, 2, \ldots$:

1. $$x = a \text{ is a root of } f(x) = 0, \quad \text{that is} \quad f(a) = 0$$

$$\Longleftrightarrow$$

$$(x - a) \text{ is a factor to } f(x), \text{ that is } f(x) = (x - a)g(x), \text{ where } \deg g(x) = n - 1.$$

$$\tag{2.27}$$

2. $$f(a) = f'(a) = f''(a) = \ldots = f^{(m-1)}(a) = 0$$

$$\Longleftrightarrow$$

$$f(x) = (x - a)^m g(x), \quad \text{where } \deg g(x) = \deg f(x) - m = n - m.$$

Theorem 2.7 (Complex conjugate roots). *If a polynomial* $f(x)$ *has only real coefficients and if* $x = \alpha + i\beta$ *is a complex root of* $f(x) = 0$, *then also* $\bar{x} = \alpha - i\beta$ *is a root of* $f(x) = 0$. *If* $\beta \neq 0$, $f(x) = (x - \alpha - i\beta)(x - \alpha + i\beta)g(x)$, *where degree* $g =$ *degree* $f - 2$.

Theorem 2.8 (The theorem of rational roots). *If, on the LHS of the polynomial equation* (2.18), *page* 19, *all coefficients* $a_0, \ldots, a_n$

are integers and if the equation has a rational root $x = \frac{s}{t}$, simplified as far as possible, then s is a divisor of a_0 and t, a divisor of a_n.

Theorem 2.9. *Every real polynomial $q(x)$ of degree n (that is with only real coefficients) can uniquely be factorized into real polynomials of the highest degree 2.*

$$q(x) = A\prod_{i=1}^{n_1}(x - a_i)^{k_i} \cdot \prod_{j=1}^{n_2}(x^2 + b_j x + c_j)^{l_j}, \tag{2.28}$$

where all a_i are real and different for $i = 1, 2, \ldots, n_1$, all pairs (b_j, c_j) are real and different for $j = 1, 2, \ldots, n_2$, and all $x^2 + b_j x + c_j$ are irreducible over $\mathbb{R}$. (That is they are not factorized into real polynomials of degree one). Further, k_i and l_j are positive integers such that

$$k_1 + k_2 + \cdots + k_{n_1} + 2(l_1 + l_2 + \cdots + l_{n_2}) = n.$$

Each complex (and thus real) polynomial $q(x)$ of degree n can be uniquely factorized into complex polynomials of degree one, that is of type $x - a_i$. More specifically, for an $A \neq 0$,

$$q(x) = A\prod_{i=1}^{m}(x - a_i)^{k_i}, \quad k_1 + k_2 + \cdots + k_m = n, \tag{2.29}$$

where $k_1, k_2, \ldots, k_m \in \mathbb{Z}_+$ and a_i are different complex numbers and the zeros a_i are of multiplicity k_i.

2.4 Rational Expression

Definition 2.5.

(i) A rational expression r is a ratio between two polynomials $p(x)$ and $q(x)$,

$$r(x) = \frac{p(x)}{q(x)}. \tag{2.30}$$

The expression is valid/defined for all x for which $q(x) \neq 0$.

(ii) (a) A polynomial $q(x)$ is a factor or divisor of the polynomial $p(x)$ if the ratio (2.30) is a polynomial (see polynomial division which follows).
This is written as $q(x)|p(x)$.

(b) If

$$(x-a)^k|p(x) \text{ but } (x-a)^{k+1} \not| p(x),$$

then $x = a$ is a zero of multiplicity k for the polynomial $p(x)$.

2.4.1 *Expansion of rational expression*

If the degree of the numerator $\geq$ degree of the denominator in (2.30), a polynomial division can be performed (see the following example).

Example 2.1. Make the division $\dfrac{2\,x^3 - x^2 - 6\,x + 14}{x^2 + x - 2}$.

Solution:

One uses the successive division algorithm

$2x-3$ *(ratio)*		
$2\,x^3 - x^2 - 6\,x + 14$	$x^2 + x - 2$	Numerator/Denominator
$-(2x^3 + 2x^2 - 4x)$		The product $2x \cdot (x^2 + x - 2)$.
$-3x^2 - 2x + 14$		Subtraction of the two sides.
$-(-3x^2 - 3x + 6)$		The product $-3 \cdot (x^2 + x - 2)$.
$x + 8$ *(remainder term)*		The subtraction of the two sides.

$x+8$ is *remainder term*. Because its degree ("1") is less than that of the denominator (= 2), the algorithm stops at this step. Then, the division means that

$$\frac{2\,x^3 - x^2 - 6\,x + 14}{x^2 + x - 2} = 2x - 3 + \frac{x+8}{x^2 + x - 2}.$$

If the degree of the denominator is > that of the numerator, (as is the case with the residual term after polynomial division), the so-called *partial fraction division (PF)* may be performed.

The following example is a continuation of the former one highlighting the meaning of the whole procedure.

Example 2.2. PF of $\dfrac{x+8}{x^2+x-2}$.

Solution:

One can factorize the denominator as $x^2+x-2=(x-1)(x+2)$. Then, the first step is making the split (Ansatz) to separate first order denominators:

$$\frac{x+8}{(x-1)(x+2)}=\frac{A}{x-1}+\frac{B}{x+2},$$

where A and B are constants, to be determined. Putting the two terms on the RHS with the same denominator yields

$$\frac{x+8}{(x-1)(x+2)}=\frac{A(x+2)+B(x-1)}{(x-1)(x+2)}.$$

Then, the numerators must be equal, hence,

$$x+8=(A+B)x+2A-B.$$

Identifying the coefficients, we end up with

$$\begin{array}{ccc} \text{LHS} & & \text{RHS} \\ x:1 & = & A+B \\ 1:8 & = & 2A-B \end{array}$$

having the unique solution $A=3$ and $B=-2$. Along with the result in Example 2.1, we finally get

$$\frac{2\,x^3-x^2-6\,x+14}{x^2+x-2}=2x-3+\frac{3}{x-1}-\frac{2}{x+2}.$$

Theorem 2.10 (Expansion of rational expression). *Assume that $p(x)$ and $q(x)$ are two real polynomials with* degree $p=m$ *and* degree $q=n$. *Then $q(x)$ can be factorized as*

$$q(x)=A\prod_{i=1}^{n_1}(x-a_i)^{k_i}\cdot\prod_{j=1}^{n_2}(x^2+b_jx+c_j)^{l_j}, \tag{2.31}$$

according to (2.28). The ratio $p(x)/q(x)$ can be expanded as

$$\frac{p(x)}{q(x)} = k(x) + \underbrace{\sum_{i=1}^{n_1}\sum_{j=1}^{k_i}\frac{A_{ij}}{(x-a_i)^j} + \sum_{i=1}^{n_2}\sum_{j=1}^{l_j}\frac{B_{ij}x+C_{ij}}{(x^2+b_ix+c_i)^j}}_{=:\, R(x),\ a\ partial\ fraction}, \tag{2.32}$$

where $k(x)$ is a polynomial of degree $m-n$, if $m \geq n$ or $k(x) \equiv 0$, if $m < n$.

Flowchart for expansion of rational expression

(i) degree $p(x) \geq$ degree $q(x) \longrightarrow$ polynomial division $\frac{p(x)}{q(x)} = k(x) + \frac{r(x)}{q(x)}$, where degree $k(x) = m - n$ and degree $r(x) <$ degree $q(x) = n$.

(ii) degree $r(x) <$ degree $q(x) \longrightarrow \frac{r(x)}{q(x)}$ is treated as the second term $R(x)$ in (2.32).

Remark. You can skip polynomial division even if $m \geq n$ and only use substitution. For $m - n =: l$, substitute the ratio by

$$\frac{p(x)}{q(x)} = \underbrace{a_lx^l + a_{l-1}x^{l-1} + \cdots + a_1x + a_0}_{=k(x)} + R(x)$$

and consider $R(x)$ as in (2.32).

With a complex polynomial $q(x)$, as in (2.29) page 23, the PF can be set as

$$\frac{r(x)}{q(x)} = \sum_{i=1}^{n_1}\sum_{j=1}^{k_i}\frac{A_{ij}}{(x-a_i)^j},$$

where all a_i are different.

2.5 Inequalities

Definition 2.6. Assume a and b are real numbers.

(i) $a \geq b$ $(a > b)$ reads a is (strictly) greater than b.
(ii) $a \leq b$ $(a < b)$ reads a is (strictly) smaller than b.
(iii) $a \leq b$ and $b \leq c \Longrightarrow a \leq c$.

Theorem 2.11. *For real numbers a, b, c, and d, the following hold true:*

$$a < b \Longleftrightarrow a + c < b + c, \qquad a < b \Longleftrightarrow ad < bd, \text{ if } d > 0.$$
$$0 \leq a < b \Longleftrightarrow \sqrt{a} < \sqrt{b}, \qquad a > b > 0 \Longleftrightarrow 0 < 1/a < 1/b.$$
$$a < 0 < b \Longleftrightarrow 1/a < 0 < 1/b, \quad 0 < a < b \Longleftrightarrow 0 < a^c < b^c, \text{ if } c > 0.$$

For each pair of real numbers a and b, $a \leq b$, $a > b$, or $a < b$.

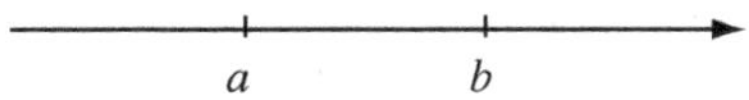

Theorem 2.12 (Arithmetic-geometric inequality). *Assume that $a_i > 0$ and that $\lambda_i > 0$ for $i = 1, 2, \ldots, n$ and $\sum_{i=1}^{n} \lambda_i = 1$. Then*

$$\sum_{i=1}^{n} \lambda_i a_i \geq \prod_{i=1}^{n} a_i^{\lambda_i}. \tag{2.33}$$

$$\begin{aligned} e^x &> x^e && \text{for all } x > 0, \quad x \neq e. \\ 3^n &> n^3 && \text{for all } n \in \mathbb{Z} \setminus \{3\}. \\ a^b &> b^a && \text{for } e \leq a < b \text{ or } b < a < e. \\ a^x &\geq x^b && \text{for } x > 0, \quad \text{if} \quad a^e \geq e^b, b > 0 \text{ and } a > 1. \end{aligned}$$

2.5.1 *Absolute value*

Definition 2.7. Let x be a real number (i.e., $x \in \mathbb{R}$). The absolute value of x means

$$|x| = \begin{cases} x, & \text{if } x \geq 0, \\ -x, & \text{if } x < 0. \end{cases} \tag{2.34}$$

$$\begin{aligned} &|a| \cdot |b| = |a \cdot b|, \quad \frac{|a|}{|b|} = \left|\frac{a}{b}\right|, \\ &|a+b| \leq |a| + |b|, \quad ||a| - |b|| \leq |a-b|, \\ &|a-b| = |b-a| = \begin{cases} b-a, & \text{if } b \geq a, \\ a-b, & \text{if } b \leq a. \end{cases} \end{aligned} \tag{2.35}$$

Remarks.

- $|x_1 - x_2|$ is the distance between the points x_1 and x_2 on the number line (the real axis).
- $\sqrt{a^2} = |a|$ for each real number a.
- $\{x : |x - x_0| = r\}$ is the set of all real numbers x with distance $r \geq 0$ to x_0. These are the two points $x = x_0 - r$ and $x = x_0 + r$.
- $\{x : |x - x_0| \leq r\}$ is the set of all real x with distance $d \leq r$ to x_0. This set can also be written as the closed interval $[x_0 - r, x_0 + r]$.
- For $|x - a|$, where x is a real variable, $x = a$ is called *breaking point.*
- $|a+b| \leq |a| + |b|$ is the *triangle inequality.*
- For real A and B, the equivalence holds

$$|A - B| < \varepsilon \iff -\varepsilon < A - B < \varepsilon.$$

2.6 Complex Numbers

From the figure to the left, some basic concepts for a complex number are defined as follows. $z = 3 + 2i$, with length $r = |z| = \sqrt{3^2 + 2^2} = \sqrt{13}$ and the argument $\theta = \arg z = \arctan(2/3)$, the angle between the positive real axis and the vector representation of the complex number, counted with positive (counterclockwise) orientation.

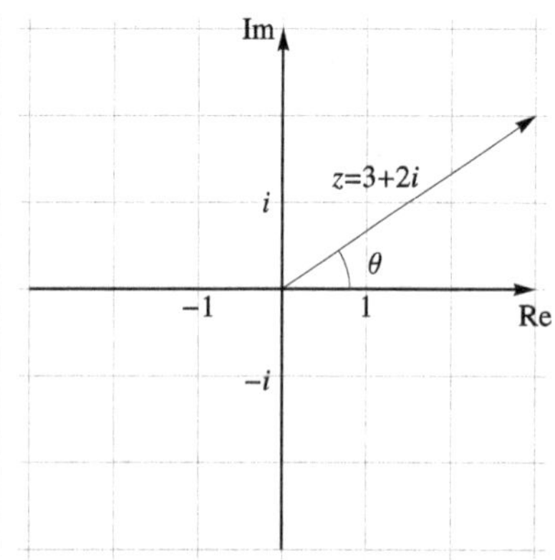

Definition 2.8. A complex number can be written as

$$z = x + i \cdot y = x + iy, \tag{2.36}$$

where x and y are real numbers (cartesian coordinates).

- The number i, with $i \cdot i = i^2 = -1$, is called the imaginary unit (in the literature related to electric engineering, one uses j instead of i, since i denotes instantaneous current).
- The form $x + iy$ is the Cartesian form of the complex number.
- x reads on the horizontal axis, the real axis.
- iy reads on the vertical axis, imaginary axis.
- x is called the real part of z and is denoted by $x = \mathrm{Re}(z)$.
- y is called the imaginary part of z and is denoted by $y = \mathrm{Im}(z)$.
- If the imaginary part is zero ($y = 0$), so $z = x$ is then (pure) *real.*
- If the real part is zero ($x = 0$), then $z = iy$ is (pure) *imaginary.*
- *The complex conjugate* $\overline{z}$ of a complex number $z = x + iy$ is the complex number $x - iy$ (the mirror image of z in the real axis).
- The absolute value of z is $|z| = \sqrt{x^2 + y^2}$ and is the length of z, seen as a vector. Alternatively, it is the distance between z and the origin.
- Some arithmetic with complex numbers can be seen in Figure 2.1.

$$\arg z = \begin{cases} \arctan(y/x), & \text{if } x > 0, \\ \arctan(y/x) + \pi, & \text{if } x < 0, \\ \dfrac{\pi}{2}, & \text{if } x = 0 \text{ and } y > 0, \\ -\dfrac{\pi}{2}, & \text{if } x = 0 \text{ and } y < 0. \end{cases}$$

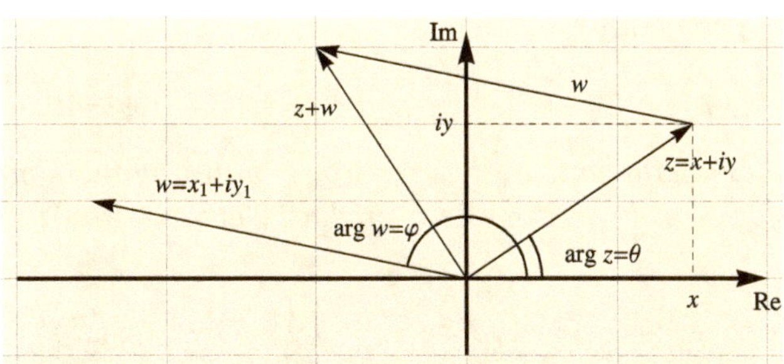

Figure 2.1: Addition of two complex numbers as a *vector*. Note, the product of two complex numbers z and w yields the vector zw with the argument $\arg(zw) = \arg z + \arg w$ and length $|z||w| = |zw|$.

The argument is the angle of the complex number, seen as a vector, with the positive real axis.

- $|z-w|$ is the distance between two complex numbers, z and w.

Theorem 2.13 (Rules of complex numbers). *Complex numbers follow the laws/rules (2.2)–(2.5) page 15. For absolute values and conjugates, the following rules apply:*

$$\begin{aligned} &|zw| = |z||w|, \quad \left|\frac{z}{w}\right| = \frac{|z|}{|w|}, \quad |z|^2 = z\cdot\overline{z}, \\ &\overline{z+w} = \overline{z}+\overline{w}, \quad \overline{zw} = \overline{z}\,\overline{w}, \quad \overline{\left(\frac{z}{w}\right)} = \frac{\overline{z}}{\overline{w}}. \end{aligned} \tag{2.37}$$

$$\begin{aligned} &2|z|^2 + z^2 + \overline{z}^2 = 4(Re\ z)^2, \\ &|z+w|^2 = |z|^2 + |w|^2 + 2\,\mathrm{Re}\,(z\overline{w}), \\ &\frac{z+\overline{z}}{2} = Re\ z, \quad \frac{z-\overline{z}}{2i} = Im\ z, \\ &|z+w| \le |z| + |w| \ (\textit{triangle inequality}). \end{aligned} \tag{2.38}$$

Let $z_1, z_2, \ldots, z_n$ be complex numbers.

Then, there is a subset $S \subseteq \{1, 2, \ldots, n\}$ such that

$$\left|\sum_{k\in S} z_k\right| \ge \frac{1}{6}\sum_{k=1}^{n} |z_k|. \tag{2.39}$$

Theorem 2.14 (Fundamental theorem of elementary algebra). *Any polynomial*

$$a_n z^n + a_{n-1}z^{n-1} + \cdots + a_1 z + a_0, \quad a_n \neq 0 \tag{2.40}$$

with complex coefficients a_k has n zeros counted by their multiplicity and therefore can be written as a product (factorization):

$$a_n z^n + a_{n-1}z^{n-1} + \cdots + a_1 z + a_0 = a_n \prod_{r=1}^{m} (z - z_r)^{k_r}, \quad a_n \neq 0, \tag{2.41}$$

where all z_r are different and $k_1, k_2, \ldots, k_m$ are positive integers with sum n.

2.6.1 *To solve equations of second degree with complex coefficients*

Example 2.3. Solve the equation $z^2 + (1+i)\, z - 4 + 8\,i = 0$.

Solution:

First, *completing the square*:

$$\begin{aligned} & z^2 + (1+i)\, z - 4 + 8i \\ &= z^2 + 2\cdot\frac{1+i}{2} z + \left(\frac{1+i}{2}\right)^2 - \left(\frac{1+i}{2}\right)^2 - 4 + 8i \\ &= \left(z + \frac{1+i}{2}\right)^2 - \left(\frac{1+i}{2}\right)^2 - 4 + 8i = 0 \\ &\iff \left(z + \frac{1+i}{2}\right)^2 = \left(\frac{1+i}{2}\right)^2 + 4 - 8i = 4 - \frac{15\,i}{2}. \end{aligned}$$

Then setting RHS: $4 - \frac{15\,i}{2} = (a + ib)^2$, the numbers (real) a and b are determined. This gives rise to the following three equations:

$$a^2 - b^2 = 4, \quad 2ab = -\frac{15}{2}, \quad a^2 + b^2 = \sqrt{4^2 + \left(-\frac{15}{2}\right)^2} = \frac{17}{2}.$$

Adding the first and last equation, yields

$$a^2 - b^2 + a^2 + b^2 = \frac{17}{2} + 4 \iff a^2 = 2 + \frac{17}{4} = \frac{25}{4} \iff a = \pm\frac{5}{2}.$$

Inserting these values for a in the second equation, we get

$$a = \frac{5}{2} \iff b = -\frac{3}{2}, \quad a = -\frac{5}{2} \iff b = \frac{3}{2}.$$

Hence,

$$\left(z + \frac{1+i}{2}\right)^2 = \left(-\frac{5}{2} + \frac{3i}{2}\right)^2,$$

or equivalently

$$z + \frac{1+i}{2} = -\frac{5}{2} + \frac{3i}{2}, \quad z + \frac{1+i}{2} = \frac{5}{2} - \frac{3i}{2}.$$

Here, the first equation gives $z = -3 + i$ and the second yields $z = 2 - 2i$. Thus, these are the two roots. By factorization we get

$$z^2 + (1+i)\,z - 4 + 8i = (z + 3 - i)(z - 2 + 2i).$$

2.6.2 *Complex numbers in polar form*

Definition 2.9.

(i) θ is an angle in radians. Then

$$e^{i\theta} := \cos\theta + i\sin\theta. \quad (2.42)$$

(ii) The polar coordinates for a complex number are (r, θ) where $|z| = r$ is its length/ modolus and $\theta = \arg z$, its angle with the positive real axis (formula (2.43)).

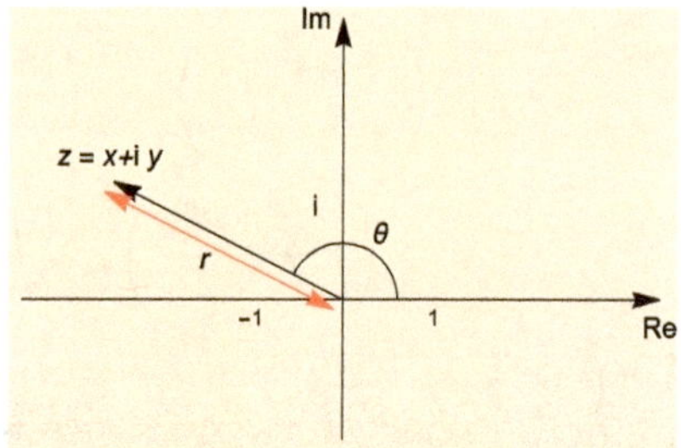

Remarks. θ is an angle in the interval $[0, 2\pi)$ or $(-\pi, \pi]$. In some applications, the angle is given in a different interval of length 2π.

Theorem 2.15.

$$x + iy = z = r(\cos\theta + i\sin\theta) = re^{i\theta}. \tag{2.43}$$

The last two expressions are called polar forms.

Any complex number in Cartesian form can be written in polar form and vice versa.

The Relation between these two coordinate forms is given by

$$\begin{cases} x = r\cos\theta, \\ y = r\sin\theta, \end{cases} \quad \textit{and} \quad x^2 + y^2 = r^2. \tag{2.44}$$

$$\arg(z \cdot w) = \arg z + \arg w + 2n\pi, \tag{2.45}$$

for some integer n.

de Moivre's formula

$$(\cos\alpha + i\sin\alpha)^n = \cos(n\alpha) + i\sin(n\alpha), \quad n \in \mathbb{Z}. \tag{2.46}$$

Expressed with (2.42), this becomes

$$\left(e^{i\alpha}\right)^n = e^{in\alpha}. \tag{2.47}$$

Euler's formulas

$$\cos\alpha = \frac{e^{i\alpha} + e^{-i\alpha}}{2}, \quad \sin\alpha = \frac{e^{i\alpha} - e^{-i\alpha}}{2i}, \quad \tan\alpha = i \cdot \frac{1 - e^{2i\alpha}}{1 + e^{2i\alpha}}. \tag{2.48}$$

Definition 2.10. A *binom* is a polynomial with exact two terms.

A *binomic* equation is an equation in the variable z, which equivalently can be written as $z^n = w$, where $n \in \mathbb{Z}_+ = \{1, 2, \ldots\}$ and $w \in \mathbb{Z}$.

Theorem 2.16.

$$z^n = w = re^{i\theta} \textit{ is a binomial equation which has } n \textit{ roots, viz.}$$
$$z = z_k = r^{1/n} e^{i(\theta + k2\pi)/n}, \quad k = 0, 1, 2, \ldots, n-1. \tag{2.49}$$

Remarks. In (2.49), k can vary through n consecutive integers, e.g., $k = 0, 1, 2, \ldots, n-1$. In the figure on the right, the seven roots of the binomial equation $z^7 = 2i$ $z_k = 2^{1/7} e^{i\pi/14 + k \cdot 2\pi i/7}$, $k = 0, 1, \ldots, 6$, are drawn as location vectors.

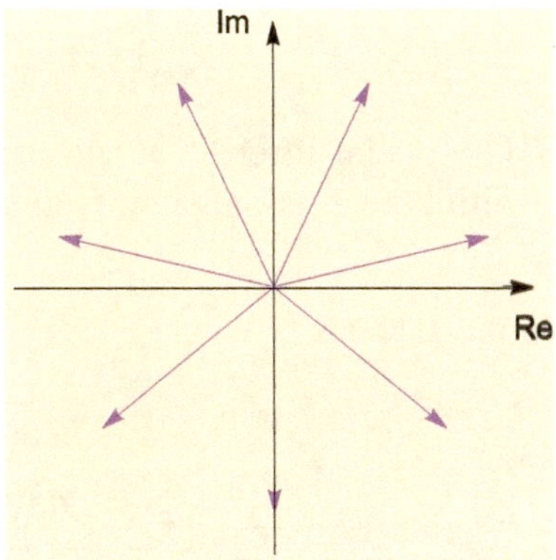

Theorem 2.17 (Complex conjugate roots). *If $f(x) = a_n x^n + a_{n-1}x^{n-1} + \cdots + a_1 x + a_0$ is a polynomial having only real or pure imaginary coefficients a_k and if $x = a+ib$ is a zero of $f(x)$, with real a, b, then $\overline{x} = a - ib$ is also a zero of $f(x)$. This in turn means that, if $b \neq 0$, then*

$$(x - (a+ib))(x - (a-ib)) = x^2 - 2ax + a^2 + b^2$$

is a factor of the polynomial $f(x)$.

2.7 Powers and Logarithms

2.7.1 *Powers*

Definition 2.11.

(i) A power is an expression of the form

$$a^b, \text{ where } a \text{ is called } \textit{base}, \text{ and } b, \textit{ exponent}. \tag{2.50}$$

(ii) Powers are well-defined in the following cases:

(a) b is an integer or $b = \frac{1}{n}$, where n is an odd integer and a, an arbitrary real number, except for $a = 0$ and $b < 0$.

(b) $a > 0$ and b an arbitrary real number.

(c) In particular, $0^0 = 1$.

(iii) For a positive integer n, the nth power of a real number a is defined as

$$a^n = \underbrace{a \cdot a \cdot \ldots \cdot a}_{n \text{ factors}}. \tag{2.51}$$
$$a^{1/n} = \sqrt[n]{a}, \quad n = 1, 2, \ldots$$

(iv) Of special interest is the base $e \approx 2.71728$ in calculus. For real numbers x, e^x, also written as $\exp(x)$, is an *exponential function.*

Theorem 2.18.

$$a^{x+y} = a^x \cdot a^y, \quad a^{x-y} = \frac{a^x}{a^y}, \quad (a^x)^y = a^{x \cdot y},$$
$$(ab)^x = a^x b^x, \quad \left(\frac{a}{b}\right)^x = \frac{a^x}{b^x}. \tag{2.52}$$

In particular,

$$a^0 = 1, \quad a^1 = a, \quad a^{-x} = \frac{1}{a^x}.$$

Prefix

Name	Meaning	Name	Symbol	Meaning	Name	Symbol
One thousand trillion	10^{15}	*peta*	P	10^{-15}	*femto*	f
One trillion	10^{12}	*tera*	T	10^{-12}	*piko*	p
One billion	10^9	*giga*	G	10^{-9}	*nano*	n
One million	10^6	*mega*	M	10^{-6}	*mikro*	μ
One thousand	10^3	*kilo*	k	10^{-3}	*milli*	m
One hundred	10^2	*hekto*	h	10^{-2}	*centi*	c
Ten	10^1	*deka*	da	10^{-1}	*deci*	d

2.7.2 *Logarithms*

Definition 2.12.

(i) Assume that $b > 0$ and $b \neq 1$. Then, the b-logarithm for a positive a is defined as the exponent $x = \log_b a$ such that $a = b^x$, e.g., $a = b^{\log_b a}$.
(ii) $\log_{10} a =: \lg a$ (10-logarithm).
(iii) $\log_e a =: \ln a$ (e-logarithm), where $e = \lim_{n \longrightarrow \infty} (1 + 1/n)^n \approx 2.71828$.

Theorem 2.19. *The laws of logarithms with 10-base. If $a > 0$, $b > 0$, and x is a real number, then*

$$\lg(ab) = \lg a + \lg b, \quad \lg(a^x) = x \lg a, \quad \lg(a/b) = \lg a - \lg b. \quad (2.53)$$

Rules (2.53) also hold for arbitrary base, and hence even for the base e.

Remarks. In particular,

$$\log_a a = 1, \quad \log_a 1 = 0, \quad \log_a(1/a) = -1.$$

$\ln := \log_e$ is called the natural logarithm and corresponds to the base e, (Euler's or Napier's constant).

Both base 10 and natural logarithms are implemented by calculators. A connection between these two logarithms is given by

$$x = e^{\ln x} = 10^{\lg x} = e^{\ln 10 \cdot \lg x} \iff \begin{cases} \ln x = \ln 10 \cdot \lg x, \\ \lg x = \lg e \cdot \ln x. \end{cases}$$

Theorem 2.20.

$$\frac{\lg x}{\lg y} = \frac{\ln x}{\ln y} = \frac{\log_a x}{\log_a y}, \qquad \text{if } a > 0, \quad a \neq 1, \text{ and } x > 0, y > 0. \tag{2.54}$$

For $a, b, c, d > 0$, all are $\neq 1$, and $x \neq 0$, the following hold true:

$$\frac{\ln a}{\ln b} = \log_b a, \quad \frac{\log_a b}{\log_c d} = \frac{\log_d c}{\log_b a}, \quad \frac{1}{\log_a b} = \log_b a, \qquad \frac{\log_a b}{x} = \log_{a^x} b, \quad a^{\log_b c} = c^{\log_b a}. \tag{2.55}$$

Chapter 3

Geometry and Trigonometry

3.1 Plane Geometry

3.1.1 *Angle*

Definition 3.1.

(i) An angle is defined by two intersecting lines on a plane (see Figure (a)).
(ii) On a plane circle, peripheral angle v and central angle w are defined as in (b).
(iii) Circle sector and arc (or circle arc) are defined as in (c).

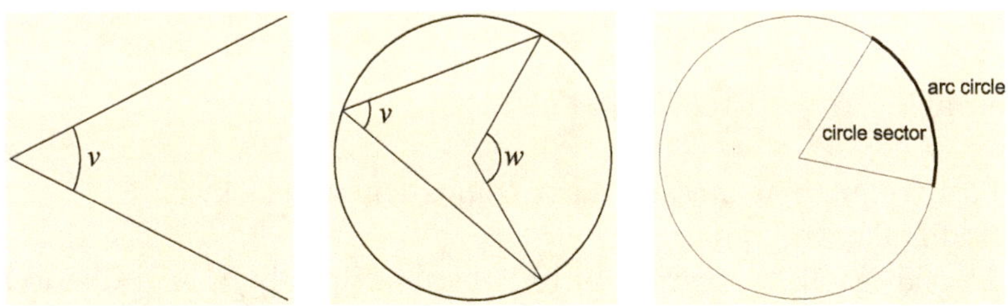

(a) (b) (c)

Theorem 3.1. *In Figure (b) above, $2v = w$. This holds even when one of the angular sides of v is tangent to the circle.*

Thus, two angles with the same angular-arc, i.e., arc of the circle cut by the sides, are equal.

Special case: *A circular angle v (with vertex on circle), opposite to a diameter of the circle, is a right angle:* $v = \frac{\pi}{2}$ $(90°)$.

For two line segments AB and CD, with their endpoints on a circle and intersecting at the point M within the circle, we obtain

$$AM \cdot MB = CM \cdot MD,$$

where XY means the distance between points X and Y.

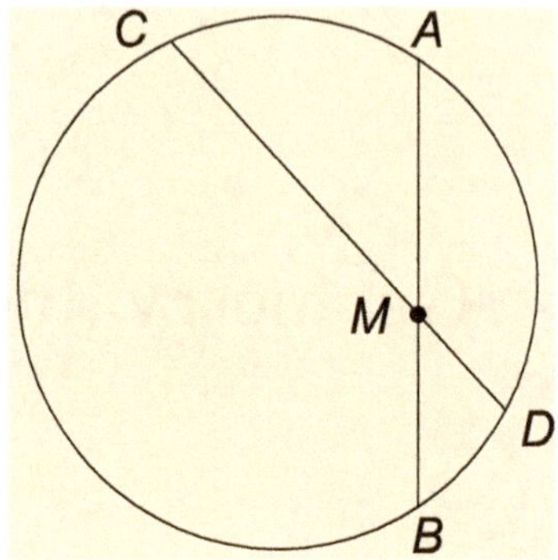

Definition 3.2. Given a circle with radius $r = 1$, *a unit circle.* Two radii build a circle sector (the smaller region inside the circle cut by the two radii). The angle ϕ, between two radii, in *radians*, is equal to the sector's arc length.

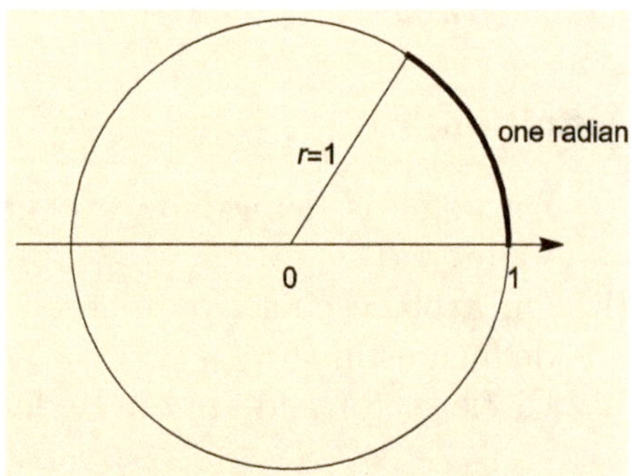

Definition of 1 radian.

Remark.

- Positive angular measurement counts counterclockwise.
- Radians have no unit.
- To convert from degrees to radians: multiply by π and divide by $180°$.
- To change from radians to degrees: multiply by $180°$ and divide by π.
- The concept of angle is generalized in mathematical analysis to angles with arbitrary value and is then specified in radians.
- One radian corresponds exactly to $\frac{180°}{\pi} \approx 57.3°$.

Conversion table: Degrees and radians

Degrees	0°	30°	45°	60°	90°	120°	150°	180°	360°
Radians	0	$\frac{\pi}{6}$	$\frac{\pi}{4}$	$\frac{\pi}{3}$	$\frac{\pi}{2}$	$\frac{2\pi}{3}$	$\frac{5\pi}{6}$	π	2π

(3.1)

3.1.2 *Units of different angular measurements*

Let x be a real number. The relationships between radians, degrees, and gon are given in the following table.
400^{g} (400 gon) corresponds to 360°.

Radian	Degree	Gon
x	$\frac{180^\circ \cdot x}{\pi}$	$\frac{200^{\mathrm{g}} \cdot x}{\pi}$
$\frac{\pi \cdot x}{180}$	x°	$\frac{10}{9} \cdot x^{\mathrm{g}}$
$\frac{\pi \cdot x}{200}$	$0.9 \cdot x^\circ$	x^{g}

(3.2)

Degree, arc minute, and arc second

An arc minute is written as $1'$ and equals $\frac{1}{60}$ degree.

An arc second is written as $1''$ and equals $\frac{1}{3600}$ degree.

For a real number x,

$$x^\circ = 60 \cdot x' = 3600 \cdot x''. \tag{3.3}$$

(a) (b) (c) (d)

Figure 3.1: (Uniformity and congruence) figures (a) and (b) are congruent, (a) and (c) are upright congruent, whereas, (a) and (d) are uniform but not congruent.

Uniformity and congruence

Two objects that are similar in shape but not necessarily of the same size are called *uniform.* If they are of the same size, they are called *congruent.* Further, there are upright and mirrored congruences. For a visualization, see Figure 3.1.

3.1.3 *Reflection in point and line*

For the figures in the plane, the following applies: reflection in a point yields a twist by 180° (an upside-down image). The reflection on a straight line yields a mirror congruent image.

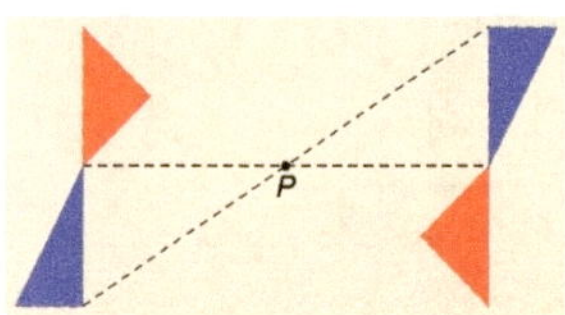

Reflection in a point, P

(a)

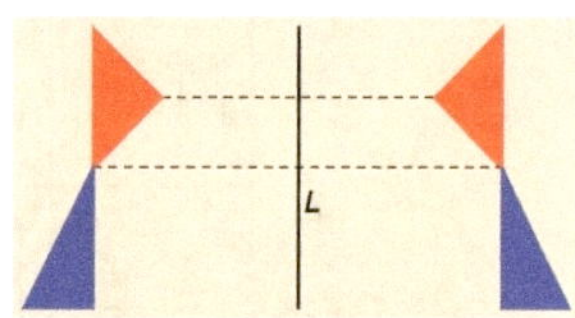

Reflection in a line, L

(b)

Theorem 3.2. *The mirror image of a line of slope k_i, reflected on a line with slope k_s, is a new line with slope, (with equal angles i).*

$$k_r = \frac{k_s^2 k_i + 2k_s - k_i}{1 - k_s^2 + 2k_s k_i}. \quad (3.4)$$

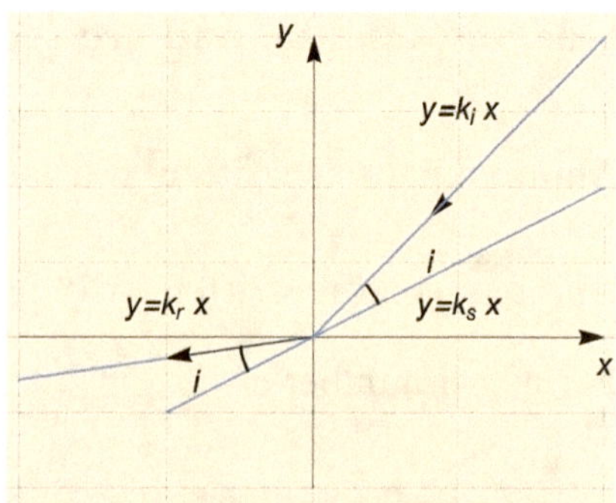

Figure corresponding to formula (3.4).

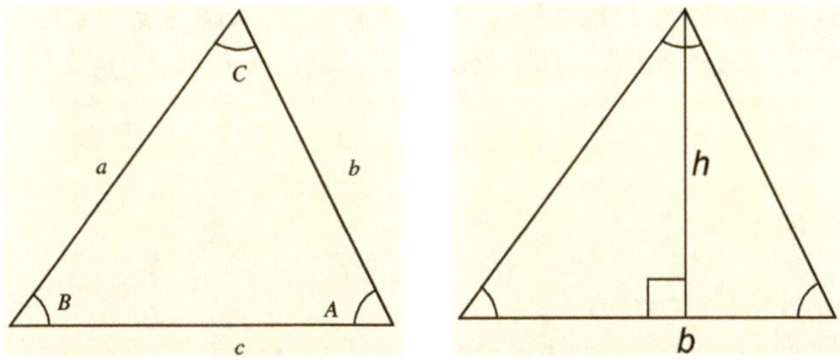

Figure 3.2: The symbols in a triangle. Triangle with base and height.

3.1.4 *Polygon*

Definition 3.3.

(i) For a triangle, the sides and angles are denoted by small and the corresponding capital letters, respectively (see the LHS triangle in Figure 3.2).

(ii) The angle A is opposite to the side a and vice versa, etc. In other words, the angle A and the side a are opposites.

(iii) The angle A and the side b are adjacent, etc. The angle A is intermediate to the sides b and c, etc.

(iv) A *pointy* angle is an angle between 0 and 90°. An *obtuse* angle is an angle between 90° and 180°. The angle 90° is called *right angle*.

(v) The *circumference* of a triangle is the sum of its sides: $a+b+c = 2s$, i.e., s is half of the circumference.

The parallel axiom:
Given any (straight) line and a point not on it, there exists one and only one (straight) line which passes through the point not intersecting the first line.

Theorem 3.3.

(i) *The sum of angles in a triangle is* $180°$ *or* π, *i.e.*, $A + B + C = 180° \ (= \pi)$. *Thus, there is, at most, one obtuse angle in a triangle.*

(ii) *The sum of two side lengths must be longer than the third side. Thus, with symbols as above* $a+b > c$, $b+c > a$, *and* $c+a > b$.

(iii) *For triangles, a large (small) side is opposite to a large (small) angle. In particular, the following equivalence applies:*

$$a \leq b \leq c \Longleftrightarrow A \leq B \leq C.$$

Theorem 3.4.

Top triangle theorem
A parallel transverse is a line parallel to a side of the triangle. The parallel transverse in the figure gives an upper triangle T_1, which is uniform with the triangle ABC. Two triangles are uniform if they have equal angles.

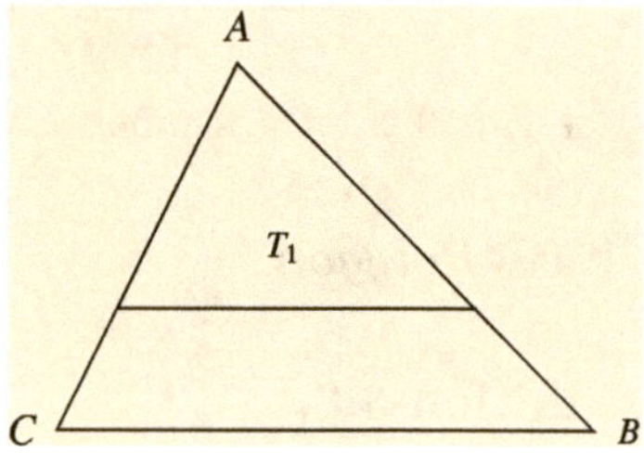

Exterior angle theorem
$A + C = D$.

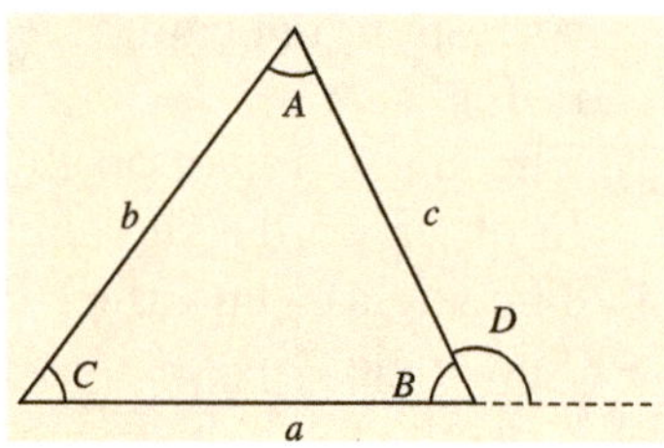

3.1.5 *Types of triangles*

Definition 3.4.

(i) In a right-angled triangle, an angle is 90°. The two sides that are perpendicular are called catheters and the third side, hypotenuse.
(ii) An isosceles triangle has (at least) two sides of equal length and thus (at least) two equal angles.
(iii) In an equilateral triangle, all sides have equal length and all angles $v = 60°$.

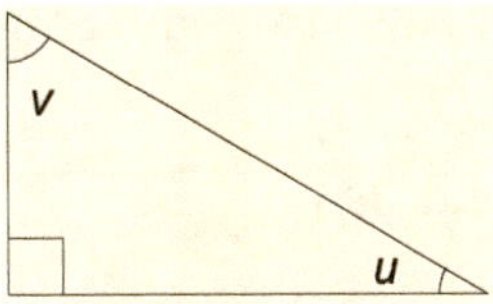

1. Right triangle

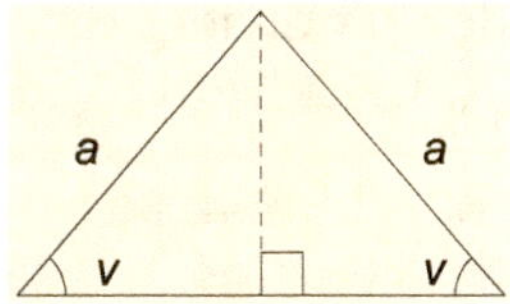

2. Isosceles triangle

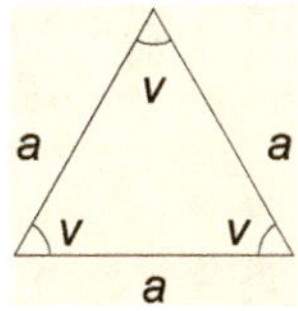

3. Equilateral triangle

Midpoint normal, height, median, and bisector

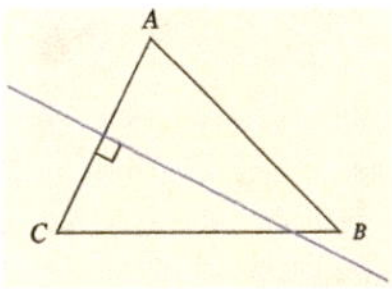

Midpoint normal

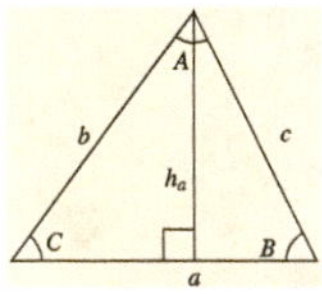

Height

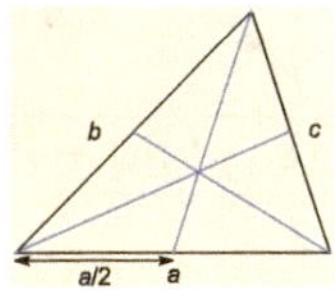

Median

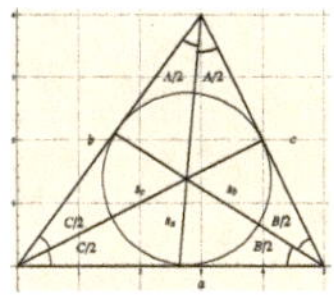

Bisectors

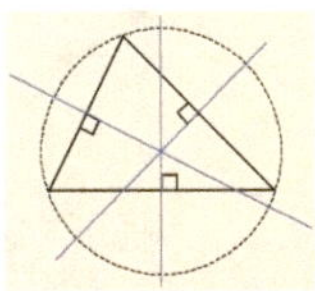

The three midpoint normals intersect in a point, which is the center of the surrounding circle of the triangle.

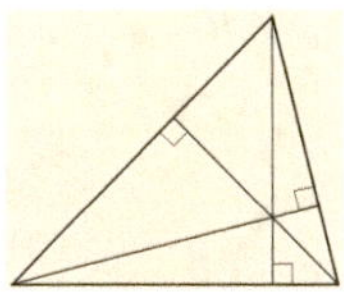

The three heights intersect in a point (might lie outside the triangle).

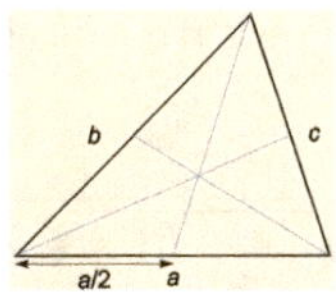

The three medians intersect each other in the centroid T of the triangle.

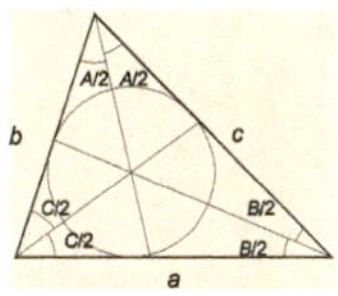

The three bisectors intersect at the center of the inscribed circle.

Definition 3.5 (Refers to the above figures). The line that is perpendicular to a side at the middle is called *midpoint normal.*

On a triangle, a line drawn from a vertex

(a) Perpendicular to the opposite side is called *height* (h_a in the left figure).
(b) To the midpoint of the opposite side is called *median* (the middle figure).

(c) To the opposite side that divides the angle into two equal angles is called *bisector* (s_a right figure).

With $T = \frac{ah_a}{2}$ denoting the triangle's area and $2s = a + b + c$, the following relations hold true.
The radius r of the inscribed circle and that of the circumscribed circle R are given by

$$r = \frac{\sqrt{(a+b-c)(a-b+c)(-a+b+c)(a+b+c)}}{2(a+b+c)}$$

$$= \frac{\sqrt{(s-a)(s-b)(s-c)s}}{s} = \frac{T}{s}$$

and

$$R = \frac{abc}{\sqrt{(-a+b+c)(a-b+c)(a+b-c)(a+b+c)}} \tag{3.5}$$

$$= \frac{abc}{4\sqrt{(s-a)(s-b)(s-c)s}},$$

respectively.

With notations as in the figures on page 43:

$$h_a = b \sin C = c \sin B$$

$$= \frac{\sqrt{(a+b+c)(-a+b+c)(a+b-c)(a-b+c)}}{4a}$$

$$= \frac{\sqrt{(s-a)(s-b)(s-c)s}}{a}. \tag{3.6}$$

$$m_a = \frac{1}{2}\sqrt{2b^2 + 2c^2 - a^2}.$$

$$s_a = \frac{\sqrt{bc\left((b+c)^2 - a^2\right)}}{b+c} = \sqrt{bc\left(1 - \frac{a^2}{(b+c)^2}\right)}.$$

Theorem 3.5. *The three midpoint normals in a triangle intersect at a point P, which is the center of the circle that circumscribes the triangle.*

The three heights of a triangle intersect at a point. This point need not be within the triangle.
The three medians of a triangle intersect at the point T; the centroid of the triangle.
The three bisectors of a triangle intersect at a point P, which is the center of the circle inscribed in the triangle.

Theorem 3.6 (The bisector theorem). *Let AD be the bisector opposite to the side BC in the triangle ABC (see the accompanying figure), then*

$$\frac{BD}{AB} = \frac{DC}{AC}\left(= \frac{\sin(A/2)}{\sin D_1}\right). \tag{3.7}$$

Furthermore, all points on AD have equal perpendicular distances to AB and AC.

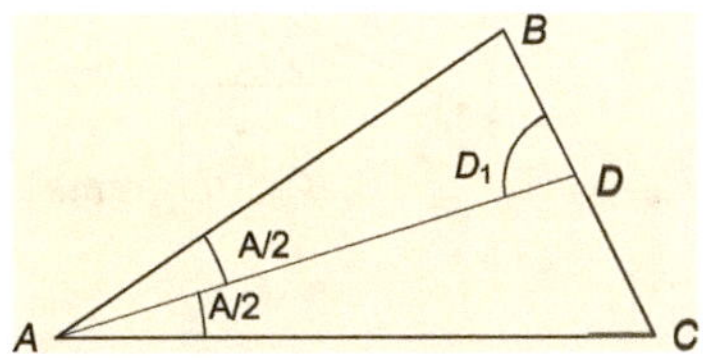

Theorem 3.7 (Some theorems in geometry).

Ceva's theorem:

Three lines between vertices and opposite sides in a triangle, intersecting in a common point, inside the triangle, divide the sides according to

$$a_1b_1c_1 = a_2b_2c_2. \tag{3.8}$$

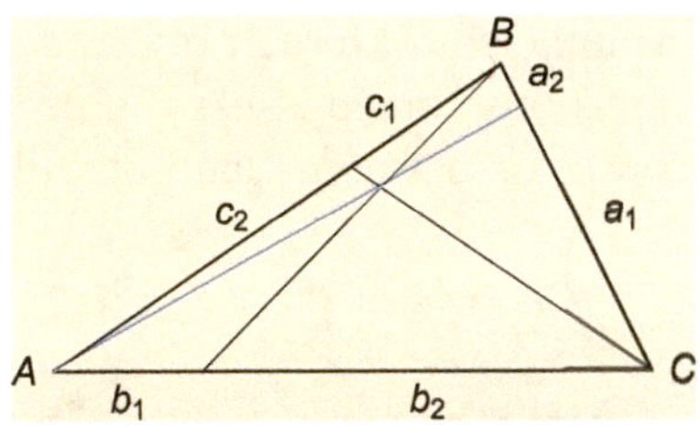

Menelao's theorem: *Let XY denote the distance between two points X and Y.*

For a line L that intersects the sides AB and AC and the extension of BC, of a triangle ABC, at the points R, Q, and P, respectively (see the following figure), yields

$$\frac{PB}{PC}\cdot\frac{QC}{QA}\cdot\frac{RA}{RB} = 1. \tag{3.9}$$

Ptolemaios' theorem: *For the line-segments obtained by diameter intersections and the sides of a quadrilateral inscribed in a circle, the following apply (the following figure to the right):*

$$mn = ac + bd, \qquad \frac{m}{n} = \frac{ad + bc}{ab + cd}. \tag{3.10}$$

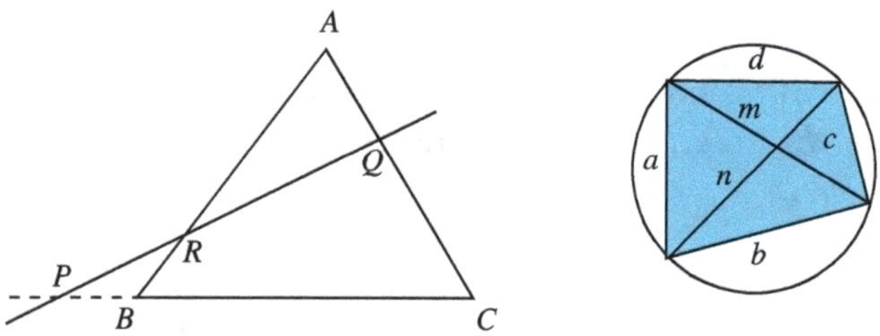

The theorems of Menelaos and Ptolemaios.

3.1.6 *Regular polygons*

Definition 3.6.

(i) In an equilateral polygon, all side lengths and angles are equal.
(ii) Mosaic equilateral polygons completely fill out the plane.

Among equilateral polygons, there are only triangles, squares, and hexagons that possess the property (ii). For example, regular penta-, septa, or octagons cannot fully cover the plane (see Figure 3.3).

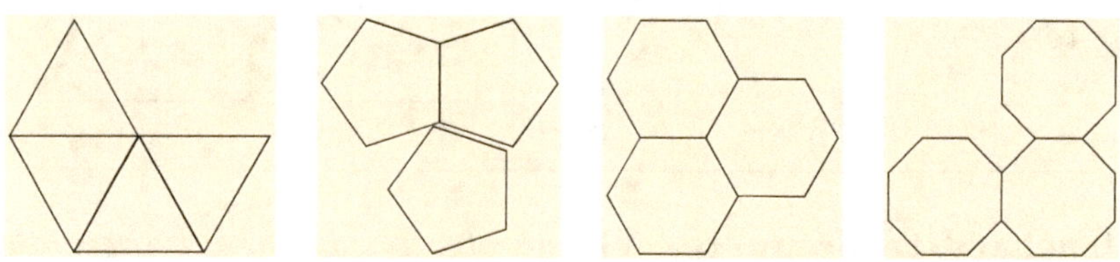

Triangles *Pentagons* *Hexagons* *Octagons*

Figure 3.3: Some equilateral polygons.

Theorem 3.8.

(i) *In an n-polygon (not necessarily equilateral), the sum of vertex angles is*

$$(n - 2) \cdot 180^\circ. \tag{3.11}$$

(ii) (a) *The angle between two adjacent sides/edges in a regular n-polygon is*

$$\frac{n-2}{n} \cdot 180^\circ. \tag{3.12}$$

(b) *The area A of an equilateral n-polygon with side length d is given by*

$$A = \frac{nd^2}{4} \cdot \tan\left(\frac{n-2}{n} \cdot 90^\circ\right). \tag{3.13}$$

Theorem 3.9 (Pythagorean and similar theorems (figure on page 56)).

(i) *In a right angled triangle with $C = 90^\circ$*

$$c^2 = a^2 + b^2. \tag{3.14}$$

(ii) *Heron's formula gives the area T, of a triangle, in terms of the size of its sides a, b, c, as follows:*

$$\begin{aligned} T &= \frac{1}{4}\sqrt{(a+b+c)\,(-a+b+c)\,(a-b+c)\,(a+b-c)} \\ &= \sqrt{s(s-a)(s-b)(s-c)}, \quad \textit{where } 2s = a+b+c. \end{aligned} \tag{3.15}$$

(iii) *Figure* 3.4:

(a) *The parallelogram law: In a parallelogram,*

$$c^2 + d^2 = 2(a^2 + b^2),$$

where a and b are the side lengths and c and d are the lengths of its diagonals.

(b) *The diagonals in a rhombus are orthogonal.*

(c) *The area of a trapezoid which is not a parallelogram is*

$$\frac{b\sqrt{(a-b+c+d)\,(a+b-c-d)\,(-a+b+c-d)\,(a+b+c-d)}}{2(b-d)}, \tag{3.16}$$

$b > d$ with b and d parallel (Figure 3.4 *(right), page* 49*).*

Remark. The converse of Pythagorean theorem:
If $c^2 = a^2 + b^2$, then $C = 90°$, which is true as well.

A Pythagorean integer triple consists of three positive integers (a, b, c) satisfying (3.14). All Pythagorean integers can be generated by

$$\begin{cases} a = x^2 - y^2, \\ b = 2xy, \\ c = x^2 + y^2, \end{cases} \tag{3.17}$$

where x, y, with $x > y$, are positive integers.

- The integers a and b can be chosen so that $\frac{a}{b}$ is (arbitrarily) close to 1, and satisfy Pythagorean theorem (3.14). This is obtained choosing integers x and y in (3.17) with properties $x > y > 0$ and $\frac{x}{y} \approx 1 + \sqrt{2}$.

Triangles with integer sides and a 60° angle between a and b (up to uniformity) are generated by

$$\begin{cases} a = (x + y)\,(3x - y)\,, \\ b = 4xy, \\ c = 3x^2 + y^2, \end{cases}$$

where $3x > y > 0$.

Similar relations for triangles with one 120° angle are (3.18)

$$\begin{cases} a = (x - y)(3x + y), \\ b = 4xy, \\ c = 3x^2 + y^2, \end{cases}$$

where $x > y > 0$.

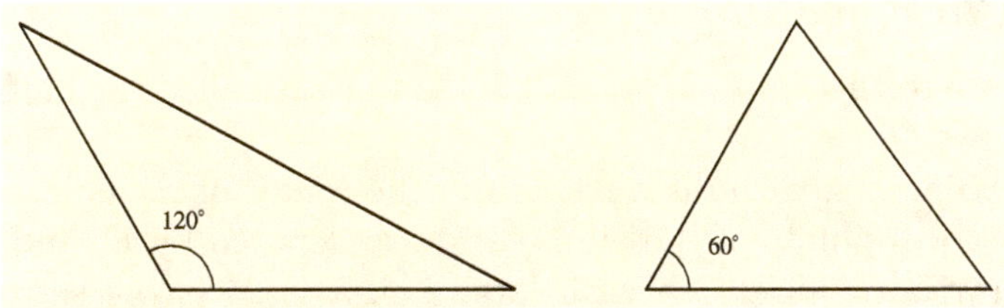

Triangles with one angle equal to 120° and 60°, respectively.

The sides of some triangles where the angle between a and b are 60°, 90°, and 120°, respectively:

60°			90°			120°		
a	b	c	a	b	c	a	b	c
1	1	1	3	4	5	3	5	7
5	8	7	5	12	13	7	8	13
16	21	19	20	21	29	11	24	31

(3.19)

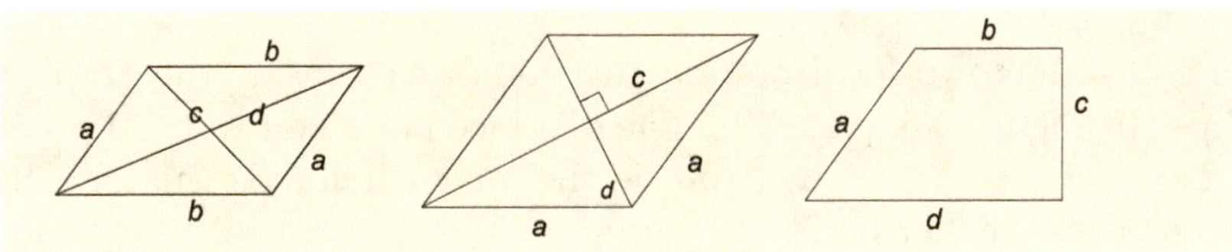

Figure 3.4: Left: parallelogram. Middle: rhombus (an equal-sided parallelogram with side length a). Right: parallel-trapezoid (not a parallelogram).

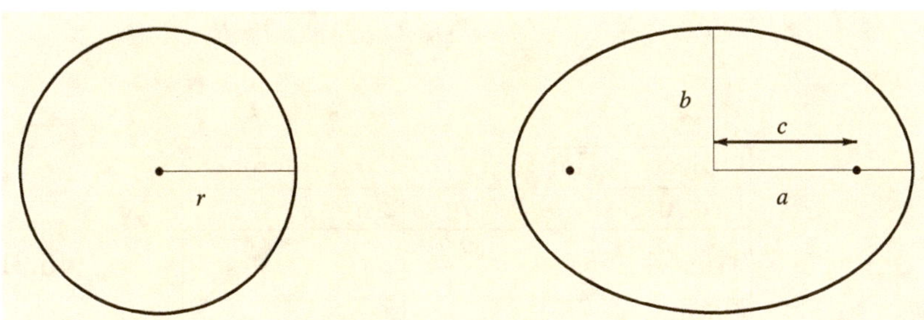

Figure 3.5: LHS: circle. A circle is a special case of an ellipse. RHS ellipse.

3.1.7 *Circle and ellipse*

In the figure to the right in Figure 3.5, a and b are called the half-axis of the ellipse.

The two points, with distance c from the center of the ellipse, are called the focal points of the ellipse and are denoted by F_1 and F_2.

Geometrically, an ellipse with large half-axis $2a$ is the set of all points P such that

$$PF_1 + PF_2 = 2a.$$

If the center of the ellipse has the coordinates $C = (x_0, y_0)$, then its equation is given by

$$(x, y) : \quad \frac{(x - x_0)^2}{a^2} + \frac{(y - y_0)^2}{b^2} = 1. \tag{3.20}$$

In particular, if the center of the ellipse coincides with the origin $x_0 = y_0 = 0$, the equation (3.20) becomes

$$\frac{x^2}{a^2} + \frac{y^2}{b^2} = 1.$$

If $a = b$, we get corresponding equations for circles with the center $C = (x_0, y_0)$.
An ellipse has two focal points (the points seen inside the graph of ellipse in Figure 3.5, page 49). The distance between two focal points is $2c$ and $a^2 = b^2 + c^2$, where a is the longer half-axis (major axis) and b is the smaller half-axis (minor axis).
There is no simple expression for perimeter of an ellipse $\mathcal{O}$, but if $a \approx b$, then $\mathcal{O} \approx \pi(a + b)$. An exact expression can be given with an *elliptic* integral:

$$\mathcal{O} = 4 \int_0^{\pi/2} \sqrt{a^2 + (b^2 - a^2)\cos^2 t}\, dt.$$

Object	Entities	Area	Circumference
Rectangle	a, b	ab	$2(a + b)$
Triangle	b, h	$\frac{b\,h}{2}$	
Circle	r	πr^2	$2\pi\, r$
Ellipse	a, b	$\pi a\, b$	

(3.21)

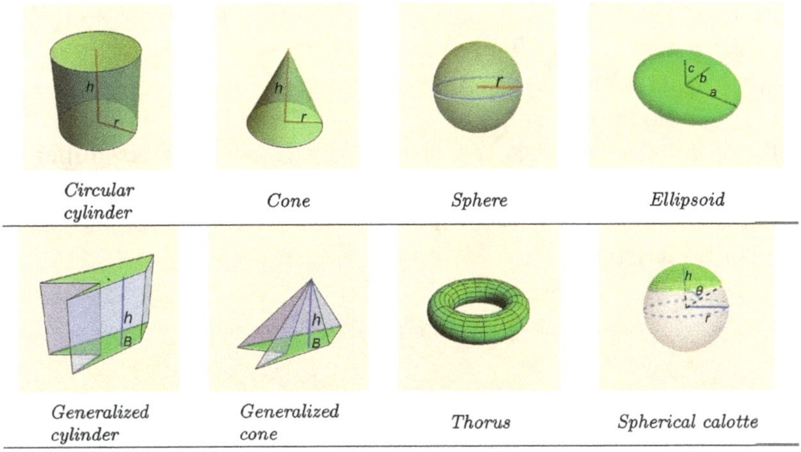

Figure 3.6: Names of some bodies.

3.2 Space Geometry

3.2.1 *Names and volumes of some common bodies*

Name, volume, and area of the bodies in Figure 3.6.

Body	Volume	Area of enclosing surface	
(1) Circular cylinder	$\pi r^2 h$	$2\pi r^2 + 2\pi r h$	
(2) Generalized cylinder	Bh		
(3) Circular cone	$\dfrac{\pi r^2 h}{3}$	$\pi r^2 + 2\pi r\sqrt{r^2 + h^2}$	
(4) Generalized cone	$\dfrac{Bh}{3}$		
(5) Globe	$\dfrac{4\pi r^3}{3}$	$4\pi r^2$	
(6) Ellipsoid	$\dfrac{4\pi abc}{3}$		
(7) Circular torus	$2R(\pi r)^2$	$4\pi^2 r\, R$	
(8) Spherical calotte	$\dfrac{\pi h^2}{3}(3r - h)$	$\pi(4\, r\, h - h^2)$	Including the disk

A circular cylinder has a mantle surface with area $2\pi rh$ (see (1)).

The area of a circular cone's mantle surface is $2\pi r\sqrt{r^2+h^2}$ (see (3)). a, b, and c are the half-axis of the ellipsoid.

There is no simple expression for an ellipsoid's surface area.

Torus volume in (7) is the product $2\pi R \cdot \pi r^2$ (the circumference of the mantle surface times the area of the cross-section).

The height of the calotte in (8) is h. The volume of the calotte can also be written as

$$V = \frac{\pi r^3}{3}(1-\cos\theta)^2(2+\cos\theta).$$

Its area is $A = 2\pi rh = 2\pi r^2(1-\cos\theta)$, including (the area of) the disk.

3.2.2 *Parallelepiped and the five regular polyhedra*

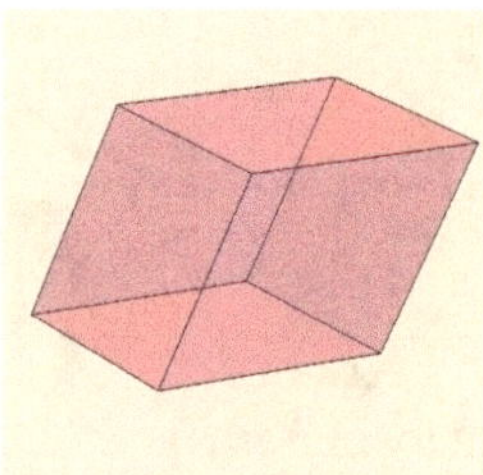

Parallelepiped

Tetrahedron

Cube

Octahedron

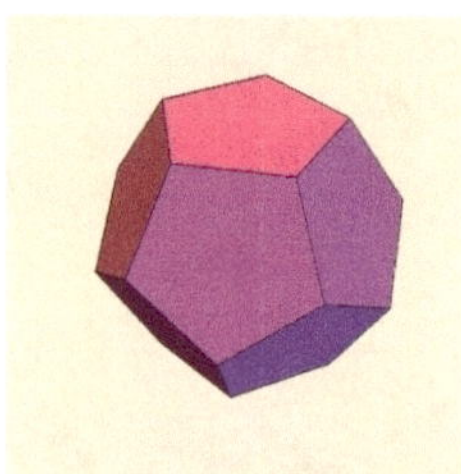

Dodecahedron

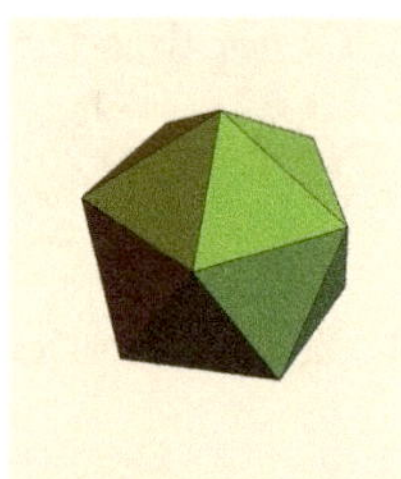

Isocahedron

Polyeder	Number of nodes N	Number of edges K	Number of sides S	Volume for edge d	Area of surface
Tetrahedron	4	6	4	$\frac{1}{6\sqrt{2}}d^3$	$\sqrt{3}\,d^2$
Cube	8	12	6	d^3	$6d^2$
Octahedron	6	12	8	$\frac{\sqrt{2}}{3}d^3$	$2\sqrt{3}\,d^2$
Dodecahedron	20	30	12	$\frac{15+7\sqrt{5}}{4}d^3$	$3\sqrt{5(5+2\sqrt{5})}\,d^2$
Icosahedron	12	30	20	$\frac{5\left(3+\sqrt{5}\right)}{12}d^3$	$5\sqrt{3}\,d^2$

(3.22)

Theorem 3.10 (The Euler relation). *The following relationship applies between N = number of corners, K = number of egdes, and S = number of sides, which also applies to irregular polyhedra.*

$$S - K + N = 2. \tag{3.23}$$

3.3 Coordinate System ($\mathbb{R}^2$)

A coordinate system (in two dimensions) consists of one flat surface and two perpendicular axes (coordinate axes).

(i) A two-dimensional coordinate system is spanned by two perpendicular coordinate axes, which we can call x-axis (horizontal) and y-axis (vertical), respectively.

(ii) A point $P = (x, y)$ in such a coordinate system has an x-coordinate, and is read from the point perpendicular to x-axis, see Figure 3.3.
The y-coordinate reads similarly as perpendicular to y-axis. x and y related to the axis, are called *Cartesian coordinates.*

(iii) The point $(0, 0)$ is called the origin of the (coordinate system).

(iv) The distance d between two points $P_1 = (x_1, y_1)$ and $P_2 = (x_2, y_2)$ in the plane is

$$d = \sqrt{(x_1 - x_2)^2 + (y_1 - y_2)^2}.$$

(v) The equation for an ellipse with axes parallel to the coordinate axes and the equation of a circle, both centered at (x_0, y_0), are given by

$$(x, y): \quad \left(\frac{x - x_0}{a}\right)^2 + \left(\frac{y - y_0}{b}\right)^2 = 1 \quad \text{and}$$

$$(x - x_0)^2 + (y - y_0)^2 = r^2, \tag{3.24}$$

respectively. The radius of the circle is r. If $a = b =: r$ in the top equation in (3.24), the equation of the circle is obtained.

Coordinate system with x- and y-axis, and the distance d between the points P_1 and P_2.

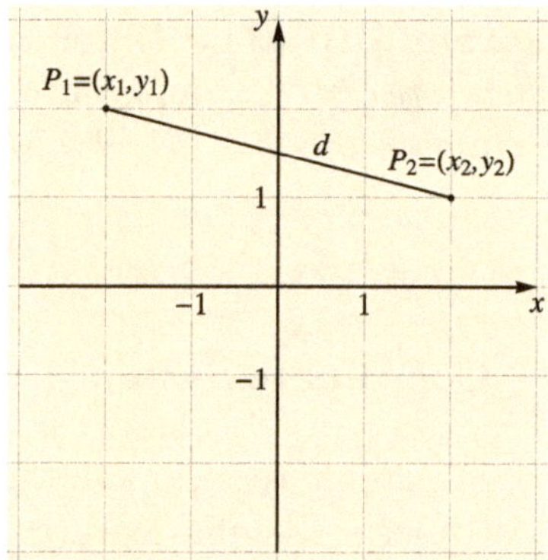

3.4 Trigonometry

Definition 3.7 (Definition of trigonometric functions).

(i) The unit circle in a Cartesian coordinate system is a circle with the radius 1 and the center at the origin $(0, 0)$. (Here cos-axis and x-axis and similarly sin-axis and y-axis coincide.)

(ii) For (x, y) on the circle, we consider the vector u from $(0, 0)$ to (x, y). The angle v between the positive x-axis and v counts positive counterclockwise and negative clockwise.

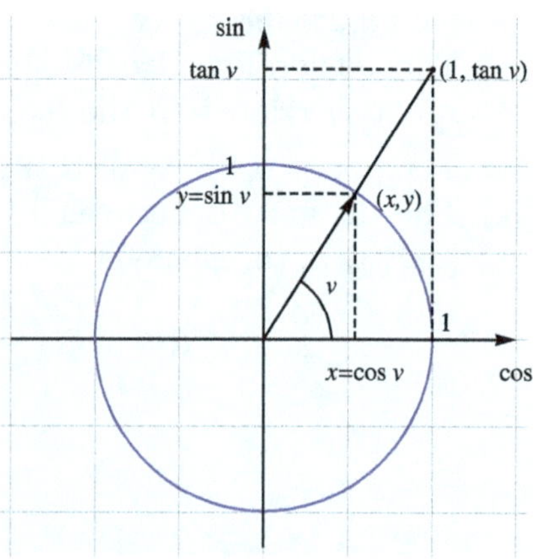

(iii) With notation as in the figure

$$\cos v = x, \quad \sin v = y, \quad \tan v = \frac{y}{x}, \quad \cot v = \frac{x}{y}. \tag{3.25}$$

In this way, parts of x- and y-axes, that are inside the unit circle, are considered cos- and sin-axes, respectively. Further, *tangent*-axis is the (vertical), oriented, real-line passing through the point $(1, 0)$ on the circle, parallel to the sin (y-axis). Likewise *co-tangent*-axis is the (horizontal), oriented, real-line passing through the point $(0, 1)$ on the circle, parallel to the cos (x-axis). In this setting, $\tan v$ is the length of the line-segment between the point $(1, 0)$ and the intersection of the *trace* of v with the tangent-axis. Similarly, the $\cot v$ is the length of the line-segment between the point $(0, 1)$ and the intersection of the *trace* of v with the co-tangent-axis.

(iv) Two angles with sums equal to $\frac{\pi}{2}$ (90°) are called *complementary* (Figure on page 56).

(v) Two angles are *supplementary* to each other if their sum is π (180°).

Amplitude A, angular velocity ω, frequency ν, period T

For the function $f(t)$, with $A > 0$, defined by

$$f(t) := A\sin(\omega\, t + \alpha), \tag{3.26}$$

A is called the *amplitude*, ω, angular velocity or angular frequency, and α, the phase (related to the function $g(t) = A\sin\omega\, t$, see also (3.38), page 60).

Period	Angular speed	Frequency
$T = \frac{1}{\nu} = \frac{2\pi}{\omega}$	$\omega = 2\pi\,\nu = \frac{2\pi}{T}$	$\nu = \frac{\omega}{2\pi} = \frac{1}{T}$

The period T of a real function f is the smallest positive number, such that

$$f(t+T) = f(t), \quad t \in \mathbb{R}.$$

Definition 3.8 (Further trigonometric functions). csc reads "cosecant" and sec reads "secant" functions, which are defined as

$$\csc v := \frac{1}{\sin v} \quad \text{and} \quad \sec v := \frac{1}{\cos v}, \qquad \text{respectively.} \tag{3.27}$$

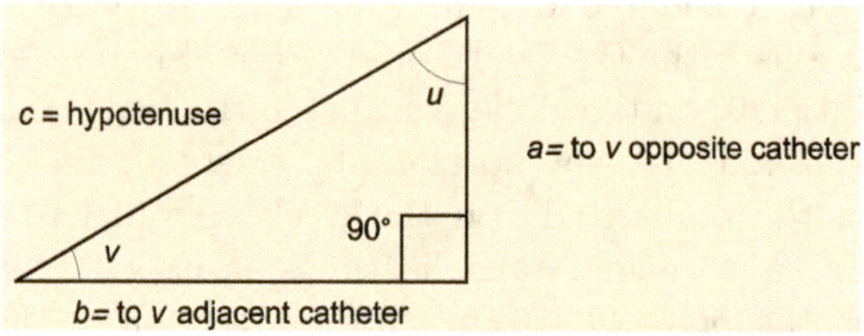

In a right angled triangle, that is with a right angled 90° or $\frac{\pi}{2}$, the Pythagorean theorem: $a^2 + b^2 = c^2$ holds. Since $u + v = \frac{\pi}{2}$, u and v are complements of each other.

In a right angled triangle for a pointy angle v, the trigonometric functions are defined as

$$\begin{aligned}
\sin v &:= \frac{\text{opposite catheter}}{\text{hypotenuse}} = \frac{a}{c},\\
\cos v &:= \frac{\text{adjacent catheter}}{\text{hypotenuse}} = \frac{b}{c},\\
\tan v &:= \frac{\text{opposite catheter}}{\text{adjacent catheter}} = \frac{a}{b},\\
\cot v &:= \frac{\text{adjacent catheter}}{\text{opposite catheter}} = \frac{b}{a},\\
\sec v &:= \frac{\text{hypotenuse}}{\text{adjacent catheter}} = \frac{c}{b},\\
\csc v &:= \frac{\text{hypotenuse}}{\text{opposite catheter}} = \frac{c}{a}.
\end{aligned} \tag{3.28}$$

Some trigonometric relations

$$
\begin{aligned}
&\sin(v + n \cdot 2\pi) = \sin v, \ \cos(v + n \cdot 2\pi) = \cos v, \quad n \in \mathbb{Z}, \\
&\tan(v + n \cdot \pi) = \tan v, \ \cot(v + n \cdot \pi) = \cot v, \quad n \in \mathbb{Z}, \\
&\sin(-v) = -\sin v, \qquad \tan(-v) = -\tan v, \\
&\cos(-v) = \cos v, \qquad \cot(-v) = -\cot v, \\
&\sin v = \cos(\tfrac{\pi}{2} - v), \qquad \tan v = \cot(\tfrac{\pi}{2} - v), \\
&\cos v = \sin(\tfrac{\pi}{2} - v), \qquad \cot v = \tan(\tfrac{\pi}{2} - v), \\
&\tan v = \frac{\sin v}{\cos v}, \qquad \cot v = \frac{1}{\tan v}, \\
&\sin^2 v + \cos^2 v = 1 \qquad \text{(The trigonometric identity)}, \\
&\sin^2 v = 1 - \cos^2 v, \qquad \cos^2 v = 1 - \sin^2 v, \\
&\sin(\pi - v) = \sin v, \qquad \cos(\pi - v) = -\cos v, \\
&\tan(\pi - v) = -\tan v, \ \cot(\pi - v) = -\cot v.
\end{aligned}
\tag{3.29}
$$

Remarks. In the above identities, the angles are in radians. Conversion between degrees and radians can be found on the page 39.

To convert an angle v from radian to degree, replace $\frac{\pi}{2}$ by $90°$, and π by $180°$.

The first two rows of (3.29) indicate that $\sin x$ and $\cos x$ have the period 2π and $\tan x$ and $\cot x$ has the period π.

The third and fourth rows in (3.29) indicate that sin, tan, and cot are odd functions, while cos is even.

3.4.1 *Basic theorems*

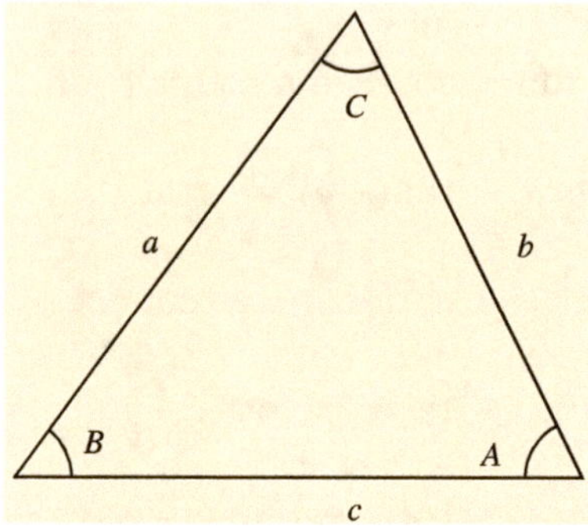

Figure 3.7: Side- and angle-notations in a triangle.

Theorem 3.11. *With notations as in Figure 3.7 and T the area of the triangle, following hold true:*

$$\text{Area theorem: } T = \frac{ab\sin C}{2}. \tag{3.30}$$

$$\text{Sine theorem: } \frac{\sin A}{a} = \frac{\sin B}{b} = \frac{\sin C}{c} = \frac{1}{2R}. \tag{3.31}$$

$$\text{Cosine theorem: } a^2 + b^2 - 2ab\cos C = c^2. \tag{3.32}$$

$$\text{Tangent theorem: } \frac{a-b}{a+b} = \frac{\tan\left(\frac{1}{2}(A-B)\right)}{\tan\left(\frac{1}{2}(A+B)\right)}. \tag{3.33}$$

R in (3.31) is the radius of the circumscribed circle of the triangle.

3.5 Addition Formulas

Theorem 3.12. *The following identities* (3.34)–(3.37) *are true for all angles α and β.*

3.5.1 *Addition formulas for sine and cosine functions*

$$\begin{aligned}\cos(\alpha-\beta) &= \cos\alpha\cos\beta+\sin\alpha\sin\beta,\\ \sin(\alpha+\beta) &= \sin\alpha\cos\beta+\cos\alpha\sin\beta,\\ \sin(\alpha-\beta) &= \sin\alpha\cos\beta-\cos\alpha\sin\beta,\\ \cos(\alpha+\beta) &= \cos\alpha\cos\beta-\sin\alpha\sin\beta.\end{aligned} \tag{3.34}$$

$$\begin{aligned}\sin(\alpha+\beta)+\sin(\alpha-\beta) &= 2\sin\alpha\cos\beta,\\ \sin(\alpha+\beta)-\sin(\alpha-\beta) &= 2\,\cos\alpha\,\sin\beta,\\ \cos(\alpha-\beta)+\cos(\alpha+\beta) &= 2\,\cos\alpha\,\cos\beta,\\ \cos(\alpha-\beta)-\cos(\alpha+\beta) &= 2\,\sin\alpha\,\sin\beta.\end{aligned} \tag{3.35}$$

$$\begin{aligned}\sin\alpha+\sin\beta &= 2\sin\frac{\alpha+\beta}{2}\cos\frac{\alpha-\beta}{2},\\ \sin\alpha-\sin\beta &= 2\sin\frac{\alpha-\beta}{2}\cos\frac{\alpha+\beta}{2},\\ \cos\alpha+\cos\beta &= 2\cos\frac{\alpha+\beta}{2}\cos\frac{\alpha-\beta}{2},\\ \cos\alpha-\cos\beta &= -2\sin\frac{\alpha+\beta}{2}\sin\frac{\alpha-\beta}{2}.\end{aligned} \tag{3.36}$$

3.5.2 *Addition formulas for tangent*

$$\tan(\alpha+\beta)=\frac{\tan\alpha+\tan\beta}{1-\tan\alpha\tan\beta}\quad \tan(\alpha-\beta)=\frac{\tan\alpha-\tan\beta}{1+\tan\alpha\tan\beta},$$

$$\cot(\alpha+\beta)=\frac{\cot\alpha\cot\beta-1}{\cot\alpha+\cot\beta}\quad \cot(\alpha-\beta)=\frac{\cot\alpha\cot\beta+1}{\cot\alpha-\cot\beta}. \tag{3.37}$$

3.5.3 *Phase–amplitude form*

Assume $a \neq 0$ or $b \neq 0$. Then

$$a \sin x + b \cos x = A \sin(x + \alpha),$$

$$\text{where } A = \sqrt{a^2 + b^2}, \text{ and } \begin{cases} \alpha = \arctan(b/a), & a > 0, \\ \alpha = \pi + \arctan(b/a), & a < 0, \\ \alpha = \dfrac{\pi}{2}, & a = 0,\ b > 0, \\ \alpha = -\dfrac{\pi}{2}, & a = 0,\ b < 0. \end{cases} \tag{3.38}$$

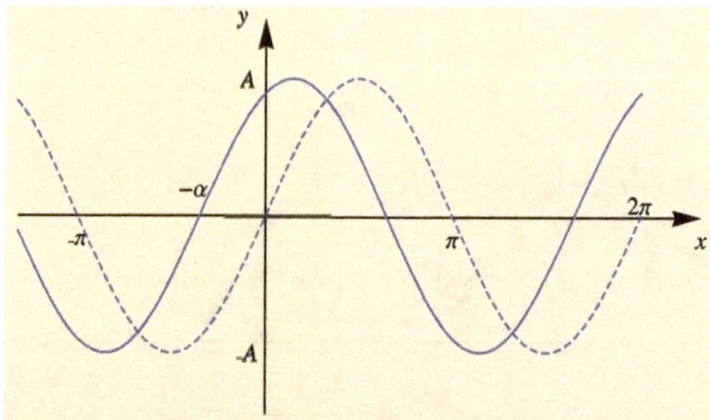

The graph of a phase-shifted function (solid) $y = A\sin(x + \alpha)$ intersects the x-axis at $x = -\alpha$. Dashed curve is $y = A \sin x$.
Note: A graph $y(t) = A\sin(\omega t + \alpha)$ intersects the t- (time) axis in $t = -\alpha/\omega$, where $y(t) = 0$.

3.5.4 *Identities for double and half angles*

$$\sin 2\alpha = 2 \sin\alpha \cos\alpha, \quad \tan 2\alpha = \frac{2\tan\alpha}{1 - \tan^2\alpha}.$$

$$\begin{aligned} \cos 2\alpha &= \cos^2\alpha - \sin^2\alpha \\ &= 2\cos^2\alpha - 1 = 1 - 2\sin^2\alpha = \cos^4\alpha - \sin^4\alpha. \end{aligned} \tag{3.39}$$

$$\cos^2\left(\frac{x}{2}\right) = \frac{1 + \cos x}{2}, \quad \sin^2\left(\frac{x}{2}\right) = \frac{1 - \cos x}{2}. \tag{3.40}$$

$$\begin{aligned} \sin^2\alpha &= \frac{1 - \cos 2\alpha}{2} = \frac{\tan^2\alpha}{1 + \tan^2\alpha} = \frac{1}{1 + \cot^2\alpha}, \\ \cos^2\alpha &= \frac{1 + \cos 2\alpha}{2} = \frac{\cot^2\alpha}{1 + \cot^2\alpha} = \frac{1}{1 + \tan^2\alpha}, \\ \tan\alpha &= \frac{\sin 2\alpha}{1 + \cos 2\alpha}, \quad \cot\alpha = \frac{\cos 2\alpha}{1 - \cos 2\alpha}. \end{aligned} \tag{3.41}$$

Remark. Observe that in (3.38), for $b > 0$ and $a < 0$, α should be chosen in the second quadrant. Then, the RHS in (3.38) is known as "phase–amplitude form". $A(> 0)$ is the amplitude and the phase constant is $-\alpha$. It is important to note that, on the LHS, the argument for sine and cosine is the same.

3.5.5 *Some exact values*

x (degree)	x (rad)	$\sin x$	$\cos x$	$\tan x$	$\cot x$
$0°$	0	0	1	0	$-$
$15°$	$\pi/12$	$\frac{\sqrt{6}-\sqrt{2}}{4}$	$\frac{\sqrt{6}+\sqrt{2}}{4}$	$2-\sqrt{3}$	$2+\sqrt{3}$
$30°$	$\pi/6$	$\frac{1}{2}$	$\frac{\sqrt{3}}{2}$	$\frac{1}{\sqrt{3}}$	$\sqrt{3}$
$45°$	$\pi/4$	$\frac{1}{\sqrt{2}}$	$\frac{1}{\sqrt{2}}$	1	1
$60°$	$\pi/3$	$\frac{\sqrt{3}}{2}$	$\frac{1}{2}$	$\sqrt{3}$	$\frac{1}{\sqrt{3}}$
$75°$	$5\pi/12$	$\frac{\sqrt{6}+\sqrt{2}}{4}$	$\frac{\sqrt{6}-\sqrt{2}}{4}$	$2+\sqrt{3}$	$2-\sqrt{3}$
$90°$	$\pi/2$	1	0	$-$	0

(3.42)

3.6 Inverse Trigonometric Functions

Definition 3.9. The inverse trigonometric functions or *arcus functions* are defined as

$$\begin{aligned}
&\begin{cases} y = \arcsin x \\ -1 \leq x \leq 1 \end{cases} \iff \begin{cases} x = \sin y \\ -\pi/2 \leq y \leq \pi/2 \end{cases} \\
&\begin{cases} y = \arccos x \\ -1 \leq x \leq 1 \end{cases} \iff \begin{cases} x = \cos y \\ 0 \leq y \leq \pi \end{cases} \\
&\begin{cases} y = \arctan x \\ -\infty < x < \infty \end{cases} \iff \begin{cases} x = \tan y \\ -\pi/2 < y < \pi/2 \end{cases} \\
&\begin{cases} y = \operatorname{arccot} x \\ -\infty < x < \infty \end{cases} \iff \begin{cases} x = \cot y \\ 0 < y < \pi. \end{cases}
\end{aligned} \tag{3.43}$$

Relations between the arcus functions

$$\arcsin(-x) = -\arcsin x, \quad \arccos(-x) = \pi - \arccos x$$

$$\arctan(-x) = -\arctan x, \quad \operatorname{arccot}(-x) = \pi - \operatorname{arccot} x$$

$$\arcsin x + \arccos x = \frac{\pi}{2}, \quad \arctan x + \operatorname{arccot} x = \frac{\pi}{2}$$

$$\arctan\frac{1}{x} = \begin{cases} -\dfrac{\pi}{2} - \arctan x, \text{ if } x < 0 \\[2ex] \dfrac{\pi}{2} - \arctan x, \text{ if } x > 0. \end{cases}$$

$$\arcsin x = \arctan\frac{x}{\sqrt{1-x^2}} = \operatorname{arccot}\frac{\sqrt{1-x^2}}{x} =$$

$$\begin{cases} -\arccos\sqrt{1-x^2}, & x < 0 \\[2ex] \arccos\sqrt{1-x^2}, & x > 0. \end{cases}$$

$$\arccos x = \begin{cases} \pi + \arctan\dfrac{\sqrt{1-x^2}}{x} = \pi - \arcsin\sqrt{1-x^2} = & \\[2ex] \pi + \operatorname{arccot}\dfrac{x}{\sqrt{1-x^2}}, & x < 0 \\[2ex] \arctan\dfrac{\sqrt{1-x^2}}{x} = \arcsin\sqrt{1-x^2} = & \\[2ex] \operatorname{arccot}\dfrac{x}{\sqrt{1-x^2}}, & x > 0. \end{cases}$$

$$\arctan x = \begin{cases} \arcsin\dfrac{x}{\sqrt{1+x^2}} - \pi/2 = \arccos\dfrac{1}{\sqrt{1+x^2}} = \operatorname{arccot}\dfrac{1}{x}, & x < 0 \\[2ex] \arcsin\dfrac{x}{\sqrt{1+x^2}} = \arccos\dfrac{1}{\sqrt{1+x^2}} = \operatorname{arccot}\dfrac{1}{x}, & x > 0. \end{cases}$$

$$\operatorname{arccot} x = \begin{cases} -\arcsin\dfrac{1}{\sqrt{1+x^2}} = \arccos\dfrac{x}{\sqrt{1+x^2}} - \pi = \arctan\dfrac{1}{x}, & x < 0 \\[2ex] \arcsin\dfrac{1}{\sqrt{1+x^2}} = \arccos\dfrac{x}{\sqrt{1+x^2}} = \arctan\dfrac{1}{x}, & x > 0. \end{cases} \tag{3.44}$$

3.7 Trigonometric Equations

(i) $\cos\alpha = \cos\beta \iff \alpha = \pm\beta + 2n\pi,\ n \in \mathbb{Z}$.
(ii) $\sin\alpha = \sin\beta \iff \alpha = \beta + 2\pi n$ or $\alpha = \pi - \beta + 2n\pi,\ n \in \mathbb{Z}$.
(iii) $\tan\alpha = \tan\beta \iff \alpha = \beta + n\pi,\ n \in \mathbb{Z}$.
Note that for $\tan\alpha + \tan\beta = 0$, one moves over one of the terms to the other side:

$$\tan\alpha = -\tan\beta = \tan(-\beta).$$

(iv) $\sin\alpha = \cos\beta$: For example, to write in the cosine-form only:

$$\sin\alpha = \cos(\pi/2 - \alpha) = \cos\beta.$$

The last two expressions are equal if
(a) $\pi/2 - \alpha = \beta + 2\pi n$ or
(b) $\alpha - \pi/2 = \beta + 2\pi n$ since $\cos(-x) = \cos x$.

(v) For equations as $-\cos 3x = \sin x$, multiplying by -1, one gets

$$-\sin x = \sin(-x) = \cos(\pi/2 - (-x)) = \cos(\pi/2 + x).$$

(vi) Equations as $\sin^2 x = 2\cos x$ can be written as second-degree equations by using $\sin^2 x = 1 - \cos^2 x$. Then substituting $\cos x = t$ yields a second-degree equation in t. Note that here $|t| \le 1$. Expressions with "linear" terms in sine and cosine with the same angular velocity ω, like $a\cos\omega t + b\sin\omega t$, can be rewritten as $A\sin(\omega t + \alpha)$, see (3.38) page 60.
(vii) In connection with studies in *Signal and System*, A and ω are called *amplitude* and *angular-frequency*, respectively.

3.8 Solving Triangles

This means that given some entities of the triangle one can determine all its sides angles (see Figure 3.7).

Number of *congruence cases* means the number of non-congruent triangles.

In a triangle the angles are in the range $(0, 180°)$ and the sum of two side lengths is longer than the third side's length (see page 41).

Following hints are useful in solving a triangle.

Known entities	Numberof congruence cases	Begin by determining
$a, A, B, \quad A+B<180°$	1	b with Sine theorem
a, b, A	$0, 1,$ or 2	B with Sine theorem
a, b, C	1	c with Cosine theorem
a, b, c	1	C with Cosine theorem

(3.45)

The following table gives formulas for the sides of a triangle in terms of the other known entities. Then, the angles are determined using the Cosine theorem ((3.33) page 58).

Solving triangles, continuation

(i) Given the circumference of the triangle: $\mathcal{O}$, one side, say a, and a nearby angle B. Then the other sides c and b are given by

$$c = \frac{(\mathcal{O} - 2\,a)\,\mathcal{O}}{2\,(\mathcal{O} - a(1+\cos B))}$$
$$b = \mathcal{O} - (a+c). \tag{3.46}$$

(ii) Given the circumference of the triangle: $\mathcal{O}$, and a side a and its opposite angle A, then

$$b, c = \frac{\mathcal{O} - a \pm \sqrt{a^2 + (2a\mathcal{O} - \mathcal{O}^2)\tan^2(A/2)}}{2}. \tag{3.47}$$

(iii) Given two angles and the area T, then

$$c = \sqrt{2T(\cot A + \cot B)}. \tag{3.48}$$

(iv) Given the circumference of the triangle: $\mathcal{O}$, one side c, and the area T, then

$$a, b = \frac{1}{2}\left(\mathcal{O} - c \pm \sqrt{\frac{c^2\,\mathcal{O}\,(\mathcal{O} - 2\,c) - 16\,T^2}{\mathcal{O}\,(\mathcal{O} - 2\,c)}}\right). \tag{3.49}$$

3.9 Coordinate System ($\mathbb{R}^3$)

A coordinate system in $\mathbb{R}^3$ (in three dimensions) consists of three mutually perpendicular axes (coordinate axes). A point $P = (x, y, z) \in \mathbb{R}^3$ has three coordinates.

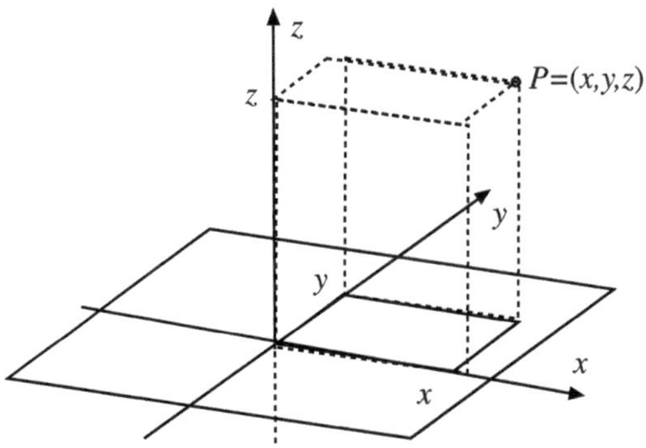

A plane perpendicular to z-axis is the $x\,y$-plane.

The axes intersect in origin $\mathcal{O} = (0, 0, 0)$.
The distance between $\mathcal{O}$ and P is given by

$$||\mathcal{O} - P|| = \sqrt{x^2 + y^2 + z^2}. \tag{3.50}$$

The distance between two points $P = (x, y, z)$ and $Q = (x_1, y_1, z_1)$ is

$$||\mathcal{P} - Q|| = \sqrt{(x - x_1)^2 + (y - y_1)^2 + (z - z_1)^2}. \tag{3.51}$$

Definition 3.10 (Spherical trigonometry).

- A sphere in $\mathbb{R}^3$ with radius $r = 1$ and center $(0, 0, 0)$ is called *unit sphere.* Its equation is then $x^2 + y^2 + z^2 = r^2 = 1$, where (x, y, z) are Cartesian coordinates.
- A midpoint plane is a plane passing through the center of the sphere. The plane's intersection with the sphere is called *great circle.*
- If a plane, intersecting the sphere, does not pass through the center of the sphere, then the intersection is called a *parallel* circle.
- A spherical triangle is the part of the surface of a sphere cut by three great circles.
- The sides of a spherical triangle are parts of large circle arcs, assigned in the arc dimensions, a, b, c. The angles are assigned the angular dimensions A, B, C (see the following figure).

- We use the notations $s = \dfrac{a+b+c}{2}$ and $S = \dfrac{A+B+C}{2}$.

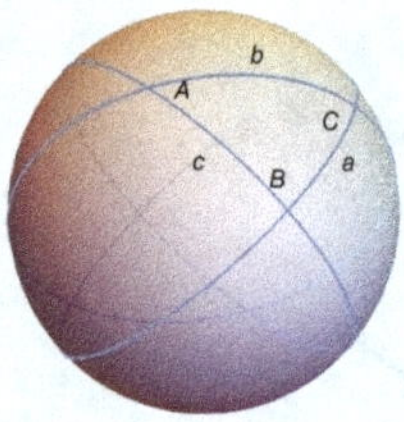

Spheric triangle on a unit sphere.

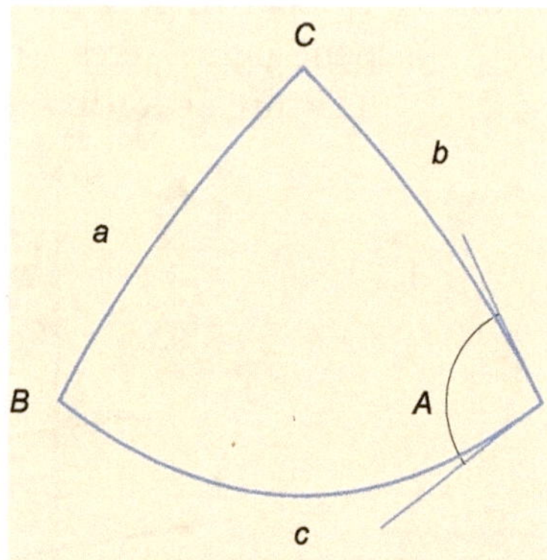

The angle A is between the tangents of circular arcs b and c.

Theorem 3.13. (Also applies to permutation of angles and sides). *All angles a, b, c, A, B, and C are assumed to be in the interval $(0, 180°)$. With notations as mentioned in the above figures:*

Inequalities

$$\begin{aligned} &0° < a+b+c < 360°,\ 180° < A+B+C < 540°, \\ &a < b < c \quad \Longleftrightarrow \quad A < B < C, \\ &a+b > c \quad \Longleftrightarrow \quad A+B > C+180°. \end{aligned} \tag{3.52}$$

Spherical excess is defined as the angle $E := A+B+C-180°$.

Sine theorem

$$\frac{\sin A}{\sin a} = \frac{\sin B}{\sin b} = \frac{\sin C}{\sin c}. \tag{3.53}$$

"The third formula"

$$\sin a \cos B = \cos b \sin c - \sin b \cos c \cos A. \tag{3.54}$$

Cosine theorem

$$\cos a = \cos b \, \cos c + \sin b \, \sin c \, \cos A. \tag{3.55a}$$

$$\cos A = -\cos B \, \cos C + \sin B \, \sin C \, \cos a. \tag{3.55b}$$

The area T of a spherical triangle with R as radius of the sphere

$$T = \frac{\pi R^2 E}{180}. \tag{3.56}$$

3.9.1 *Identities in spherical trigonometry*

Gauss' identities

$$\frac{\sin\frac{1}{2}(a-b)}{\sin\frac{1}{2}c} = \frac{\sin\frac{1}{2}(A-B)}{\cos\frac{1}{2}C}, \quad \frac{\cos\frac{1}{2}(a-b)}{\cos\frac{1}{2}c} = \frac{\sin\frac{1}{2}(A+B)}{\cos\frac{1}{2}C},$$

$$\frac{\sin\frac{1}{2}(a+b)}{\sin\frac{1}{2}c} = \frac{\cos\frac{1}{2}(A-B)}{\sin\frac{1}{2}C}, \quad \frac{\cos\frac{1}{2}(a+b)}{\cos\frac{1}{2}c} = \frac{\cos\frac{1}{2}(A+B)}{\sin\frac{1}{2}C}. \tag{3.57}$$

---

d'Alembert's identities

$$\begin{aligned}
\sin\frac{A}{2}\sin\frac{b+c}{2} &= \sin\frac{a}{2}\cos\frac{B-C}{2},\\
\sin\frac{A}{2}\cos\frac{b+c}{2} &= \cos\frac{a}{2}\cos\frac{B+C}{2},\\
\cos\frac{A}{2}\sin\frac{b-c}{2} &= \sin\frac{a}{2}\sin\frac{B-C}{2},\\
\cos\frac{A}{2}\cos\frac{b-c}{2} &= \cos\frac{a}{2}\sin\frac{B+C}{2}.
\end{aligned} \tag{3.58}$$

Napier's identities

$$\tan\frac{b+c}{2}\cos\frac{B+C}{2} = \tan\frac{a}{2}\cos\frac{B-C}{2}, \tag{3.59a}$$

$$\tan\frac{b-c}{2}\sin\frac{B+C}{2} = \tan\frac{a}{2}\sin\frac{B-C}{2}, \tag{3.59b}$$

$$\tan\frac{B+C}{2}\cos\frac{b+c}{2} = \cot\frac{A}{2}\cos\frac{b-c}{2}, \tag{3.59c}$$

$$\tan\frac{B-C}{2}\sin\frac{b+c}{2}=\cot\frac{A}{2}\sin\frac{b-c}{2}. \tag{3.59d}$$

Identities for half angles

$$\begin{aligned}\sin^2\frac{A}{2}&=\frac{\sin(s-b)\,\sin(s-c)}{\sin b\,\sin c},\\ \cos^2\frac{A}{2}&=\frac{\sin s\,\sin(s-a)}{\sin b\,\sin c},\\ \sin^2\frac{a}{2}&=\frac{\sin S\,\cos(S-A)}{\sin B\,\sin C},\\ \cos^2\frac{a}{2}&=\frac{\cos(S-B)\,\cos(S-C)}{\sin B\,\sin C}.\end{aligned} \tag{3.60}$$

3.9.2 *Triangle solution of spheric triangle*

Solution process for six different cases. See conditions for $a+b+c$ and $A+B+C$, page 66.

	Known parameters	Method
(1)	a, b, c	A, B, C using (3.55a).
(2)	A, B, C	a, b, c using (3.55b).
(3)	a, b, C	c using (3.55a). Then A and B using (3.55a).
(4)	B, C, a	A using (3.55b). Then b and c using (3.55b).
(5)	a, A, B	b using (3.53). Then c and C using (3.59a) and (3.59c), respectively.
(6)	b, c, B	C using (3.53). Then a and A using (3.59a), and (3.59c), respectively.

Items (5) and (6) may concern two congruent cases.

Chapter 4

Vector Algebra

4.1 Basic Concepts

Definition 4.1. A geometric vector is represented by an arrow (see the following figure). A vector is presented in $\mathbb{R}^2$ (the plane), but the notion can be generalized to $\mathbb{R}^n$.

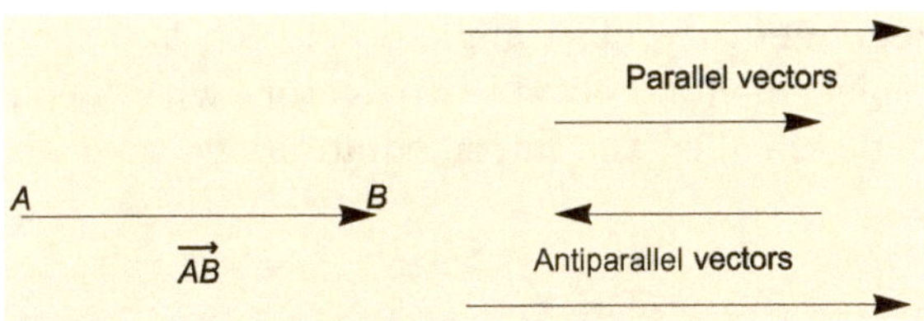

LHS: Vector with starting-point A and endpoint B. RHS: Parallel and anti-parallel vectors.

A vector is denoted by a letter, either with line on top: $\overline{a}$, below $\underline{a}$, or in boldface: $\boldsymbol{a}$. If the starting- and endpoints are A and B, respectively, then the vector is written as $\overrightarrow{AB}$.

Two vectors $\boldsymbol{a}$ and $\boldsymbol{b}$ are *parallel* if they are parallel directed-lines with the same directions. This is denoted $\boldsymbol{a} \parallel \boldsymbol{b}$. If the lines are parallel and directions opposite, then the vectors are called *antiparallel.*

Two vectors $\boldsymbol{a}$ and $\boldsymbol{b}$ are *equal*: $\boldsymbol{a} = \boldsymbol{b}$ if either one can be *parallel-moved* so that they coincide. Thus, in the figure, the vectors on the right top and bottom are equal.

Definition 4.2.

Length of a vector $\boldsymbol{a}$ is the length of the arrow (its line-segment) in a suitable unit of length and is denoted $a = |\boldsymbol{a}|$ (≥ 0), i.e., either with absolute-value sign or only with a regular lowercase a.
The *zero vector*, $\mathbf{0}$, has length 0 and, graphically, represents a point.
Multiplication with scalar: Multiplication of $\boldsymbol{a}$ with a real number (scalar) k yields a vector $k \cdot \boldsymbol{a}$ with the same direction as $\boldsymbol{a}$ if $k \geq 0$ or opposite direction if $k < 0$. In either case, its length is $|k\boldsymbol{a}| = |k||\boldsymbol{a}| = |k|a$.

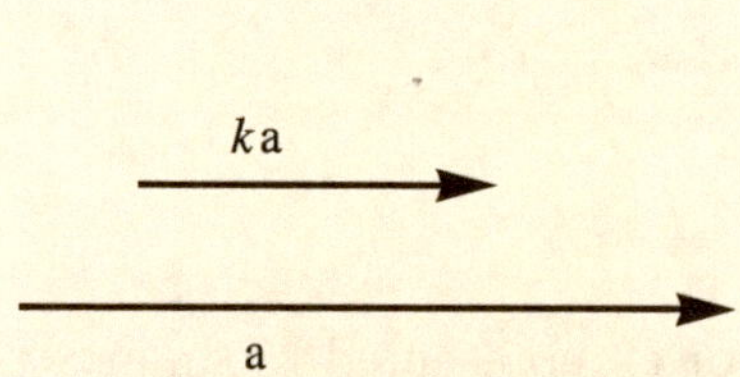

Multiplication with scalar (k, $0 < k < 1$) of the vector $\boldsymbol{a}$.

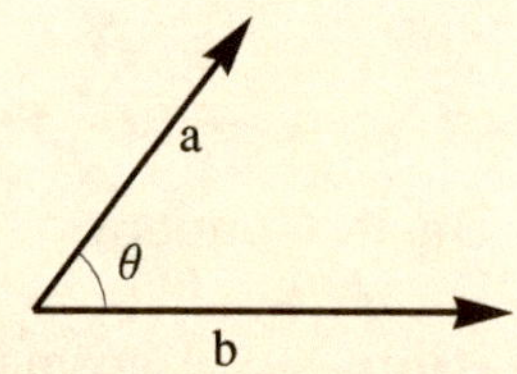

Angle between vectors.

Angle between vectors: The angle θ between two vectors is built by parallel displacement to obtain two vectors with a common start point. This angle is called the intermediate angle to $\boldsymbol{a}$ and $\boldsymbol{b}$.

$$0^\circ \leq \theta \leq 180^\circ.$$

If $\theta = 90^\circ$, the vectors are *orthogonal.* This is denoted $\boldsymbol{a} \perp \boldsymbol{b}$.
Two vectors $\boldsymbol{a}$ and $\boldsymbol{b}$ are normal if they are orthogonal.
Addition of vectors: To add (or sum) two vectors $\boldsymbol{a}$ and $\boldsymbol{b}$, one may parallel move, e.g., $\boldsymbol{a}$ so that its endpoint coincides with the starting point of $\boldsymbol{b}$ (see Figure (a)).
Sum of the vectors (Figure 4.1) $\boldsymbol{a} + \boldsymbol{b}$ is the vector that has the same starting point as $(\boldsymbol{a})$ and the same endpoint as $(\boldsymbol{b})$.
In particular, $\boldsymbol{a} + (-\boldsymbol{a}) = \mathbf{0}$: the zero vector.
The sum of two vectors $\boldsymbol{a}$ and $\boldsymbol{b}$: $\boldsymbol{r} := \boldsymbol{a} + \boldsymbol{b}$ is called *resultant* and the two vectors $\boldsymbol{a}$ and $\boldsymbol{b}$ are called *composants* of $\boldsymbol{r}$.
The inner product between $\boldsymbol{a}$ and $\boldsymbol{b}$ is defined as

$$\boldsymbol{a} \cdot \boldsymbol{b} =: |\boldsymbol{a}| \cdot |\boldsymbol{b}| \cos\theta = ab\cos\theta. \tag{4.1}$$

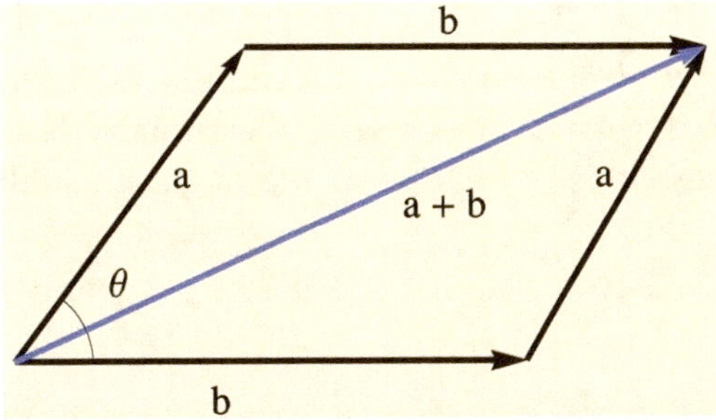

Figure 4.1: Addition of vectors: $\boldsymbol{a} + \boldsymbol{b} = \boldsymbol{b} + \boldsymbol{a} = \boldsymbol{r}$.

Remark. If two vectors are (anti-) parallel with the same (opposite) directions, their intermediate angle is 0° (180°).

In particular,
$\boldsymbol{a} \cdot \boldsymbol{b} = 0$ if $\boldsymbol{a} \perp \boldsymbol{b}$, since then the intermediate angle is 90° and $\cos 90° = 0$.
For two parallel vectors $\boldsymbol{a}$ and $\boldsymbol{b}$, $\boldsymbol{a} \parallel \boldsymbol{b}$

$$\boldsymbol{a} \cdot \boldsymbol{b} = |\boldsymbol{a}||\boldsymbol{b}|.$$

As a special case $\boldsymbol{a} \cdot \boldsymbol{a} = |\boldsymbol{a}| \cdot |\boldsymbol{a}| \cos 0 = |\boldsymbol{a}|^2$.
Addition is both commutative and associative (Figure 4.1).
Inner product is a measure of the interaction of two vectors:

$$\boldsymbol{a} \cdot \boldsymbol{b} = |\boldsymbol{a}||\boldsymbol{b}| \cos \theta,$$

where θ is the angle between $\boldsymbol{a}$ and $\boldsymbol{b}$, thus $|\boldsymbol{b}| \cos \theta$ is the projection of $\boldsymbol{b}$ on $\boldsymbol{a}$.

Elementary calculus with vectors: Let $\boldsymbol{a}$, $\boldsymbol{b}$, and $\boldsymbol{c}$ be vectors and α and β scalars (real numbers). Then

For addition and multiplication with scalar,

$$\begin{aligned} \boldsymbol{a} + \boldsymbol{b} &= \boldsymbol{b} + \boldsymbol{a} && \text{(commutative law)}, \\ \boldsymbol{a} + (\boldsymbol{b} + \boldsymbol{c}) &= (\boldsymbol{a} + \boldsymbol{b}) + \boldsymbol{c} && \text{(associative law)}, \\ \alpha(\boldsymbol{a} + \boldsymbol{b}) &= \alpha\, \boldsymbol{a} + \alpha\, \boldsymbol{b} && \text{(distributive law/vectors)}, \\ (\alpha + \beta)\boldsymbol{a} &= \alpha\, \boldsymbol{a} + \beta\, \boldsymbol{a} && \text{(distributive law/scalars)}. \end{aligned} \tag{4.2}$$

For inner product,

$$\begin{aligned} \boldsymbol{a}\cdot\boldsymbol{b} &= \boldsymbol{b}\cdot\boldsymbol{a} && \text{(commutative law)},\\ \boldsymbol{a}\cdot(\boldsymbol{b}+\boldsymbol{c}) &= \boldsymbol{a}\cdot\boldsymbol{b}+\boldsymbol{a}\cdot\boldsymbol{c} && \text{(distributive law)},\\ |\boldsymbol{a}+\boldsymbol{b}| &\le |\boldsymbol{a}|+|\boldsymbol{b}| && \text{(triangle inequality)}. \end{aligned} \tag{4.3}$$

Definition 4.3. Given vectors $\boldsymbol{u}_1$, $\boldsymbol{u}_2, \ldots,$ $\boldsymbol{u}_m$ and scalars $\alpha_1, \alpha_2, \ldots, \alpha_m$.
The vector

$$\alpha_1\boldsymbol{u}_1 + \alpha_2\boldsymbol{u}_2 + \cdots + \alpha_m\boldsymbol{u}_m = \sum_{k=1}^{m} \alpha_k\boldsymbol{u}_k$$

is a *linear combination* of these vectors.

The vectors $\boldsymbol{u}_1$, $\boldsymbol{u}_2, \ldots,$ $\boldsymbol{u}_m$ are called *linearly independent* if

$$\alpha_1\boldsymbol{u}_1+\alpha_2\boldsymbol{u}_2+\cdots+\alpha_m\boldsymbol{u}_m = 0 \quad\Longrightarrow\quad \alpha_1 = \alpha_2 = \ldots = \alpha_m = 0,$$

otherwise they are called *linearly dependent.*

In each set of linearly dependent vectors, at least one of the vectors can be written as a linear combination of the remaining vectors.

Definition 4.4 Vectors in coordinate system (component form in $\mathbb{R}^2$).

(i) A vector with starting point at the origin of the coordinate system $O = (0;0)$ and endpoint $P = (x;y)$: a *position vector* is denoted $\overrightarrow{OP} = (x,y)$. Coordinates of the endpoint are *components* of the vector.

(ii) In a two-dimensional *Cartesian* coordinate system, the basis vectors $\boldsymbol{e}_x := (1,0)$ and $\boldsymbol{e}_y := (0,1)$ are orthogonal and have length 1. This is generalized to $\mathbb{R}^n$.

(iii) The length of the vector $\overrightarrow{OP} = (x,y)$ in two-dimensional Cartesian coordinate system is $|\overrightarrow{OP}| = \sqrt{x^2+y^2}$.

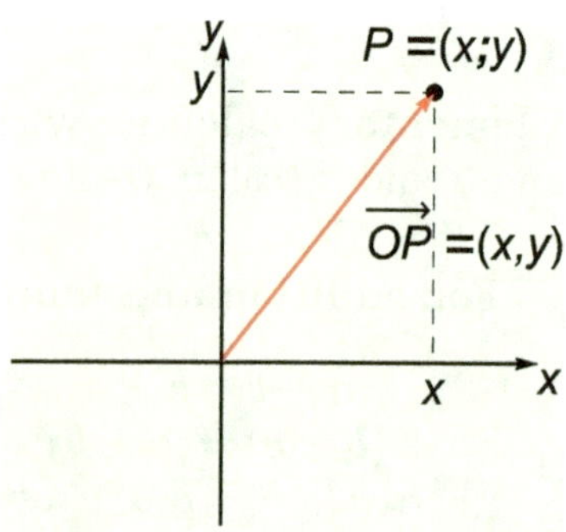

Vector in coordinate system ($\mathbb{R}^2$).

Theorem 4.1. *Addition and multiplication by scalars are performed component-wise:*

(i) *For two points* $P = (x_1; y_1)$ *and* $Q = (x_2; y_2)$ *with position vectors* $\overrightarrow{OP} = (x_1, y_1)$ *and* $\overrightarrow{OQ} = (x_2, y_2)$*, their sum is the vector*

$$\overrightarrow{OP} + \overrightarrow{OQ} = (x_1 + x_2, y_1 + y_2)$$

and their difference (in that order) is

$$\overrightarrow{PQ} = \overrightarrow{OQ} - \overrightarrow{OP} = (x_2 - x_1, y_2 - y_1).$$

(ii) *For a scalar* c*,*

$$c\overrightarrow{OP} = c(x_1, y_1) = (cx_1, cy_1).$$

Definition 4.5. The distance between two *points* $P = (x_1, y_1)$ and $Q = (x_2, y_2)$ is defined as the length of the vector $\overrightarrow{PQ}$, i.e.,

$$|\overrightarrow{PQ}| = \sqrt{(x_2 - x_1)^2 + (y_2 - y_1)^2}. \tag{4.4}$$

There is basically no difference between a point's coordinates and the components of its position vector.

Vectors $\boldsymbol{e}_x := (1, 0)$ and $\boldsymbol{e}_y := (0, 1)$ are unit vectors along respective coordinate axis with properties:

$$\begin{cases} \boldsymbol{e}_x \cdot \boldsymbol{e}_x = |\boldsymbol{e}_x||\boldsymbol{e}_x| \cos 0 = |\boldsymbol{e}_x|^2 = 1, \\ \boldsymbol{e}_y \cdot \boldsymbol{e}_y = |\boldsymbol{e}_y|^2 = 1, \\ \boldsymbol{e}_x \cdot \boldsymbol{e}_y = \boldsymbol{e}_y \cdot \boldsymbol{e}_x = |\boldsymbol{e}_x||\boldsymbol{e}_y| \cos \frac{\pi}{2} = 0. \end{cases}$$

Each vector $\boldsymbol{u} = (x, y)$ (in a coordinate system) can be expressed as a linear combination of $\boldsymbol{e}_x$ and $\boldsymbol{e}_y$:

$$\boldsymbol{u} = (x, y) = x(1, 0) + y(0, 1) = x\boldsymbol{e}_x + y\boldsymbol{e}_y.$$

A vector in $\mathbb{R}^n$ is written as $\boldsymbol{u} = (x_1, x_2, \ldots x_n)$ with component-wise addition and multiplication by scalar. In particular, the length of $\boldsymbol{u}$ is

$$|\boldsymbol{u}| = \sqrt{x_1^2 + x_2^2 + \cdots + x_n^2}.$$

When adding two vectors $\overrightarrow{OP}$ and $\overrightarrow{OQ}$, the latter is moved in parallel so that its endpoint coincides with the starting point of the first.

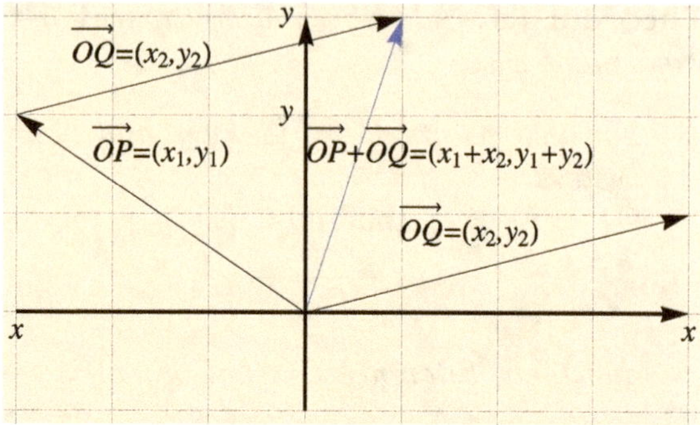

Theorem 4.2.

(i) *Let* $\boldsymbol{a}$ *and* $\boldsymbol{b}$ *be two vectors in* $\mathbb{R}^2$, *which are neither parallel nor antiparallel. Then, they may serve as basis vectors for* $\mathbb{R}^2$, *i.e., any vector* $\boldsymbol{v} \in \mathbb{R}^2$ *can, uniquely, be expressed as a linear combination of these two vectors. More specifically, there are uniquely determined scalars* x *and* y *such that*

$$\boldsymbol{v} = x\boldsymbol{a} + y\boldsymbol{b}.$$

(ii) *Inner product in component form: If* $\boldsymbol{a} = (x_1, y_1)$ *and* $\boldsymbol{b} = (x_2, y_2)$, *then*

$$\boldsymbol{a} \cdot \boldsymbol{b} = x_1x_2 + y_1y_2. \tag{4.5}$$

(iii) *The unit vector parallel to the vector* $\boldsymbol{a} = (x_1, y_1) \neq \boldsymbol{0}$ *is*

$$\frac{1}{|\boldsymbol{a}|}\,\boldsymbol{a} = \frac{1}{\sqrt{x_1^2 + y_1^2}}\,(x_1, y_1). \tag{4.6}$$

4.1.1 *Line in* $\mathbb{R}^2$

Theorem 4.3.

(i) *General form of equation of a line in* $\mathbb{R}^2$ *as a set is as follows:*

$$\{(x, y) : \quad ax + by + c = 0\}, \quad \text{where } (a, b) \neq (0, 0). \tag{4.7}$$

(ii) *For an arbitrary* $\boldsymbol{r} = (x, y)$, *and fixed* $\boldsymbol{r}_0 = (x_0, y_0)$, *on a line, with the trace (direction) vector* $\boldsymbol{v} = \overrightarrow{P_0P_1} = (\alpha, \beta)$. *The parameter form of the line is given as in* (4.8).

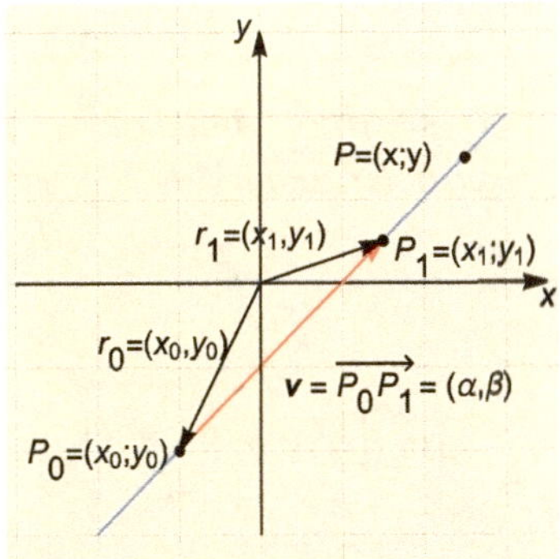

$$\begin{cases} x = \alpha t + x_0 \\ y = \beta t + y_0 \end{cases} \quad (x, y) = t(\alpha, \beta) + (x_0, y_0) \text{ or } \quad \boldsymbol{r} = t\boldsymbol{v} + \mathbf{r}_0,$$

where $t \in \mathbb{R}$.

$$\tag{4.8}$$

(iii) *Relation between* (4.7) *and* (4.8):

$$\begin{cases} x = -bt + x_0, \\ y = at + y_0, \end{cases} \quad \text{where} \quad \begin{cases} ax + by + c = 0, \\ ax_0 + by_0 + c = 0. \end{cases} \tag{4.9}$$

(iv) *The Intercept form of a line that does not pass through origin (nor parallel to coordinate axes for* a, $b \neq \infty$*) is*

$$\frac{x}{a} + \frac{y}{b} = 1. \tag{4.10}$$

The points $(x_1; y_1) = (a; 0)$ *and* $(x_2; y_2) = (0; b)$ *are intersection points with the respective axes.*

Theorem 4.4.

(i) ***Distance between point and line:*** *Given a point* $P = (x_1; y_1)$ *and a line with equation* (4.7). *Their distance* d *is then*

$$d = \frac{|ax_1 + by_1 + c|}{\sqrt{a^2 + b^2}}. \quad (4.11)$$

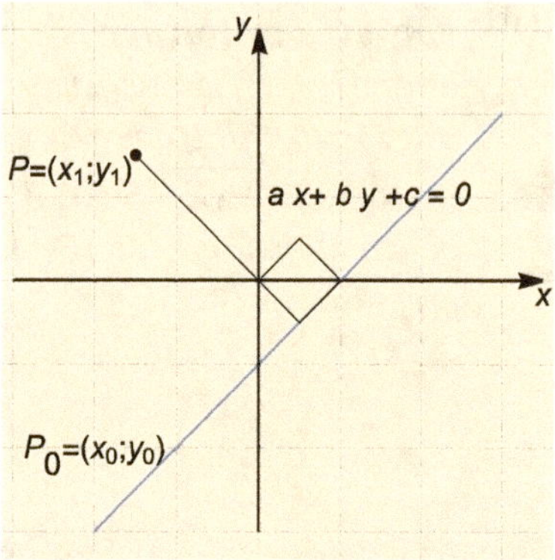

(ii) *The area* T *of the triangle with vertices* $(0, 0)$, (x_1, y_1), *and* (x_2, y_2) *is*

$$T = \frac{1}{2}|x_2\, y_1 - x_1\, y_2|\,. \quad (4.12)$$

4.2 Vectors in $\mathbb{R}^3$

Vectors in $\mathbb{R}^3$ follow the same calculation rules as in $\mathbb{R}^2$. An additional concept is the *cross product* of two vectors in $\mathbb{R}^3$.

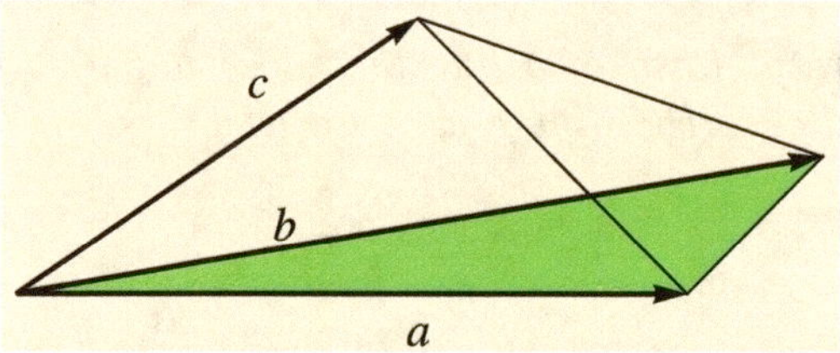

Tetrahedron spanned by three vectors.

(i) The unit vectors in $\mathbb{R}^3$, along the axes, are as follows:

$$\mathbf{e}_x := \boldsymbol{i} = (1, 0, 0), \quad \mathbf{e}_y := \boldsymbol{j} = (0, 1, 0), \quad \mathbf{e}_z := \boldsymbol{k} = (0, 0, 1). \quad (4.13)$$

(ii) A vector can be written in component form as a position vector, as follows: $\boldsymbol{a} = (x, y, z) = x\mathbf{e}_x + y\mathbf{e}_y + z\mathbf{e}_z$.

(iii) Addition and multiplication by scalar (α) for $\boldsymbol{a} = (x_1, y_1, z_1)$ and $\boldsymbol{b} = (x_2, y_2, z_2)$ is component-wise:

$$\boldsymbol{a} + \boldsymbol{b} = (x_1 + x_2, y_2 + y_2, z_1 + z_2)$$

$$\alpha\, \boldsymbol{a} = (\alpha\, x_1, \alpha\, y_1, \alpha\, z_1).$$

(iv) The inner product and length in $\mathbb{R}^3$ are as follows:

$$\boldsymbol{a} \cdot \boldsymbol{b} = (x_1, y_1, z_1) \cdot (x_2, y_2, z_2) = x_1x_2 + y_1y_2 + z_1z_2. \tag{4.14}$$

$$|\boldsymbol{a}| = \sqrt{\boldsymbol{a} \cdot \boldsymbol{a}} = \sqrt{x_1^2 + y_1^2 + z_1^2}. \tag{4.15}$$

(v) An equation of a line in $\mathbb{R}^3$, in parametric form, is

$$(x, y, z) = t(\alpha, \beta, \gamma) + (x_0, y_0, z_0) \text{ or } \boldsymbol{r} = t \cdot \boldsymbol{v} + \mathbf{r}_0 \tag{4.16}$$

with $\boldsymbol{v} = (\alpha, \beta, \gamma)$ as *direction vector*. Alternatively,

$$\begin{cases} x = \alpha t + x_0, \\ y = \beta t + y_0, \\ z = \gamma t + z_0. \end{cases} \quad t \in \mathbb{R}.$$

(vi) This is generalized to line in $\mathbb{R}^n$:

$$\begin{aligned} \boldsymbol{r} &= (x_1, x_2, \ldots, x_n) \in \mathbb{R}^n && \text{arbitrary point on the line,} \\ \boldsymbol{r}_0 &= (x_{0,1}, x_{0,2}, \ldots, x_{0,n}) \in \mathbb{R}^n && \text{starting point on the line,} \\ \boldsymbol{v} &= (v_1, v_2, \ldots, v_n) \in \mathbb{R}^n && \text{direction vector.} \end{aligned}$$

The corresponding line has the parameter form

$$\boldsymbol{r} = t\,\boldsymbol{v} + \boldsymbol{r}_0, \quad t \in \mathbb{R}. \tag{4.17}$$

(vii) Consider a Cartesian coordinate system with the axes as in the figure and the three basis vectors $(\mathbf{e}_x, \mathbf{e}_y, \mathbf{e}_z)$. In this order, they constitute a *Right-handed Coordinate System* (RHC-system), whereas $(\mathbf{e}_x, \mathbf{e}_z, \mathbf{e}_y)$ is a *Left-handed Coordinate System* (LHC-system).

4.2.1 *Cross product and scalar triple product*

Definition 4.6. Let $\boldsymbol{a}$ and $\boldsymbol{b}$ be two vectors in $\mathbb{R}^3$ with the angle θ between.

The cross product of $\boldsymbol{a}$ and $\boldsymbol{b}$ is a vector, denoted by $\boldsymbol{a} \times \boldsymbol{b}$ and with the properties

(i) $(\boldsymbol{a} \times \boldsymbol{b}) \perp \boldsymbol{a}, \qquad (\boldsymbol{a} \times \boldsymbol{b}) \perp \boldsymbol{b}$,
(ii) $|\boldsymbol{a} \times \boldsymbol{b}| = |\boldsymbol{a}| \cdot |\boldsymbol{b}| \cdot \sin\theta$,
(iii) $(\boldsymbol{a}, \boldsymbol{b}, \boldsymbol{a} \times \boldsymbol{b})$ forms a right-oriented system, as in the following figure.
The cross product can alternatively be written as $\boldsymbol{a} \times \boldsymbol{b} = |\boldsymbol{a}| \cdot |\boldsymbol{b}| \cdot \sin\theta\, \boldsymbol{n}$,
where $|\boldsymbol{n}| = 1$ och $\boldsymbol{a}, \boldsymbol{b}, \boldsymbol{n}$ form a right-oriented system.
(iv) Triple scalar product between the vectors $\boldsymbol{a}$, $\boldsymbol{b}$, and $\boldsymbol{c}$ is defined as

$$[\boldsymbol{a}, \boldsymbol{b}, \boldsymbol{c}] = (\boldsymbol{a} \times \boldsymbol{b}) \cdot \boldsymbol{c}. \tag{4.18}$$

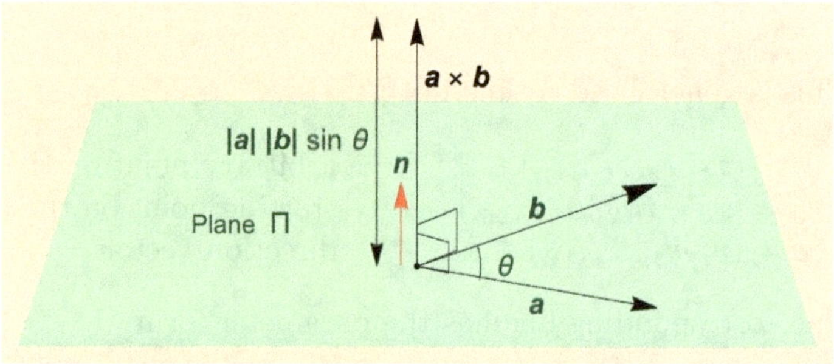

The cross product $\boldsymbol{a} \times \boldsymbol{b} := (|\boldsymbol{a}| \cdot |\boldsymbol{b}| \cdot \sin\theta)\, \boldsymbol{n}$, with length $|\boldsymbol{a} \times \boldsymbol{b}| = |\boldsymbol{a}| \cdot |\boldsymbol{b}| \cdot \sin\theta$. The vectors $\boldsymbol{a}$ and $\boldsymbol{b}$ lie in a plane Π. $\boldsymbol{n}$ is a normal vector to Π, it is also the unit vector, parallel to $\boldsymbol{a} \times \boldsymbol{b}$.

Calculation rules of inner and cross product

Theorem 4.5.

$$\boldsymbol{a} \cdot \boldsymbol{b} = \boldsymbol{b} \cdot \boldsymbol{a} \qquad \textit{(commutativity)}$$

$$\boldsymbol{a} \cdot (\boldsymbol{b} + \boldsymbol{c}) = \boldsymbol{a} \cdot \boldsymbol{b} + \boldsymbol{a} \cdot \boldsymbol{c} \qquad \textit{(distributivity)}$$

$$\boldsymbol{a} \times \boldsymbol{b} = -\boldsymbol{b} \times \boldsymbol{a} \qquad \textit{(anticommutativity)}$$

$$\boldsymbol{a} \times (\boldsymbol{b} + \boldsymbol{c}) = \boldsymbol{a} \times \boldsymbol{b} + \boldsymbol{a} \times \boldsymbol{c} \qquad \textit{(distributivity)}$$

$$(\boldsymbol{a} \times \boldsymbol{b}) \cdot \boldsymbol{c} = \boldsymbol{a} \cdot (\boldsymbol{b} \times \boldsymbol{c})$$

$$[\boldsymbol{a}, \boldsymbol{b}, \boldsymbol{c}] = -[\boldsymbol{b}, \boldsymbol{a}, \boldsymbol{c}]$$

$$[\boldsymbol{a}, \boldsymbol{b}, \boldsymbol{c} + \boldsymbol{d}] = [\boldsymbol{a}, \boldsymbol{b}, \boldsymbol{c}] + [\boldsymbol{a}, \boldsymbol{b}, \boldsymbol{d}]$$

$$\boldsymbol{a} \times (\boldsymbol{b} \times \boldsymbol{c}) = (\boldsymbol{a} \cdot \boldsymbol{c})\boldsymbol{b} - (\boldsymbol{a} \cdot \boldsymbol{b})\boldsymbol{c}$$

$$(\boldsymbol{a} \times \boldsymbol{b}) \times (\boldsymbol{c} \times \boldsymbol{d}) = [\boldsymbol{a}, \boldsymbol{c}, \boldsymbol{d}]\,\boldsymbol{b} - [\boldsymbol{b}, \boldsymbol{c}, \boldsymbol{d}]\,\boldsymbol{a}$$

$$(\boldsymbol{a} \times \boldsymbol{b}) \cdot (\boldsymbol{c} \times \boldsymbol{d}) = (\boldsymbol{a} \cdot \boldsymbol{c})(\boldsymbol{b} \cdot \boldsymbol{d}) - (\boldsymbol{a} \cdot \boldsymbol{d})(\boldsymbol{b} \cdot \boldsymbol{c}). \qquad (4.19)$$

Theorem 4.6.

(i) $V = |\,[\boldsymbol{a}, \boldsymbol{b}, \boldsymbol{c}]\,|$ *is the volume of the parallelepiped spanned by the vectors* $\boldsymbol{a}, \boldsymbol{b}, \boldsymbol{c}$ *oriented as in Figure* 4.2.

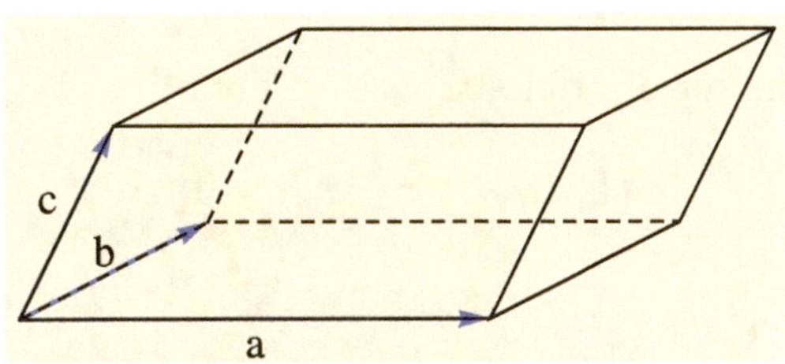

Figure 4.2: Parallelepiped spanned by three vectors.

(ii) $T = \frac{1}{6}|\,[\boldsymbol{a}, \boldsymbol{b}, \boldsymbol{c}]\,|$ *is the volume of the tetrahedron spanned by the vectors* $\boldsymbol{a}, \boldsymbol{b}, \boldsymbol{c}$.

Theorem 4.7. *Put* $\boldsymbol{a} = a_1 \boldsymbol{e}_x + a_2 \boldsymbol{e}_y + a_3 \boldsymbol{e}_z$, $\boldsymbol{b} = b_1 \boldsymbol{e}_x + b_2 \boldsymbol{e}_y + b_3 \boldsymbol{e}_z$ *and* $\boldsymbol{c} = c_1 \boldsymbol{e}_x + c_2 \boldsymbol{e}_y + c_3 \boldsymbol{e}_z$, *where* $\{\boldsymbol{e}_x, \boldsymbol{e}_y, \boldsymbol{e}_z\}$ *is an RHC-base. Then the cross product is*

$$\boldsymbol{a} \times \boldsymbol{b} = \det \begin{bmatrix} \boldsymbol{e}_x & \boldsymbol{e}_y & \boldsymbol{e}_z \\ a_1 & a_2 & a_3 \\ b_1 & b_2 & b_3 \end{bmatrix} \tag{4.20}$$

$$= (a_2 b_3 - b_2 a_3)\boldsymbol{e}_x + (a_3 b_1 - b_3 a_1)\boldsymbol{e}_y + (a_1 b_2 - b_1 a_2)\boldsymbol{e}_z.$$

Triple scalar product is given by

$$(\boldsymbol{a} \times \boldsymbol{b}) \cdot \boldsymbol{c} = \det \begin{bmatrix} a_1 & a_2 & a_3 \\ b_1 & b_2 & b_3 \\ c_1 & c_2 & c_3 \end{bmatrix} \tag{4.21}$$

$$= a_1 b_2 c_3 + a_2 b_3 c_1 + a_3 b_1 c_2 - (a_1 b_3 c_2 + a_2 b_1 c_3 + a_3 b_2 c_1).$$

4.2.2 *Plane in* $\mathbb{R}^3$

Definition 4.7. Let $\boldsymbol{n} = (A, B, C) \neq \boldsymbol{0}$ be a normal vector to a plane Π, $\boldsymbol{r}_0 = (x_0, y_0, z_0)$, a fixed vector, and $\boldsymbol{r} = (x, y, z)$, an arbitrary vector, considered as points on the plane. Then Π's equation can be written as

$$\boldsymbol{n} \cdot (\boldsymbol{r} - \boldsymbol{r}_0) = 0. \tag{4.22}$$

The plane is the set of points

$$\{\boldsymbol{r} : \quad \boldsymbol{n} \cdot (\boldsymbol{r} - \boldsymbol{r}_0) = 0\}.$$

In coordinate form:

$$\{(x, y, z) : \quad Ax + By + Cz + D = 0\}, \tag{4.23}$$

where $-D = Ax_0 + By_0 + Cz_0$.

The distance between two objects, e.g., a plane and a point, is referred to the shortest distance, hence the "orthogonal" distance between them.

Two lines in $\mathbb{R}^3$ are parallel if they have (anti-)parallel direction vectors.

Two planes in $\mathbb{R}^3$ are parallel if they have (anti-)parallel normal vectors.

4.2.3 *Distance between some objects in* $\mathbb{R}^3$

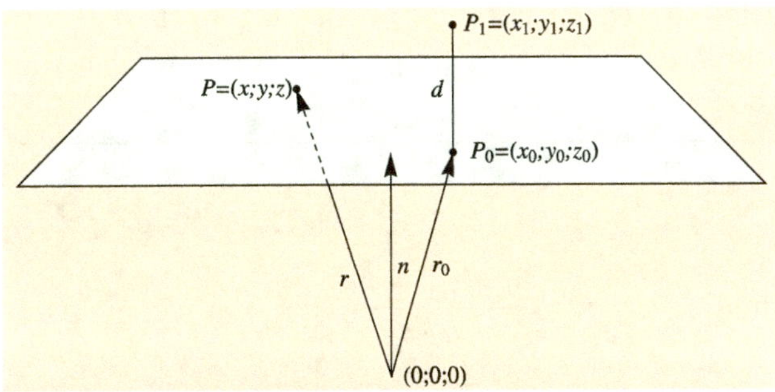

Distance d between plane and point.

Theorem 4.8.

(i) **Distance between plane and point**
The distance d between the plane $\Pi : Ax + By + Cz + D = 0$ *and a point* $\boldsymbol{r}_1 = (x_1, y_1, z_1)$:

$$d = \frac{|Ax_1 + By_1 + Cz_1 + D|}{\sqrt{A^2 + B^2 + C^2}} = \frac{|\boldsymbol{n} \cdot (\boldsymbol{r}_1 - \boldsymbol{r}_0)|}{|\boldsymbol{n}|}, \tag{4.24}$$

where $\boldsymbol{r}_0 \in \Pi$ *and* $\boldsymbol{n}$ *is a normal vector to* Π.

(ii) **Distance between line and point**
The distance d between a line: $\boldsymbol{r} = t\boldsymbol{v} + \boldsymbol{r}_0$ *with* $\boldsymbol{r}_0 = (x_0, y_0, z_0)$ *and a point/vector* $\boldsymbol{r}_1 := (x_1, y_1, z_1)$ *is*

$$d = \frac{|\boldsymbol{v} \times (\boldsymbol{r}_1 - \boldsymbol{r}_0)|}{|\boldsymbol{v}|}. \tag{4.25}$$

(iii) **Distance between two lines**
Let $\boldsymbol{v}_0$ *and* $\boldsymbol{v}_1$ *be the directional vectors for two lines:* L_0 *and* L_1, $\boldsymbol{r}_0 \in L_0$ *and* $\boldsymbol{r}_1 \in L_1$, *then*

(a) *If the lines are parallel, the distance d is obtained by* (4.25), *where $\boldsymbol{v}$ can be chosen as $\boldsymbol{v}_0$ or $\boldsymbol{v}_1$.*

(b) *If the lines are not parallel, then the distance d is given by*

$$d = \frac{|(\boldsymbol{v}_0 \times \boldsymbol{v}_1) \cdot (\boldsymbol{r}_1 - \boldsymbol{r}_0)|}{|\boldsymbol{v}_0 \times \boldsymbol{v}_1|}. \tag{4.26}$$

4.2.4 *Intersection, projection, lines, and planes*

Given a point $\boldsymbol{p}$, a line in parameter form $\boldsymbol{r} = t\,\boldsymbol{v} + \boldsymbol{r}_0$, and a plane with equation $\boldsymbol{n} \cdot ((x, y, z) - \boldsymbol{r}_1) = 0$. Letters in bold are position vectors/points, e.g., with $\boldsymbol{r} = (x, y, z)$. The following projections are meant orthogonal.

Intersection between line and plane	$\boldsymbol{r} = \dfrac{\boldsymbol{n} \cdot (\boldsymbol{r}_1 - \boldsymbol{r}_0)}{\boldsymbol{n} \cdot \boldsymbol{v}} \boldsymbol{v} + \boldsymbol{r}_0$
Projection of point $\boldsymbol{p}$ on line	$\boldsymbol{r} = \dfrac{\boldsymbol{v} \cdot (\boldsymbol{p} - \boldsymbol{r}_0)}{\lvert\boldsymbol{v}\rvert^2} \boldsymbol{v} + \boldsymbol{r}_0$
Reflection of point $\boldsymbol{p}$ in line	$\boldsymbol{r} = 2\,\dfrac{\boldsymbol{v} \cdot (\boldsymbol{p} - \boldsymbol{r}_0)}{\lvert\boldsymbol{v}\rvert^2} \boldsymbol{v} + 2\boldsymbol{r}_0 - \boldsymbol{p}$
Projection of point $\boldsymbol{p}$ on plane	$\boldsymbol{r} = \boldsymbol{p} + \dfrac{\boldsymbol{n} \cdot (\boldsymbol{r}_1 - \boldsymbol{p})}{\lvert\boldsymbol{n}\rvert^2} \boldsymbol{n}$
Reflection of point $\boldsymbol{p}$ in plane	$\boldsymbol{r} = \boldsymbol{p} + 2\,\dfrac{\boldsymbol{n} \cdot (\boldsymbol{r}_1 - \boldsymbol{r}_0)}{\lvert\boldsymbol{n}\rvert^2} \boldsymbol{n}$
Projection of line on plane, not parallel. Note! $(\boldsymbol{n} \cdot \boldsymbol{v})\,\boldsymbol{n} - \lvert\boldsymbol{n}\rvert^2\,\boldsymbol{v} = \boldsymbol{n} \times (\boldsymbol{n} \times \boldsymbol{v})$.	$\boldsymbol{r} = t\left((\boldsymbol{n} \cdot \boldsymbol{v})\,\boldsymbol{n} - \lvert\boldsymbol{n}\rvert^2\,\boldsymbol{v}\right) + \dfrac{\boldsymbol{n} \cdot (\boldsymbol{r}_1 - \boldsymbol{r}_0)}{\boldsymbol{n} \cdot \boldsymbol{v}} \boldsymbol{v} + \boldsymbol{r}_0,\ t \in \mathbb{R}$

Projection of line on plane, parallel
$$r = t\,v + \frac{n\cdot(r_1 - r_0)}{|n|^2}\,n + r_0,\ t\in\mathbb{R}$$

Reflection of line in plane, not parallel
$$r = t\left(2(n\cdot v)n - |n|^2\,v\right) + \frac{n\cdot(r_1 - r_0)}{n\cdot v}\,v + r_0,\ t\in\mathbb{R} \tag{4.27}$$

Reflection of line in plane, parallel
$$r = t\,v + 2\,\frac{n\cdot(r_1 - r_0)}{|n|^2}\,n + r_0,\ t\in\mathbb{R}.$$

Projection and mirroring of point, $\boldsymbol{r}_1$, in plane (below). Projection, $\boldsymbol{r}_1$, and mirroring of point in line (right).

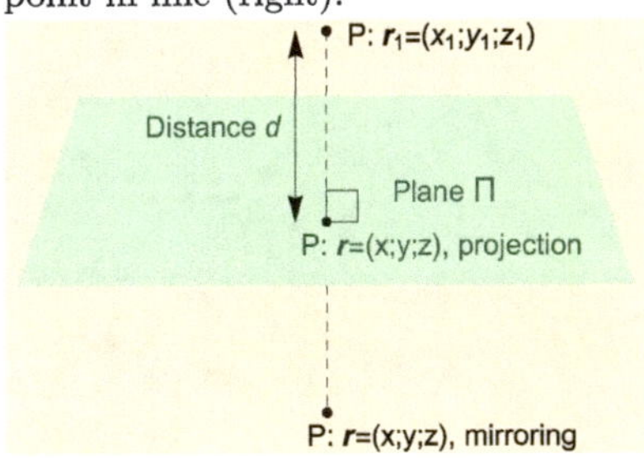

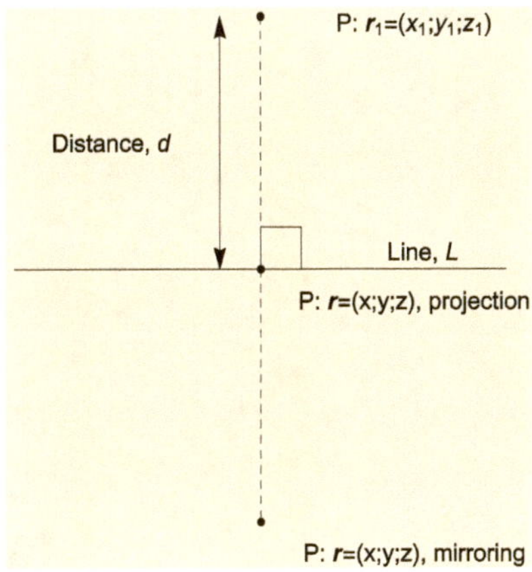

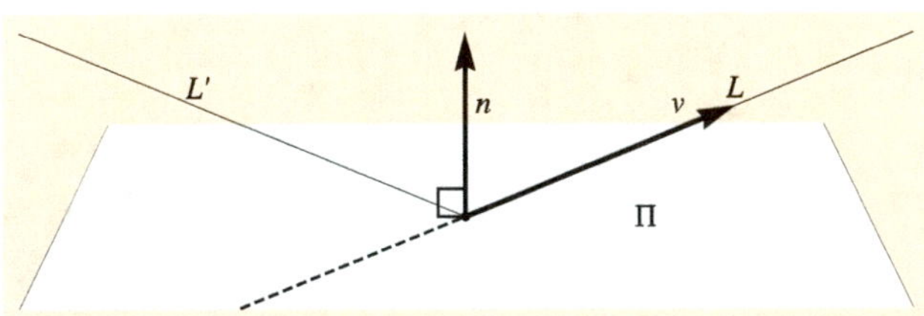

Reflection of line $L \nparallel \Pi$ in the plane Π gives the line L'.

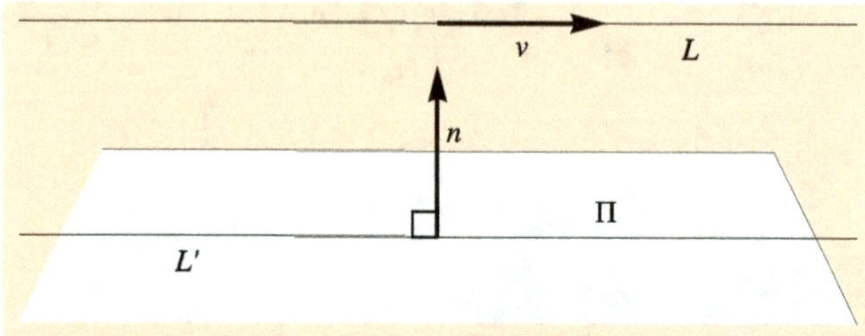

Reflection of line $L \parallel \Pi$ on the plane Π gives the line L'.

Chapter 5

Linear Algebra

5.1 Linear Equation Systems

Definition 5.1.

$$
\begin{aligned}
a_{11}x_1 + a_{12}x_2 + \cdots + a_{1n}x_n &= y_1, \\
a_{21}x_1 + a_{22}x_2 + \cdots + a_{2n}x_n &= y_2, \\
&\ddots \\
a_{m1}x_1 + a_{m2}x_2 + \cdots + a_{mn}x_n &= y_m
\end{aligned} \tag{5.1}
$$

is a linear system of equations, in short ES, with m equations and n variables (unknowns) $x_1, x_2, \ldots, x_n$.

Definition 5.2.

(i) A matrix $\boldsymbol{A}$ of type $m \times n$, with *element* a_{ij}, is given by

$$
\boldsymbol{A} = \begin{bmatrix} a_{11} & a_{12} & \dots & a_{1n} \\ a_{21} & a_{22} & \dots & a_{2n} \\ & & \ddots & \\ a_{m1} & a_{m2} & \dots & a_{mn} \end{bmatrix}. \tag{5.2}
$$

(ii) In matrix form, (5.1) is written as

$$[\boldsymbol{A}|\boldsymbol{Y}] := \underbrace{\underbrace{\left[\begin{array}{cccc|c} a_{11} & a_{12} & \dots & a_{1n} & y_1 \\ a_{21} & a_{22} & \dots & a_{2n} & y_2 \\ & & \ddots & & \vdots \\ a_{m1} & a_{m2} & \dots & a_{mn} & y_m \end{array}\right]}_{\text{Coefficient matrix (type } m\times n).}}_{\text{Augmented matrix (type } m\times(n+1)).} \tag{5.3}$$

One can have several RHS in (5.3), which correspond to several equation systems with the same coefficient matrix.

Definition 5.3. The transpose of the matrix $\boldsymbol{A}$ in (5.2) is the matrix, putting the element $a_{j,k}$ in (5.2) in a matrix of type $n \times m$ in position (k, j)

$$\boldsymbol{A}^T = \begin{bmatrix} a_{11} & a_{21} & \dots & a_{m1} \\ a_{12} & a_{22} & \dots & a_{m2} \\ \vdots & \vdots & \ddots & \vdots \\ a_{1n} & a_{2n} & \dots & a_{mn} \end{bmatrix}. \tag{5.4}$$

$\boldsymbol{A}$ is called quadratic if $m = n$.
$\boldsymbol{A}$ is symmetric if $\boldsymbol{A}^T = \boldsymbol{A}$, i.e., $a_{ij} = a_{ji}$.
$\boldsymbol{A}$ is anti-symmetric if $\boldsymbol{A}^T = -\boldsymbol{A}$, i.e., $a_{ij} = -a_{ji}$ and $a_{ii} = 0$.

Theorem 5.1. *Both symmetric and anti-symmetric matrices are quadratic. I, $\boldsymbol{A}^T \cdot \boldsymbol{A}$, $\boldsymbol{A} \cdot \boldsymbol{A}^T$ are quadratic and symmetric (multiplication of matrices on page* 92).

$$\mathit{type}\,(\boldsymbol{A}^T) = n \times m \iff \mathit{type}\,(\boldsymbol{A}) = m \times n.$$

Definition 5.4.

- The Matrix $\boldsymbol{A}$ in (5.2) is also written as $(a_{ij})_{m\times n}$. $\begin{bmatrix} a_{i1} & a_{i2} & \dots & a_{in} \end{bmatrix}$ is the ith row and $\begin{bmatrix} a_{1j} \\ a_{2j} \\ \vdots \\ a_{mj} \end{bmatrix}$ the jth column. These two considered as vectors are called row and column vectors, respectively.

- An $n \times n$ square matrix $\boldsymbol{A}$ has order n.
- In a matrix of order n, the sequence of elements $a_{ii}, i = 1, 2, \ldots, n$, is called the *main diagonal.*
- The sum of the diagonal elements is called the *trace* of $\boldsymbol{A}$ and is denoted

$$\operatorname{tr}(\boldsymbol{A}) := \sum_{i=1}^{n} a_{ii}.$$

Theorem 5.2. *The trace is a linear operator, i.e., for two square matrices of same order n*

$$tr(x\,\boldsymbol{A} + y\,\boldsymbol{B}) = x\, tr\boldsymbol{A} + y\, tr\boldsymbol{B}, \qquad (\textit{square matrices}).$$

$$tr(\boldsymbol{A} \cdot \boldsymbol{B}) = tr(\boldsymbol{B} \cdot \boldsymbol{A}), \qquad (\textit{type}\, \boldsymbol{A} = m \times n, \quad \textit{type}\, \boldsymbol{B} = n \times m).$$

$$tr(\boldsymbol{A}) = \sum_{i=1}^{n} \lambda_i \quad \textit{where } \lambda_i \textit{ are the eigenvalues of } \boldsymbol{A}.$$

(*Eigenvalues; see page* 105.)

Definition 5.5.

- An upper triangular matrix of type $m \times n$ is of the form

$$\boldsymbol{A} = \begin{bmatrix} a_{11} & a_{12} & \ldots & a_{1m} & \cdots & a_{1n} \\ 0 & a_{22} & \ldots & a_{2m} & \cdots & a_{2n} \\ & & \ddots & & \ddots & \vdots \\ 0 & 0 & \ldots & a_{mm} & \ldots & a_{mn} \end{bmatrix} \qquad (m \le n), \tag{5.5}$$

or

$$\boldsymbol{A} = \begin{bmatrix} a_{11} & a_{12} & \ldots & a_{1n} \\ 0 & a_{22} & \ldots & a_{2n} \\ \vdots & \vdots & \ddots & \vdots \\ 0 & 0 & \ldots & a_{mn} \\ 0 & 0 & \ldots & 0 \\ \vdots & \vdots & \ddots & \vdots \\ 0 & 0 & \ldots & 0 \end{bmatrix} \qquad (m \ge n), \tag{5.6}$$

where the elements below the main diagonal $= 0$.
Lower triangular matrix is defined similarly.

- A diagonal matrix is a square matrix where $a_{ij} \equiv 0$ for every $i \neq j$.
- The unit matrix of order n denoted by $\boldsymbol{I} = \boldsymbol{I}_n$ is a square matrix (of order n) with $a_{ij} = 0$, for $i \neq j$, and $a_{ii} = 1,\ i, j = 1, 2, \ldots, n$. Example

$$\boldsymbol{I}_1 = 1, \quad \boldsymbol{I}_2 = \begin{bmatrix} 1 & 0 \\ 0 & 1 \end{bmatrix}, \quad \boldsymbol{I}_3 = \begin{bmatrix} 1 & 0 & 0 \\ 0 & 1 & 0 \\ 0 & 0 & 1 \end{bmatrix}$$

$$\text{and generally} \quad \boldsymbol{I}_n = \begin{bmatrix} 1 & 0 & \ldots & 0 \\ 0 & 1 & \ldots & 0 \\ & & \ddots & \\ 0 & 0 & \ldots & 1 \end{bmatrix} \text{ of type } \boldsymbol{I}_n = n \times n. \qquad (5.7)$$

5.1.1 *Solution of linear system of equations with matrices*

To solve a linear system of equations (5.1) in matrix form, the most common *elimination method*, also known as *Gaussian elimination*, yields a simple algorithm, see what follows.

Definition 5.6. The three *elementary row operations*, for matrices, are

- **R1:** Multiplication of one row by a number, which then is element-wise added to another row.
- **R2:** Interchanging two rows.
- **R3:** Multiplication of one row by a number $\neq 0$.

Remarks. Two matrices $\boldsymbol{A}$ and $\boldsymbol{A}'$, such that one is transferred to the other by a sequence of elementary row operations, are called *row equivalent*. This is written as $\boldsymbol{A} \sim \boldsymbol{A}'$.

- Evidently, conversing the order of row operations $\boldsymbol{A} \sim \boldsymbol{A}'$ yields $\boldsymbol{A}' \sim \boldsymbol{A}$, so $\sim$ is a kind of equivalence.
- Row operations on matrices need not be associated to linear equation systems.

Definition 5.7. In this definition, one considers a matrix $\boldsymbol{A}$ given as (5.2) page 85 and that $a_{1,1} \neq 0$.

A row (column) in a matrix where at least one element is $\neq 0$ is a non-zero row (non-zero-column).

If all elements in the row (column) are zero, the row (column) is called a *zero row (zero column)*.

The first non-zero element $a_{j,k}$ in a row, counted from the left, is called *pivot element*, that is $a_{j',k} = 0$ for $j' = 1, 2, \ldots, j-1$.

A matrix with pivot element in positions (j,k) and $(j+1,k')$ with $k' > k$ is on *Echelon form* and the corresponding position (j,k) is a *pivot position*.

A column in a matrix with pivot position is called *pivot column*.

A matrix on echelon form with all its pivot elements equal to 1, *and all other elements in the same column* equal to zero, is on *(row) reduced echelon form*, see (5.9) page 90.

The rank of a matrix is the number of non-zero rows in a row equivalent matrix in echelon form, i.e., the number of pivot positions.

Theorem 5.3. *Applying* **R1, R2, R3** *on page* 88, *to the matrix* $\boldsymbol{A}$, *one eventually reaches a row equivalent unique row reduced echelon matrix* $\boldsymbol{A}'$.

The rank is unique (due to the proposition above). The definition above implies that all zero rows in $\boldsymbol{A}'$ *are the rows with highest row indices i.e., are at the bottom in* $\boldsymbol{A}'$.

Theorem 5.4. *Let* $\boldsymbol{A}$ *be a coefficient matrix,* $[\boldsymbol{A}|\,\boldsymbol{Y}]$ *an augmented matrix, and the number of variables is* n. *Then*

$$\begin{aligned} \textit{Rank } \boldsymbol{A} = \textit{Rank } [\boldsymbol{A}|\,\boldsymbol{Y}] = n &\iff \textit{Number of solutions } = 1, \\ \textit{Rank } \boldsymbol{A} = \textit{Rank } [\boldsymbol{A}|\,\boldsymbol{Y}] < n &\iff \textit{Number of solutions } = \infty, \\ \textit{Rank } \boldsymbol{A} < \textit{Rank } [\boldsymbol{A}|\,\boldsymbol{Y}] &\iff \textit{Number of solutions } = 0. \end{aligned} \tag{5.8}$$

$$
A \sim A' = \begin{bmatrix}
\boxed{1} & b_{12}\dots b_{1k_1} & 0 & b_{1,(k_1+2)}\cdots b_{1k_2} & 0 & \dots & & \\
0 & \dots\dots\dots & \boxed{1} & b_{2,(k_1+2)}\cdots b_{2k_2} & 0 & & & \\
0 & \dots\dots\dots & 0 & 0\dots\dots\dots 0 & \boxed{1} & \dots & & \\
0 & \dots\dots\dots & 0 & 0\dots\dots\dots 0 & 0 & \ddots & & \\
0 & \dots\dots\dots & 0 & 0\dots\dots\dots 0 & 0 & \dots & \boxed{1} & b_{rk_r}\dots b_{rn} \\
0 & \dots\dots\dots & 0 & 0\dots\dots\dots 0 & 0 & \dots & 0 & 0\dots\dots 0 \\
\vdots & \vdots & \vdots & \vdots & \vdots & \vdots & \vdots & \vdots \\
0 & \dots\dots\dots & 0 & 0\dots\dots\dots 0 & 0 & \dots & 0 & 0\dots\dots 0
\end{bmatrix}. \tag{5.9}
$$

A': *the equivalent* row reduced *matrix for* A (5.2) *is a result of a finite number of elementary row operations on* A.

The dimension of the solution space (the space of all solutions), for an equation system with coefficient matrix A, *is*

$$n - r = n - \operatorname{Rank} A.$$

5.1.2 *Column, row, and null-spaces*

Definition 5.8. Let A be a matrix of type $m \times n$.

(i) The column space $\mathcal{K}_A$ of A is the space of all linear combinations of its columns.
(ii) The row space $\mathcal{R}_A$ of A is the space of all linear combinations of its rows.
(iii) The null space $\mathcal{N}_A$ (null-space) of A is the space of all vectors $x \in \mathbb{R}^n$ such that $Ax = \mathbf{0}$.

Theorem 5.5. *If* A, X, *and* Y *are matrices of type* $m \times n$, $n \times 1$, *and* $m \times 1$, *respectively,* A_j *are the columns of* A, $j = 1, 2, \dots, n$, A^i *are the rows of* A, $X = [x_1, x_2, \dots, x_n]^T$, *and* $Y = [y_1, y_2, \dots, y_m]$, *then*

$$
\begin{aligned}
AX &= x_1 A_1 + x_2 A_2 + \cdots + x_n A_n, \\
YA &= y_1 A^1 + x_2 A^2 + \cdots + y_m A^m.
\end{aligned} \tag{5.10}
$$

Theorem 5.6. *Let* $\boldsymbol{A}$ *be a matrix of type* $m \times n$*. Then*

$$\operatorname{Rank} \boldsymbol{A} = \dim(\mathcal{K}_{\boldsymbol{A}}) = \dim(\mathcal{R}_{\boldsymbol{A}})\,. \tag{5.11}$$

Theorem 5.7 (The dimension theorem).

$$\dim(\mathcal{N}_{\boldsymbol{A}}) + \operatorname{Rank} \boldsymbol{A} = n = \textit{number of columns of } \boldsymbol{A}. \tag{5.12}$$

For a $n \times n$ *matrix* $\boldsymbol{A}$ *with full rank: Rank* $\boldsymbol{A} = n, \dim(\mathcal{N}_{\boldsymbol{A}}) = 0$, *i.e.,* $\mathcal{N}_{\boldsymbol{A}} = \{\boldsymbol{0}\}$.

5.2 Matrix Algebra

Definition 5.9 (Addition and multiplication by scalars). Only matrices of same type can be added (so-called element-wise addition). Let

$$\boldsymbol{A} = (a_{ij})_{m\times n} \text{ and } \boldsymbol{B} = (b_{ij})_{m\times n},$$

then

$$\boldsymbol{A} + \boldsymbol{B} = (a_{ij} + b_{ij})_{m\times n}. \tag{5.13}$$

Multiplication by a scalar (real/complex number) c is performed element-wise.

$$c\boldsymbol{A} = c(a_{ij})_{m\times n} = (c\, a_{ij})_{m\times n}. \tag{5.14}$$

Definition 5.10. Multiplication of two matrices $\boldsymbol{A}$ and $\boldsymbol{B}$ (in this order) is possible only if the number of columns in $\boldsymbol{A}$ is equal to the number of rows in $\boldsymbol{B}$.

More specifically, for type $\boldsymbol{A} = m \times n$ and type $\boldsymbol{B} = n \times p$

$$\boldsymbol{A} \cdot \boldsymbol{B} = \boldsymbol{C} = (c_{ij})_{m\times p}, \tag{5.15}$$

where $c_{ij} = a_{i1}b_{1j} + a_{i2}b_{2j} + \cdots + a_{in}b_{nj}$, $i = 1, 2, \ldots, m$, and $j = 1, 2, \ldots, p.$, i.e., c_{ij} in $\boldsymbol{AB}$ is a result of (vector) multiplication of ith row in $\boldsymbol{A}$ with jth column in $\boldsymbol{B}$.

$$\begin{bmatrix} i\text{th row} \end{bmatrix} \cdot \begin{bmatrix} j\text{th column} \end{bmatrix} = \begin{matrix}\text{element at}\\ \text{position } (i,j)\end{matrix}$$

Figure 5.1: Matrix multiplication.

Theorem 5.8 (Matrix operations).

Associative addition

$$(\boldsymbol{A}+\boldsymbol{B})+\boldsymbol{C}=\boldsymbol{A}+(\boldsymbol{B}+\boldsymbol{C}),$$

Commutative addition

$$\boldsymbol{A}+\boldsymbol{B}=\boldsymbol{B}+\boldsymbol{A}, \quad \textit{if type } \boldsymbol{A} = \textit{type } \boldsymbol{B} = \textit{type } \boldsymbol{C}.$$

Associative multiplication (guideline via Figure 5.1)

$$(\boldsymbol{A}\cdot\boldsymbol{B})\cdot\boldsymbol{C}=\boldsymbol{A}\cdot(\boldsymbol{B}\cdot\boldsymbol{C}) \qquad \textit{for} \begin{cases} \textit{type } \boldsymbol{A} = m\times n, \\ \textit{type } \boldsymbol{B} = n\times p, \ \textit{and} \\ \textit{type } \boldsymbol{C} = p\times r. \end{cases}$$

The left distributive law (guideline via Figure 5.1)

$$\boldsymbol{A}\cdot(\boldsymbol{B}+\boldsymbol{C})=\boldsymbol{A}\cdot\boldsymbol{B}+\boldsymbol{A}\cdot\boldsymbol{C} \quad \textit{for} \begin{cases} \textit{type } \boldsymbol{A} = m\times n, \ \textit{and} \\ \textit{type } \boldsymbol{B} = \textit{type } \boldsymbol{C} = n\times p. \end{cases}$$

The right distributive law (guideline via Figure 5.1)

$$(\boldsymbol{A}+\boldsymbol{B})\cdot\boldsymbol{C}=\boldsymbol{A}\cdot\boldsymbol{C}+\boldsymbol{B}\cdot\boldsymbol{C} \quad \textit{for} \begin{cases} \textit{type } \boldsymbol{A} = \textit{type } \boldsymbol{B} = m\times n, \ \textit{and} \\ \textit{type } \boldsymbol{C} = n\times p. \end{cases} \tag{5.16}$$

The equation system (5.1) in matrix form (5.3) can be written as a matrix multiplication

$$\begin{bmatrix} a_{11} & a_{12} & \dots & a_{1n} \\ a_{21} & a_{22} & \dots & a_{2n} \\ & & \ddots & \\ a_{m1} & a_{m2} & \dots & a_{mn} \end{bmatrix} \cdot \begin{bmatrix} x_1 \\ x_2 \\ \vdots \\ x_n \end{bmatrix} = \begin{bmatrix} y_1 \\ y_2 \\ \vdots \\ y_m \end{bmatrix}. \tag{5.17}$$

Remarks. The matrices $\boldsymbol{A} \cdot \boldsymbol{B}$ och $\boldsymbol{B} \cdot \boldsymbol{A}$ are generally not equal, i.e., the multiplication is not commutative.

A *necessary* condition for two matrices to commute, that is $\boldsymbol{A} \cdot \boldsymbol{B} = \boldsymbol{B} \cdot \boldsymbol{A}$, is that $\boldsymbol{A}$ and $\boldsymbol{B}$ are quadratic matrices of the same order.

Theorem 5.9. *For unit matrices* $\boldsymbol{I}$

$$\boldsymbol{I} \cdot \boldsymbol{A} = \boldsymbol{A} \quad \textit{and} \quad \boldsymbol{A} \cdot \boldsymbol{I} = \boldsymbol{A}. \tag{5.18}$$

For the transpose, the following hold true:

$$(\boldsymbol{A} + \boldsymbol{B})^T = \boldsymbol{A}^T + \boldsymbol{B}^T, \textit{ if type } \boldsymbol{A} = \textit{type } \boldsymbol{B}.$$

$$(\boldsymbol{A} \cdot \boldsymbol{B})^T = \boldsymbol{B}^T \cdot \boldsymbol{A}^T, \quad \textit{if type } \boldsymbol{A} = m \times n, \quad \textit{type } \boldsymbol{B} = n \times p. \tag{5.19}$$

5.2.1 *Inverse matrix*

Definition 5.11. If for a square matrix $\boldsymbol{A}$, there exists a matrix $\boldsymbol{A}^{-1}$ such that

$$\boldsymbol{A}^{-1} \cdot \boldsymbol{A} = \boldsymbol{A} \cdot \boldsymbol{A}^{-1} = \boldsymbol{I}, \tag{5.20}$$

then the matrix $\boldsymbol{A}^{-1}$ is called the inverse of $\boldsymbol{A}$. Then we say that $\boldsymbol{A}$ is invertible.

Theorem 5.10. *Suppose that* $\boldsymbol{A}$ *and* $\boldsymbol{B}$*: are invertible of the same order.*

Then the following relations hold true:

$$\begin{gathered}(\boldsymbol{A} \cdot \boldsymbol{B})^{-1} = \boldsymbol{B}^{-1} \cdot \boldsymbol{A}^{-1}, \qquad (\boldsymbol{A}^T)^{-1} = (\boldsymbol{A}^{-1})^T \\ \boldsymbol{A} \cdot \boldsymbol{X} = \boldsymbol{C} \Longleftrightarrow \boldsymbol{X} = \boldsymbol{A}^{-1} \boldsymbol{C}.\end{gathered} \tag{5.21}$$

Remarks. Let type $\boldsymbol{A} = m \times n$ in (5.18). Then its left and right unit matrices are of type $m \times m$ and $n \times n$, respectively.

For an invertible matrix $\boldsymbol{A}$, using row reducing, the equation systems $\boldsymbol{AX} = \boldsymbol{I}$ (all matrices are of order n) yields the solution (matrix) $\boldsymbol{X} = \boldsymbol{A}^{-1}$. This is called *Jacobi's method.*

5.2.2 *Elementary matrices*

The row operations **R1–R3**, page 88, may be performed using *elementary matrices*:

Theorem 5.11. *Let $\boldsymbol{I} = \boldsymbol{I}_n$ be the unit matrix of order n and $\boldsymbol{A}$ a matrix of type $\boldsymbol{A} = n \times p$.*

- **R1:** *Multiplying row i by a scalar c and adding (element-wise) to the row j yields $\boldsymbol{A}' = \boldsymbol{E}(1) \cdot \boldsymbol{A}$, where $\boldsymbol{E}(1)$ is $\boldsymbol{I}$ with c at position (j, i).*
- **R2:** *Interchanging rows i and j gives a matrix $\boldsymbol{A}' = \boldsymbol{E}(2) \cdot \boldsymbol{A}$, where $\boldsymbol{E}(2)$ is the matrix $\boldsymbol{I}$ where rows i and j are interchanged.*
- **R3:** *Multiplying of row i with $c \neq 0$ gives a matrix $\boldsymbol{A}' = \boldsymbol{E}(3) \cdot \boldsymbol{A}$, where $\boldsymbol{E}(3)$ is the matrix $\boldsymbol{I}$, but with c in the position (i, i).*

5.2.3 *LU-factorization*

Definition 5.12.

- A matrix $\boldsymbol{L}$ is lower triangular, if all elements above the main diagonal are, equal to zero.
- A matrix $\boldsymbol{U}$ is upper triangular, if all elements beneath the main diagonal are zero.

Theorem 5.12. *Let $\boldsymbol{A}$ be a matrix of type $\boldsymbol{A} = m \times n$.*

Then there exist a lower triangular matrix $\boldsymbol{L}$ of type $\boldsymbol{L} = m \times m$, with only ones in the main diagonal, and an upper triangular matrix $\boldsymbol{U}$, type $\boldsymbol{U} = m \times n$ such that

$$\boldsymbol{A} = \boldsymbol{L} \cdot \boldsymbol{U}. \tag{5.22}$$

If only **R1** *and* **R3** *are used to get*

$$\boldsymbol{A}' = \left(\prod_{j=1}^{p} \boldsymbol{E}_j \right) \cdot \boldsymbol{A}$$

via elementary matrices $\boldsymbol{E}_j$, and if $\boldsymbol{A}'$ is an upper triangular matrix, then

$$\boldsymbol{L} = \left(\prod_{j=1}^{p} \boldsymbol{E}_j \right)^{-1} \quad \text{and} \quad \boldsymbol{U} = \boldsymbol{A}'.$$

Here, $\boldsymbol{E}_j$ are all, lower triangular, elementary matrices.

5.2.4 *Quadratic form*

Definition 5.13. Assume that $\boldsymbol{A}$ is a symmetric (and hence square) matrix (i.e., $a_{ij} = a_{ji}$) of order n and $\boldsymbol{x}$ is a matrix (vector) of type $n \times 1$. Then,

$$q(\boldsymbol{x}) := \boldsymbol{x}^T \cdot \boldsymbol{A} \cdot \boldsymbol{x} \tag{5.23}$$

is called a *quadratic form*.

$$\begin{array}{llll} (1) & q(\boldsymbol{x}) > 0, & \boldsymbol{x} \neq \boldsymbol{0} & q \text{ is positive definite,} \\ (2) & q(\boldsymbol{x}) < 0, & \boldsymbol{x} \neq \boldsymbol{0} & q \text{ is negative definite,} \\ (3) & q(\boldsymbol{x}) < 0, \text{ and } q(\boldsymbol{x}) > 0 & & q \text{ is indefinite,} \\ & \quad \text{for different } \boldsymbol{x}. & & \end{array} \tag{5.24}$$

If in (1) > 0 and (in (2) < 0) are changed to ≥ 0, and (≤ 0), then q is positive (negative) semi-definite, respectively.

5.3 Determinant

Definition 5.14. Consider a permutation $(k_1, k_2, \ldots, k_n)$ of $(1, 2, \ldots, n)$. The number of inversions, denoted by $|(k_1, k_2, \ldots, k_n)|$, is the number of pairs with the property $k_i > k_j$ where $i < j$. Let $\boldsymbol{A} = (a_{ij})_{n \times n}$ be a square matrix of order n. Its *determinant* is a real (complex) number given by

$$\det \boldsymbol{A} = \sum_{(k_1, k_2, \ldots, k_n)} (-1)^{|(k_1, k_2, \ldots, k_n)|} a_{1k_1} \cdot a_{2k_2} \cdot \ldots \cdot a_{nk_n}, \tag{5.25}$$

where the sum is taken over all permutations $(k_1, k_2, \ldots k_n)$ of $(1, 2, \ldots, n)$.

The determinant of a matrix $\boldsymbol{A}$ is denoted $\det \boldsymbol{A}$ or simply $|\boldsymbol{A}|$.

Example 5.1. For $(4, 2, 3, 1)$, the number of permutations is $|(4, 2, 3, 1)| = 5$, because of $(4, 1), (4, 2), (4, 3), (3, 1), (2, 1)$.

Theorem 5.13. *Let* $\boldsymbol{A}$ *and* $\boldsymbol{B}$ *be square matrices of the same type/order. Then*

$$\det(\boldsymbol{A}\cdot\boldsymbol{B}) = \det\boldsymbol{A}\cdot\det\boldsymbol{B}, \qquad \det\boldsymbol{A} = \det(\boldsymbol{A}^T). \tag{5.26}$$

$$\det(\boldsymbol{A}^{-1}) = (\det\boldsymbol{A})^{-1} \quad \textit{(if } \boldsymbol{A} \textit{ is invertible, i.e., } \det\boldsymbol{A}\neq 0\textit{)}. \tag{5.27}$$

The determinant of an upper or lower triangular matrix is the product of the elements on the main diagonal.

The determinant of all identity matrices is $\det\boldsymbol{I} = 1$.

$$\det\begin{bmatrix} a & b \\ c & d \end{bmatrix} =: \begin{vmatrix} a & b \\ c & d \end{vmatrix} = ad - bc \quad \textit{and}$$

$$\det\begin{bmatrix} a_{11} & a_{12} & a_{13} \\ a_{21} & a_{22} & a_{23} \\ a_{31} & a_{32} & a_{33} \end{bmatrix} = \begin{array}{l} a_{11}a_{22}a_{33} + a_{12}a_{23}a_{31} + a_{13}a_{21}a_{32} + \\ -(a_{11}a_{23}a_{32} + a_{12}a_{21}a_{33} + a_{13}a_{22}a_{31}). \end{array} \tag{5.28}$$

Sarrus' rule

Only for matrices of type 3×3, *Sarrus' rule* makes sense. Putting the two first columns to the right of the matrix, the following diagonal multiplication procedure applies, where product by blue colored arrows are taken with a minus sign. (The red a_{11} to a_{22} is not counted).

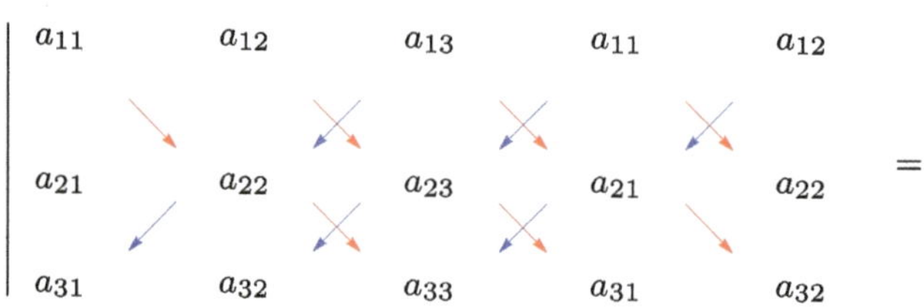

$$a_{11}a_{22}a_{33} + a_{12}a_{23}a_{31} + a_{13}a_{21}a_{32} \quad - \quad (a_{13}a_{22}a_{31} + a_{11}a_{23}a_{32} + a_{12}a_{21}a_{33})$$

The inverse of a matrix $\boldsymbol{A}$ of order 2 (type 2×2) exists precisely when $\det\boldsymbol{A} = ad - bc \neq 0$, see (5.28). Then the inverse is

$$\begin{bmatrix} a & b \\ c & d \end{bmatrix}^{-1} = \frac{1}{ad-bc}\begin{bmatrix} d & -b \\ -c & a \end{bmatrix}. \tag{5.29}$$

Definition 5.15. The *sub-matrix* $\boldsymbol{A}_{ij}$ of $\boldsymbol{A}$ of order n is the square matrix of order $n-1$ obtained when the ith row and jth column of $\boldsymbol{A}$ are removed.

The corresponding *sub-determinant* is $d_{ij} := \det \boldsymbol{A}_{ij}$.

The inverse of a matrix $\boldsymbol{A}$ exists if and only if $\det \boldsymbol{A} \neq 0$. The determinant of a 3×3-matrix,

$$\boldsymbol{A} = \begin{bmatrix} a_{11} & a_{12} & a_{13} \\ a_{21} & a_{22} & a_{23} \\ a_{31} & a_{32} & a_{33} \end{bmatrix},$$

here denoted by D, is given by (5.28). The inverse of $\boldsymbol{A}$ is then

$$\boldsymbol{A}^{-1} = \frac{1}{D} \begin{bmatrix} d_{11} & -d_{21} & d_{31} \\ -d_{12} & d_{22} & -d_{32} \\ d_{13} & -d_{23} & d_{33} \end{bmatrix},$$

where the d_{ij}s are defined as in (5.30). Note the index-shift on ds!

More generally, the following rule holds.

Theorem 5.14. *Given a square matrix* $\boldsymbol{A}$ *of type* $\boldsymbol{A} = n \times n$ *(order* n*) with* $\det \boldsymbol{A} = D \neq 0$.

Let $\boldsymbol{A}_{ij}$ *be the matrix, of type* $(n-1) \times (n-1)$*, obtained from* $\boldsymbol{A}$ *by removing row* i *and column* j *and set* $d_{ij} = \det \boldsymbol{A}_{ij}$*. Then*

$$\boldsymbol{A}^{-1} = \frac{1}{D} \begin{bmatrix} d_{11} & -d_{21} & \ldots & (-1)^{1+n} d_{n1} \\ -d_{12} & d_{22} & \ldots & (-1)^{2+n} d_{n2} \\ \vdots & \vdots & \ddots & \vdots \\ (-1)^{n+1} d_{1n} & (-1)^{n+2} d_{2n} & \ldots & d_{nn} \end{bmatrix}. \tag{5.30}$$

The element in position (i,j) *i* $\boldsymbol{A}^{-1}$ *is thus* $(-1)^{i+j} \cdot \dfrac{d_{ji}}{D} = (-1)^{i+j} \cdot \dfrac{\det \boldsymbol{A}_{ij}}{D}$.

5.3.1 Number of solutions for ES, determinant, and rank

Theorem 5.15. *Let* $\boldsymbol{A}$ *be a matrix type* $\boldsymbol{A} = n \times n$, *i.e., a square matrix of order* n. *Then the following four statements are equivalent.*

$$\begin{array}{l} \bullet \ \det \boldsymbol{A} \neq 0. \\ \bullet \ \boldsymbol{A}^{-1} \textit{ exists.} \\ \bullet \ \boldsymbol{A} \cdot \boldsymbol{X} = \boldsymbol{B} \textit{ has a unique solution } \boldsymbol{X}. \\ \bullet \ \operatorname{Rank} \boldsymbol{A} = n. \end{array} \tag{5.31}$$

5.3.2 Computing the determinant using sub-determinants

The determinant of

$$\boldsymbol{A} := \begin{bmatrix} a_{11} & a_{12} & \dots & a_{1n} \\ a_{21} & a_{22} & \dots & a_{2n} \\ & & \ddots & \\ a_{n1} & a_{n2} & \dots & a_{nn} \end{bmatrix},$$

can be obtained *expanding along row number* i, i.e.,

$$\det \boldsymbol{A} = \sum_{j=1}^{n} (-1)^{i+j} a_{ij} \det \boldsymbol{A}_{ij}. \tag{5.32}$$

Likewise, one gets the determinant of $\boldsymbol{A}$ expanding with respect to column j:

$$\det \boldsymbol{A} = \sum_{i=1}^{n} (-1)^{i+j} a_{ij} \det \boldsymbol{A}_{ij}. \tag{5.33}$$

5.3.3 Cramer's rule

If in the equation system (5.17) page 92, $m = n$, one gets

$$\underbrace{\begin{bmatrix} a_{11} & a_{12} & \dots & a_{1n} \\ a_{21} & a_{22} & \dots & a_{2n} \\ & & \ddots & \\ a_{n1} & a_{n2} & \dots & a_{nn} \end{bmatrix}}_{=\boldsymbol{A}} \cdot \begin{bmatrix} x_1 \\ x_2 \\ \vdots \\ x_n \end{bmatrix} = \begin{bmatrix} y_1 \\ y_2 \\ \vdots \\ y_n \end{bmatrix}. \tag{5.34}$$

Assume that $\det \boldsymbol{A} \neq 0$. By $\boldsymbol{A}^{(j)}$ we mean the matrix obtained from $\boldsymbol{A}$ when substituting column j by the RHS: $(y_1, y_2, \ldots, y_n)^T$. Then

$$x_j = \frac{\det \boldsymbol{A}^{(j)}}{\det \boldsymbol{A}}, \qquad j = 1, 2, \ldots, n. \tag{5.35}$$

Example 5.2. We use Cramer's rule to solve the equation system

$$\begin{cases} 2x + 3y &= 5, \\ -x + 2y &= 1. \end{cases} \tag{5.36}$$

Solution. The coefficient matrix and the matrices in nominator in Cramer's rule are

$$\boldsymbol{A} = \begin{bmatrix} 2 & 3 \\ -1 & 2 \end{bmatrix} \qquad \boldsymbol{A}^{(1)} = \begin{bmatrix} 5 & 3 \\ 1 & 2 \end{bmatrix} \qquad \boldsymbol{A}^{(2)} = \begin{bmatrix} 2 & 5 \\ -1 & 1 \end{bmatrix}.$$

Thus,

$$x = \frac{\det \boldsymbol{A}^{(1)}}{\det \boldsymbol{A}} = \frac{7}{7} = 1 \quad \text{and} \quad y = \frac{\det \boldsymbol{A}^{(2)}}{\det \boldsymbol{A}} = \frac{7}{7} = 1,$$

which is the exact solution of the equation system (5.36).

5.3.4 *Determinant and row operations*

Theorem 5.16. *Given a square matrix* $\boldsymbol{A}$, *and a row-equivalent matrix* $\boldsymbol{A}'$ *of* $\boldsymbol{A}$, *obtained by elementary row operation on* $\boldsymbol{A}$.

- **R1** *Multiplying a row by a number and then adding (element-wise) to another row does not change the value of the determinant, i.e.,*

$$\det \boldsymbol{A}' = \det \boldsymbol{A}.$$

- **R2** *The change on two rows changes the sign of the determinant:*

$$\det \boldsymbol{A}' = -\det \boldsymbol{A}.$$

- **R3** *Multiplying a row or column by a number* $c \neq 0$ *means the determinant is multiplied by* c*:*

$$\det \boldsymbol{A}' = c \det \boldsymbol{A}.$$

- *If all elements in a row (column) are equal to zero, then* $\det \boldsymbol{A} = 0$.
- *If two rows (columns) are equal, then* $\det \boldsymbol{A} = 0$.
- *For a matrix* $\boldsymbol{A}$ *of order* n, *multiplied by a number* c,

$$\det(c \cdot \boldsymbol{A}) = c^n \cdot \det \boldsymbol{A}.$$

5.3.5 *Pseudoinverse*

There are several definitions of pseudoinverse. What follows is the most common one: the Moore–Penrose inverse.

Given a matrix $\boldsymbol{A} = (a_{jk})_{m \times n}$ of type $m \times n$ with real entries (elements). Its *pseudoinverse* is defined as the matrix $\boldsymbol{A}^+$ satisfying

$\boldsymbol{A} \cdot \boldsymbol{A}^+ \cdot \boldsymbol{A} = \boldsymbol{A}$.
$\boldsymbol{A}^+ \cdot \boldsymbol{A} \cdot \boldsymbol{A}^+ = \boldsymbol{A}^+$.
$(\boldsymbol{A} \cdot \boldsymbol{A}^+)^T = \boldsymbol{A} \cdot \boldsymbol{A}^+$, that is $\boldsymbol{A} \cdot \boldsymbol{A}^+$ is symmetric.
$(\boldsymbol{A}^+ \cdot \boldsymbol{A})^T = \boldsymbol{A}^+ \cdot \boldsymbol{A}$, that is $\boldsymbol{A}^+ \cdot \boldsymbol{A}$ is also symmetric.

Properties

If $\boldsymbol{A}$ has linearly independent columns, then $m \geq n$, $\boldsymbol{A}^T \cdot \boldsymbol{A}$ is invertible, and

$$\boldsymbol{A}^+ = (\boldsymbol{A}^T \cdot \boldsymbol{A})^{-1} \cdot \boldsymbol{A}^T \text{ implying } \boldsymbol{A}^+ \cdot \boldsymbol{A} = \boldsymbol{I}_n,$$

that is $\boldsymbol{A}^+$ is a left inverse of $\boldsymbol{A}$.

If $\boldsymbol{A}$ has linearly independent rows, then $m \leq n$, $\boldsymbol{A} \cdot \boldsymbol{A}^T$ is invertible, and

$$\boldsymbol{A}^+ = \boldsymbol{A}^T \cdot (\boldsymbol{A} \cdot \boldsymbol{A}^T)^{-1} \text{ implying } \boldsymbol{A} \cdot \boldsymbol{A}^+ = \boldsymbol{I}_m,$$

that is $\boldsymbol{A}^+$ is a right inverse of $\boldsymbol{A}$.

Remarks. The notion of pseudoinverse is often defined for complex-valued matrices. In that case, $\boldsymbol{A}^T$ is substituted by a hermitian matrix, i.e., a matrix $\boldsymbol{A}^*$ with entries $a_{kj} = \overline{a_{jk}}$.

To solve a linear system of equations: $\boldsymbol{A} \cdot \boldsymbol{X} = \boldsymbol{Y}$ with $\boldsymbol{A}^T \cdot \boldsymbol{A}$ invertible, the best solution in *Least square (LS) terms*, see the following, is

$$\boldsymbol{X} = \hat{\boldsymbol{X}} = \boldsymbol{A}^+ \cdot \boldsymbol{A} \cdot \boldsymbol{Y}.$$

With $\boldsymbol{X} = \hat{\boldsymbol{X}}$ the value

$$||\boldsymbol{A} \cdot \boldsymbol{X} - \boldsymbol{Y}||$$

is the smallest possible one.

Even a singular square matrix has a psuedoinverse, for example,

$$\boldsymbol{A} := \begin{bmatrix} 1 & 2 \\ -2 & -4 \end{bmatrix} \text{ with } \boldsymbol{A}^+ = \frac{1}{25}\begin{bmatrix} 1 & -2 \\ 2 & -4 \end{bmatrix}.$$

Even if a linear equation system $\boldsymbol{Ax} = \boldsymbol{y}$, type $\boldsymbol{A} = m \times n$, has no solution $\boldsymbol{x}$, its *reduced equation system*: $\boldsymbol{A}^T\boldsymbol{A}\cdot\boldsymbol{x} = \boldsymbol{A}^T\cdot\boldsymbol{y}$ has indeed a solution, which is an approximate solution of $\boldsymbol{Ax} = \boldsymbol{y}$.

Definition 5.16. The matrix equation $\boldsymbol{A}\cdot\boldsymbol{x} = \boldsymbol{y} \Longrightarrow \boldsymbol{A}^T\boldsymbol{A}\cdot\boldsymbol{x} = \boldsymbol{A}^T\cdot\boldsymbol{y}$, where the latter is called *reduced.*

The LS method is about finding a solution $x_1, x_2, \ldots, x_n$ so that the "LS-error"

$$\eta := \sqrt{\frac{1}{m}(\varepsilon_1^2 + \varepsilon_2^2 + \cdots + \varepsilon_n^2)},$$

with

$$\begin{cases} \varepsilon_1 & = a_{11}\,x_1 + a_{12}\,x_2 + \cdots + a_{1n}\,x_n - y_1, \\ \ldots = \ldots \\ \varepsilon_m & = a_{m1}\,x_1 + a_{m2}\,x_2 + \cdots + a_{mn}\,x_n - y_m, \end{cases}$$

becomes minimal.

Note that if $\boldsymbol{x} = (x_1, x_2, \ldots, x_n)^T$ is an exact solution $\varepsilon_i = 0$, $i = 1, \ldots, m$. Then the LS-error $\eta = 0$.

The terms ε_i, $i = 1, \ldots, m$ are the differences between LHS and RHS of the ES $\boldsymbol{Ax} = \boldsymbol{y}$.

Theorem 5.17. *Consider the linear equation system* (5.1) *page* 85

$$\boldsymbol{A}\cdot\boldsymbol{x} = \boldsymbol{y}. \tag{5.37}$$

The norm of $\boldsymbol{x}$ *(the Euclidean distance between* $\boldsymbol{x}$ *and origin) is defined as*

$$\|\boldsymbol{x}\| = \|(x_1, x_2, \ldots, x_n)\| = \sqrt{x_1^2 + x_2^2 + \cdots + x_n^2}.$$

There is at least one solution $\boldsymbol{x} = \boldsymbol{x}_0$*, which minimizes the LS-error* $\eta\ \|\boldsymbol{A}\cdot\boldsymbol{x} - \boldsymbol{y}\|$*. Furthermore, if* $\boldsymbol{x}$ *is an approximate solution of* (5.37)*, then*

$$\|\boldsymbol{A}\cdot\boldsymbol{x} - \boldsymbol{y}\| \text{ minimal} \Longleftrightarrow \boldsymbol{A}^T\boldsymbol{A}\cdot\boldsymbol{x} = \boldsymbol{A}^T\cdot\boldsymbol{y}. \tag{5.38}$$

The equation on the RHS (5.38) *always has a solution that is the best approximate solution in the sense of the LS method.*

In the case that $\boldsymbol{A}^T \cdot \boldsymbol{A}$ *is invertible, the best solution of* (5.37) *in LS terms is given by*

$$\boldsymbol{x} = (\boldsymbol{A}^T \cdot \boldsymbol{A})^{-1} \cdot \boldsymbol{A}^T \cdot \boldsymbol{y},$$

where the matrix $(\boldsymbol{A}^T \cdot \boldsymbol{A})^{-1} \cdot \boldsymbol{A}^T$ *is the (left) pseudoinverse of* $\boldsymbol{A}$.

Given coordinates $(x_1, y_1), (x_2, y_2), \ldots, (x_m, y_m)$ *for* m *points in* $\mathbb{R}^2$.

To adjust a line of the form $y = ax + b$ *to these points, one gets the equation system (ES)*

$$\begin{aligned} ax_1 + b &= y_1 \\ ax_2 + b &= y_2 \\ &\ddots \\ ax_m + b &= y_m \end{aligned} \quad \text{or} \quad \begin{bmatrix} x_1 & 1 \\ x_2 & 1 \\ \vdots & \vdots \\ x_m & 1 \end{bmatrix} \cdot \begin{bmatrix} a \\ b \end{bmatrix} = \begin{bmatrix} y_1 \\ y_2 \\ \vdots \\ y_m \end{bmatrix}. \tag{5.39}$$

By writing the matrix-equation in compact form

$$\boldsymbol{X} \cdot \begin{bmatrix} a \\ b \end{bmatrix} = \boldsymbol{Y}, \tag{5.40}$$

the best solution in LS terms is given by

$$\boldsymbol{X}^T \cdot \boldsymbol{X} \cdot \begin{bmatrix} a \\ b \end{bmatrix} = \boldsymbol{X}^T \cdot \boldsymbol{Y}.$$

This matrix-equation has a solution. In the case $\boldsymbol{X}^T \cdot \boldsymbol{X}$ *is invertible, the best LS solution is*

$$\begin{bmatrix} a \\ b \end{bmatrix} = (\boldsymbol{X}^T \cdot \boldsymbol{X})^{-1} \cdot \boldsymbol{X}^T \cdot \boldsymbol{Y}. \tag{5.41}$$

To adjust a polynomial of (at most) second degree, $y = a_2x^2 + a_1x + a_0$, *the corresponding matrix-equation is*

$$\begin{bmatrix} x_1^2 & x_1 & 1 \\ x_2^2 & x_2 & 1 \\ \vdots & \vdots & \vdots \\ x_m^2 & x_m & 1 \end{bmatrix} \cdot \begin{bmatrix} a_2 \\ a_1 \\ a_0 \end{bmatrix} = \begin{bmatrix} y_1 \\ y_2 \\ \vdots \\ y_m \end{bmatrix},$$

or in short terms, similar to (5.40)

$$\boldsymbol{X} \cdot \begin{bmatrix} a_2 \\ a_1 \\ a_0 \end{bmatrix} = \boldsymbol{Y}.$$

In the case $\boldsymbol{X}^T \cdot \boldsymbol{X}$ *is invertible, the best LS solution is*

$$\begin{bmatrix} a_2 \\ a_1 \\ a_0 \end{bmatrix} = (\boldsymbol{X}^T \cdot \boldsymbol{X})^{-1} \cdot \boldsymbol{X}^T \cdot \boldsymbol{Y}. \tag{5.42}$$

5.3.6 *Best LS solutions for some common functions*

To adjust other functions than polynomials to points, logarithms are used.

Below, the base of 10 and e are used, i.e., the logarithm $\log_{10} = \lg$ and $\log_e = \ln$.

$$y = C\,x^a \iff \lg C + a \lg x = \lg y.$$

Now a plays the same role as a in (5.40) and $\lg C = b$.

The corresponding matrix-equation is

$$\begin{bmatrix} \lg x_1 & 1 \\ \lg x_2 & 1 \\ \vdots & \vdots \\ \lg x_m & 1 \end{bmatrix} \cdot \begin{bmatrix} a \\ \lg C \end{bmatrix} = \begin{bmatrix} \lg y_1 \\ \lg y_2 \\ \vdots \\ \lg y_m \end{bmatrix}.$$

In the case the matrix on the left-hand side, $\boldsymbol{X}$, is invertible, The LS solution is given by (5.41).

For an exponential relation $y = Ca^x$, using logarithms,

$$y = Ca^x \iff x \lg a + \lg C = \lg y.$$

For adjustment of the type in $y = a \ln x + b$, one considers $X := \ln x$ as a new variable and applies (5.40), but with x_j replaced by

$X_j = \ln x_j$, so the matrix $\boldsymbol{X}$ is

$$\boldsymbol{X} = \begin{bmatrix} \ln x_1 & 1 \\ \ln x_2 & 1 \\ \vdots & \vdots \\ \ln x_m & 1 \end{bmatrix}, \quad \text{and} \quad \boldsymbol{X} \cdot \begin{bmatrix} a \\ b \end{bmatrix} = \boldsymbol{Y},$$

with $\boldsymbol{Y}$ as in (5.40).

Example 5.3. Only the linear case is addressed here. Following points are given: $\begin{array}{c|ccccc} x & 1 & 2 & 3 & 4 & 5 \\ y & 4 & 3 & 3 & 1 & 2 \end{array}$. For the line, one uses (5.39). This gives the matrix equation

$$\begin{bmatrix} 1 & 1 \\ 2 & 1 \\ 3 & 1 \\ 4 & 1 \\ 5 & 1 \end{bmatrix} \cdot \begin{bmatrix} a \\ b \end{bmatrix} = \begin{bmatrix} 4 \\ 3 \\ 3 \\ 1 \\ 2 \end{bmatrix} \quad \text{or} \quad \boldsymbol{X} \cdot \begin{bmatrix} a \\ b \end{bmatrix} = \boldsymbol{Y}.$$

Multiplying by $\boldsymbol{X}^T$ from the left, one gets

$$\begin{bmatrix} 55 & 15 \\ 15 & 5 \end{bmatrix} \cdot \begin{bmatrix} a \\ b \end{bmatrix} = \begin{bmatrix} 33 \\ 13 \end{bmatrix} \Longleftrightarrow \begin{bmatrix} a \\ b \end{bmatrix} = \frac{1}{10} \cdot \begin{bmatrix} 1 & -3 \\ -3 & 11 \end{bmatrix} \cdot \begin{bmatrix} 33 \\ 13 \end{bmatrix} = \begin{bmatrix} -0.6 \\ 4.4 \end{bmatrix}.$$

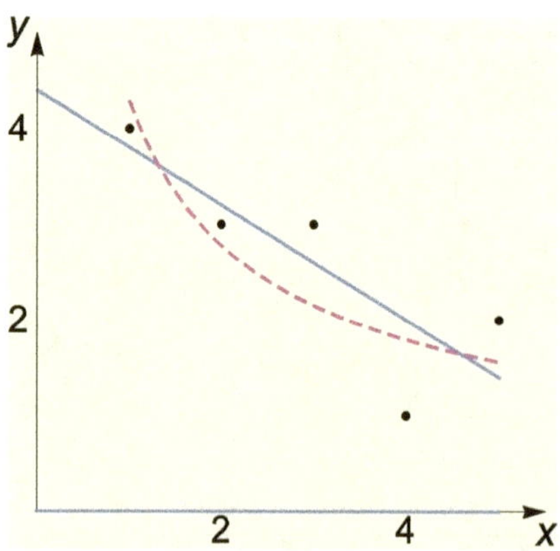

Dashed and purple colored: Power function $y = 0.6 \cdot x^{-0.6}$.

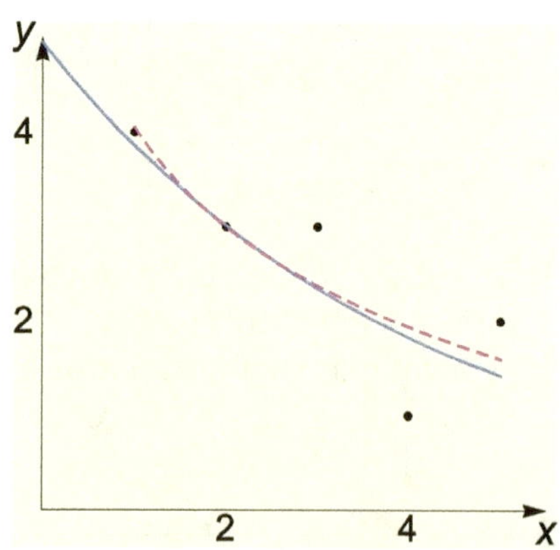

Dashed and purple colored: Logarithmic function $y = 4.25 - 1.9 \ln x$.

5.3.7 *Eigenvalues and eigenvectors*

Definition 5.17. A square matrix $\boldsymbol{A}$ has an *eigenvector* $\boldsymbol{x} \neq \mathbf{0}$ if the equation

$$\boldsymbol{A} \cdot \boldsymbol{x} = \lambda \boldsymbol{x}, \tag{5.43}$$

has a solution for some scalar λ. Then λ is called an *eigenvalue.*

Theorem 5.18.

(i) *Eigenvalues λ are roots of the* secular equation

$$s(\lambda) := \det(\boldsymbol{A} - \lambda \boldsymbol{I}) = 0. \tag{5.44}$$

(ii) *Eigenvalues of a symmetric (real) matrix are real.*
(iii) *For two different eigenvalues, the corresponding eigenvectors are orthogonal.*

Determining eigenvalues and eigenvectors

(i) The λs are obtained solving the polynomial equation (5.44).
(ii) For each λ the corresponding eigenvector $\boldsymbol{x}$ is obtained solving (5.43) the homogenous matrix equation

$$(\boldsymbol{A} - \lambda \boldsymbol{I})\boldsymbol{x} = \mathbf{0}, \text{ or } [\boldsymbol{A} - \lambda \boldsymbol{I}],$$

the augmented matrix. In the case $\boldsymbol{A} = (a_{jk})_{3\times 3}$, the augmented matrix becomes

$$\begin{bmatrix} a_{11} - \lambda & a_{12} & a_{13} \\ a_{21} & a_{22} - \lambda & a_{23} \\ a_{31} & a_{32} & a_{33} - \lambda \end{bmatrix},$$

where the RHS, $\mathbf{0} = [0 \quad 0 \quad 0]^T$, does not need to be put out.

Properties of eigenvalues

(i) An eigenvalue λ of multiplicity k (as a root for the polynomial equation (5.44)) yields k linearly independent eigenvectors, spanning the corresponding eigenspace E_λ.

(ii) The sum of the eigenvalues and the diagonal elements are equal:

$$\lambda_1 + \lambda_2 + \cdots + \lambda_n = a_{11} + a_{22} + \cdots + a_{nn},$$

the trace of $\boldsymbol{A}$.

(iii) The product of all eigenvalues is the determinant of $\boldsymbol{A}$:

$$\lambda_1 \cdot \lambda_2 \cdot \ldots \cdot \lambda_n = \det\ \boldsymbol{A}.$$

Thus, if an eigenvalue of the matrix $\boldsymbol{A}$ is $= 0$, then det $\boldsymbol{A} = 0$, and hence $\boldsymbol{A}$ is not invertible.

5.3.8 *Diagonalization of matrix*

Definition 5.18. Diagonalizing a matrix $\boldsymbol{A}$ means to find an orthogonal matrix $\boldsymbol{P}$, such that

$$\boldsymbol{P}^{-1}\boldsymbol{A}\boldsymbol{P} = \boldsymbol{D}, \tag{5.45}$$

where $\boldsymbol{D}$ is a diagonal matrix, i.e., $d_{ij} = 0$ for all $i \neq j$.

Theorem 5.19 (The spectral theorem).

(i) *A diagonalizable matrix* $\boldsymbol{A}$ *is a quadratic one, here of type* $\boldsymbol{A} = n \times n$.

(ii) $\boldsymbol{A}$ *is diagonalizable* $\iff$ *Its* n *eigenvectors are linearly independent.*

(iii) *If* $\boldsymbol{A}$ *is diagonalizable, the columns in* $\boldsymbol{P}$ *are the normalized eigenvectors and the diagonal elements of* $\boldsymbol{D}$*:* $d_{ii} = \lambda_i$, *are the corresponding eigenvalues. More specifically, if* $\{\mathbf{v}_1, \mathbf{v}_2, \ldots, \mathbf{v}_n\}$ *are the normalized, linearly independent eigenvectors, and* $\{\lambda_1, \lambda_2, \ldots, \lambda_n\}$ *are the corresponding eigenvalues, then*

$$\boldsymbol{P} = \begin{bmatrix}\mathbf{v}_1\ \mathbf{v}_2\ \ldots\ \mathbf{v}_n\end{bmatrix},\ \textit{and}\ \boldsymbol{D} = \begin{bmatrix} \lambda_1 & 0 & 0 & \ldots & 0 \\ 0 & \lambda_2 & 0 & \ldots & 0 \\ \vdots & \vdots & \vdots & \ddots & \vdots \\ 0 & 0 & 0 & \ldots & \lambda_n \end{bmatrix}. \tag{5.46}$$

(iv) *For a diagonalizable matrix* $\boldsymbol{A}$,

$$\boldsymbol{A}^n = \boldsymbol{P}\boldsymbol{D}^n\boldsymbol{P}^{-1}, \tag{5.47}$$

where the elements of $\boldsymbol{D}^n$ *are given by* $d_{ij}^n = 0$, *for* $i \neq j$ *and* $d_{ii}^n = \lambda_i^n$, $n = 0, 1, 2, \ldots$

Orthogonal matrix

Definition 5.19. δ_{ij}, Kronecker's delta, is defined as

$$\delta_{ij} = \begin{cases} 0 \text{ if } i \neq j, \\ 1 \text{ if } i = j. \end{cases} \tag{5.48}$$

A matrix $\boldsymbol{P}$ is called orthogonal if

(i) $\boldsymbol{P}$ is quadratic (of order n) and
(ii) the columns of $\boldsymbol{P}$ are orthonormal:

$$P_i^T \cdot P_j = \delta_{ij}, \quad i, j = 1, 2, \ldots, n,$$

where P_i is the ith column in $\boldsymbol{P}$, i.e., $\boldsymbol{P}$ is orthogonal if $\boldsymbol{P}^T \boldsymbol{P} = I$.

Theorem 5.20.

(i)

$$\begin{cases} \lambda_{\min} \leq \dfrac{\boldsymbol{x}^T A \boldsymbol{x}}{\|\boldsymbol{x}\|} \leq \lambda_{\max}, \quad (\boldsymbol{x} \neq \boldsymbol{0}), \\ \\ \textit{equality} \iff \textit{only if } \boldsymbol{x} = \textit{ the corresponding eigenvector.} \end{cases}$$

(ii) *Let $\boldsymbol{P}$ be a square matrix. Then the following holds true:*

$$\boldsymbol{P} \textit{ is orthogonal} \iff P_i^T \cdot P_j = \delta_{ij}.$$

(iii) $\boldsymbol{P}$ *is orthogonal* $\Longrightarrow$

$$\begin{aligned} &(\text{a})\ \boldsymbol{P}^T \textit{ is orthogonal}, && (\text{b})\ \det \boldsymbol{P} = \pm 1, \\ &(\text{c})\ \boldsymbol{P}^T = \boldsymbol{P}^{-1}, && (\text{d})\ (\boldsymbol{P} \cdot \boldsymbol{u})^T \cdot (\boldsymbol{P} \cdot \boldsymbol{v}) = \boldsymbol{u}^T \cdot \boldsymbol{v}, \quad \boldsymbol{u}, \boldsymbol{v} \in \mathbb{R}^n. \\ &(\text{e})\ \boldsymbol{u} \perp \boldsymbol{v} \iff \boldsymbol{P}\boldsymbol{u} \perp \boldsymbol{P}\boldsymbol{v}, && (\text{f})\ \|\boldsymbol{P}\boldsymbol{u}\| = \|\boldsymbol{u}\|, \qquad \boldsymbol{u} \in \mathbb{R}^n. \end{aligned} \tag{5.49}$$

(iv) *If $\boldsymbol{P}$ and $\boldsymbol{R}$ are orthogonal of order n, then $\boldsymbol{P}^{-1}$ and $\boldsymbol{PR}$ are orthogonal.*

(v) *The eigenvalues λ of an orthogonal matrix have absolute value 1, i.e., $|\lambda| = 1$.*

Theorem 5.21.

(i) *An orthogonally diagonalizable matrix $\boldsymbol{A}$ is symmetric.*
(ii) *(The Spectral theorem) for a (real) matrix $\boldsymbol{A}$, the following two statements are equivalent:*
   (a) *$\boldsymbol{A}$ is symmetric.*
   (b) *$\boldsymbol{A}$ is orthogonally diagonalizable.*

5.3.9 *Matrices with complex elements*

Definition 5.20. A matrix $\boldsymbol{A}$ with complex elements is called a *complex matrix.*

The complex conjugate of an $m \times n$ matrix $\boldsymbol{A} = (a_{jk})$ is the $m \times n$ matrix $\overline{\boldsymbol{A}} = (\overline{a}_{jk})$, i.e., the matrix $\boldsymbol{A}$ with elements that are *complex conjugated.*

The conjugate transpose (Adjoint Hermitian matrix) of $\boldsymbol{A}$ is given by

$$\boldsymbol{A}^* := (\overline{\boldsymbol{A}})^T.$$

A quadratic complex matrix $\boldsymbol{U}$ is called *unitary* if its column vectors are orthonormal, i.e., if $\boldsymbol{U}^*\boldsymbol{U} = \boldsymbol{I}$, $(\boldsymbol{U}^* = \boldsymbol{U}^{-1})$.

A quadratic complex matrix $\boldsymbol{H}$ is called *Hermitian* if $\boldsymbol{H}^* = \boldsymbol{H}$, i.e., $\boldsymbol{H}$ is equal to the transpose of its conjugate.

Theorem 5.22 (Properties of conjugated transpose). *Let $\boldsymbol{A}$ and $\boldsymbol{B}$ be $m \times n$ (complex) matrices. Then the following hold true:*

(i) $(\boldsymbol{A}^*)^* = \boldsymbol{A}$.
(ii) $(\boldsymbol{A} + \boldsymbol{B})^* = \boldsymbol{A}^* + \boldsymbol{B}^*$.
(iii) $(z\boldsymbol{A})^* = \overline{z}\boldsymbol{A}^*, \quad z \in \mathbb{C}$.
(iv) *Furthermore, if $\boldsymbol{A}$ and $\boldsymbol{B}$ are quadratic matrices of the same order, then $(\boldsymbol{A}\boldsymbol{B})^* = \boldsymbol{B}^*\boldsymbol{A}^*$.*

Finite-dimensional linear space

Definition 5.21. A linear space is a set M whose elements, also called vectors, have the following properties:

(i) Addition $(+)$ is a commutative and associative binary operation.

(ii) There is an element (vector) $\mathbf{0} \in M$, such that $\boldsymbol{x} + \mathbf{0} = \boldsymbol{x}$. Further, for every $\boldsymbol{x} \in M$ there exists a $-\boldsymbol{x} \in M$, such that $\boldsymbol{x} + (-\boldsymbol{x}) = \mathbf{0}$.

(iii) Let K be a (number) field, e.g., $K = \mathbb{R}$ or $K = \mathbb{C}$. For every $k \in K$ and $\boldsymbol{x} \in M$ $k\boldsymbol{x} \in M$ (k is referred as *scalar*).

(iv) A set $\{\boldsymbol{v}_1, \boldsymbol{v}_2, \ldots, \boldsymbol{v}_k\} \subseteq M$ is called linearly independent if

$$a_1\mathbf{v}_1 + a_2\boldsymbol{v}_2 + \cdots + a_k\boldsymbol{v}_k = \mathbf{0} \Longrightarrow a_1 = a_2 = \ldots = a_k = 0,$$

otherwise it is linearly dependent.

(v) $\{\boldsymbol{v}_1, \boldsymbol{v}_2, \ldots, \boldsymbol{v}_n\} \subseteq M$ spans M if every $\boldsymbol{x} \in M$ can be written as a linear combination of $\{\boldsymbol{v}_1, \boldsymbol{v}_2, \ldots, \boldsymbol{v}_n\}$. More specifically, if there are scalars $a_1, a_2, \ldots, a_n$ such that

$$\boldsymbol{x} = a_1\boldsymbol{v}_1 + a_2\boldsymbol{v}_2 + \cdots + a_n\boldsymbol{v}_n.$$

If in addition $\boldsymbol{v}_1, \boldsymbol{v}_2, \ldots, \boldsymbol{v}_n$ are linearly independent, the set $\{\boldsymbol{v}_j, j = 1, 2, \ldots, n\}$ is a basis of M and M has dimension n: $\dim M = n$.

5.3.10 *Base*

Definition 5.22. A finite set of vectors $\{\boldsymbol{e}_i\}_{i=1}^n$ in a vector space M is a basis for M if every vector $\boldsymbol{a} \in M$ can be uniquely represented as a linear combination of $\boldsymbol{e}_i$s. In other words, there are unique scalars $a_i \in K$ such that

$$\boldsymbol{a} = a_1\boldsymbol{e}_1 + \cdots + a_n\boldsymbol{e}_n. \tag{5.50}$$

Hence, the vector space M is n-dimensional.

The basis is orthonormal if

$$\boldsymbol{e}_i \cdot \boldsymbol{e}_j = \delta_{ij}, \quad i, j = 1, 2, \ldots, n. \tag{5.51}$$

Definition 5.23.

(i) An inner product $\langle \boldsymbol{u}\,,\, \boldsymbol{v}\rangle$, also written as $\boldsymbol{u} \cdot \boldsymbol{v}$, is a binary function/operation on a linear space L, satisfying the following properties

$$
\begin{aligned}
\boldsymbol{u}\cdot\boldsymbol{v} &= \overline{\boldsymbol{v}\cdot\boldsymbol{u}} \qquad \text{(complex conjugate)}\\
\boldsymbol{u}\cdot(\boldsymbol{v}+\boldsymbol{w}) &= \boldsymbol{u}\cdot\boldsymbol{v}+\boldsymbol{u}\cdot\boldsymbol{w}\\
\boldsymbol{u}\cdot\boldsymbol{u} &\geq 0 \quad \text{equality only if} \quad \boldsymbol{u}=\boldsymbol{0}\\
\sqrt{\boldsymbol{u}\cdot\boldsymbol{u}} &= \|\boldsymbol{u}\|\\
k(\boldsymbol{u}\cdot\boldsymbol{v}) &= (k\boldsymbol{u})\cdot\boldsymbol{v}, \qquad k \text{ scalar.}
\end{aligned}
\tag{5.52}
$$

(ii) Two vectors $\boldsymbol{u}$ och $\boldsymbol{v}$ are orthogonal if $\boldsymbol{u}\cdot\boldsymbol{v}=0$.

Theorem 5.23.

$$
\begin{aligned}
&|\boldsymbol{u}\cdot\boldsymbol{v}| \leq \|\boldsymbol{u}\|\cdot\|\boldsymbol{v}\| \qquad \textit{(Cauchy–Schwarz inequality)}\\
&\|\boldsymbol{u}+\boldsymbol{v}\| \leq \|\boldsymbol{u}\|+\|\boldsymbol{v}\| \qquad \textit{(The triangle inequality, which follows from the C–S inequality)}
\end{aligned}
\tag{5.53}
$$

If $\{\mathbf{e}_k\}_{k=1}^{n}$ *is an orthonormal basis for* M, *then*

$\forall\, \boldsymbol{a} \in M, \quad \boldsymbol{a} = a_1\mathbf{e}_1 + a_2\mathbf{e}_2 + \cdots + a_n\mathbf{e}_n$ *and* $a_i = \boldsymbol{a}\cdot\mathbf{e}_i$. $i = 1, 2, \ldots, n$.

Theorem 5.24.

(i) *Assume that the vector space* M *has* $\dim M = n$ *and let* $\boldsymbol{u}_1, \boldsymbol{u}_2, \ldots, \boldsymbol{u}_m \in M$.

  (a) *If* $\boldsymbol{u}_1, \boldsymbol{u}_2, \ldots, \boldsymbol{u}_m$ *are linearly independent, so is* $m \leq n$.
  (b) *If* $\boldsymbol{u}_1, \boldsymbol{u}_2, \ldots, \boldsymbol{u}_m$ *are linearly independent and* $m = n$, *then* $\boldsymbol{u}_1, \boldsymbol{u}_2, \ldots, \boldsymbol{u}_n$ *is a basis for* M.

(ii) **Gram–Schmidt Orthogonalization procedure**
An arbitrary basis $\boldsymbol{v}_1, \boldsymbol{v}_2, \ldots, \boldsymbol{v}_n$ can be orthogonalized so that an orthonormal basis $\mathbf{e}_1, \mathbf{e}_2, \ldots, \mathbf{e}_n$ is obtained:

$$\boldsymbol{e}_i \cdot \boldsymbol{e}_j = \delta_{ij}, \quad i, j = 1, 2, \ldots, n. \tag{5.54}$$

To this end, first an orthogonal basis $\{\boldsymbol{u}_i\}_{i=1}^n$ is constructed setting

$$\begin{aligned}
\boldsymbol{u}_1 &= \boldsymbol{v}_1 \\
\boldsymbol{u}_2 &= \boldsymbol{v}_2 - \frac{\boldsymbol{v}_2 \cdot \boldsymbol{u}_1}{\boldsymbol{u}_1 \cdot \boldsymbol{u}_1} \boldsymbol{u}_1 \\
\vdots\ &\quad \vdots \\
\boldsymbol{u}_n &= \boldsymbol{v}_n - \sum_{j=1}^{n-1} \frac{\boldsymbol{v}_n \cdot \boldsymbol{u}_j}{\boldsymbol{u}_j \cdot \boldsymbol{u}_j} \boldsymbol{u}_j .
\end{aligned}$$

Finally, normalizing yields the desired orthonormal basis:

$$\mathbf{e}_j = \frac{\mathbf{u}_j}{\|\mathbf{u}_j\|}, \quad j = 1, 2, \ldots, n .$$

5.3.11 *Basis and coordinate change*

Theorem 5.25. *Let $\{\mathbf{e}_i, i = 1, 2, \ldots, n\}$ and $\{\mathbf{f}_j, j = 1, 2, \ldots, n\}$ be two bases of the same linear space. Then there are scalars b_{ji}, such that*

$$\mathbf{f}_j = \sum_{i=1}^{n} b_{ji} \mathbf{e}_i, \tag{5.55}$$

or in matrix form

$$\begin{bmatrix} \mathbf{f}_1 \\ \vdots \\ \mathbf{f}_n \end{bmatrix} = \underbrace{\begin{bmatrix} b_{11}\ b_{12}\ \ldots\ b_{1n} \\ \ddots \\ b_{n1}\ b_{n2}\ \ldots\ b_{nn} \end{bmatrix}}_{\textit{Transformation matrix } B^T} \cdot \begin{bmatrix} \mathbf{e}_1 \\ \vdots \\ \mathbf{e}_n \end{bmatrix} . \tag{5.56}$$

Let $\boldsymbol{v}$ be an arbitrary vector. Then there are scalars x_i och y_j, $i, j = 1, 2, \ldots, n$ such that

$$\boldsymbol{v} = \sum_i^n x_i \mathbf{e}_i = \sum_j^n y_j \mathbf{f}_j.$$

Let a point P have coordinates $[x_1, \ldots, x_n]^T$ and $[y_1, \ldots, y_n]^T$ with respect to coordinate systems $(O, x_1, \ldots, x_n)$ and $(\Omega, y_1, \ldots, y_n)$, respectively. Then

$$\begin{bmatrix} x_1 \\ \vdots \\ x_n \end{bmatrix} = \begin{bmatrix} x_{01} \\ \vdots \\ x_{0n} \end{bmatrix} + \boldsymbol{B} \begin{bmatrix} y_1 \\ \vdots \\ y_n \end{bmatrix}, \tag{5.57}$$

where $\Omega = (x_{01}; \ldots; x_{0n})$ is the $x-$coordinates of the origin of the $y-$system, and $\boldsymbol{B}$ is the transformation matrix in (5.57).

If the coordinate systems have the same origin, then $x_{01} = \cdots = x_{0n} = 0$.

Theorem 5.26. *If both bases are orthogonal and have a common origin, then $\boldsymbol{B}$ is an orthogonal matrix and*

$$\boldsymbol{B}^T = \boldsymbol{B}^{-1} \text{ hence } \begin{bmatrix} y_1 \\ \vdots \\ y_n \end{bmatrix} = \boldsymbol{B}^T \begin{bmatrix} x_1 \\ \vdots \\ x_n \end{bmatrix}. \tag{5.58}$$

Theorem 5.27 (Rotation of coordinate systems). *If the Cartesian system (O, x, y) rotates, counterclockwise (about origin) with a rotation angle α to the coordinate system (O, x_1, y_1), then the coordinates in the two systems are related as*

$$\begin{cases} x = x_1 \cos\alpha - y_1 \sin\alpha \\ y = x_1 \sin\alpha + y_1 \cos\alpha \end{cases} \Longleftrightarrow \begin{cases} x_1 = x\cos\alpha + y\sin\alpha \\ y_1 = -x\sin\alpha + y\cos\alpha. \end{cases}$$

Or in matrix representation

$$\begin{bmatrix} x \\ y \end{bmatrix} = \begin{bmatrix} \cos\alpha & -\sin\alpha \\ \sin\alpha & \cos\alpha \end{bmatrix} \begin{bmatrix} x_1 \\ y_1 \end{bmatrix} \Longleftrightarrow \begin{bmatrix} x_1 \\ y_1 \end{bmatrix} = \begin{bmatrix} \cos\alpha & \sin\alpha \\ -\sin\alpha & \cos\alpha \end{bmatrix} \begin{bmatrix} x \\ y \end{bmatrix}.$$

5.4 The Quaternion Ring

The Quaternion- or the Hamilton ring, denoted by $\mathbb{H}$, is a four-dimensional algebraic structure.

Definition 5.24. The numbers $\mathbf{i}, \mathbf{j}, \mathbf{k}$ satisfy

$$\begin{aligned} &\mathbf{i}^2 = \mathbf{j}^2 = \mathbf{k}^2 = \mathbf{ijk} = -1 \\ &\text{and} \\ &\mathbf{ij} = -\mathbf{ji}, \quad \mathbf{jk} = -\mathbf{kj}, \quad \mathbf{ki} = -\mathbf{ik}. \end{aligned} \tag{5.59}$$

A number of the form $q = x + y\mathbf{i} + z\mathbf{j} + t\mathbf{k}$, where $x, y, z, t \in \mathbb{R}$ is called a *quaternion.*

Theorem 5.28 (General quaternion properties).

$$\begin{aligned} &(q_1 + q_2) + q_3 = q_1 + (q_2 + q_3),\ q_1 + q_2 = q_2 + q_1, \\ &(q_1 \cdot q_2) \cdot q_3 = q_1 \cdot (q_2 \cdot q_3), \qquad \begin{cases} q_1(q_2 + q_3) = q_1q_2 + q_1q_3, \\ (q_1 + q_2)q_3 = q_1q_3 + q_2q_3. \end{cases} \end{aligned} \tag{5.60}$$

Multiplication is not commutative i.e., in general $q_1 \cdot q_2 \neq q_2 \cdot q_1$.

For every $q \neq 0$, *there is a multiplicative inverse* q^{-1} *such that*

$$q \cdot q^{-1} = q^{-1} \cdot q = 1. \tag{5.61}$$

Definition 5.25. For a quaternion $q = x + y\mathbf{i} + z\mathbf{j} + t\mathbf{k}$, its conjugate and norm $|\cdot|$ are defined as

$$\overline{q} = x - (y\mathbf{i} + z\mathbf{j} + t\mathbf{k}) \quad \text{and} \quad |q| = \sqrt{q\overline{q}} \geq 0, \text{ respectively.} \tag{5.62}$$

5.4.1 *Splitting a quaternion q in its scalar and vector parts*

One may split

$$q = \underbrace{x}_{\text{scalar part}} + \underbrace{y\mathbf{i} + z\mathbf{j} + t\mathbf{k}}_{\text{vector part}}.$$

The vector part may be denoted by $\boldsymbol{u}$ or $\boldsymbol{v}$.

Theorem 5.29.

$$\begin{aligned} &|q|^2 = x^2 + \boldsymbol{v}^2 = x^2 + y^2 + z^2 + t^2, \textit{ where } \boldsymbol{v} = y\mathbf{i} + z\mathbf{j} + t\mathbf{k} \\ &\overline{q_1 \cdot q_2} = \overline{q_2} \cdot \overline{q_1}. \end{aligned} \tag{5.63}$$

If $\boldsymbol{u}$ and $\boldsymbol{v}$ are two vector parts, then

$$\boldsymbol{u}\boldsymbol{v} = -\boldsymbol{u}\cdot\boldsymbol{v} + \boldsymbol{u}\times\boldsymbol{v}, \tag{5.64}$$

where $\boldsymbol{u}\boldsymbol{v}$ is the usual multiplication in $\mathbb{H}$, $\cdot$ is the scalar product, and $\times$ is the vector product.

5.4.2 *Matrix representation*

Theorem 5.30. *With the matrices*

$$E = \begin{bmatrix} 1 & 0 \\ 0 & 1 \end{bmatrix}, \quad I = \begin{bmatrix} i & 0 \\ 0 & 1 \end{bmatrix}, \quad J = \begin{bmatrix} 0 & 1 \\ -1 & 0 \end{bmatrix}, \quad K = \begin{bmatrix} 0 & i \\ i & 0 \end{bmatrix}. \tag{5.65}$$

$$Q := xE + yI + zJ + tK = \begin{bmatrix} x+iy & z+it \\ -z+it & x-iy \end{bmatrix} \tag{5.66}$$

is a complex matrix representation of a quaternion. Setting $u = x+iy$ and $v = z+it$,

$$\det Q = |u|^2 + |v|^2 = x^2 + y^2 + z^2 + t^2. \tag{5.67}$$

Theorem 5.31 (Basis and dual basis in $\mathbb{R}^3$).

(i) *Given three vectors $\underset{1}{\mathbf{e}}$, $\underset{2}{\mathbf{e}}$, and $\underset{3}{\mathbf{e}}$ i $\mathbb{R}^3$, with their triple scalar product satisfying*

$$[\underset{1}{\mathbf{e}}, \underset{2}{\mathbf{e}}, \underset{3}{\mathbf{e}}] = \underset{1}{\mathbf{e}} \cdot (\underset{2}{\mathbf{e}} \times \underset{3}{\mathbf{e}}) \neq 0. \tag{5.68}$$

The set $\{\underset{1}{\mathbf{e}}, \underset{2}{\mathbf{e}}, \underset{3}{\mathbf{e}}\}$ serves as a basis for $\mathbb{R}^3$.

(ii) *The dual basis is defined as*

$$\overset{1}{\mathbf{e}} = \frac{\underset{2}{\mathbf{e}} \times \underset{3}{\mathbf{e}}}{[\underset{1}{\mathbf{e}}, \underset{2}{\mathbf{e}}, \underset{3}{\mathbf{e}}]}, \quad \overset{2}{\mathbf{e}} = \frac{\underset{3}{\mathbf{e}} \times \underset{1}{\mathbf{e}}}{[\underset{1}{\mathbf{e}}, \underset{2}{\mathbf{e}}, \underset{3}{\mathbf{e}}]}, \quad \overset{3}{\mathbf{e}} = \frac{\underset{1}{\mathbf{e}} \times \underset{2}{\mathbf{e}}}{[\underset{1}{\mathbf{e}}, \underset{2}{\mathbf{e}}, \underset{3}{\mathbf{e}}]}. \tag{5.69}$$

In particular, $[\overset{1}{\mathbf{e}}, \overset{2}{\mathbf{e}}, \overset{3}{\mathbf{e}}] = \dfrac{1}{[\underset{1}{\mathbf{e}}, \underset{2}{\mathbf{e}}, \underset{3}{\mathbf{e}}]}$ *and* $\underset{i}{\mathbf{e}} \cdot \overset{j}{\mathbf{e}} = \delta_i^j$.

(iii) *Every vector in* $\mathbb{R}^3$ *can then be uniquely written as*

$$\boldsymbol{v} = v^1 \underset{1}{\mathbf{e}} + v^2 \underset{2}{\mathbf{e}} + v^3 \underset{3}{\mathbf{e}} = v_1 \overset{1}{\mathbf{e}} + v_2 \overset{2}{\mathbf{e}} + v_3 \overset{3}{\mathbf{e}}. \tag{5.70}$$

$v^j = \boldsymbol{v} \cdot \overset{j}{\mathbf{e}} = \dfrac{[\boldsymbol{v}, \underset{k}{\mathbf{e}}, \underset{l}{\mathbf{e}}]}{[\underset{1}{\mathbf{e}}, \underset{2}{\mathbf{e}}, \underset{3}{\mathbf{e}}]}$, $i = 1, 2, 3$, *are the contra-variant components,*

$v_j = \boldsymbol{v} \cdot \underset{j}{\mathbf{e}} = \dfrac{[\boldsymbol{v}, \overset{k}{\mathbf{e}}, \overset{l}{\mathbf{e}}]}{[\overset{1}{\mathbf{e}}, \overset{2}{\mathbf{e}}, \overset{3}{\mathbf{e}}]}$ *are the covariant components.*

$(j, k, l) = (1, 2, 3), \quad (2, 3, 1), \quad (3, 1, 2).$

(iv) *The scalar product of* $\boldsymbol{u}$ *and* $\boldsymbol{v}$ *can be written as*

$$\boldsymbol{u} \cdot \boldsymbol{v} = \sum_{i,j=1}^{3} u^i v^j (\underset{i}{\mathbf{e}} \cdot \underset{j}{\mathbf{e}}) = \sum_{i,j=1}^{3} u_i v_j (\overset{i}{\mathbf{e}} \cdot \overset{j}{\mathbf{e}}). \tag{5.71}$$

(v) *Letting* $g^{ij} = \overset{i}{\mathbf{e}} \cdot \overset{j}{\mathbf{e}}$, *and* $g_{ij} = \underset{i}{\mathbf{e}} \cdot \underset{j}{\mathbf{e}}$

$$v^i = \sum_{j=1}^{3} v_j g^{ij}, \qquad v_j = \sum_{i=1}^{3} v^i g_{ij}, \qquad i, j = 1, 2, 3.$$

5.5 Optimization

5.5.1 *Linear optimization*

Definition 5.26.

Notations

$$\boldsymbol{b}^T = [b_1\, b_2 \ldots b_n], \quad \mathbf{c}^T = [c_1\, c_2 \ldots c_m],$$

$$\boldsymbol{x}^T = [x_1\, x_2 \ldots x_n],\ \boldsymbol{y}^T = [y_1\, y_2 \ldots y_m]\,.$$

$$\boldsymbol{A}, \text{ a real matrix:} \quad \text{type}\, \boldsymbol{A} = m \times n\,.$$

A function $f : \mathbb{R}^n \to \mathbb{R}$,

$$f(\boldsymbol{x}) := \boldsymbol{b}^T \cdot \boldsymbol{x} = b_1 x_1 + b_2 x_2 + \cdots + b_n x_n \tag{5.72}$$

is called a *target function.*

A vector-inequality $\boldsymbol{x} \geq \boldsymbol{x}'$ means

$$x_1 \geq x_1',\, x_2 \geq x_2',\, \ldots, x_n \geq x_n'\,.$$

Similarly, $\boldsymbol{x} > \boldsymbol{x}'$ means $x_1 > x_1'$, $x_2 > x_2'$, $\ldots$, $x_n > x_n'$.

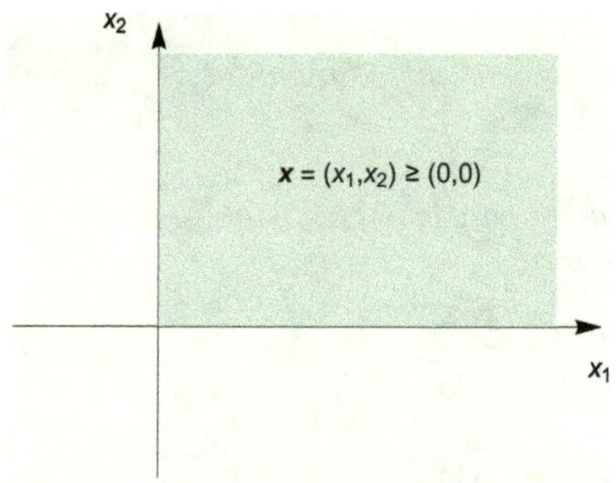

The subset $\{x \geq 0\} \subset \mathbb{R}^2$

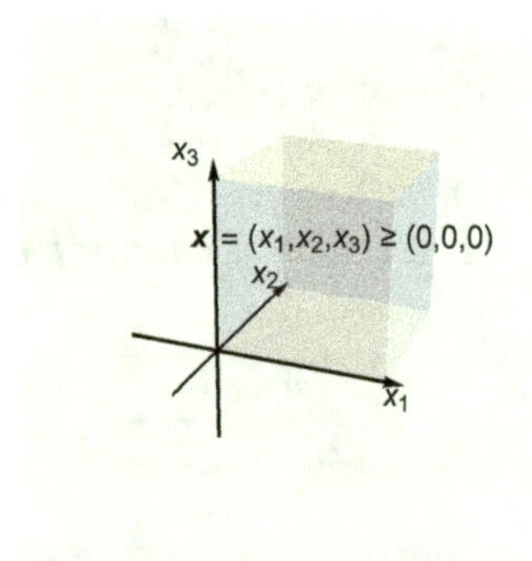

The subset $\{x \geq 0\} \subset \mathbb{R}^3$

A *side-condition* or *constraint* is formulated as

$$\boldsymbol{A} \cdot \boldsymbol{x} \geq \boldsymbol{c}, \quad \boldsymbol{A} \cdot \boldsymbol{x} \leq \boldsymbol{c} \text{ or } \boldsymbol{A} \cdot \boldsymbol{x} = \boldsymbol{c}\,. \tag{5.73}$$

$\boldsymbol{x}$ satisfying a constraint, like in (5.73), is called *an acceptance or a valid point.*

A *linear programming (LP)* is the task of optimizing (maximizing/minimizing) a target function (5.72).

A valid vector $\boldsymbol{x}$ that corresponds to the optimal value of the target function is denoted $\hat{\boldsymbol{x}}$. The optimal value is then $f(\hat{\boldsymbol{x}})$. The problems

$$(1)\begin{cases} \max(\boldsymbol{b}^T\,\boldsymbol{x}), \\ \boldsymbol{x} \geq 0, \\ \boldsymbol{A}\,\boldsymbol{x} \leq \boldsymbol{c}, \end{cases} \quad \text{and} \quad (2)\begin{cases} \min(\boldsymbol{c}^T\,\boldsymbol{y}), \\ \boldsymbol{y} \geq 0, \\ \boldsymbol{A}^T\,\boldsymbol{y} \geq \boldsymbol{b}, \end{cases} \tag{5.74}$$

are *dual* (LP duals).

Remarks. A constraint $\boldsymbol{A} \cdot \boldsymbol{x} = \boldsymbol{c}$ can be returned to the constraint $\begin{cases} \boldsymbol{A}\,\boldsymbol{x} \le \boldsymbol{c} \\ -\boldsymbol{A}\,\boldsymbol{x} \le -\boldsymbol{c} \end{cases}$, which in turn, can be written as

$$\begin{bmatrix} \boldsymbol{A} \\ -\boldsymbol{A} \end{bmatrix} \cdot \boldsymbol{x} \le \begin{bmatrix} \boldsymbol{c} \\ -\boldsymbol{c} \end{bmatrix},$$

i.e., the same form as in (1) (5.74).

Theorem 5.32 (The duality theorem). *Consider the two dual LPs in* (5.74).

The following hold true:

(i) *If* (1) *has valid points,* $\boldsymbol{x}$, *then* (2) *has an optimal solution,* $\hat{\boldsymbol{y}}$.
(ii) *If* (2) *has valid points,* $\boldsymbol{y}$, *then* (1) *has an optimal solution,* $\hat{\boldsymbol{x}}$.
(iii) *If* (1) *and* (2) *have acceptance/valid points, then both* (1) *and* (2) *are optimal solutions with same optimal value, that is* $\boldsymbol{b}^T \cdot \hat{\boldsymbol{x}} = \boldsymbol{c}^T \cdot \hat{\boldsymbol{y}}$.

Theorem 5.33 (The complementarity theorem). *Assume that* $\boldsymbol{x}$ *and* $\boldsymbol{y}$ *are valid solutions to* (1) *and* (2) *in* (5.74), *respectively. Then the following three statements are equivalent*:

(i) $\boldsymbol{x}$ *and* $\boldsymbol{y}$ *are optimal solutions,*
(ii) $\boldsymbol{b}^T\,\boldsymbol{x} = \boldsymbol{c}^T\,\boldsymbol{y}$.
(iii)

$$\begin{aligned} \boldsymbol{x} > \boldsymbol{0} &\Longrightarrow a_{1j}y_1 + a_{2j}y_2 + \cdots + a_{mj}y_m = b_j,\ j = 1, 2, ..., n. \\ \boldsymbol{y} > \boldsymbol{0} &\Longrightarrow a_{i1j}x_1 + a_{i2}x_2 + \cdots + a_{in}x_n = c_i,\ i = 1, 2, ..., m. \end{aligned} \tag{5.75}$$

Remarks. The duality theorem on page 117 applies unrestricted for

$$(1')\begin{cases} \max(\boldsymbol{b}^T\,\boldsymbol{x}), \\ \boldsymbol{x} \ge 0, \\ \boldsymbol{A}\,\boldsymbol{x} = \boldsymbol{c}, \end{cases} \quad \text{and} \quad (2')\begin{cases} \min(\boldsymbol{c}^T\,\boldsymbol{y}), \\ \boldsymbol{A}^T\,\boldsymbol{y} \ge \boldsymbol{b}. \end{cases} \tag{5.76}$$

Statements (iii) in (5.75) can, alternatively, be formulated as

$$\boldsymbol{x} = \mathbf{0} \Longleftarrow a_{1jj}y_1 + a_{2j}y_2 + \cdots + a_{mj}y_m > b_j,\ j = 1, 2, \ldots, n.$$

$$\boldsymbol{y} = \mathbf{0} \Longleftarrow a_{i1j}x_1 + a_{i2}x_2 + \cdots + a_{in}x_n < c_i,\ i = 1, 2, \ldots, m.$$

Theorem 5.34 (Farkas' lemma).

$$\text{Either } \exists \quad \boldsymbol{x} : \begin{cases} \boldsymbol{A}\,\boldsymbol{x} = \boldsymbol{b}, \\ \\ \boldsymbol{x} \geq 0, \end{cases} \qquad \text{or} \qquad \exists \quad \boldsymbol{y} : \begin{cases} \boldsymbol{A}^T\,\boldsymbol{y} \geq \boldsymbol{b}, \\ \\ \boldsymbol{b}^T\,\boldsymbol{y} < 0, \end{cases} \tag{5.77}$$

that is, exactly one of the systems has a solution (but not both).

5.5.2 *Convex optimization*

Definition 5.27. A subset $K \subseteq \mathbb{R}^n$ is convex if for each pair of points $\boldsymbol{x}$ and $\boldsymbol{x}'$ in K, $t\boldsymbol{x} + (1-t)\boldsymbol{x}' \in K$ for all $t : 0 < t < 1$.

A function $f : K \to \mathbb{R}$ is *convex* if

$$f(t\,\boldsymbol{x} + (1-t)\boldsymbol{x}') \leq t\,f(\boldsymbol{x}) + (1-t)f(\boldsymbol{x}').$$

With $\leq$ replaced by $<$, the function is *strongly convex*.

Let f, g_1, g_2, ..., g_m be convex and differentiable functions on $K \subseteq \mathbb{R}^n$.

A *convex program* is

$$\text{(CP)} \quad \begin{cases} \min f(\boldsymbol{x}) & & \\ & & \\ g_1(\boldsymbol{x}) & \leq c_1 & \\ g_2(\boldsymbol{x}) & \leq c_2 & \boldsymbol{x} \in K, \\ & \vdots & \\ g_m(\boldsymbol{x}) & \leq c_m & \end{cases} \tag{5.78}$$

where $g_j(\boldsymbol{x}) \leq c_j$, $j = 1, 2, \ldots, m$ are m constraints.

A vector $\boldsymbol{x}$ which satisfies all constraints is said to be a point (vector) of acceptance or a valid point.

Remarks. Convexity interprets as if $\boldsymbol{x}$ and $\boldsymbol{x}'$ are in K, then all points on the line segment between the points also belongs to K, see the following figure on the left.

A constraint $g(\boldsymbol{x}) \leq c$, where g is convex, interpreted as a set,

$$\{\boldsymbol{x} \in \mathbb{R}^n : g(\boldsymbol{x}) \leq c\} \quad \text{is convex.}$$

All constraints together in (5.78), page 118, constitute an intersection of sets:

$$\bigcap_{j=1}^{m} \{\boldsymbol{x} : g_j(\boldsymbol{x}) \leq c_j\},$$

which is also a convex set.

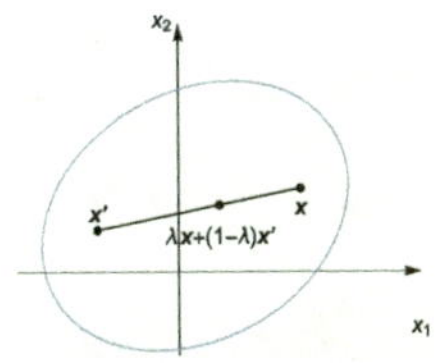

Convex set in $\mathbb{R}^2$ *containing* $\boldsymbol{x}$, $\boldsymbol{x}'$, *and the line between.*

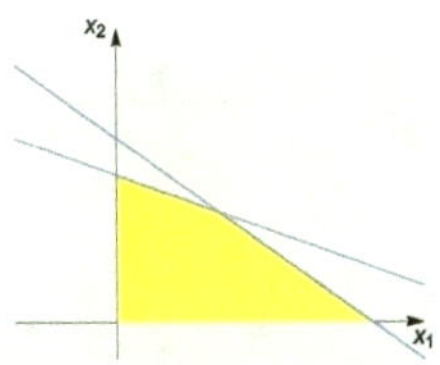

Convex domain, bounded by $\boldsymbol{x} = (x_1, x_2) \geq 0$, $g_1(\boldsymbol{x}) = 2x_1 + x_2 \leq 5$ *and* $g_2(\boldsymbol{x}) = x_1 + x_2 \leq 4$.

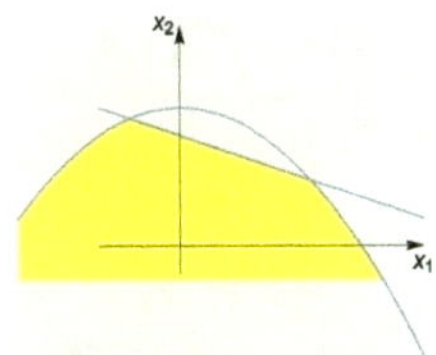

Convex domain, bounded by $g_1(\boldsymbol{x}) = x_1^2 + x_2 \leq 5$ *and* $g_2(\boldsymbol{x}) = x_1 + x_2 \leq 4$.

Theorem 5.35. *If K is an open interval and $f : K \to \mathbb{R}$ is convex, then f is continuous. Let K be an interval, a, b, $c \in K$ satisfy $a < b < c$, and f be a convex on K. Then the following inequality holds:*

$$\frac{f(b) - f(a)}{b - a} \leq \frac{f(c) - f(b)}{c - b}.$$

Theorem 5.36. *Assume that the convex program* (5.78), *page* 118, *has at least one solution.*

(i) *If the function f is strictly convex,* (5.78) *has a unique solution, $\hat{\boldsymbol{x}} \in K$.*
(ii) *If* (5.78) *has two solutions, $\hat{\boldsymbol{x}}_1$ and $\hat{\boldsymbol{x}}_2$, then also all $\hat{\boldsymbol{x}} = \lambda\hat{\boldsymbol{x}}_1 + (1 - \lambda)\hat{\boldsymbol{x}}_2$ for all $\lambda : 0 < \lambda < 1$ are solutions. (This means, that all points on the line segment between the points are solutions).*

Theorem 5.37 (The convex Kuhn–Tucker theorem). *Suppose that in the convex program (5.78) all constraints are fulfilled for a vector/point* $\hat{\boldsymbol{x}}$, *(i.e., an acceptance point). Furthermore, assume that there is a vector* $\hat{\boldsymbol{y}} \geq \mathbf{0}$, *that is*

$$\hat{y}_1 \geq 0,\ \hat{y}_2 \geq 0, ..., \hat{y}_m \geq 0,$$

that satisfies the constraints

(i) $\hat{y}_j(g_j(\hat{\boldsymbol{x}}) - c_j) = 0, \quad j = 1, 2, ..., m.$

(ii) $\dfrac{\partial f}{\partial x_k}(\hat{\boldsymbol{x}}) + \hat{y}_1 \dfrac{\partial g_1}{\partial x_k}(\hat{\boldsymbol{x}}) + \hat{y}_2 \dfrac{\partial g_2}{\partial x_k}(\hat{\boldsymbol{x}}) + \cdots + \hat{y}_m \dfrac{\partial g_m}{\partial x_k}(\hat{\boldsymbol{x}}) = 0,\ k = 1, 2, ..., n.$

Then $\hat{\boldsymbol{x}}$ *is an optimal solution, to (5.78), i.e., f assumes a minimum:* $\min f(\boldsymbol{x}) = f(\hat{\boldsymbol{x}})$.

With the restriction that

$$\boldsymbol{x} \geq \mathbf{0}$$

(i.e., $x_k \geq 0$ *for* $k = 1, 2, ..., n$*) and*

(i′) $\hat{y}_j(g_j(\hat{\boldsymbol{x}}) - c_j) = 0, \qquad j = 1, 2, ..., m.$

(ii′) $\dfrac{\partial f}{\partial x_k}(\hat{\boldsymbol{x}}) + \hat{y}_1 \dfrac{\partial g_1}{\partial x_k}(\hat{\boldsymbol{x}}) + \cdots + \hat{y}_m \dfrac{\partial g_m}{\partial x_k}(\hat{\boldsymbol{x}}) \geq 0, \qquad k = 1, 2, ..., n.$

(iii′) $\hat{\boldsymbol{x}}_k \left[\dfrac{\partial f}{\partial x_k}(\hat{\boldsymbol{x}}) + \hat{y}_1 \dfrac{\partial g_1}{\partial x_k}(\hat{\boldsymbol{x}}) + \cdots + \hat{y}_m \dfrac{\partial g_m}{\partial x_k}(\hat{\boldsymbol{x}})\right] = 0,\ k = 1, 2, ..., n.$

$\hat{\boldsymbol{x}}$ *is an optimal solution of (5.78).*

Chapter 6

Algebraic Structures

A binary operation $*$ defined on a set M is a function $* : M \times M \to M$. On an element level this is represented as $M \times M \ni (a, b) \mapsto a * b \in M$. The operation is commutative if $a * b = b * a$ and associative if $(a * b) * c = a * (b * c)$.

6.1 Overview

Algebraic structure	Description
Semi group $(M, *)$	$*$Associative.
Monoid $(M, *) = (M, *, e)$	Semi group with identity element e: $e * a = a * e = a$.
Group $(M, *) = (M, *, e)$	Monoid for which every $a \in M$ has a unique inverse $a^{-1} \in M$ with property $a^{-1} * a = a * a^{-1} = e$.
Abelian group or	Group such that $*$ is commutative.
Ring $(M, +, \circ)$	$(M, +)$ is an Abelian group and $(M, \circ)$ is a semigroup and $a \circ (b + c) = a \circ b + a \circ c$, $(b + c) \circ a = b \circ a + c \circ a$.
Field $(M, +, \circ)$	Ring with the property that $M \setminus \{0\}$ an Abelian group under multiplication $\circ$.
Lattice $(M, \leq)$	$(S, \leq)$ is a partially ordered set (poset) such that each pair a, b of elements in M has a largest lower limit, LLL and a lowest upper limit LUL.
Boolean algebra $(M, +, \circ, ^{-}, 0, 1)$	The binary operations $+$ and $\circ$ are commutative, associative, and distributive over the underlying scalar fields. The elements 0 and 1 are identity elements of $+$ and $\circ$, respectively. $\overline{a}$ is the complement of a, if $a + \overline{a} = 1$ and $a \circ \overline{a} = 0$.

(6.1)

Algebraic structure	Example
Semi group	The set of integers under addition (multiplication). The set of binary relations on a set under composition.
Monoid	The set of real numbers under addition and with 0 as identity. The real numbers under multiplication and with 1 as identity. The power set $\mathcal{P}(\Omega)$ under union and with $\emptyset$ as identity.
Group	The set of permutations of a set under composition. The set of symmetries of a regular polygon. $(\mathbb{Z}_n, +_n)$ where $+_n$ is addition modulo n.
Abelian group	The integers under addition. The set of rational numbers $\neq 0$ under multiplication.
Ring	The set of integers (rational numbers, even numbers, real numbers, complex numbers) under addition and multiplication. $(\mathbb{Z}_n, /, +_n, \times_n)$; where $+_n$ and $\times_n$ are addition and multiplication modulo n.
Field	The set of rational, real, or complex numbers under addition and multiplication. $(\mathbb{Z}_n, +_n, \times_n)$, n is prime number.
Lattice	Power set $\mathcal{P}(\Omega)$ under inclusion: $A, B \in \mathcal{P}(\Omega)$; $A \subset B$. The set of positive integers under D, where D is the relation "divide" $a\,D\,b$ $(a \mid b)$ if a divides b. Supremum and infimum of a and b is defined as GCD and LCM of a and b.
Boolean algebra	$(\mathcal{P}(\Omega), \cup, \cap,$ "complements", $\emptyset, \Omega)$. $(S, \wedge, \vee, \neg, F, T)$ where S is the set of equivalence classes of expression forms n expressions and F and T are contradiction and tautology, respectively. $S = \{0, 1\}$ with the common definition for addition and multiplication with the exception that $1 + 1 = 1$. (Boolean addition and multiplication) and with $\overline{0} = 1$, $\overline{1} = 0$.

6.2 Homomorphism and Isomorphism

Homomorphism

Assume that M_1 and M_2 are algebraic structures of the same type. A map $\varphi : M_1 \to M_2$ is called a *homomorphism* if φ preserves the algebraic structure.

Semigroup

Let $(M_1, *_1)$ and $(M_2, *_2)$ be semigroups. A map $\varphi : M_1 \to M_2$ is called *semigroup-homomorphism* if for every a, $b \in M_1$,

$$\varphi(a *_1 b) = \varphi(a) *_2 \varphi(b).$$

Monoid

Let $(M_1, *_1)$ och $(M_2, *_2)$ be monoids with identity elements e_1 and e_2, respectively. A semigroup-homomorphism $\varphi : M_1 \to M_2$ is called *monoid-homomorphism* if

$$\varphi(e_1) = e_2.$$

Group

Let $(M_1, *_1)$ och $(M_2, *_2)$ be two groups. A map $\varphi : M_1 \to M_2$ such that for all a, $b \in M_1$,

$$\varphi(a *_1 b) = \varphi(a) *_2 \varphi(b)$$

is called a *group-homomorphism*. Group properties yield

$$\varphi(e_1) = \varphi(e_2), \quad \text{and} \quad \varphi(a^{-1}) = \varphi(a)^{-1}, \quad \forall a \in M_1.$$

Ring

Let $(M_1, +_1, \circ_1)$ and $(M_2, *_2, \circ_2)$ be rings. A map $\varphi : M_1 \to M_2$ such that for all a, $b \in M_1$

$$\varphi(a +_1 b) = \varphi(a) +_2 \varphi(b) \quad \text{and} \quad \varphi(a \circ_1 b) = \varphi(a) \circ_2 \varphi(b)$$

is called *ring-homomorphism*.

Lattice
Let $(M_1, \leq_1)$ and $(M_2, \leq_2)$ be lattices. A map $\varphi : M_1 \to M_2$ such that for all $a, b \in M_1$

$$\varphi(\text{GLB}(a, b)) = \text{GLB}(\varphi(a), \varphi(b))$$
$$\varphi(\text{LUB}(a, b)) = \text{LUB}(\varphi(a), \varphi(b))$$

is called *lattice-homomorphism.*

Due to the properties of lattices

$$a \leq_1 b \Longrightarrow \varphi(a) \leq_2 \varphi(b).$$

Boolean algebra (For definition, see (6.1) page 121).

Let $(M_1, +_1, \circ_1, ^-, 0_1, 1_1)$ and $(M_2, +_2, \circ_2, ^-, 0_2, 1_2)$ be Boolean algebras.

A map $\varphi : M_1 \to M_2$ is called *Boolean homomorphism* if for every $a, b \in M_1$.

$$\varphi(a +_1 b) = \varphi(a) +_2 \varphi(b), \quad \varphi(a \circ_1 b) = \varphi(a) \circ_2 \varphi(b),$$
$$\varphi(0_1) = 0_2, \quad \varphi(1_1) = 1_2, \quad \varphi(\overline{a}) = \overline{\varphi(a)}.$$

To show that $\varphi : M_1 \to M_2$ is a Boolean homomorphism, it suffices to show

$$\varphi(a +_1 b) = \varphi(a) +_2 \varphi(b) \quad \text{and} \quad \varphi(\overline{a}) = \overline{\varphi(a)}.$$

A bijective homomorphism is called *isomorphism.*

The inverse of a bijective homomorphism, i.e., of an isomorphism, is an isomorphism.

Two algebraic structures are called *isomorphic* if there exists an isomorphism between them.

6.3 Groups

Definition 6.1.

(i) A group consists of a set $G = \langle \cdot \rangle \neq \emptyset$ and a binary operation $*$ on $G \times G \xrightarrow{*} G$, such that

(a) $a, b \in G \Longrightarrow a * b \in G$, i.e., closed under the operation $*$.
(b) For $a, b, c \in G$, $\quad (a * b) * c = a * (b * c)$ (associativity).
(c) There is an identity element $e \in G$ with property $a * e = e * a = a$.
(d) For every $a \in G$, there is an inverse $a^{-1} \in G$, such that $a * a^{-1} = a^{-1} * a = e$.

(ii) A group is commutative or Abelian if $a * b = b * a$ for each pair of elements.
(iii) The group is finite if G contains only a finite number of elements, i.e., $|G| \in \mathbb{Z}_+$.
(iv) A group generated by an element $a \in G$ is called cyclic and a is called *generator*. That is:

$$\langle a, a^2, a^3, \ldots, * \rangle = G \quad a^2 := a * a, \quad a^3 := a * a * a, \ldots$$

(v) A subset H of G is a subgroup of G if H fulfills the criteria for a group.

In the following, the operation $*$ is suppressed, one writes $a\,b$ instead of $a * b$, and $a^n = \underbrace{a * a * a * \ldots * a}_{n \text{ factors}}$.

(vi) For a subset $H \subseteq G$ we define $aH = \{ah; h \in H\}$ and $Ha = \{ha : h \in H\}$ as the left and right coset of H, respectively.
(vii) If H is a subgroup of G, the relation $a \equiv b \mod H$ is defined if $ab^{-1} \in H$. This can equivalently be written as $a \in Hb$.
(viii) The period of an element $a \in G$ is the least positive integer m satisfying $a^m = e$. If no such m exists, then the period of a is infinite. The period of a is written as $m =: o(a)$ (the order of a).

Remarks. Summing up: We only write ab. For Abelian groups, it is customary to write $*$ as $+$, i.e., use the presentation $a + b$ instead of $a * b$.

A group is actually $\langle G, * \rangle$, but for simplicity, it is only presented by G.

Theorem 6.1. *Let G be a group.*

(i) *A subset $H \neq \emptyset$ of G is a subgroup if and only if*
(a) $a, b \in H \Longrightarrow ab \in H$ *and*
(b) $a \in H \Longrightarrow a^{-1} \in H$

or alternatively if and only if

$$a, b \in H \Longrightarrow ab^{-1} \in H.$$

(ii) *If H is a finite subset of G, to be a subgroup, it suffices that H is closed under multiplication.*
(iii) $a \equiv b \bmod H$ *defines an equivalence relation on G and thus creates a partition of G.*
(iv) *If H is a finite subset of G and $a \in G$, then the left and right cosets aH and Ha have the same cardinality.*
(v) *(Lagrange's theorem) For a subgroup H, of a finite group G, the number of elements of H is a divisor to the number of elements of G, i.e., $\frac{|G|}{|H|}$ is an integer.*

Theorem 6.2. *Let G be a group.*

(i) *Suppose that G is finite and $a \in G$ has the period m. Then*

$$\langle a, a^2, a^3, \ldots, a^m, * \rangle =: H,$$

is a subgroup of G and m is a divisor of $|G|$.
(ii) *If G is finite, then $a^{|G|} = e$, the identity element of G.*
(iii) *Assume that $|G| = p$ is a prime number. Then G is cyclic.*
(iv) *Let H and K be subgroups of G. Then*
  (a) *$H \cap K$ is a subgroup of G (and of H and K).*
  (b) *Set $HK = \{hk : h \in H,\ k \in K\}$. Then HK is a subgroup of G if and only if $HK = KH$.*

Definition 6.2.

(i) A subgroup N of G is *normal*, if $aN = Na$ for each $a \in G$, i.e., if each right and left coset of an element are equal. That N is normal is sometimes written as $N \lhd G$.
(ii) $G/N = \{aN, a \in G\}$ is quotient group, where N is a normal subgroup of G.

(iii) A mapping $\phi : G \to G'$ between two groups is a homomorphism if

$$\phi(ab) = \phi(a)\phi(b).$$

(a) $\ker \phi = \{x \in G : \phi(x) = e'\}$, where e' is the identity element of G'.
(b) An injective homomorphism ($\phi(a) = \phi(b) \implies a = b$) is called monomorphism.
(c) An invertible homomorphism ϕ is called isomorphism.
(d) An isomorphism $\phi : G \to G$ is called automorphism.

Theorem 6.3.

(i) *Assume that $N \lhd G$. Then the quotient group G/N defines a group. The group operation is defined as $aN * bN := \{an_1bn_2 : n_1, n_2 \in N\}$.*
(ii) *Assume that $\phi : G \to G'$ is a group homomorphism. Then the following hold true:*
  (a) $\phi(e) = e'$, $\phi(a^{-1}) = (\phi(a))^{-1}$.
  (b) $\phi(G)$ *is a subgroup of* G'.
  (c) $\ker \phi \lhd G$.
  (d) $\ker \phi = \{e'\} \Leftrightarrow \phi$ *is a monomorphism.*

Theorem 6.4. *Let p be a prime number and G, a finite group.*

(i) *If $|G| = p^2$, then G is Abelian.*
(ii) *(Sylow's theorem) If p^α is a divisor of $|G|$, then G has a subgroup H such that $|H| = p^\alpha$, i.e., H is a subgroup of order p^α.*

6.3.1 *Examples of groups*

In the following, $\boldsymbol{A}$ is a real (or complex) matrix.

$$\begin{array}{ll} \text{Commutative} & \begin{cases} \langle \mathbb{Z}, + \rangle, \quad \langle a \in \mathbb{Z}_n : \text{GCD}(a,n) = 1, \cdot \rangle, \; n = 1, 2, \ldots \\ \langle \mathbb{Z}_n, + \rangle, n = 1, 2, \ldots, \quad \langle \mathbb{R}_+, \cdot \rangle, \quad \langle \mathbb{C} \setminus \{0\}, \cdot \rangle \\ \langle \mathbb{R}_+ \setminus \{1\}, x * y := x^{\log_a y}, a \in \mathbb{R}_+ \setminus \{1\} \rangle. \end{cases} \\ \text{Non-commutative} & \begin{cases} \langle \mathbf{A}, \text{ type } \boldsymbol{A} = n \times n : \det \mathbf{A} \neq 0, \cdot \rangle, \; n = 2, 3, \ldots \\ \langle \mathbf{A}, \text{ type } \boldsymbol{A} = n \times n : \det \mathbf{A} = 1, \cdot \rangle, \; n = 2, 3, \ldots \\ \langle f : A \to A, \quad f \text{ bijection}, \circ \rangle. \end{cases} \end{array} \tag{6.2}$$

6.4 Rings

Definition 6.3.

(i) A set R with two operations $+$ and $*$, written as $\langle R; +, *\rangle$, is called a *ring*, if (a)–(c) hold true

(a) $\langle R; +\rangle$ is an Abelian group,
(b) R is closed under the associative operation $*$,
(c) $a * (b + c) = a * b + a * c$, $(b + c) * a = b * a + c * a$ (left- and right distributive law, respectively).

(ii) $S \subseteq R$ is a *subring* of R if S itself is a ring. (The notation for operations $*$ and $\cdot$ are usually suppressed.)
(iii) R is a commutative ring if $ab = ba$ for all $a, b \in R$.
(iv) R is a ring with identity element (neutral element) (e or even 1), if $1 \cdot a = a \cdot 1 = a$ for all $a \in R$.
(v) A commutative ring R with the identity element e that lacks zero divisors, i.e., fulfills the cancellation law

$$ab = ac \Longrightarrow b = c, \text{ if } a \neq 0 \quad \text{is called } \textit{integral domain.}$$

(vi) A unit u has the property that there is an element v such that $uv = vu = 1$. This is written $u|1$ (u divides 1), i.e., u has a multiplicative inverse: $v = u^{-1}$. (Note that unit is not the same as unit element!)
(vii) An element p is irreducible if $p = ab \Longrightarrow$ implies that a or b is a unit.
(viii) A ring is a *unique factorization domain* (UFD) if every element which is not a unit can be uniquely factorized into irreducible elements.
(ix) If $\langle R \setminus \{0\}, \cdot\rangle$ is a group, then R is called a division ring, or even a skew-field. If the group is commutative, it is called *field.*

Definition 6.4.

(i) $a^k = \underbrace{a \cdot a \cdot \cdots \cdot a}_{k \text{ factors}}.$
(ii) An element $a \neq 0$ is called *nilpotent* if $a^k = 0$ for some integer $k > 1$.

(iii) An element a is called *idempotent* if $a^2 = a$.
(iv) A zero divisor is an element $a \neq 0$ such that there exists an element $b \neq 0$ such that $ab = 0$ or $ba = 0$.

Definition 6.5.

(i) An ideal $I \subseteq R$ is a subring such that if $r \in R$ and $i \in I$, then $ri,\ ir \in I$.
(ii) $I + r = \{i + r : i \in I\}$ is a (right-)coset of R.
(iii) An ideal I of R is maximal if $I \subset R$, $I \neq R$ and there is no other ideal $J \neq R$ such that $I \subset J$. In other words: For an ideal J such that $I \subseteq J \subseteq R$, either $I = J$ or $J = R$.
(iv) $\phi : R \to S$ is a ring homomorphism, if R and S are rings such that $\phi(a+b) = \phi(a)+\phi(b)$, and $\phi(ab) = \phi(a)\phi(b)$, for all $a, b \in R$.
   (a) ϕ is called monomorphism if ϕ is injective.
   (b) ϕ is called isomorphism if ϕ is bijective. The rings R and S are then called isomorph.

Definition 6.6.

(i) A commutative ring is an integral domain if there is no zero divisors.
(ii) An Euclidean ring R is an integral domain if there is a map $d : R \to \mathbb{N}$ such that
   (a) For every pair $a, b \in R \setminus \{0\}$, $d(a)d(b) \leq d(ab)$ and
   (b) There exist s, $t \in R$ such that $a = tb + r$, where $r = 0$ or $d(r) < d(b)$.
(iii) *A principal ideal domain* (PID) is an integral domain, such that each ideal is generated by only one element.

Theorem 6.5.

(i) *If there exists an integer $k > 1$ such that for every $a \in R$, $a^k = a$, then the ring is commutative.*
(ii) *A finite integral domain is a field.*
(iii) *An Euclidean ring is a principal ideal ring.*
(iv) *An Euclidean ring has a unit-element* 1.

6.4.1 *Examples of rings*

Ring	Comment
$\langle \mathbb{Z}, +, \cdot \rangle$	UFD, PID, Eucl. ring
$\langle \boldsymbol{A} = (a_{ij})_{n\times n},\ a_{ij} \in \mathbb{C},\ +, \cdot \rangle$	Ring with unit
$\langle \mathbb{C}, +, \cdot \rangle$	Field
$\langle \mathbb{R}, +, \cdot \rangle$	Field
$\langle \mathbb{H}, +, \cdot \rangle$	Division ring
$\langle \mathbb{Z}_p, +, \cdot \rangle$, p prime number	Field

(6.3)

For every integer d with no square factor, $\langle \{x + y\sqrt{d}, \quad x, y \in \mathbb{Q}\}, +, \cdot \rangle$ is a commutative ring.

- The only integers $d < 0$ for which the ring has unique prime factorization are

$$d = -1, -2, -3, -7, -11, -19, -43, -67, -163.$$

- The only integers d for which the ring is Euclidean are

$$d \left| \begin{array}{ccccccccccc} -11, & -7, & -3, & -2, & -1, & 2, & 3, & 5, & 6, & 7, & 11 \\ 11, & 13, & 17, & 19, & 21, & 29, & 33, & 37, & 41, & 57, & 73. \end{array} \right.$$

Chapter 7

Logic and Number Theory

7.1 Combinatorics

7.1.1 *Sum and product*

Definition 7.1. The sum $a_1 + a_2 + \cdots + a_n$ and the product $a_1 \cdot a_2 \cdot \ldots \cdot a_n$ are written as

$$\sum_{k=1}^{n} a_k \quad \text{and} \quad \prod_{k=1}^{n} a_k, \text{ respectively.} \tag{7.1}$$

$\sum$ is called the sum symbol and $\prod$ is called the product symbol. The summation and the product can have other *indices* than 1 and n. k is called *dummy index* and can be replaced by any other symbol which is not included in the expression of a_k:s.

Sequences and sums are introduced on page 307 and subsequent pages.

7.1.2 *Factorials*

Example 7.1. The product $1 \cdot 2 \cdot 3 \cdot 4$ is written $\prod_{k=1}^{4} k$. This product of consecutive integers is also written 4! (reads "four-factorial").

In the following definitions, n represents a non-negative integer.

Definition 7.2.

$$\begin{aligned} 0! &= 1, \quad \text{zero-factorial} \\ 1 \cdot 2 \cdot \ldots \cdot n &= n!, \quad n \text{ factorial} \\ 1 \cdot 3 \cdot 5 \cdot \ldots \cdot (2n-1) &= (2n-1)!!, \quad 2n-1 \text{ semi factorial} \\ 2 \cdot 4 \cdot 6 \cdot \ldots \cdot (2n) &= (2n)!!, \quad 2n \text{ semi factorial.} \end{aligned} \tag{7.2}$$

Theorem 7.1.

$$\begin{aligned} &(2n)!! = 2^n \cdot n!, \qquad 2^n n!(2n-1)!! = (2n)! \\ &n! \approx \left(\frac{n}{e}\right)^n \sqrt{2\pi n}, \quad \textit{with asymptotic equivalence} \\ &\qquad\qquad \text{(Stirling's formula).} \end{aligned} \tag{7.3}$$

7.1.3 *Permutations and combinations*

The principle of multiplication says that for n numbers where each can be chosen in r_k, $k = 1, 2, \ldots, n$ ways, the total number of choices is

$$r_1 \cdot r_2 \cdot \ldots \cdot r_n = \prod_{k=1}^{n} r_k \text{ ways.} \tag{7.4}$$

It follows that

(i) The number of different ways where n different elements can be arranged is $n!$

(ii) Let A be a set containing n elements.

  (a) The number of different ways to choose k elements of A, with respect to mutual order (without replacement), is

$$n(n-1)(n-2)\ldots(n-k+1).$$

  This is the number of *permutations* and is denoted by $(n)_k$.

  (b) The number of different ways to choose k elements from A ($|A| = n$) with no reference to their order of appearance is $n(n-1)(n-2)\ldots(n-k+1)/k!$ This is the number of *combinations* denoted by $\binom{n}{k}$, which is also the number of subsets of A having k elements.

$$\binom{n}{k} = \frac{n(n-1)(n-2)\ldots(n-k+1)}{k!} = \frac{n!}{k!(n-k)!}. \tag{7.5}$$

This number is called *binomial coefficient.*

Theorem 7.2.

$$\binom{n}{k} = \binom{n}{n-k}, \quad k = 0, 1, \ldots, n \quad n = 0, 1, \ldots \tag{7.6}$$

$$\binom{n}{k-1} + \binom{n}{k} = \binom{n+1}{k}, \quad k = 1, \ldots, n \quad n = 1, 2, \ldots \tag{7.7}$$

$$\binom{n}{0} = \binom{n}{n} = 1, \quad n = 0, 1, \ldots \tag{7.8}$$

To choose elements, k times, out of a set with n elements

	With regard to order	With no regard to order
With replacement	n^k	$\binom{n+k-1}{k}$
Without replacement	$\frac{n!}{(n-k)!} = (n)_k$	$\binom{n}{k}$

(7.9)

7.2 Proof by Induction

The *induction principle* is used to verify a given equation, $P(n)$, $n \in \mathbb{Z}$, for integers. The procedure goes as follows:

Step 1. Show that $P(n_0)$ is true (n_0 is the starting value for n in $P(n)$).
Step 2. Show that if $P(n)$ is true, then ($\Longrightarrow$) $P(n+1)$ is true.
Steps 1 and 2 imply that $P(n)$ is true for all $n = n_0, n_0 + 1, \ldots$, according to the induction principle.

7.2.1 *Strong induction*

The strong induction has an altered form of the Step 2:

Step 1. Show that $P(n_0)$ is valid.
Step 2. Show that $P(k)$ is valid for all $n_0 \leq k \leq n \Longrightarrow P(n+1)$.
By the induction principle, Steps 1 and 2 yield $P(n)$ for all $n = n_0, n_0 + 1, \ldots$

7.3 Relations

Definition 7.3.

(i) Let $A_1, A_2, \ldots, A_n$ be n number of sets. An $n-$relation R on $A_1, A_2, \ldots, A_n$ is a subset of $\prod_{k=1}^{n} A_k = A_1 \times A_2 \times \cdots \times A_n$. That $(x_1, x_2, \ldots, x_n) \in R$ is denoted by $x_1 \, R \, x_2 \, R \ldots R \, x_n$.

(ii) If all A_k are equal $(= A)$, then R is called an $n-$relation on A.

(iii) A relation R on two sets A and B, i.e., $R \subset A \times B$, is called a binary relation and is written as $x \, R \, y$; $(x \in A,\ y \in B)$, and thus means that $(x, y) \in R$.

Definition 7.4 (Some different types of binary relations $x \, R \, y$).

(i) R is reflexive if $x \, R \, x$ for all $x \in R$.

(ii) R is symmetric if $x \, R \, y \Longrightarrow y \, R \, x$ for all $x, y \in R$.

(iii) R is anti-symmetric if $x \, R \, y$ and $y \, R \, x \Longrightarrow x = y$.

(iv) R is transitive if

$$x \, R \, y \quad \text{and} \quad y \, R \, z \Longrightarrow x \, R \, z.$$

(v) A binary relation on A is an equivalence relation if it is reflexive, symmetric, and transitive.

(vi) A partially ordered relation R is a binary relation on a set A, which is reflexive, anti-symmetric, and transitive. $x \, R \, y$ is written $x \preceq y$. The corresponding set A is a partially ordered set (a poset).

(vii) **Composition of two relations**
For two binary relations R on $A \times B$ and S on $B \times C$, a binary relation $S \circ R : A \times C$ is defined by $x \, S \circ R \, z$, where $x \in A$, $z \in C$, $\exists y \in B$ such that $x \, R \, y$ and $y \, S z$.

Definition 7.5.

(i) Assume that

$$\boldsymbol{x} = (x_1, \ldots, x_m) \in A_1 \times A_2 \times \cdots \times A_m$$

and

$$\boldsymbol{y} = (y_1, \ldots, y_n) \in B_1 \times B_2 \times \cdots \times B_n.$$

A relation R on $A_1 \times A_2 \times \cdots \times A_m \times B_1 \times B_2 \times \cdots \times B_n$ is called a *function* if for every $\boldsymbol{x} \in \prod_{i=1}^{m} A_i$ there exists a unique $\boldsymbol{y} \in \prod_{j=1}^{n} B_j$. This is then written as $R : \prod_{i=1}^{m} A_i \to \prod_{j=1}^{n} B_j$, in short $\boldsymbol{x}\, R\, \boldsymbol{y}$ or $R(\boldsymbol{x}) = \boldsymbol{y}$.
In particular, a binary relation $\boldsymbol{x}\, R\, \boldsymbol{y}$ is a *function* on $A \times B$ if for every $\boldsymbol{x} \in A$ there exists a *unique* $\boldsymbol{y} \in B$, such that $\boldsymbol{x}\, R\, \boldsymbol{y}$.

(ii) The inverse relation R^{-1} corresponding to R is defined as

$$\boldsymbol{y}\, R^{-1}\, \boldsymbol{x} \Leftrightarrow \boldsymbol{x}\, R\, \boldsymbol{y}.$$

Remarks. Let $\cup_i A_i = A$ be a partition of a set A. The relation $\boldsymbol{x} R \boldsymbol{y}$ on A, defined as $\boldsymbol{x} R \boldsymbol{y} \Leftrightarrow \boldsymbol{x}, \boldsymbol{y} \in A_i$, is called an equivalence relation. This is an alternative definition of equivalence relation. For a function R, the relation $\boldsymbol{x}\, R\, \boldsymbol{y}$ on $X \times Y$ is written $\boldsymbol{y} = R(\boldsymbol{x})$.

A function f is

- Injective if $f(\boldsymbol{x}_1) = f(\boldsymbol{x}_2) \Longrightarrow \boldsymbol{x}_1 = \boldsymbol{x}_2$.
- Surjective if for every $\boldsymbol{y}$ there exists an $\boldsymbol{x}$, such that $\boldsymbol{y} = f(\boldsymbol{x})$.
- Bijective, if f is both injective and surjective.
  $f : X \to Y$ is bijective if and only if the inverse relation $f^{-1} : Y \to X$ is a function. f^{-1} is called the inverse function of f, and is defined by

$$f^{-1}(\boldsymbol{y}) = \boldsymbol{x} \Leftrightarrow \boldsymbol{y} = f(\boldsymbol{x}) \quad \text{for every} \quad (\boldsymbol{x}, \boldsymbol{y}) \in X \times Y.$$

An equivalence relation on A implies a partition of A and vice versa: $\boldsymbol{x}$ and $\boldsymbol{y}$ belong to the same $A_i \Leftrightarrow \boldsymbol{x} R \boldsymbol{y}$.

Definition 7.6. Assume that $f : X \to Y$ is a function, A a subset of X, and B a subset of Y. Then

$$\begin{aligned} &X = D_f, \quad \text{is called the domain of } f \text{ and} \\ &f(X) = f(D_f) = V_f, \quad \text{is the range of } f. \\ &\qquad \text{Further, the following sets} \\ &f(A) := \{f(x) : x \in A\} \text{ and } f^{-1}(B) = \{x : f(x) \in B\} \\ &\qquad \text{are well defined.} \end{aligned} \tag{7.10}$$

Theorem 7.3. *Under the same conditions as in* (7.10) ($A_i \subseteq X$ *and* $B_i \subseteq Y$), *the following relations hold true*

$$f(\cap_i A_i) \subseteq \cap_i f(A_i), \quad f(\cup_i A_i) = \cup_i f(A_i). \tag{7.11}$$

$$f^{-1}(\cap_i B_i) = \cap_i f^{-1}(B_i), \quad f^{-1}(\cup_i B_i) = \cup_i f^{-1}(B_i)$$

$$f^{-1}(Y \setminus B) = X \setminus f^{-1}(B)$$

$$f(f^{-1}(B)) \subseteq B, \quad A \subseteq f^{-1}(f(A)) \tag{7.11a}$$

$$f(A) = \emptyset \Leftrightarrow A = \emptyset.$$

$$B = \emptyset \Longrightarrow f^{-1}(B) = \emptyset. \tag{7.11b}$$

$$A_1 \subseteq A_2 \Longrightarrow f(A_1) \subseteq f(A_2). \tag{7.11c}$$

$$B_1 \subseteq B_2 \Longrightarrow f^{-1}(B_1) \subseteq f^{-1}(B_2).$$

(i) *The inclusion in* (7.11) *is an equality for each class* $A_i \Leftrightarrow f$ *is injective.*
(ii) *The first inclusion in* (7.11a) *is an equality for each* $B \Leftrightarrow f$ *is surjective.*
(iii) *The second inclusion in* (7.11a) *is an equality for all* $A \Leftrightarrow f$ *is injective.*
(iv) *Equation* (7.11b) *is an equivalence* (*reversible*) *if* f *is surjective.*
(v) *The first* (*second*) *implication in* (7.11c) *is an equivalence, if* f *is injective* (*surjective*).

Definition 7.7. The reflective, symmetrical, and transitive *cover* of a relation R is the smallest relation $r(R)$, $s(R)$, och $t(R)$, with R being reflexive, symmetric, and transitive, respectively.

Theorem 7.4.

(i) $R \circ R \subseteq R \Leftrightarrow R$ *is transitive.*
(ii) *The composition rule for binary relations.*
  (a) $\circ$ *is associative, that is* $R \circ (T \circ S) = (R \circ T) \circ S$.
  (b) *Following distributive laws and inclusions are valid*

$$\begin{aligned} &(R \circ T) \cup (S \circ T) = (R \cup S) \circ T \\ &(T \cup R) \circ (T \cup S) = T \circ (R \cup S) \\ &(R \cap S) \circ T \subseteq (R \circ T) \cap (S \circ T) \\ &T \circ (R \cap S) \subseteq (T \circ R) \cap (T \circ S). \end{aligned} \tag{7.12}$$

(iii) *Let $\mathcal{E}(A)$ be the set of equivalence relations on A. Furthermore, assume that $R, S \in \mathcal{E}(A)$. Then*

$$\mathcal{R} = \mathcal{R}^{-1} \quad \text{and} \quad R \circ S \in \mathcal{E}(A) \Leftrightarrow R \circ S = S \circ R. \tag{7.13}$$

(iv) **Rules for covers (For definition, see page 136).** *Let $s(R)$ and $t(R)$ be the symmetric- and transitive covers of R, respectively. Then,*

   (a) *R reflexive $\Longrightarrow s(R)$ and $t(R)$ reflexive.*
   (b) *R symmetric $\Longrightarrow s(R)$ and $t(R)$ symmetric.*
   (c) *R transitive $\Longrightarrow r(R)$ transitive.*

(v) *The reflexivity, symmetry, and transitivity properties of R are inherited from $R^2 := R \circ R$.*

Definition 7.8. Lattices and posets (partially ordered set).

(i) A transitive and anti-symmetric binary relation on a set A is called a partially ordered relation R on A, or in short a "poset" of A. The anti-symmetry and transitivity hence implies that

$$x \preceq y \text{ and } y \preceq x \Longrightarrow x = y, \quad x \preceq y \text{ and } y \preceq z \Longrightarrow x \preceq z.$$

   (a) The relation $x\,R\,y$ is written $x \preceq y$ or $y \succeq x$. That $x \preceq y$ but $x \neq y$ is written as $x \prec y$.
   (b) If $x \prec y$, then x is a "precursor" of y and y is a "follower" of x.
   (c) If $x \prec y$ and there is no other element z, such that $x \prec z \prec y$, then x is called "immediate" or "direct" precursor of y. The direct follower is defined analogously.

(ii) x and y are *comparable*, if $x \preceq y$ or $x \succeq y$.
(iii) If all pairs x and y in a poset are comparable, then the poset is said to be a totally ordered set (chain).
(iv) A sequence $\ldots \preceq x_n \ldots \preceq x_1 \preceq x_0$ is called a descending chain.
(v) A poset such that each descending chain has a smallest element is called well ordered.
(vi) An element $x \in A$ in a poset is maximal (minimal) if no other element $y \in A$ has the property $x \prec y$ $(y \prec x)$.
(vii) x is a largest (smallest) element of a poset A if $y \preceq x$ $(x \preceq y)$ for all $y \in A$. It is clear that a largest (smallest) element is unique and maximal (minimal).

(viii) Let $B \subseteq A$. An element $m \in A$ for which $b \preceq m$ ($b \succeq m$) for each $b \in B$ is called a majorant (minorant) of B. A majorant (minorant) m_0 of B, such that $m_0 \preceq m$ ($m_0 \succeq m$) for each other majorant (minorant) m of B, is called the "supremum ('infimum') of B" , and is denoted by $\sup B$ ($\inf B$).

Example 7.2. The Relation $\preceq$ defined on the set of positive integers $a \preceq b \Leftrightarrow a|b$ constitutes $\{1, 2, 3, \ldots\}$ a poset. The following is a so-called Hasse diagram considering the relation on $\{1, 2, 3, \ldots, 23, 24\}$.

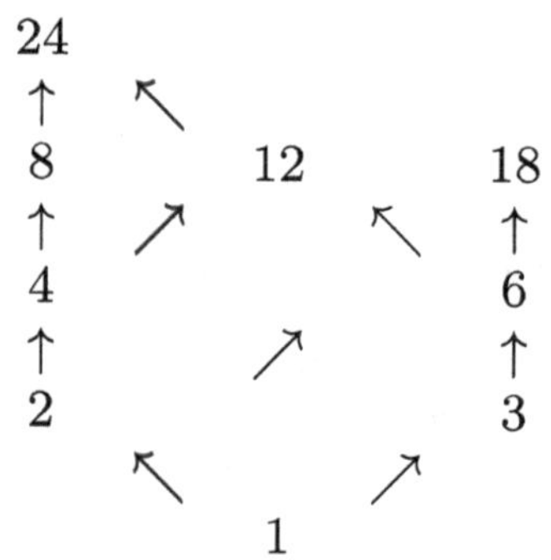

In the above Hasse diagram: The Maximal elements are 18 and 24. The minimal element likewise the smallest is 1.

Definition 7.9. A lattice L is a set which is closed under the binary operations $\vee$ and $\wedge$.

Let $a, b, c \in L$, then

Commutative laws: $a \vee b = b \vee a, \quad a \wedge b = b \wedge a$
Associative laws: $(a \vee b) \vee c = a \vee (b \vee c), \quad (a \wedge b) \wedge c = a \wedge (b \wedge c)$
Absorption laws: $a \vee (a \wedge b) = a, \quad a \wedge (a \vee b) = a$

(7.14)

The dual to a proposition P for a lattice L containing $\vee$ and $\wedge$ is the expression P' given by P through shifting $\vee$ to $\wedge$ and $\wedge$ to $\vee$.

Theorem 7.5.

(i) *Every finite poset A of a lattice L can be numbered by means of a function $f : A \to \{1, 2, 3, \ldots\}$, with property $a \preceq b \Longrightarrow f(a) \leq f(b)$.*

(ii) *For a lattice, the following properties hold:*

(a) *P true $\Leftrightarrow$ P' true.*
(b) *(Idempotent laws) $a \vee a = a$ and (thus as above) $a \wedge a = a$.*
(c) *$a \wedge b = a \Leftrightarrow a \vee b = b$.*
(d) *The relation $a \preceq b$ defined as $a \wedge b = a$ is a poset on L.*

(iii) *Assume that P is a poset such that for each pair $a, b \in P$, both* $\inf(a, b)$ *and* $\sup(a, b)$ *exist. Denoting* $\inf(a, b) = a \wedge b$ *and* $\sup(a, b) = a \vee b$, *we have*

(a) *$(P, \wedge, \vee)$ is a lattice.*
(b) *The partial order on P induced by a lattice is the same as the original partial order on P.*

(iv) *(Hausdorff's maximality theorem) Every poset A contains a maximal complete ordered subset B. This means that B is not a proper subset of a poset $B' \subseteq A$.*

Remarks. (iii)(a) in the previous theorem states that a lattice becomes a poset if $a \wedge b \equiv \inf(a, b)$ and $a \vee b \equiv \sup(a, b)$ exist for each pair a, b.

Definition 7.10.

(i) (a) A lattice L has a lower bound d if $d \preceq x$ for every $x \in L$.
(b) A lattice L has an upper bound u if $x \preceq u$ for every $x \in L$.
(c) A lattice is bounded above (below) if there exists an upper (lower) bound. A lattice is bounded if it has both an upper and a lower bound.

(ii) A lattice is distributive if

$$a \wedge (b \vee c) = (a \wedge b) \vee (a \wedge c) \quad \text{and}$$
$$a \vee (b \wedge c) = (a \vee b) \wedge (a \wedge c). \tag{7.15}$$

(iii) Let L be a lattice. An element $a \in L$ is irreducible if

$$a = x \wedge y \Longrightarrow a = x \text{ or } a = y. \tag{7.16}$$

(iv) The irreducible elements which directly precede d are called atoms or prime elements.

(v) An element y is a complement of an element x if

$$x \vee y = u \text{ and } x \wedge y = d.$$

(vi) A bounded lattice, where every element has a complement, is called a complement-lattice.

Theorem 7.6. *Let x be an element of a lattice L.*

(i) *A lattice is non-distributive if and only if it contains some of the following sublattices (i) or (ii) in the figure on the right.*

(ii) *x has a unique, direct precursor $\Leftrightarrow$ x is irreducible.*

(iii) *Assume that L is a finite, distributive lattice. Then, for each element x, there exist unique irreducible elements a_i, $i = 1, 2, \ldots, n$, such that*

$$x = a_1 \vee a_2 \vee \ldots \vee a_n. \quad (7.17)$$

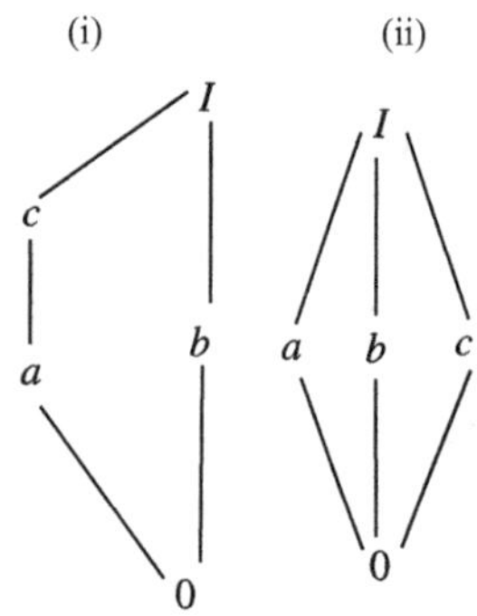

(iv) *Assume that L is a bounded distributive lattice, then x has at most one complement.*

(v) *Assume that L is a finite and distributive complement-lattice. Then each element x can be uniquely expressed as in (7.17), where a_i are atoms of L.*

7.4 Expressional Logic

Definition 7.11 (Logical symbols).

(i) Truth symbols: TRUE (t), FALSE (f).

(ii) Logical connectives.

(a) $\wedge$ also written AND.
(b) $\vee$ also written OR.
(c) $\neg$ reads "not".
(d) $\rightarrow$ reads "implies".
(e) $\longleftrightarrow$ reads "equivalent to" and means both $\rightarrow$ and $\leftarrow$.
(f) For variables uppercase $P, Q, R, S, \dots$, is used.

Logical connectives operate in algebraic order of priority, otherwise from left to right with or without parentheses. For instance, $\neg P \vee Q = (\neg P) \vee Q$ and $P \vee Q \wedge R = P \vee (Q \wedge R)$.

(iii) A proposition is built up by a sequence of logical connectives on variables which is also reffered as a *well-formed formula: wff.*
(iv) Two equivalent wff:s V and W (like $P \wedge Q$ and $Q \wedge P$) are written as $V \equiv W$. One writes $W(P, Q, R)$ if W is defined by only these variables.
(v) A proposition is
  (a) A tautology if it is true for all admissible values of its variables.
  (b) A contradiction if it is false for at least one value of one of its variables.

7.4.1 *Tautology and contradiction*

To determine whether a statement is a tautology or not, its truth value table is constructed. One may also use a contradiction argument.

Tables

The truth value tables for $P \rightarrow Q$ and $P \longleftrightarrow Q$ are

P	Q	$P \rightarrow Q$
t	t	t
t	f	f
f	t	t
f	f	t

and

P	Q	$P \longleftrightarrow Q$
t	t	t
t	f	f
f	t	f
f	f	t

, respectively. (7.18)

The truth value tables for $P \vee Q$ and $P \wedge Q$ are

P	Q	$P \vee Q$
t	t	t
t	f	t
f	t	t
f	f	f

and

P	Q	$P \wedge Q$
t	t	t
t	f	f
f	t	f
f	f	f

, respectively. (7.19)

Example 7.3. To determine whether $((P \longleftrightarrow Q) \wedge Q) \rightarrow P$ is a tautology or not, the following truth value table (which includes all steps) is used. (The table is read from left to right.)

Truth (value) table							
	$(P$	$\longleftrightarrow$	$Q)$	$\wedge$	$Q)$	$\rightarrow$	P
	t	t	t	t	t	t	t
	t	f	f	f	f	t	t
	f	f	t	f	t	t	f
	f	t	f	f	f	t	f
steps	1	2	1	3	1	4	1

Alternatively, one can use contradiction, to verify whether the statement is a tautology or not.

By (7.18) an implication $P \rightarrow Q$ is false only if P and Q have values t and f, respectively.

Arguing with contradiction							
step	$(P$	$\longleftrightarrow$	$Q)$	$\wedge$	Q	$\rightarrow$	P
1						f	
2			t				f
3	f	$\boxed{t}$			t		
4	f		t				
5		$\boxed{f}$					

The t in row 2 refers to the whole expression $(P \longleftrightarrow Q) \wedge Q$.

The whole contradiction is false, due to $\boxed{\text{f}}$ in row 5.

So $((P \longleftrightarrow Q) \wedge Q) \rightarrow P$ is a tautology.

Example 7.4. Consider the statement

$$W = (P \to \neg Q) \to ((P \vee Q) \to Q).$$

We try with truth-value table to determine whether W is a tautology or not.

$(P \to$	$\neg$	$Q)$	$\to$	$[(P$	$\vee$	$Q)$	$\to$	$Q]$
t	f	f	t	t	t	t	t	t
t	t	t	t	t	t	f	t	t
f	t	f	$\boxed{f}$	f	t	t	f	f
f	t	t	t	f	f	f	t	f
(1)		(2)						

Numbered columns (1) and (2) contain all possible combinations for (P, Q) values. In implication column we get an f (third row, fourth column) (boxed). Thus, the implication is indeed a contradiction, whence the statement is not a tautology.

Alternatively, as in the previous example, one can, through contradiction, decide whether W is a tautology or not. We then assume that LHS is true and RHS is false, i.e.,

$$\begin{array}{ccccccccc} (P \to & \neg & Q) & \to & [(P & \vee & Q) & \to & Q]. \\ & t & & & & & f & & \end{array}$$

Because now RHS contains an implication thus according to (7.18) $P \vee Q$ is true (for P true and Q false). Then $\neg Q$ is true, so P is true. Thus, $\underset{t}{P} \to \underset{t}{(\neg Q)}$ in LHS, which is true. So this does not lead to a contradiction, hence W is not a tautology.

Quine's method.

To check whether a wff claim: $W(P, P_1, P_2, \ldots)$ is a tautology or not, take P as true t, and false f, respectively, and see whether a tautology is obtained. More precisely

Theorem 7.7.

$$W(P, P_1, P_2, \ldots) \text{ is a tautology}$$
$$\Leftrightarrow \begin{cases} W(t, P_1, P_2, \ldots) & \text{are both} \\ W(f, P_1, P_2, \ldots) & \text{tautologies.} \end{cases} \tag{7.20}$$

Normal forms

Definition 7.12. The expressions

$$\bigvee_{j=1}^{n} (\wedge_{m_j=1}^{n_j} * P_{m_j}), \quad \bigwedge_{j=1}^{n} (\vee_{m_j=1}^{n_j} * P_{m_j}), \tag{7.21}$$

with $*$ as $\neg$ (or nothing), are called a disjunctive (DNF) and a conjunctive (CNF) normal form, respectively.

A DNF or a CNF is *complete*, if for each $\wedge_{m_j=1}^{n_j} * P_{m_j}$ or $\vee_{m_j=1}^{n_j} * P_{m_j}$, respectively, found the same variables possibly included $*$ in front.

Theorem 7.8.

(i) *Each wff is equivalent to a CNF or a DNF.*
(ii) *Each wff which is not a contradiction is equivalent to a complete DNF.*
(iii) *Every wff which is not a tautology is equivalent to a complete CNF.*

7.4.2 *Methods of proofs*

Theorem 7.9. *To prove a claim $V \to W$, or a theorem, the following methods are used*

(i) *Direct proof: $V \to W$.*
(ii) *Indirect proof (proof by negation): $\neg W \to \neg V$, i.e., to prove the (equivalent) contra-positive statement.*
(iii) *Proof by contradiction: $(V \wedge \neg W) \to f$, i.e., false ("reduction ad absurdum").*

Remarks. Solving an equation $f(x) = 0$ (real or complex) yields a number of x-values, say $x_1, x_2, \ldots, x_n$, that might satisfy $f(x_i) = 0$, $i = 1, 2, \ldots, n$ or not! To see this, we note that

(i) The implication

$$f(x) = 0 \Longrightarrow x = x_1, x = x_2, \ldots, x = x_n$$

means that the x which satisfy $f(x) = 0$ are found among $x_1, x_2, \ldots, x_n$ but that does not rule out that some of these x_i may be false solutions of $f(x) = 0$.

(ii) The implication

$$f(x) = 0 \Longleftarrow x = x_1, x = x_2, \ldots, x = x_n$$

means that $x = x_1, x = x_2, \ldots, x = x_n$ solve $f(x) = 0$, but that does not rule out that there might be other solutions for $f(x) = 0$.

7.5 Predicate Logic

Definition 7.13.

(i) $\exists$ is the existence quantifier and reads "exists".
(ii) $\forall$ is the universal quantifier and reads "for all".

Definition 7.14. The first-order predicate logic has the following ingredients:

(i) Individual variables x, y, z
(ii) Individual constants a, b, c
(iii) Function constants f, g, h
(iv) Predicate constants p, q, r
(v) Connectivity symbols $\neg$, $\rightarrow$, $\vee$, $\wedge$
(vi) Quantifiers $\exists$, $\forall$
(vii) Parentheses ()

In the expression $\exists x$, so that the wff, W is valid, and is said to be *the frame of the quantifier* $\exists x$. If W contains x, then x is a bound of W, otherwise x is free.

7.6 Boolean Algebra

Definition 7.15. Assume that B contains (at least) two elements, denoted 0 and 1. B i called a boolean algebra if there exist two binary operations $+$ and $*$, such that $x + y \in B$ and $x * y \in B$ for all pairs $x, y \in B$. Furthermore, $+$ and $*$ shall fulfill the following properties:

$$\begin{aligned}
&(x+y)+z = x+(y+z), && (x*y)*z = x*(y*z),\\
&x+y = y+x, && x*y = y*x,\\
&x+(y*z) = (x+y)*(x+z), && x*(y+z) = x*y+x*z,\\
&x+0 = 0, && x*1 = x,\\
&x+x' = 0, && x*x' = 1.
\end{aligned} \tag{7.22}$$

The last two expressions say that there is a complement x' for each x.

For each statement containing elements from B, $+$, $*$, 0, and 1, the dual statement is defined interchanging all $+$ and $*$ as well as all 0 and 1. (For instance, $(1 + x) * (y + 0) = y$ has the dual $(0 * x) + (y * 1) = y$.)

Theorem 7.10. *The following identities are derived from* (7.22).

$$\begin{aligned}
&x+x = x, && x*x = x,\\
&x+1 = 1, && x*0 = 0,\\
&x+(x*y) = x, && x*(x+y) = x.
\end{aligned} \tag{7.23}$$

7.6.1 *Graph theory*

Definition 7.16. A multi-graph consists of a number of nodes/vertices $\{v_i\}$ and a number of edges $\{e_i\}$ (Figure 7.1).

(i) Two nodes u and v are (directly) connected if there is an edge e between them. It is then written $e = (u, v)$.
(ii) For a graph, there is at most one edge between two different nodes or there is no edge $e = (u, u)$ (a so-called loop).
(iii) A multi-graph is connected if there is an edge between each pair of nodes.
(iv) A connected graph is complete if for each pair of nodes u and v there is an edge (u, v).
(v) A path between $u = e_0$ and $v = e_n$ is a sequence of directly connected nodes;

$$u, e_1, v_2, e_2, \ldots, e_{n-1}, v \quad \text{where} \quad e_k = (v_k, v_{k+1}).$$

(vi) The length of the path is $n - 1$ as in notations above.
(vii) A node is isolated if there is no edge connecting it to other nodes (e.g., node 1 in Figure 7.1 is isolated).

(viii) A path in a multi-graph is

(a) A trail if all its edges are different.
(b) A path if all nodes are different.
(c) A cycle, if all nodes are different except for the first and the last (which coincide). A cycle is a k−cycle if it has length k.

(ix) A minimal connection between two nodes is the shortest path between them.
(x) The distance between two nodes is the length of the shortest trail between them.
(xi) The degree of a node is the number of its edges (e.g., node 4 in Figure 7.1 has degree 4).
(xii) The diameter of a graph is the maximal length of its minimal connections.
(xiii) A multi-graph is traversable, if there is a path which uses each edge exactly once (a traversable trail).
(xiv) An Euclidean graph is a traversable multi-graph.
(xv) A multi-graph is a planar graph if it can be drawn so that no edge crosses the other.

Theorem 7.11.

(i) *Every path is a trail.*
(ii) *There is a path between two nodes* $\Longleftrightarrow$ *there is a trail between them.*
(iii) *The Relation "u and v are connected by a path" is an equivalence relation. Corresponding equivalence class $\{u\}$ is the largest connected subgraph, containing u.*
(iv) *(Euler) A multi-graph is traversable* $\Longleftrightarrow$ *At most two nodes have odd degree.*
(v) *(Euler) For connected multi-graphs $V - E + R = 2$, where V, E, and R are the number of nodes (vertices), edges, and domains (regions), respectively.*
(vi) *(Four color problem) For a connected graph, the different domains are colored with only one of the four colors, so that the adjacent domains get different colors.*
(vii) *(Kuratowski's theorem) A graph is plane if it does not contain any partial graphs, see Figure 7.3.*

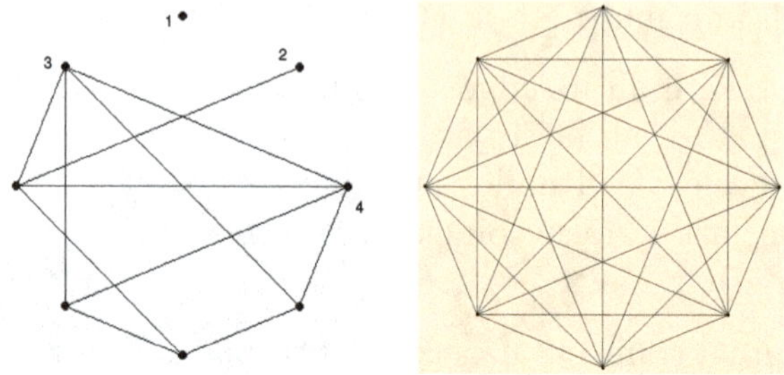

Figure 7.1: LHS: A graph with 8 nodes. RHS: A complete graph.

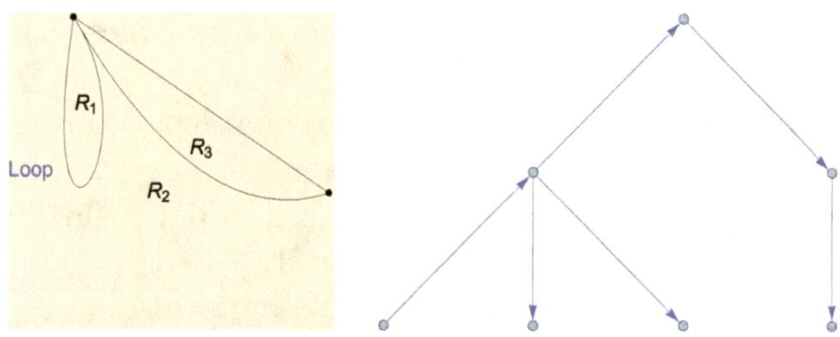

Figure 7.2: LHS: A multi-graph, dividing the plane in three regions. RHS: A tree.

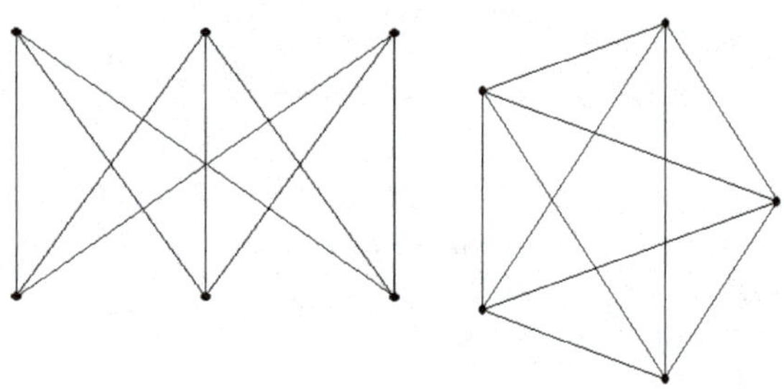

Figure 7.3: Subgraphs which are not plane.

7.6.2 *Trees*

Definition 7.17. A tree is a cycle free connected graph (Figure 7.2).

Theorem 7.12. *Let G be a graph with more than one node. Then (i)–(iv) are equivalent.*

(i) *G is a tree.*
(ii) *Every pair of nodes is connected with precisely one edge.*
(iii) *G is connected, but if an edge is removed, then the resulting graph is not connected.*
(iv) *G contains no cycles, but if a (non-isolated) edge is added to G, then the resulting graph will have exactly one cycle.*

Furthermore, for a finite G, with $n > 1$ nodes, the following claims are equivalent.

(i) *G is a tree.*
(ii) *G is cycle free with $n-1$ edges.*
(iii) *G is connected with $n-1$ edges.*

7.7 Difference Equations

Definition 7.18. A linear difference equation with constant coefficients, $a_0, a_1, \ldots, a_n$, in a (unknown) sequence $(r_n)_{n=1}^{\infty}$ of order n is an equation of the form

$$a_n r_n + a_{n-1} r_{n-1} + \cdots + a_1 r_1 + a_0 r_0 = g(n), \quad a_n \neq 0. \tag{7.24}$$

If $g(n) = 0$ for all n, the equation is called homogeneous. The "teristic polynomial" for the difference equation is given by

$$\begin{aligned} p(x) :&= a_n x^n + a_{n-1} x^{n-1} + \cdots + a_1 x + a_0 \\ &= a_n \prod_{r=1}^{m} (x - x_r)^{k_r}, \quad a_n \neq 0, \end{aligned} \tag{7.25}$$

where RHS in (7.25) is the unique factorization of $p(x)$ and $k_1, k_2, \ldots, k_m$ are positive integers with $k_1 + k_2 + \cdots + k_m = n$.

Theorem 7.13.

(i) *The solution $(r_n)_{n=1}^{\infty}$ to the equation (7.24) is given by $r_n = r_{n,h} + r_{n,p}$, where h stands for homogenous solution and p for a particular solution, corresponding to RHS $= 0$ and RHS $= g(n)$, respectively.*

(a) *The homogenous solution is*

$$r_{n,h} = \sum_{j=1}^{m} p_j(n) x_j^n, \tag{7.26}$$

where $p_j(n)$ is a polynomial of degree $p_j < k_j$, due to (7.25).

(b) *A particular solution is obtained by a suitable ansatz, that depends on the RHS $= g(n)$ and, in some cases, also the homogenous solution.*

(ii) *Ansatz algorithm of particular solution $r_{n,p}$ for some special RHS (right-hand sides). If*

(a) $g(n) = a \cdot n^k, \quad k = 0, 1, 2, \ldots$
Choose $r_{n,p}$, a polynomial in n of degree k.

(b) $g(n) = c\, x_0^n$.
If x_0 is not a zero of the characteristic polynomial (7.25), choose $r_{n,p} = a\, x_0^n$.
If $x_0 = x_j$, i.e., a zero of multiplicity k_j to the characteristic polynomial, choose $r_{n,p} = x_j^n p(n)$, where p is a polynomial of degree $k_j - 1$.

(c) *For each term $g_j(n)$ in $g(n) = \sum_{j=1}^{t} g_j(n)$, $j = 1, 2, \ldots, m$, on the RHS, make an ansatz r_{n,p_s}. The particular solution is then the sum of the different particular solutions.*

(iii) *Initial conditions for the difference equation (7.24) is of the form*

$$r_k = b_k, \quad k = 0, 1, 2, 3, \ldots, n-1,$$

for arbitrary real/complex numbers, $b_0, b_1, \ldots, b_{n-1}$. These conditions determine the n unknown coefficients in $r_{n,h}$. This yields a unique solution $r_n = r_{n,h} + r_{n,p}$.

7.8 Number Theory

Definition 7.19. The following a, b, c, m, n are positive integers.

7.8.1 *Introductory concepts*

(i) That a is a divisor of b means that b/a is an integer and is written $a|b$. GCD(m, n) is the Greatest Common Divisor of m and n.

(ii) A common multiple of a and b is number c, such that $a|c$ and $b|c$. LCM(m, n) is the Least Common Multiple of m and, n.

(iii) Two numbers m and n are relatively prime, if GCD$(m, n) = 1$.

(iv) (a) $\Phi(n)$ is the number of $m \in \{1, 2, \ldots, n\}$ such that GCD$(m, n) = 1$ (Φ is known as Euler Totient function).

(b) $\sigma(n)$ is the sum of the divisors of n.

(c) The Möbius μ–function is defined as

$$\mu(n) = \begin{cases} 0, & \text{if } n \text{ contains a square,} \\ (-1)^k, & \text{if } n = p_1 \cdot p_2 \cdot \ldots \cdot p_k, \\ & \text{where } p_j \text{ are distinct prime numbers,} \\ \mu(1) = 1. & \end{cases}$$

(d) $\sigma(n) = \sum_{d|n} d$ is the sum of divisors of n.

(e) $\tau(n) = \sum_{d|n} 1$ is the sum of number of divisors of n.

(v) A prime p is an integer ≥ 2, such that its only divisors are 1 and p.

(vi) A prime twin is a pair of primes p_1 and p_2 with $|p_1 - p_2| = 2$, for instance, 29 and 31.

(vii) A real number x which is a root/solution of a polynomial equation $f(x) = 0$ with only rational coefficients is called algebraic. If x is not a root of such an equation, then it is called transcendent.

(viii) **Rest class mod n:** Let a, b, n be integers and $n > 0$. Then

$$a \equiv b \mod n \Longleftrightarrow n|a - b. \tag{7.27}$$

(ix) A function f defined on $\mathbb{Z}_+$ is called multiplicative if

$$f(mn) = f(m)f(n), \quad \forall m,\, n : \gcd(m, n) = 1.$$

Theorem 7.14. *Following functions are multiplicative*

(i) *Euler's totient function* $\Phi(n)$.

(ii) *The Möbius function* $\mu(n)$.

(iii) *The function* $\sigma(n)$: *the sum of divisors of* n.

(iv) *The function* $\tau(n)$: *the sum of the numbers of divisors of* n.

Definition 7.20. For a real number x, the largest integer number $\leq x$ is denoted by $[x]$ and is called the integer part of x.

A chain fraction is obtained by setting $a_0 = [x]$ and if $x \neq a_0$, define x_1 by $x = a_0 + 1/x_1$. Inductively, x_n and a_n are determined through $x_{n-1} = a_{n-1} + 1/x_n$, if $a_{n-1} \neq x_{n-1}$, and then $a_n = [x_n]$.

The algorithm

$$x = a_0 + \cfrac{1}{a_1 + \cfrac{1}{a_2 + \ddots}} \tag{7.28}$$

is referred to as a chain fraction. This is presented in concise form

$$x = a_0 + \frac{1}{a_1+}\frac{1}{a_2+}\cdots \quad \text{or} \quad x = [a_0, a_1, \ldots].$$

An irrational number x that solves $ax^2 + bx + c = 0$, for rational a, b, c, is called quadratic irrational.

Theorem 7.15.

(i) *x is a rational number$\Longleftrightarrow$ $x_n = a_n$ for some n, i.e., the chain fraction is finite. This can be written as*

$$x = a_0 + \cfrac{1}{a_1 + \cfrac{1}{a_2 + \ddots \cfrac{1}{a_n}}} = [a_0, a_1, \ldots, a_n]. \tag{7.29}$$

(ii) *More precisely there is a bijection between x and all $[a_0, a_1, \ldots, a_n]$, where $a_n \geq 2$, $n = 1, 2, \ldots$*

(iii) *x is quadratic irrational if and only if*

$$x = [a_0, a_1, \ldots, a_{k-1}, \overline{a_k, a_{k+1}, \ldots, a_{k+m-1}}],$$

where $\overline{a_k, a_{k+1}, \ldots, a_{k+m-1}}$ means that this sequence is infinitely repeated.

(iv) *The numbers $[a_0, a_1, \ldots, a_n]$ converge to x, as $n \to \infty$.*

Remarks. The numbers $\sqrt{2}$ and $\sqrt{3}$ are algebraic.
The numbers $\ln 2$, π, and e are transcendental, likewise for the number $\sum_{k=1}^{\infty} 2^{-k!}$.

Example of a quadratic irrational representation:

$$\sqrt{2} = [1, 2, 2, 2, \ldots] = [1, \overline{2}].$$

Theorem 7.16. *Let $a_1, a_2, \ldots, a_n$ be algebraic numbers. If $b_0, b_1, b_2, \ldots, b_n$ are algebraic numbers $\neq 0$, then*

$e^{b_0} \cdot a_1^{b_1} \cdot a_2^{b_2} \cdot \cdots \cdot a_n^{b_n}$	is transcendent (Baker).	
$b_1 \ln a_1 + b_2 \ln a_2 + \cdots + b_n \ln a_n =: c$	is transcendent, if $c \neq 0$ (Baker).	(7.30)
$b_1 e^{a_1} + b_2 e^{a_2} + \cdots + b_n e^{a_n} \neq 0$	if in addition all a_i are different (Lindemann).	

The first 100 primes

2	3	5	7	11	13	17	19	23	29
31	37	41	43	47	53	59	61	67	71
73	79	83	89	97	101	103	107	109	113
127	131	137	139	149	151	157	163	167	173
179	181	191	193	197	199	211	223	227	229
233	239	241	251	257	263	269	271	277	281
283	293	307	311	313	317	331	337	347	349
353	359	367	373	379	383	389	397	401	409
419	421	431	433	439	443	449	457	461	463
467	479	487	491	499	503	509	521	523	541

The Euclidean algorithm: Let a and b be integers such that $a > b > 0$. Then there are unique k and r where $0 \leq r < b$, such that

$$\frac{a}{b} = k + \frac{r}{b}. \tag{7.31}$$

If $r > 0$, we apply (7.31) on b and $r =: r_1$. Hence, there are k_1 and r_2, where $0 \leq r_2 < r_1$. $\frac{b}{r_1} = k_1 + \frac{r_2}{r_1}$. Since $r_1 > r_2 \geq 0$ are integers, the algorithm ends after a finite number of steps. Let r_n be the last rest term > 0. Then $r_n = \text{GCD}(a, b)$.

Theorem 7.17.

(i) $\text{LCM}(a, b) \cdot \text{GCD}(a, b) = ab$.
(ii) *There are infinitely many primes.*

(iii) *Every integer $n > 0$ can be uniquely written as a product of primes:*

$$n = \prod_{j=1}^{k} p_j^{\alpha_j}, \quad p_1 < p_2 < \cdots < p_k, \qquad (7.32)$$

where p_j are different primes and α_j are positive integers.

(iv) $\Phi(n) = n - 1 \Longleftrightarrow n$ *is a prime.* (Φ : *Euler totient*).

(v) (a) *The relation* (7.27) *is an equivalence relation.*
(b) *The equivalence classes can be represented by*

$$\{0, 1, 2, \ldots, n-1\} =: \mathbb{Z}_n.$$

(c) *This set constitutes a ring $< \mathbb{Z}_n, +, \cdot >$, which becomes a field if and only if n is prime.*

(vi) *Let p be a prime number and p^j be the highest power of p, which divides n. Then*

$$\Phi(n) = n \prod_{p:\, p|n} (1 - 1/p), \quad n = \sum_{d:\, d|n} \Phi(d),$$
$$\sigma(n) = \prod_{p:\, p|n} \frac{(p^{j+1} - 1)}{p - 1}, \quad \tau(n) = \prod_{p:\, p|n} (j+1). \qquad (7.33)$$

(vii) *For a multiplicative function f*

$$f(n) = \prod_{j=1}^{k} f(p_j^{\alpha_j}), \quad \text{where } n \text{ is given by (7.32).}$$

(viii) *If f is multiplicative, then $g(n) := \sum_{d:\, d|n} f(d)$ is multiplicative.*

(ix) *Let $\{p_n\}_{n=1}^{\infty}$ be the enumeration of the prime numbers in order of their magnitude. Then, for an infinite number of indices $n, r > 0$,*

$$\liminf_{n\to\infty} \frac{p_{n+r} - p_n}{\ln n} \le 1, \quad p_{n+r} - p_n < (\ln n)^{8r/(8r+1)}. \qquad (7.34)$$

The Euler Totient function $\Phi(n)$, the number of numbers $1, 2, \ldots, n$, relative prime to n.

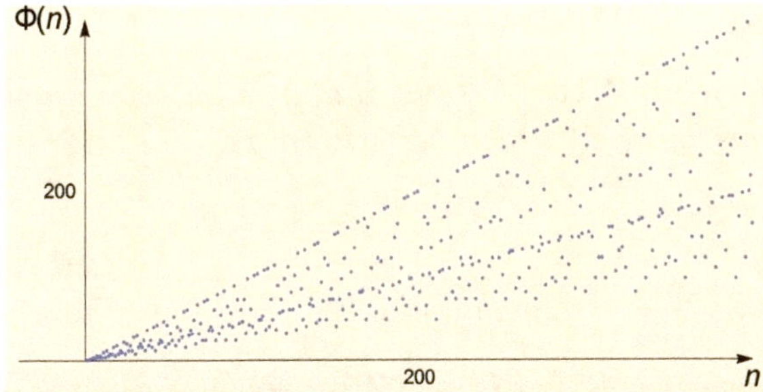

Definition 7.21.

(i) A perfect number n is a positive integer such that $n = 2\sigma(n) := 2\sum_{d|n} d$, where d:s are divisors of n.
(ii) Two numbers m and n are friendly if $2\sigma(n) = m$ och $2\sigma(m) = n$.
(iii) The numbers 6 and 28 are perfect, since $1+2+3+6 = 2 \cdot 6$ and $1+2+4+7+14+28 = 2 \cdot 28$. The first eight perfect numbers are

6	28	496	8128
33550336	8589869056	137438691328	2305843008139952128

(iv) The numbers 220 and 284 are friendly.

Riemann's z-function

Riemann's z-function is defined as

$$\zeta(z) = \sum_{n=1}^{\infty} \frac{1}{n^z},$$

where z is a complex number. The sum is absolutely convergent for $\text{Re}(z) > 1$. For this function, the following holds true:

$$\zeta(z) = \prod_{p}^{\infty} \frac{1}{(1 - 1/p^z)}, \quad \text{for prime numbers } p \text{ and } \text{Re}(z) > 1.$$

$\zeta(z)$ can analytically be extended to $\mathbb{Z} \setminus \{1\}$ and has zeros at $z = -2, -4, -6, \ldots$

Riemann's hypothesis states that the remaining zeros are lying on the line $\text{Re}(z) = 1/2$.

7.8.2 *Some results*

Theorem 7.18. *Assume that* n *is an even number. Then* n *is perfect precisely when it can be written as*

$$n = 2^{p-1}(2^p - 1), \tag{7.35}$$

where $2^p - 1$ *is prime number (and hence* p *is prime).*

Theorem 7.19 (The Chinese remainder theorem). *Assume that* $n_1, n_2, \ldots, n_k$ *are, pairwise, relatively prime numbers and* $c_1, c_2, \ldots, c_k$*, arbitrary integers.*

Then there is an integer x *which solves all equations*

$$x \equiv c_j \mod n_j, \quad j = 1, 2, \ldots, k.$$

Theorem 7.20 (Euler's theorem). *Assume* a *and* n *are relatively prime natural numbers. Then*

$$a^{\Phi(n)} \equiv 1 \mod n. \tag{7.36}$$

Fermat's little theorem. *Let* p *be a prime number, then*

$$a^p \equiv p \mod p \quad \text{and in particular}$$

$$a^{p-1} \equiv 1, \quad \text{if } (a, p) = 1, \text{ i.e., they are relatively prime.}$$

The latter follows from Euler's theorem, since $\Phi(p) = p - 1$.

Theorem 7.21 (Wilson's theorem).

$$(p-1)! \equiv -1 \mod p \Longleftrightarrow p \textit{ is a prime number.} \tag{7.37}$$

In addition, it follows that $3! = 2 \mod 4$*. More generally,*

$$(n-1)! = 0 \mod n, \quad \textit{for } n \geq 6, \textit{ and not a prime number.} \tag{7.38}$$

Theorem 7.22 (Liouville's theorem). *If* x *is algebraic of degree* $n > 1$*, then there is a constant* $c = c(x)$ *such that* $|x - p/q| < c/q^n$ *holds for all rational numbers* p/q *(*p *and* q *integers).*

For each $\alpha > 2$ *and* x*, there is a constant* $C = C(x, \alpha)$ *such that* $|x - p/q| < C(x, \alpha)/q^\alpha$ *holds for all rational numbers* p/q *(*p *and* q *integers).*

Theorem 7.23 (Fermat's last theorem (Wiles, 1994)). *There are no integers $a, b, c > 0$ such that*

$$a^n + b^n = c^n, \quad \text{for } n = 3, 4, \ldots \tag{7.39}$$

Remarks. The trivial integer solution to the above relation means that $a = 0$ or $b = 0$.

The case $n = 2$ is discussed in the chapter on geometry.

(i) Every integer $n \geq 0$ can be expressed as the sum of four integer squares (Lagrange). This means that there exist integers a, b, c, and d, such that

$$n = a^2 + b^2 + c^2 + d^2.$$

(ii) Let n be an integer.

For each prime number $p|n$ with $n = 3 \mod p$, the largest integer exponent α for which $p^\alpha|n$ is an even number.

$$\Longleftrightarrow$$

n can be expressed as the sum of two square integers (Fermat, Euler).

(iii) It is uncertain if there exist infinitely many prime twins.

(iv) Nor is the following assertion proven:
Each even positive integer can be written as the sum of two primes (Goldbach's conjecture).

(v) (a) There are an infinite number of primes p, such that $p + 2$ either is a prime number or a product of two prime numbers (Chen 1974).
(b) Every large enough odd integer > 0 can be written as the sum of at most three prime numbers (Vinogradov).

(vi) Catalan's equation

$$x^p - y^q = 1 \tag{7.40}$$

considers looking for positive integers x, p, y, q. Here, the only known non-trivial solution is

$$3^2 - 2^3 = 1.$$

In 1844, Catalan made the conjecture that there are no other solutions.

It is necessary to have $p|y$ and $q|x$. Tijdman showed in 1976 that the number of solutions is finite. Catalan's problem may possibly be solved using computers.

7.8.3 *RSA encryption*

The principle of RSA encryption is based on two, different large primes p and q, that are known only to ones who keep an encrypted message secret, while the product $p \cdot q$ is known. Here, the factorization of $p \cdot q$ is the challenge.

(i) Encryption of m is the number $m^d \mod pq$.
(ii) Decrypting of m^d is $(m^d)^e \mod pq$. By Euler's theorem (7.36) page 156, we have that

$$(m^d)^e = m^{de} = m^{k\varphi(pq)+1} = m \mod pq.$$

Chapter 8

Calculus of One Variable

8.1 Elementary Topology on $\mathbb{R}$

The set of real numbers, satisfying the inequality $-1 < x \leq 3$ is denoted by brackets: $(-1, 3]$. This is an example of an *interval.* The two-sided inequality can thus be expressed as $x \in (-1, 3]$.

Definition 8.1. Four interval types and their equivalent representations:

$$\begin{aligned} a < x \leq b &\Longleftrightarrow x \in (a, b] \\ a \leq x < b &\Longleftrightarrow x \in [a, b) \\ a \leq x \leq b &\Longleftrightarrow x \in [a, b] \\ a < x < b &\Longleftrightarrow x \in (a, b). \end{aligned} \tag{8.1}$$

Remarks. The two last intervals in the definition are called closed and open, respectively.

The points a and b are called endpoints.

Intervals are (thus) subsets of $\mathbb{R}$.

The interval $(-1, 3]$ (see Figure 8.1) is a subset (sub interval) of, for instance, $(-3, 4)$, which can be denoted by $(-1, 3] \subset (-3, 4)$.

Only in the strict inequalities, ($<$), $a = -\infty$ or $b = \infty$ is allowed.

A number x satisfying $a < x < b$ is in the interior of any of the intervals in (8.1). Such an x is called an *interior point.*

An interval (a, b) such that $a < x < b$ is called a neighborhood of x. More generally, a neighborhood to a point x is a set M, such that $x \in (a, b) \subseteq M$.

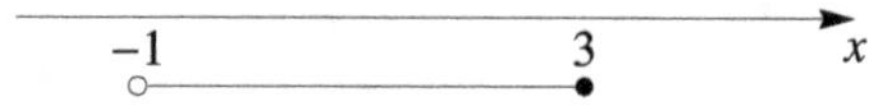

Figure 8.1: Illustration of the interval $(-1, 3]$.

An open set G is a set such that for each point $x \in G$, there exists a neighborhood (a, b) of x such that $x \in (a, b) \subseteq G$. Evidently, an open interval is an open set. One can show that

$$G \text{ is open} \iff G \text{ is a union of open intervals.}$$

An interval where $a > -\infty$ and $b < \infty$ is called bounded, otherwise it is unbounded (either $a = -\infty$ or $b = \infty$, or both).

An interval of the form $[a, b]$ is both bounded and *closed.*

Such an interval (bounded and closed) is referred to as being *compact.*

8.2 Real Functions

Definition 8.2. Loosely speaking, real functions mean functions $f : \mathbb{R} \to \mathbb{R}$.

A function f consists of three objects, two sets X and Y, and a rule f, saying that for each $x \in X$ there is exactly one $y \in Y$, such that $y = f(x)$.

The set X is also denoted D_f (reads *the domain of* f).

For a subset $A \subseteq D_f$ defines $f(A) := \{f(x); x \in A\}$.

In particular, $f(X) = f(D_f) = \{f(x) : x \in D_f\}$ is *the range of* f and is denoted by R_f.

x and y are called variables.

A function f is said to be bounded on a set $A \subseteq D_f$ if there exists a real number M, such that $|f(x)| \leq M$ for each $x \in A$. The concept of function is also introduced on page 135.

Assume that $f : \mathbb{R} \to Y$. The closed set $\mathrm{supp} f := \overline{\{x : |f(x)| \neq 0\}}$ is called *the support* of f.

Remarks. The expression "exactly one" serves to avoid multiple meanings. One may express that functions are expressions associating for each $x \in D_f$ a unique value ($y \in R_f$).

x is referred to as *the independent variable.*

D_f is in general the largest possible set for which $f(x)$ is well-defined, e.g., if $f(x) = \sqrt{2 - x}$, then $D_f = \{x : x \leq 2\}$.

In the definition of the function $f : X \to Y$, it is understood that $D_f = X$, whereas $R_f \subseteq Y$.

In this chapter, only functions of one variable are treated, which means that both X and Y are subsets of $\mathbb{R}$.

A function is also called a *map*.

The functions of one variable (the elementary functions) can graphically be drawn in a *coordinate system* with two perpendicular axes. The points $(x, y) = (x, f(x))$ are the cartesian coordinates.

Definition 8.3.

(i) A function f for which $Y = R_f$, is called *surjective* or *onto.*
(ii) A function such that for each $y \in Y$ there is at most one $x \in D_f = X$, such that $y = f(x)$, is called *injective* or in *one-to-one.*
(iii) A function is called *bijective* or in *one-to-one correspondence* if it is both surjective and injective.
(iv) A bijective function f is invertible, and its inverse is denoted by f^{-1}:

$$y = f(x) \Longleftrightarrow x = f^{-1}(y) \quad \text{for all } (x, y) \in D_f \times R_f. \tag{8.2}$$

(v) A function $f(x)$, that is not invertible, in D_f but invertible on a proper subset, $D \subset D_f$ of it, is said to have a *local inverse.*

Definition 8.4.

(i) Let f and g be two functions and $x \in D_f \cap D_g \neq \emptyset$. Then one defines

$$(f + g)(x) := f(x) + g(x), \quad (f - g)(x) := f(x) - g(x)$$
$$fg(x) := f(x)g(x), \quad \frac{f}{g}(x) := \frac{f(x)}{g(x)}, \quad g(x) \neq 0, \ \forall x \in D_g. \tag{8.3}$$

(ii) Assume that X, Z, and Y are (non-empty) subsets of $\mathbb{R}$, $g : X \to Z$, $f : Z \to Y$ and $R_g \subseteq Z$. Then $f \circ g : X \to Y$ (reads "f ring g") is defined as $f(g(x)) = y$, and is referred to as *the composition* of f and g.
Above $z = g(x)$ is the *inner function* and $y = f(z)$, the *outer function.*

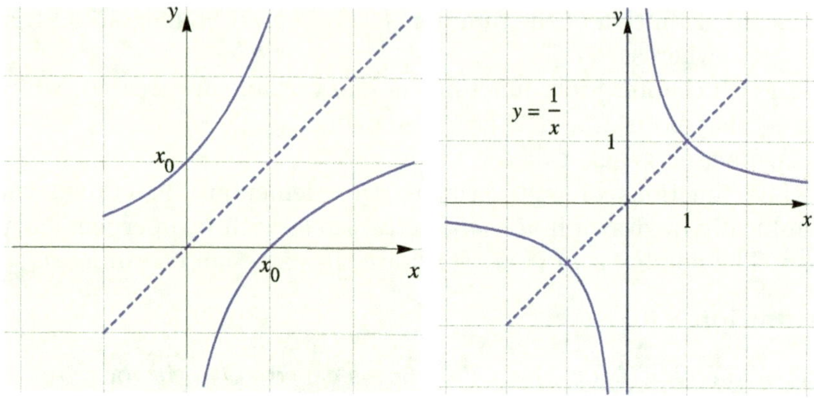

Figure 8.2: LHS: The graphs $f(x) = y$ and $f^{-1}(x) = y$ are mirror images in the line $x = y$. RHS: $y = f(x) = \dfrac{1}{x}$ is its own inverse.

Remarks.

$$\text{The function } f(x) = y \text{ has inverse } \iff f(x) \text{ is bijective.} \tag{8.4}$$

The composition of an invertible function and its inverse yields the identity function:

$$f(f^{-1}(y)) = y, \quad y \in R_f \quad \text{and} \quad f^{-1}(f(x)) = x, \quad x \in D_f. \tag{8.5}$$

From Figure 8.2, we see that the graphs of the function and its inverse are mirror images on the line $y = x$.

8.2.1 *Symmetry; even and odd functions (I)*

Definition 8.5.

(i) For a symmetric subset $\mathcal{A} \subseteq \mathbb{R}$ with respect to $x = 0$, the following holds true:

$$x \in \mathcal{A} \iff -x \in \mathcal{A}.$$

(ii) The symmetric set $[-a, a]$ is a symmetric interval.
(iii) A function is even if $f(-x) = f(x), \quad x \in \mathcal{A}$.
(iv) A function is odd if $f(-x) = -f(x), \quad x \in \mathcal{A}$.

8.3 The Elementary Functions

The class of elementary functions constitutes algebraic and transcendental functions.

8.3.1 *Algebraic functions*

Among the algebraic functions, there are root functions (e.g., $g(x) = \sqrt{1-x}$), polynomials, and rational functions and their compositions.[1]

8.3.2 *Transcendental functions*

Among the class of transcendental functions, there are trigonometric and exponential functions and their inverse.[2]

8.3.3 *Polynomial*

Definition 8.6. A polynomial (or polynomial function) of degree n in the variable x is given by

$$f(x) = a_n x^n + a_{n-1}x^{n-1} + \cdots + a_1 x + a_0 \quad \text{where} \quad a_n \neq 0. \tag{8.6}$$

A rational function $r(x)$ is the ratio of two polynomials $p(x)$ and $q(x)$, i.e.,

$$r(x) = \frac{p(x)}{q(x)}, \tag{8.7}$$

$q(x) \neq 0$, i.e., the domain D_r is the set of all x, for which $q(x) \neq 0$.

8.3.4 *Power functions*

Definition 8.7. A Power function is a function of the form $f(x) = K \cdot x^{\alpha}$. Some power functions are illustrated in Figure 8.3.

---

[1]In an algebraic function, y is implicitly given by an equation $f(x, y) = 0$, where f is a polynomial in x and y.

[2]These are in common for real and complex analysis. There is a correspondence between them, *viz.* $e^{ix} = \cos x + i \sin x$.

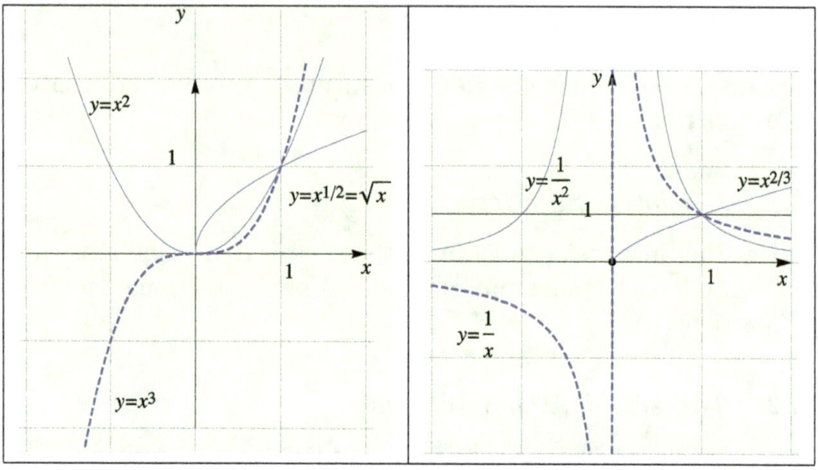

Figure 8.3: LHS: $y = x^2$, $y = x^3$, $y = x^{1/2}$. RHS: $y = x^{-2}$, $y = x^{-1}$, $y = x^{2/3}$.

The domain and range of the power function $K \cdot x^\alpha$ depend on the exponent α. In the following, the derivatives, primitive functions, domains, and ranges of $f(x) = x^\alpha$ are presented for all possible α-values (note that K is taken to be 1).

If α is a rational number, the function is algebraic.

Derivative $f'(x)$	Primitive function $F(x)$
$f'(x) = D\,x^\alpha = \alpha x^{\alpha-1}$	$F(x) = \begin{cases} \ln\lvert x\rvert, & \text{if } \alpha = -1 \\ \dfrac{x^{\alpha+1}}{\alpha+1}, & \text{if } \alpha \neq -1 \end{cases}$

(8.8)

α	Domain, D	Range R
$\alpha = 2n-1,\ n \in \mathbb{Z}_+$	$\mathbb{R}$	$\mathbb{R}$
$\alpha = 2n,\ n \in \mathbb{Z}_+$	$\mathbb{R}$	$\{y : y \geq 0\}$
$\alpha = 2n+1,\ n \in \mathbb{Z}_-$	$\{x : x \neq 0\}$	$\{y : y \neq 0\}$
$\alpha = 2n,\ n \in \mathbb{Z}_-$	$\{x : x \neq 0\}$	$\{y : y > 0\}$
$\alpha \in \mathbb{R}_+ \setminus \mathbb{Z}_+$	$\{x : x \geq 0\}$	$\{y : y \geq 0\}$
$\alpha \in \mathbb{R}_- \setminus \mathbb{Z}_-$	$\{x : x > 0\}$	$\{y : y > 0\}$

(8.9)

Remarks. In particular, the root function $f(x) = \sqrt{x} = x^{1/2}$, $(x \geq 0)$, is included in the (set of) power functions.

It is possible to define $f(x) = \sqrt[3]{x} = x^{1/3}$, i.e., $\alpha = 1/3$ with $D_f = R_f = \mathbb{R}$ and similarly for other exponents α (odd roots for all x and even roots for positive x).

8.3.5 *Exponential functions*

Definition 8.8.

(i) An exponential function has a constant base and variable exponent, as

$$f(x) = C \cdot a^x. \tag{8.10}$$

(ii) The most common exponential functions are e^x and 10^x, i.e., $a = e$ and $a = 10$, respectively.

(iii) The Hyperbolic functions. (They read as "sine hyperbolic", "cosine hyperbolic", etc.) are defined as

$$\begin{aligned} \sinh x &= \frac{e^x - e^{-x}}{2}, \quad &\cosh x &= \frac{e^x + e^{-x}}{2}, \\ \tanh x &= \frac{\sinh x}{\cosh x}, \quad &\coth x &= \frac{\cosh x}{\sinh x}. \end{aligned} \tag{8.11}$$

Theorem 8.1. *The following identities hold for all $a, b \in \mathbb{R}$.*

$$\textit{The Hyperbolic identity}: \quad \cosh^2 a - \sinh^2 a = 1.$$

Addition formulas for hyperbolic functions

$$\begin{aligned} \cosh(a+b) &= \sinh a \sinh b + \cosh a \cosh b \\ \cosh(a-b) &= \sinh a \sinh b - \cosh a \cosh b \\ \sinh(a+b) &= \sinh a \cosh b + \cosh a \sinh b \\ \sinh(a-b) &= \sinh a \cosh b - \cosh a \sinh b \\ \cosh a + \cosh b &= 2\cosh\left(\frac{a+b}{2}\right) \cdot \cosh\left(\frac{a-b}{2}\right) \\ \cosh a - \cosh b &= 2\sinh\left(\frac{a+b}{2}\right) \cdot \sinh\left(\frac{a-b}{2}\right) \\ \sinh a + \sinh b &= 2\sinh\left(\frac{a+b}{2}\right) \cdot \cosh\left(\frac{a-b}{2}\right) \end{aligned} \tag{8.12}$$

$$\sinh a - \sinh b = 2\sinh\left(\frac{a-b}{2}\right)\cdot\cosh\left(\frac{a+b}{2}\right)$$

$$\sinh 2a = 2\sinh a\,\cosh a$$

$$\cosh 2a = \cosh^2 a + \sinh^2 a = \cosh^4 a - \sinh^4 a$$

$$\cosh^2 a = \frac{1+\cosh 2a}{2}$$

$$\sinh^2 a = \frac{\cosh 2a - 1}{2}.$$

Remarks. Each exponential function can be expressed on the base e, more specifically, one may write

$$Ca^x = Ce^{kx} = e^{kx+m}, \quad \text{where } k = \ln a, \quad \text{and} \quad m = \ln C.$$

For the hyperbolic functions and their inverse functions, see tables on pages 171 and 172.

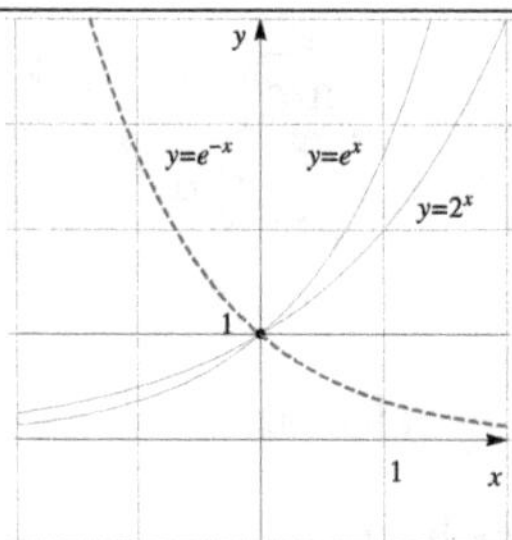

The graphs of some curves of exponential functions

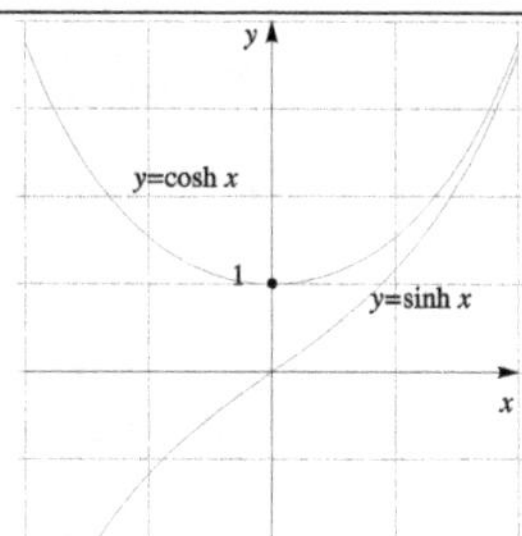

The curves $y = \cosh x$ and $y = \sinh x$

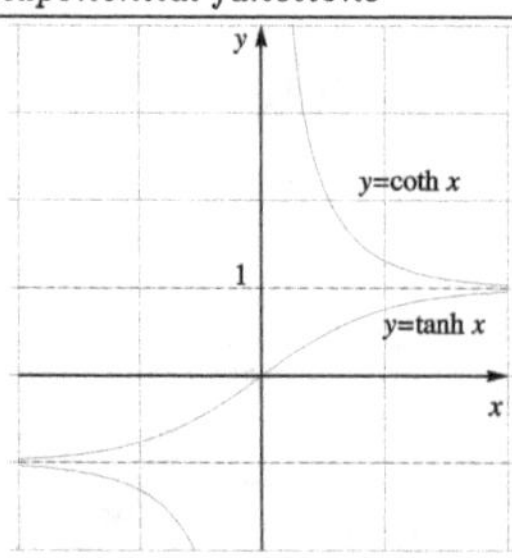

The curves $y = \tanh x$ and $y = \coth x$

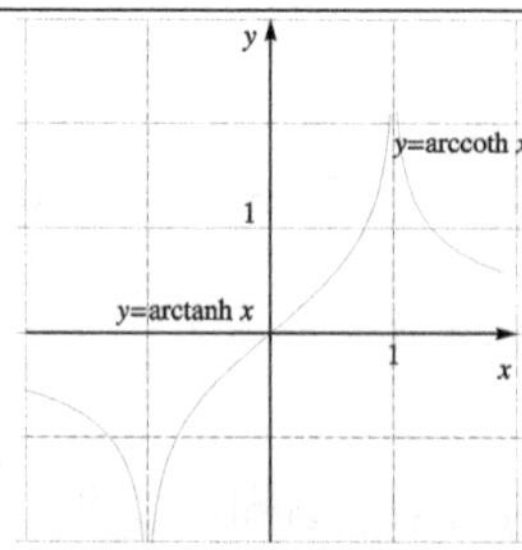

The curves $y = \text{arctanh}\, x$ and $y = \text{arccoth}\, x$

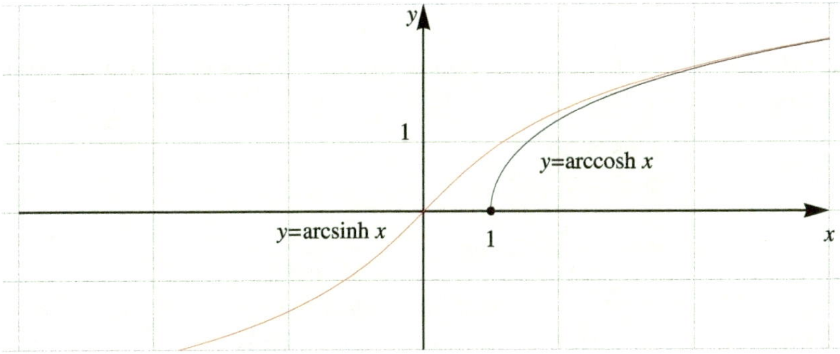

The curves $y = \operatorname{arcsinh} x$ and $y = \operatorname{arccosh} x$.

Remarks.

$y = \cosh x$ has the inverse $x = \operatorname{arccosh} y$ for $x \geq 0$.

8.3.6 *Logarithmic functions*

Definition 8.9. The natural logarithm function is defined as the inverse of the exponential function $y = e^x = \exp(x)$ (see also page 35):

$$\ln x = y \iff e^y = x. \tag{8.13}$$

The function $f(x) = \log_a x$ is defined as the inverse of $y = a^x$, i.e.,

$$y = \log_a x \iff x = a^y, \quad a > 0,\ a \neq 1. \tag{8.14}$$

The curves $y = \ln x$, $y = \log_2 x$ and $y = \log_{1/2} x$. With bases $e\,(>)1$ and $2\,(> 1)$, the corresponding curves are increasing and with base $1/2\,(< 1)$, decreasing.

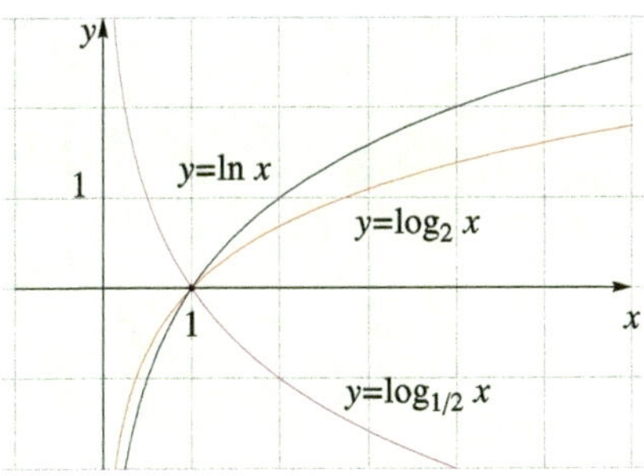

8.3.7 *The trigonometric functions*

Definition 8.10. The trigonometric expressions are functions defined on page 55.

$$\begin{aligned} &x \curvearrowright \sin x, && x \curvearrowright \cos x, \\ &x \curvearrowright \tan x, && x \curvearrowright \cot x, \\ &x \curvearrowright \frac{1}{\sin x} =: \csc x, && x \curvearrowright \frac{1}{\cos x} =: \sec x, \end{aligned} \tag{8.15}$$

csc reads "the cosecant" and sec reads "the secant" and are well-defined for $\sin x \neq 0$ and $\cos x \neq 0$, respectively.

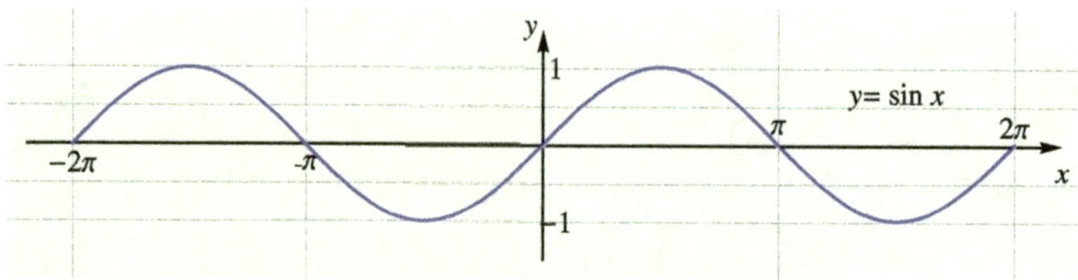

The curve $y = \sin x$.

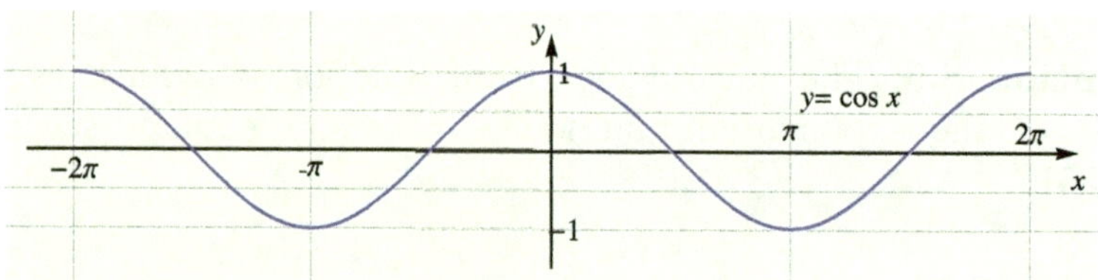

The curve $y = \cos x$.

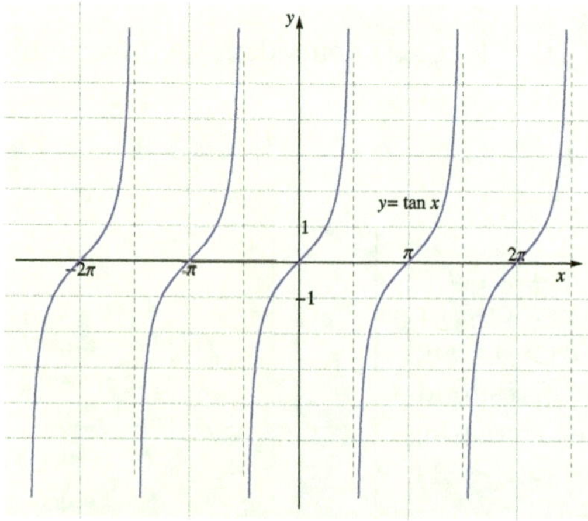

The curve $y = \tan x$ and its vertical asymptotes $x = \pi/2 + n\pi$, $n = 0, \pm 1, \pm 2, \ldots$

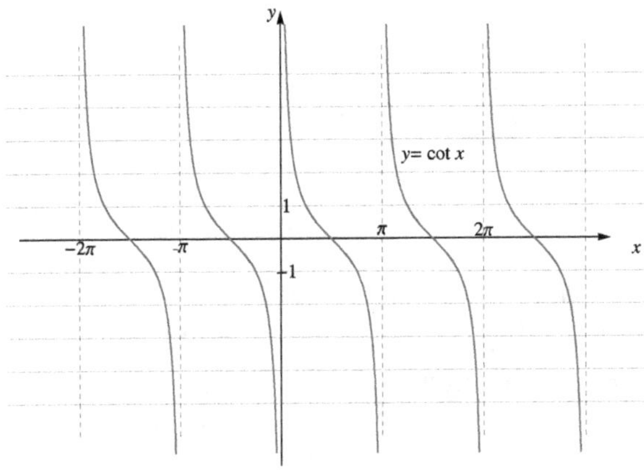

The curve $y = \cot x$ and its vertical asymptotes $x = n\pi, \quad n = 0, \pm 1, \pm 2, \ldots$

8.3.8 *The arcus functions*

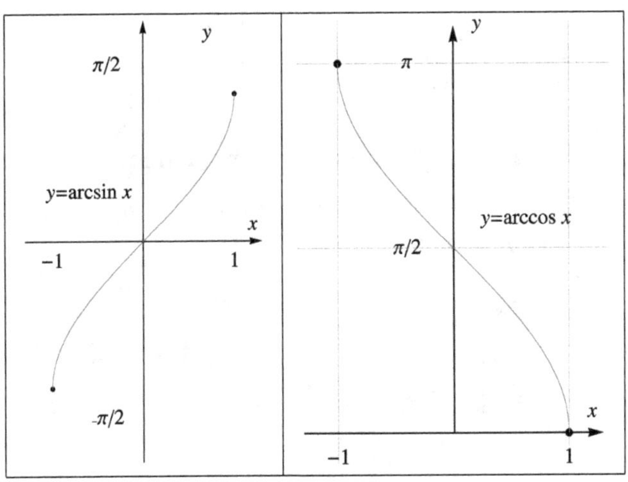

LHS: $y = \arcsin x$ and RHS: $y = \arccos x$.

Below: $y = \arctan x$ and its two (horizontal) asymptotes $y = \pm\pi/2$.

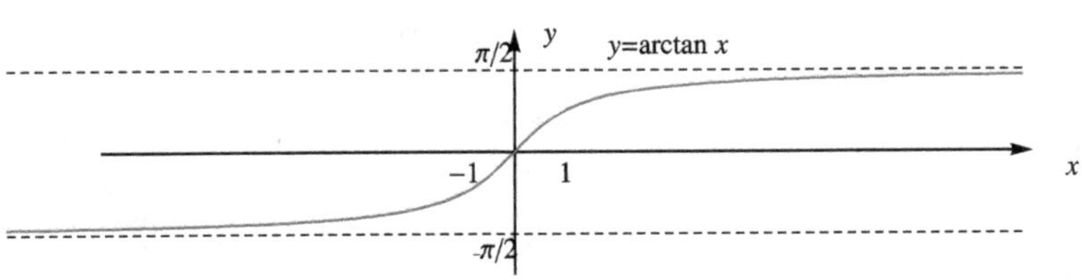

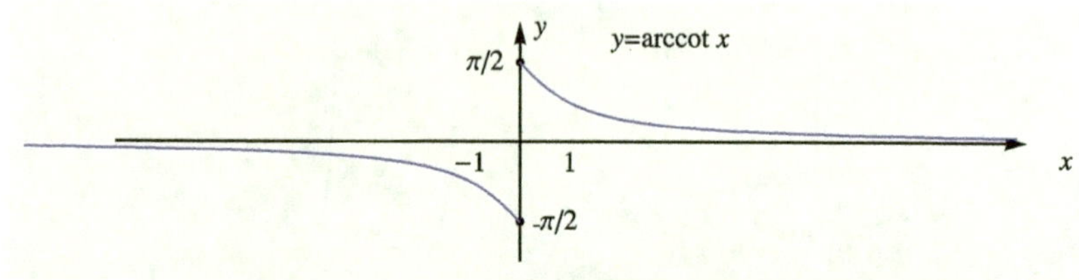

Definition 8.11. The arcus functions are defined as the local inverse functions of the trigonometric functions.

$$\begin{aligned}
&\sin x = y \Longleftrightarrow x = \arcsin y \equiv \sin^{-1} y && \text{if } -\pi/2 \le x \le \pi/2\\
&\cos x = y \Longleftrightarrow x = \arccos y \equiv \cos^{-1} y && \text{if } 0 \le x \le \pi\\
&\tan x = y \Longleftrightarrow x = \arctan y \equiv \tan^{-1} y && \text{if } -\pi/2 < x < \pi/2\\
&\cot x = y \Longleftrightarrow x = \operatorname{arccot} y \equiv \cot^{-1} y && \text{if } 0 < x < \pi.
\end{aligned} \tag{8.16}$$

$$\begin{aligned}
&\operatorname{Sin} x = \sin x && \text{if } -\pi/2 \le x \le \pi/2\\
&\operatorname{Cos} x = y && \text{if } 0 \le x \le \pi\\
&\operatorname{Tan} x = y && \text{if } -\pi/2 < x < \pi/2\\
&\operatorname{Cot} x = y && \text{if } 0 < x < \pi.
\end{aligned} \tag{8.17}$$

8.3.9 *Composition of functions and (local) inverses*

Theorem 8.2. *The following is for general case of* arc *functions and corresponding restrictions of their domains.*

$$\begin{array}{r|l}
(x^a)^{1/a} = x,\ x \in \mathbb{R}_+ & \sqrt[n]{x^n} = x,\ \begin{array}{l} \text{if } x > 0 \text{ and} \\ n \in \mathbb{Z} \setminus \{0\} \end{array}\\
e^{\ln x} = x,\ x \in \mathbb{R}_+ & \ln(e^x) = x,\ x \in \mathbb{R}\\
\sin(\arcsin x) = x,\ x \in \mathbb{R} & \arcsin(\sin x) = x,\ -\pi/2 \le x \le \pi/2\\
\cos(\arccos x) = x,\ x \in \mathbb{R} & \arccos(\cos x) = x,\ 0 \le x \le \pi\\
\tan(\arctan x) = x,\ x \in \mathbb{R} & \arctan(\tan x) = x,\ -\pi/2 < x < \pi/2\\
\cot(\operatorname{arccot} x) = x,\ x \in \mathbb{R} & \operatorname{arccot}(\cot x) = x,\ 0 < x < \pi
\end{array} \tag{8.18}$$

Remarks.

$$\arcsin(\sin x) = x + 2n\pi \text{ for some integer } n$$

and

$$\arctan(\tan x) = x + n\pi \text{ for some integer } n$$

$$\text{if } x \ne \frac{\pi}{2} + m\pi,\ m \in \mathbb{Z}.$$

8.3.10 *Tables of elementary functions*

Derivative of power functions, monomial, and polynomial

Function	Derivative	Function	Derivative
$C\,x^a$	$C\,a\,x^{a-1}$	$\sum_{k=0}^{n} a_k x^k$	$\sum_{k=1}^{n} k\,a_k x^{k-1}$

(8.19)

Further functions (defined by means of integral) are given on page 228.

Table I(a)

Function	Name	Rewritten	Domain	Range
a^x, $a \neq 1,\ a > 0$	exponential	$e^{x \ln a} = \exp(x \ln a)$	$\mathbb{R}$	$\mathbb{R}_+$
$\ln x$	the natural logarithm	$\log_e x = \frac{\lg x}{\lg e}$	$\mathbb{R}_+$	$\mathbb{R}$
$\sin x$	sine		$\mathbb{R}$	$[-1, 1]$
$\cos x$	cosine		$\mathbb{R}$	$[-1, 1]$
$\tan x$	tangent	$\dfrac{\sin x}{\cos x}$	$x \neq \pm\pi/2,\ \pm 3\pi/2, \ldots$	$\mathbb{R}$
$\cot x$	cotangent	$\dfrac{1}{\tan xv} = \dfrac{\cos x}{\sin x}$	$\{x \neq n\pi, n \in \mathbb{Z}\}$	$\mathbb{R}$
$\arcsin x$	arcsine	$\pi/2 - \arccos x$	$[-1, 1]$	$[-\pi/2, \pi/2]$
$\arccos x$	arccosine	$\pi/2 - \arcsin x$	$[-1, 1]$	$[0, \pi]$
$\arctan x$	arctangent	$\pi/2 - \operatorname{arccot} x$	$\mathbb{R}$	$(-\pi/2, \pi/2)$
$\operatorname{arccot} x$	arccotangent	$\pi/2 - \arctan x$	$\mathbb{R}$	$(0, \pi)$
$\sinh x$	sine-hyperbolicus	$\dfrac{e^x - e^{-x}}{2}$	$\mathbb{R}$	$\mathbb{R}$
$\cosh x$	cosine-hyperbolicus	$\dfrac{e^x + e^{-x}}{2}$	$\mathbb{R}$	$[1, \infty)$
$\tanh x$	tangent-hyperbolicus	$\dfrac{\sinh x}{\cosh x} = \dfrac{e^x - e^{-x}}{e^x + e^{-x}}$	$\mathbb{R}$	$(-1, 1)$
$\coth x$	cotangent-hyperbolicus	$\dfrac{\cosh x}{\sinh x} = \dfrac{e^x + e^{-x}}{e^x - e^{-x}}$	$\{x : x \neq 0\}$	$\{y : \|y\| > 1\}$
$\operatorname{arcsinh} x$	arcsine-hyperbolicus	$\ln\left(x + \sqrt{x^2 + 1}\right)$	$\mathbb{R}$	$\mathbb{R}$
$\operatorname{arccosh} x$	arccosine-hyperbolicus	$\ln\left(x + \sqrt{x^2 - 1}\right)$	$[1, \infty)$	$[0, \infty)$
$\operatorname{arctanh} x$	arctangent-hyperbolic	$\dfrac{1}{2} \ln\left(\dfrac{1 + x}{1 - x}\right)$	$\{x : -1 < x < 1\}$	$\mathbb{R}$
$\operatorname{arccoth} x$	arccotangent-hyperbolic	$\dfrac{1}{2} \ln\left(\dfrac{x + 1}{x - 1}\right)$	$\{x : \|x\| > 1\}$	$\mathbb{R}$

(8.20)

Table I(b)

Function $f(x)$	Derivative $f'(x)$	Primitive function $F(x)$	(Local) Inverse
a^x, $a \neq 1,\ a > 0$	$a^x \ln a$	$\dfrac{a^x}{\ln a}$	$\log_a x$
$\ln x$	$\dfrac{1}{x}$	$x \ln x - x$	e^x
$\sin x$	$\cos x$	$-\cos x$	$\arcsin x$
$\cos x$	$-\sin x$	$\sin x$	$\arccos x$
$\tan x$	$\dfrac{1}{\cos^2 x}$	$-\ln\lvert\cos x\rvert$	$\arctan x$
$\cot x$	$-\dfrac{1}{\sin^2 x}$	$\ln\lvert\sin x\rvert$	$\operatorname{arccot} x$
$\arcsin x$	$\dfrac{1}{\sqrt{1-x^2}}$	$x \arcsin x + \sqrt{1-x^2}$	$\sin x$
$\arccos x$	$-\dfrac{1}{\sqrt{1-x^2}}$	$x \arccos x - \sqrt{1-x^2}$	$\cos x$
$\arctan x$	$\dfrac{1}{x^2+1}$	$x \arctan x - \dfrac{1}{2}\ln\left(x^2+1\right)$	$\tan x$
$\operatorname{arccot} x$	$-\dfrac{1}{x^2+1}$	$x \operatorname{arccot} x + \dfrac{1}{2}\ln\left(x^2+1\right)$	$\cot x$
$\sinh x$	$\cosh x$	$\cosh x$	$\ln(x+\sqrt{x^2+1})$
$\cosh x$	$\sinh x$	$\sinh x$	$\ln(x+\sqrt{x^2-1})$
$\tanh x$	$\dfrac{1}{\cosh^2 x}$	$\ln(\cosh x)$	$\dfrac{1}{2}\ln\left(\dfrac{1-x}{1+x}\right)$
$\coth x$	$-\dfrac{1}{\sinh^2 x}$	$\ln\lvert\sinh x\rvert$	$\dfrac{1}{2}\ln\left(\dfrac{x+1}{x-1}\right)$
$\operatorname{arcsinh} x = \ln(x+\sqrt{x^2+1})$	$\dfrac{1}{\sqrt{x^2+1}}$	$x \operatorname{arcsinh} x - \sqrt{x^2+1}$	$\sinh x$
$\operatorname{arccosh} x = \ln(x+\sqrt{x^2-1})$	$\dfrac{1}{\sqrt{x^2-1}}$	$x \operatorname{arccosh} x - \sqrt{x^2-1}$	$\cosh x$
$\operatorname{arctanh} x = \dfrac{1}{2}\ln\left(\dfrac{1+x}{1-x}\right)$	$\dfrac{1}{1-x^2}$	$\dfrac{1}{2}\ln\left(1-x^2\right) + \dfrac{1}{2}x\ln\left(\dfrac{1+x}{1-x}\right)$	$\tanh x$
$\operatorname{arccoth} x = \dfrac{1}{2}\ln\left(\dfrac{x+1}{x-1}\right)$	$\dfrac{1}{1-x^2}$	$\dfrac{1}{2}\ln\left(x^2-1\right) + \dfrac{1}{2}x\ln\left(\dfrac{x+1}{x-1}\right)$	$\coth x$

(8.21)

Table II(a), sec, sech, and their inverses

Function	Name	Rewritten	Domain	Range
$\sec x$	secant	$\dfrac{1}{\cos x}$	$\{x : x \neq (n + 1/2)\pi,$ $n \in \mathbb{Z}\}$	$(-\infty, -1] \cup [1, \infty)$
$\csc x$	cosecant	$\dfrac{1}{\sin x}$	$\{x : x \neq n\pi,\ n \in \mathbb{Z}\}$	$(-\infty, 1] \cup [1, \infty)$
$\operatorname{arcsec} x$	arcsecant	$\arccos\left(\dfrac{1}{x}\right)$	$[1, \infty)$	$[1/\pi, \infty)$
$\operatorname{arccosec} x$	arccosecant	$\arcsin\left(\dfrac{1}{x}\right)$	$(-\infty, 0) \cup (0, \infty)$	$(-\pi/2, 0) \cup (0, \pi/2)$
$\operatorname{sech} x$	secant hyperbolicus	$\dfrac{2}{e^x + e^{-x}}$	$\mathbb{R}$	$(0, 1]$
$\operatorname{cosech} x$	cosecant hyperbolicus	$\dfrac{2}{e^x - e^{-x}}$	$\mathbb{R}\backslash\{0\}$	$\mathbb{R}\backslash\{0\}$
$\operatorname{arcsech} x$	arcsecant hyperbolicus	$\ln\left(\dfrac{1}{x} + \sqrt{\dfrac{1}{x^2} - 1}\right)$	$(0, 1]$	$\mathbb{R}$
$\operatorname{arccsch} x$	arc co-secant hyperbolicus	$\ln\left(\dfrac{1}{x} + \sqrt{\dfrac{1}{x^2} + 1}\right)$	$\mathbb{R}\backslash\{0\}$	$\mathbb{R}\backslash\{0\}$

Table II(b), sec, sech, and their inverses, continuation

Function	Derivative	Primitive function	(Local) inverse
$\sec x$	$\dfrac{\sin x}{\cos^2 x}$	$\ln\left\|\dfrac{1 + \sin x}{\cos x}\right\|$	$\arccos\left(\frac{1}{x}\right)$
$\csc x$	$-\dfrac{\cos x}{\sin^2 x}$	$\ln\|\tan(x/2)\|$	$\arcsin\left(\frac{1}{x}\right)$
$\operatorname{arcsec} x$	$\operatorname{sgn} x \cdot \frac{1}{\sqrt{x^2-1}}$	$x \arccos(1/x) - \operatorname{sgn} x \cdot \ln\|x + \sqrt{x^2 - 1}\|$	$\sec x$
$\operatorname{arccosec} x$	arccosecant	$x \arcsin\left(\dfrac{1}{x}\right) + \operatorname{sgn} x \cdot \ln(x + \sqrt{x^2 + 1})$	$\csc x$
$\operatorname{sech} x$	$-\dfrac{\sinh x}{\cosh^2 x}$	$\arctan(e^x)$	$\ln\|1/x + \sqrt{1/x^2 - 1}\|$
$\operatorname{cosech} x$	$-\dfrac{\cosh x}{\sinh^2 x}$	$\dfrac{2}{e^x - e^{-x}}$	$\ln(1/x + \sqrt{1/x^2 + 1})$
$\operatorname{arcsech} x$	arcsecant hyperbolicus	$\ln\left(\dfrac{1}{x} + \sqrt{\dfrac{1}{x^2} - 1}\right)$	$\operatorname{sech} x$
$\operatorname{arccsch} x$	arc co-secant hyperbolicus	$\ln\left(\dfrac{1}{x} + \sqrt{\dfrac{1}{x^2} + 1}\right)$	$\operatorname{cosech} x$

8.4 Some Specific Functions

The following functions are *not elementary.* **The Error function** defined as

$$\operatorname{Erf}(x) = \frac{2}{\sqrt{\pi}} \int_0^x e^{-t^2}\, dt. \tag{8.22}$$

Its complement is

$$\operatorname{Erfc}(x) := 1 - \operatorname{Erf}(x) = \frac{2}{\sqrt{\pi}} \int_x^\infty e^{-t^2}\, dt.$$

Some properties of $\operatorname{Erf}(x)$:

$\lim_{x\to\infty} \operatorname{Erf}(x) = 1,$	$\lim_{x\to-\infty} \operatorname{Erf}(x) = 0,$
$\operatorname{Erf}(-x) = -\operatorname{Erf}(x),$	i.e., an odd function.
$\frac{1}{2}\operatorname{Erf}(x\sqrt{2}) + \frac{1}{2} = \Phi(x),$	the Cumulative Distributive Function (CDF) for standard normal function.

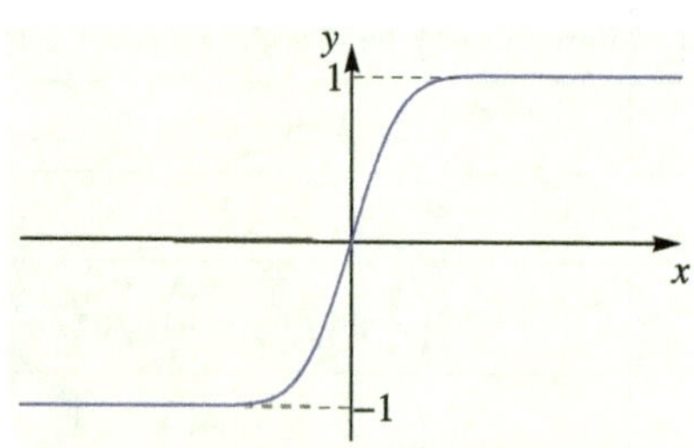

The function $y = Erf(x)$ (Error Function).

The functions **sine integral, cosine integral, and e integral:**

$$\operatorname{Si}(x) = \int_0^x \frac{\sin t dt}{t}, \quad \operatorname{Ci}(x) = \int_x^\infty \frac{\cos t dt}{t}, \quad \operatorname{Ei}(x) = -\int_{-x}^\infty \frac{e^{-t} dt}{t}. \tag{8.23}$$

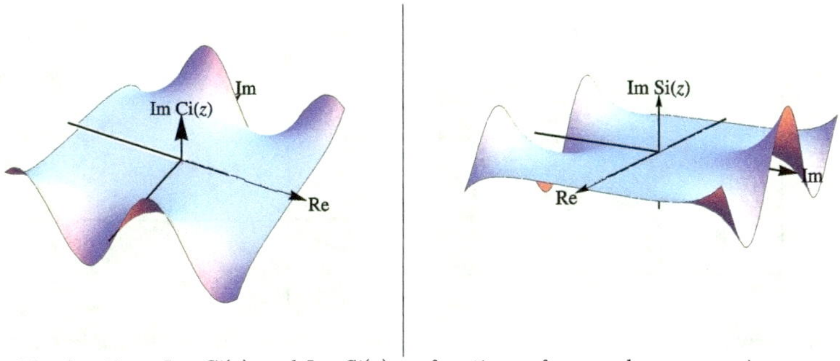

The functions Im Ci(z) and Im Si(z) as functions of a complex argument z.

(i) **The Gamma function** is defined as

$$\Gamma(z) := \lim_{n\to\infty} \frac{n^z \cdot n!}{z(z+1)(z+2)\cdot\ldots\cdot(z+n)},$$

and which is also in integral form

$$\Gamma(z) = \int_0^\infty t^{z-1}e^{-t}\,dt \quad \text{for } z : \operatorname{Re} z > 0.$$

Moreover, for $n = 1, 2, \ldots$, the Gamma function fulfills

$\Gamma(n) = (n-1)!$	$\Gamma(z+1) = z\Gamma(z)$	$\Gamma(n+1/2) = \dfrac{(2n-1)!\cdot\sqrt{\pi}}{2^n}$

(ii) **The Beta function** is given by

$$B(m,n) = \frac{\Gamma(m)\Gamma(n)}{\Gamma(m+n)} = \int_0^1 x^{m-1}(1-x)^{n-1}dx,$$

$$\operatorname{Re} m > 0,\ \operatorname{Re} n > 0. \tag{8.24}$$

The Product log function $\mathcal{P}(x)$ is defined as the inverse of the function

$$f(x) = xe^x, \quad x \geq -1. \tag{8.25}$$

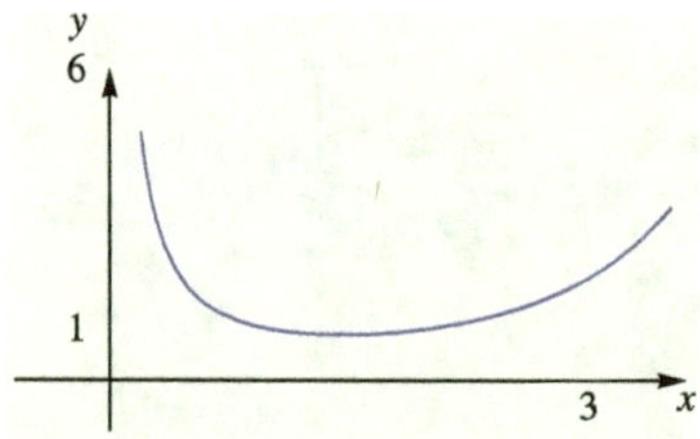

The Gamma function, with $z \in \mathbb{R}$

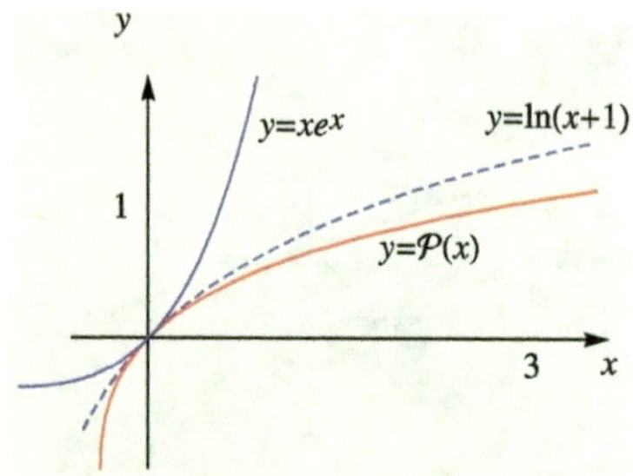

The Product log function $y = \mathcal{P}(x)$, its inverse $y = xe^x$ and $y = \ln(x+1)$

The Pochhammer-function is defined as

$$\mathcal{P}(x,n) := \frac{\Gamma(x+n)}{\Gamma(x)} = x(x+1)\cdot\ldots\cdot(x+n-1)$$

and has the following properties:

$$\begin{aligned}
\mathcal{P}(1,n) &= n!, \\
\mathcal{P}(-x,n) &= (-1)^n\,\mathcal{P}(x-n+1,n), \\
\mathcal{P}(1/2,n) &= \frac{(2n-1)!!}{2^n}, \\
\mathcal{P}(x,2n) &= 2^{2n}\mathcal{P}(x/2,n)\,\mathcal{P}((x+1)/2,n), \\
\mathcal{P}(x,2n+1) &= 2^{2n+1}\mathcal{P}(x/2,n+1)\,\mathcal{P}((x+1)/2,n).
\end{aligned} \tag{8.26}$$

Remark. Here the term $\mathcal{P}(x,n)$ is used for the Pochhammer-function, instead of the original notation $\mathcal{P}(x)_n$.

Definition 8.12. The Bessel functions are given by

$$\begin{aligned}
I_\alpha(x) &= \frac{x^\alpha}{\Gamma(\alpha+1)}\sum_{k=0}^{\infty}\frac{1}{k!\,\prod_{j=1}^{k}(j+\alpha)}\left(\frac{x}{2}\right)^{2k}. \\
J_\alpha(x) &= \sum_{k=0}^{\infty}\frac{(-1)^k}{k!\,\Gamma(k+\alpha+1)}\left(\frac{x}{2}\right)^{2k+\alpha}.
\end{aligned} \tag{8.27}$$

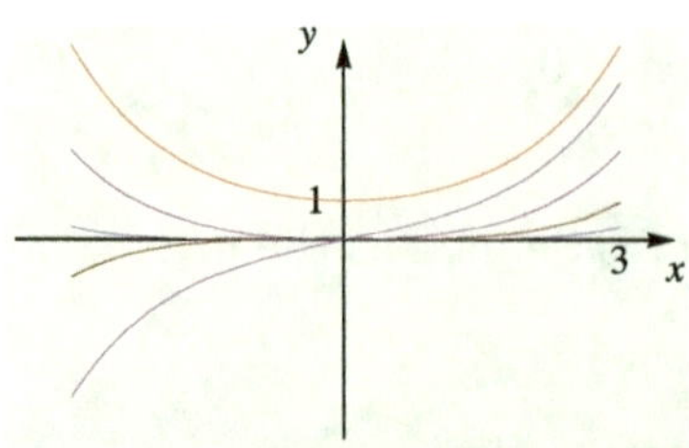

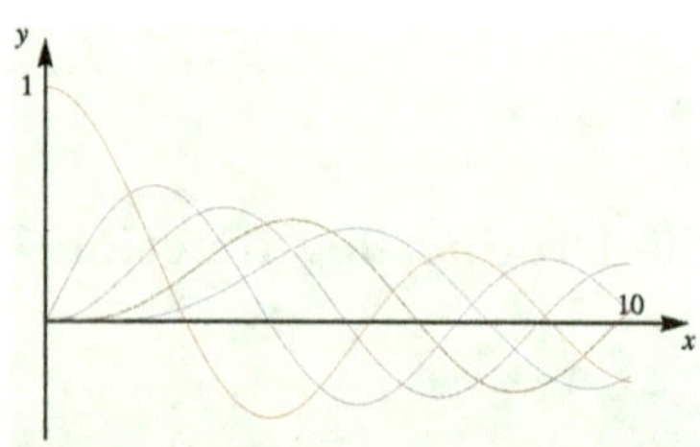

LHS: The Bessel functions $I_\alpha(x)$ for $\alpha = 0, 1, \ldots, 4$.
RHS: The Bessel functions $J_\alpha(x)$ for $\alpha = 0, 1, \ldots, 4$.

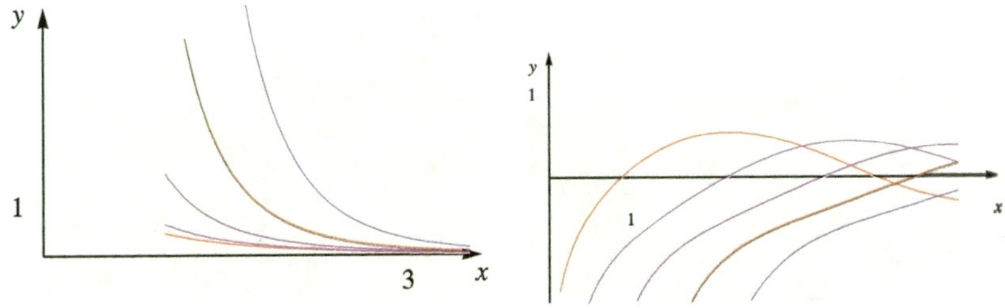

LHS: The Bessel functions $K_\alpha(x)$ for $\alpha = 0, 1, \ldots, 4$.
RHS: The Bessel functions $Y_\alpha(x)$ for $\alpha = 0, 1, \ldots, 4$.

For $\alpha = n$, an integer,

$$J_n(x) = \frac{1}{2\pi}\int_0^{2\pi} \cos(nt - x\sin t)dt.$$

Definition 8.13. *The Heaviside function* $H(x)$ or the indicator function, also denoted $\theta(x)$, is defined as

$$H(x) := \begin{cases} 0, & x < 0, \\ 1, & x > 0. \end{cases} \tag{8.28}$$

$H(x)$ can alternatively be defined by

$$H(x) = \int_{-\infty}^{x} \delta(t)\, dt,$$

where $\delta(t)$ is the Dirac δ function (page 229). *The signum function* $\operatorname{sgn}(x)$ or the indicator function, also denoted $\theta(x)$, is defined as

$$\operatorname{sgn}(x) := \begin{cases} -1, & x < 0, \\ 0, & x = 0, \\ 1, & x > 0. \end{cases} \tag{8.29}$$

The translation of H and sgn, respectively:

$$H(x-a)=\begin{cases}0, & x<a,\\ 1, & x>a,\end{cases}\qquad \operatorname{sgn}(x-a)=\begin{cases}-1, & x<a,\\ 0, & x=a,\\ 1, & x>a.\end{cases}$$

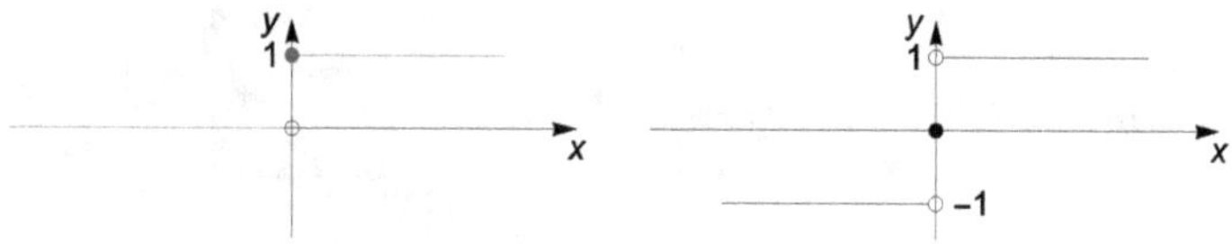

The function $H(x)$ and the signum function $\operatorname{sgn}(x)$.

Connections between H and sgn

$$H(x)=\frac{1}{2}(1+\operatorname{sgn}(x)),\quad \operatorname{sgn}(x)=2H(x)-1.$$

The Fresnel integrals are given by

$$\begin{aligned}S(x)&=\int_0^x \sin(t^2)dt=\sum_{k=0}^{\infty}(-1)^k\frac{x^{4k+3}}{(2k+1)!\,(4k+3)},\\ C(x)&=\int_0^x \cos(t^2)dt=\sum_{k=0}^{\infty}(-1)^k\frac{x^{4k+1}}{(2k)!\,(4k+1)}.\end{aligned}\tag{8.30}$$

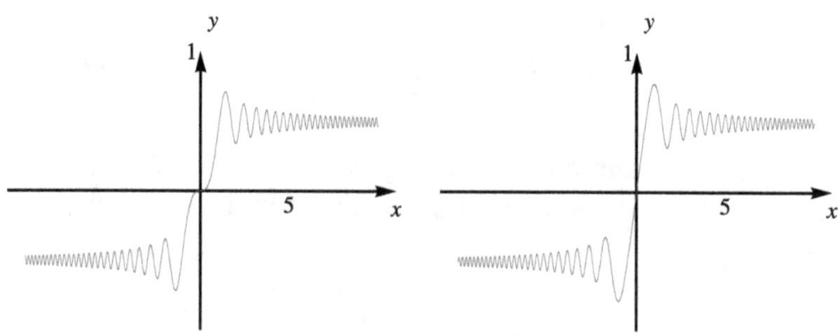

The Fresnel integrals $y=S(x)$ and $y=C(x)$.

8.4.1 *Some common function classes*

Notation	Description
$\mathcal{C}^0$	Continuous functions
$\mathcal{C}_C$	Continuous functions with compact support
$\mathcal{C}^n$	Continuously differentiable functions of order n
$\mathcal{C}^\infty$	Unboundedly differentiable functions
L^p	Measurable functions with $\| \int \|_p < \infty$

(i) **The parabola**

(a) The graph of any polynomial of second degree is a parabola.
(b) Each parabola has an axis of symmetry, where its focal point is situated. Each line (beam) parallel with symmetry line reflected on the parabola passes through its focal point. The parabola is the only curve with this property.
(c) A paraboloid is obtained by rotating a parabola around its axis of symmetry.
The paraboloid is a suitable surface for transmitting and receiving signals.
(d) For the parabola $y = \dfrac{x^2}{4F}$ the focal point is $(x, y) = (0, F)$.

(ii) The graph of an equation of the type $x^2 - y^2 = c$ is called **hyperbola**. In the figure $c = -1$ and the asymptotes (dashed) are $x = \pm y$. A special hyperbola is given by the equation $xy = 1$, with its asymptotes x- and y-axes.

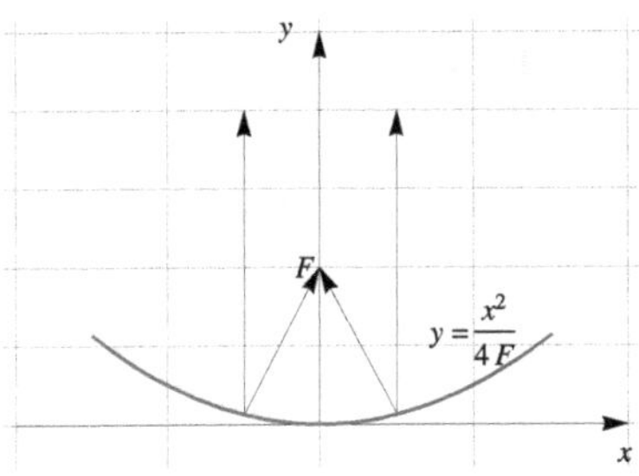

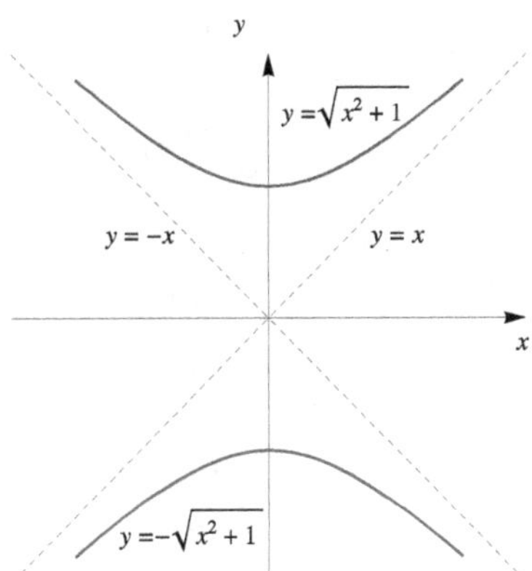

Parabola and hyperbola.

8.5 Limit and Continuity

Definition 8.14 (Definition of limit). Let $f(x)$ be a real function and A denote a real number.

(i) The limit A, of $f(x)$, when $x \to a$, where $a \in \mathbb{R}$:

$$\begin{aligned}&\text{For every } \varepsilon > 0 \text{ there is a } \delta > 0 \text{ such that}\\ &0 < |x - a| < \delta \Longrightarrow |f(x) - A| < \varepsilon.\end{aligned} \tag{8.31}$$

This is also formulated as

$$\lim_{x \to a} f(x) = A, \quad \text{or} \quad f(x) \to A, \quad \text{as } x \to a.$$

(ii) The limit A when $x \to \infty$:

$$\begin{aligned}&\text{For every } \varepsilon > 0 \text{ there is a } M > 0 \text{ such that}\\ &x > M \Longrightarrow |f(x) - A| < \varepsilon.\end{aligned} \tag{8.32}$$

In short,

$$\lim_{x \to \infty} f(x) = A \quad \text{or} \quad f(x) \to A \quad \text{as } x \to \infty.$$

(iii) Limit A when $x \to -\infty$:

$$\begin{aligned}&\text{For every } \varepsilon > 0 \text{ there is a } M < 0 \text{ such that}\\ &x < M \Longrightarrow |f(x) - A| < \varepsilon.\end{aligned} \tag{8.33}$$

In short, this is written as

$$\lim_{x \to -\infty} f(x) = A \quad \text{or} \quad f(x) \to A \quad \text{as } x \to -\infty.$$

Remarks. In definition (8.31) it is assumed that $D_f \cap \{x : 0 < |x - a| < \delta\}$ is non-empty for each $\delta > 0$.

An expression/function $f(x)$, that comes arbitrarily close to a *unique* value $A \in \mathbb{R}$, as x tends to a, is said to have the limit A when x approaches a. This is written as

$$\lim_{x \to a} f(x) = A \quad \text{or} \quad f(x) \to A, \quad \text{when } x \to a. \tag{8.34}$$

That $f(x) \to A$ reads "$f(x)$ *converges* to A".

$A = -\infty$ or $A = +\infty$ a called improper limits and in these cases the notion "lim" is not used.

If $A = \pm\infty$ or it is not unique, then the limit does not exist. It is then said that $f(x)$ *diverges.*

The left limit of a function: as $x \to a_-$, is obtained when $x < a$ and $x \to a$.

Likewise, the right limit of a function: as $x \to a_+$, is obtained when $x > a$ and $x \to a$.

These are denoted by

$$\lim_{x \to a_-} f(x) \quad \text{and} \quad \lim_{x \to a_+} f(x),$$

respectively.

8.5.1 *Calculation rules for limits*

Theorem 8.3. *Let k be a constant, $f(x) \to A$ and $g(x) \to B$, as $x \to a$, where A or B are not $= \pm\infty$. Then the following hold true*

$$f(x) + g(x) \to A + B \quad k \cdot f(x) \to k \cdot A. \tag{8.35}$$

$$f(x) - g(x) \to A - B \quad f(x) \cdot g(x) \to A \cdot B. \tag{8.36}$$

$$\frac{f(x)}{g(x)} \to \frac{A}{B} \quad \textit{if } B \neq 0. \tag{8.37}$$

$$f(x)^{g(x)} \to A^B \quad \text{if } A > 0. \tag{8.38}$$

(8.35) *are the linearity properties of the limits.*

Furthermore, if $h(y) \to C$ as $y \to B$ and $g(x) \to B$ as $x \to a$, then

$$h(g(x)) \to C \quad \text{as } x \to a. \tag{8.39}$$

The last statement means, in short, that

$$\lim_{x \to a} h(g(x)) = C. \tag{8.40}$$

In particular, the following holds true:

$$f(x)^B \to A^B \quad \text{and} \quad A^{g(x)} \to A^B \text{ as } x \to a. \tag{8.41}$$

Theorem 8.4 (The Squeeze theorem). *Assume that*

$$f(x) \le g(x) \le h(x), \quad f(x) \to B, \quad \text{and} \quad h(x) \to B \text{ as } x \to a.$$

Then even

$$g(x) \to B \quad \text{as } x \to a. \tag{8.42}$$

8.5.2 *Corollary from the limit laws*

Theorem 8.5. *Assume that* $h(x) \to 0$ *and* $0 \le g(x) \le h(x)$ *as* $x \to a$. *Then it follows that*

$$\begin{aligned} &(1)\ g(x) \to 0. \\ &(2)\ \textit{Assume further that } f(x) = h(x) \cdot k(x), \\ &\quad\ \textit{where } k(x) \textit{ is bounded, i.e., } |k(x)| \le M. \\ &\quad\ \textit{Then also } f(x) \to 0, \textit{ as } x \to a. \end{aligned} \tag{8.43}$$

8.5.3 *The size order between exp-, power-, and logarithm functions*

Theorem 8.6. *Assume that* $a > 1$ *and* $c > 0$. *Then*

$$\frac{x^b}{a^x} \longrightarrow 0 \quad \text{as } x \to \infty. \tag{8.44}$$

$$\frac{(\ln x)^b}{x^c} \longrightarrow 0 \quad \text{as } x \to \infty. \tag{8.45}$$

$$|\ln x|^b \cdot x^c \longrightarrow 0 \quad \text{as } x \to 0_+. \tag{8.46}$$

$$\begin{aligned} \frac{\mathcal{P}(x)}{\ln x} &\longrightarrow 1 \qquad \text{as } x \to \infty, \\ \frac{\mathcal{P}(x)}{\ln(x+1)} &\longrightarrow 1 \quad \text{as } x \to 0, \end{aligned} \tag{8.47}$$

where $\mathcal{P}(x)$ *is the product-log function, page* 175.

8.5.4 *Limits for the trigonometric functions*

The basic limit for trigonometric functions is as follows:

Theorem 8.7.

$$\lim_{x\to 0} \frac{\sin x}{x} = 1. \tag{8.48}$$

The limit value assumes that x is in radians. This also applies in all real analysis. The limit holds even for complex variable x.

8.5.5 *Some special limits*

Theorem 8.8.

$$\begin{array}{ll}
\lim_{n\to\pm\infty} (1+1/n)^n = e, & \lim_{n\to\pm\infty} (1+x/n)^n = e^x. \\
\lim_{h\to 0} (1+h)^{1/h} = e, & \lim_{n\to\infty} e^{-n}\left(1+n+\dfrac{n^2}{2!}+\cdots+\dfrac{n^n}{n!}\right) = \dfrac{1}{2}. \\
\lim_{n\to\infty} \dfrac{x^n}{n!} = 0, & \lim_{x\to 0_+} x^{1/x} = 0, \quad \lim_{x\to\infty} x^{1/x} = 1.
\end{array} \tag{8.49}$$

8.5.6 *Some derived limits*

$$\begin{array}{l|l}
\lim_{x\to 0} \dfrac{\tan x}{x} = 1 & \lim_{x\to 0} \dfrac{1-\cos x}{x^2} = \dfrac{1}{2} \\
\lim_{x\to 0} \dfrac{\arcsin x}{x} = 1 & \lim_{x\to} \dfrac{\arctan x}{x} = 1 \\
\lim_{x\to 0} \dfrac{\frac{\pi}{2} - \arccos x}{x} = 1 & \lim_{x\to 0} \dfrac{\sinh x}{x} = 1 \\
\lim_{x\to 0} \dfrac{\ln(x+1)}{x} = 1 & \lim_{x\to 0} \dfrac{\tanh x}{x} = 1 \\
\lim_{x\to 0} \dfrac{\cosh x - 1}{x^2} = \dfrac{1}{2} & \lim_{x\to 0} \dfrac{\operatorname{arcsinh} x}{x} = 1
\end{array}$$

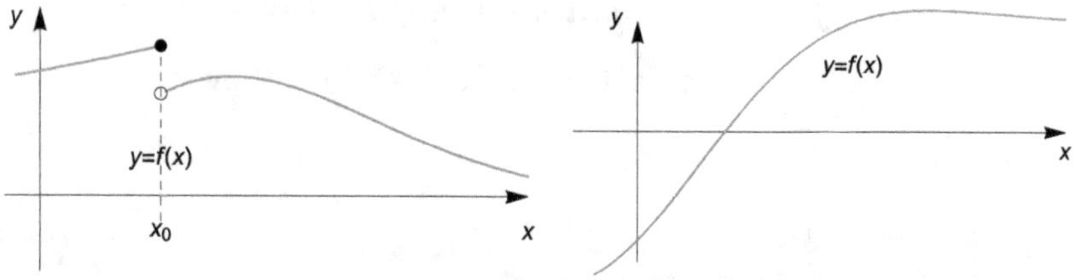

Figure 8.4: The function on the left is discontinuous at $x = x_0$ but continuous for all other x values in its domain of definition. The function on the right is continuous.

8.6 Continuity

8.6.1 *Definition*

We consider real functions, defined in an interval, or a union of intervals.

Definition 8.15. Let $f(x)$ be a real function.

(i) Definition of continuity at a point, $(a, f(a))$, where $x = a \in D_f$:
A function $f(x)$ is continuous at the point $x = a$, if

$$\begin{aligned} &\text{for each } \varepsilon > 0 \text{ there is a } \delta > 0 \text{ such that} \\ &|x - a| < \delta \Longrightarrow |f(x) - f(a)| < \varepsilon. \end{aligned} \tag{8.50}$$

This is written as

$$\begin{aligned} f(x) &\to f(a), \qquad \text{as } x \to a \\ &\text{or} \\ \lim_{x\to a} f(x) &= f(a). \end{aligned} \tag{8.51}$$

(ii) Definition of continuity on a set
A function f is continuous on a set $E \subseteq D_f$, if it is continuous at each $x \in E$.

(iii) A function which is not continuous is called *discontinuous* (Figure 8.4, left).
If this is the case for $f(x)$ at a point $x = a \in D_f$, then the function i said to be discontinuous at $x = a$, or has a discontinuity at $x = a$.

(iv) Let $E \subseteq D_f$. If for each $\varepsilon > 0$ there is a $\delta > 0$, such that

(a)

$$|x - x'| < \delta \Longrightarrow |f(x) - f(x')| < \varepsilon, \quad \forall x, x' \in E, \quad (8.52)$$

then the function is said to be *uniformly continuous* on the set E and if

(b)

$$|x - x'|^\alpha < \delta \Longrightarrow |f(x) - f(x')| < \varepsilon, \quad \forall x, x' \in E, \quad (8.53)$$

then the function i said to be Lipschitz continuous of order $\alpha(> 0)$ on the set E.

Remarks. Note that the definition (i) assumes that $x = a$ belongs to the domain of the function.

The definitions can be generalized to functions $f : \mathbb{R}^m \to \mathbb{R}^n$, where m and n are arbitrary integers.

The function $f(x) := \dfrac{\sin x}{x}$ has $D_f = \{x \in \mathbb{R} : x \neq 0\}$. Since

$$\lim_{x \to 0} \frac{\sin x}{x} = 1,$$

$f(x)$ can be *extended* to a continuous function on whole $\mathbb{R}$ if and only if one puts $f(0) = 1$.

8.6.2 *Calculus rules for continuity*

Theorem 8.9. *Sum, difference, product, ratio, and composition of two continuous functions are continuous.*

The ratio $f(x)/g(x)$ is continuous provided that $g(x) \neq 0$.

8.6.3 *Some theorems about continuity*

Definition 8.16. A real function f assumes a greatest value $f_{\max}$, in a set $A \subseteq D_f$, if there exists $x_0 \in A$ such that $f(x_0) = f_{\max} \geq f(x)$ for all $x \in A$.

A real function f assumes a smallest value $f_{\min}$ in a set $A \subseteq D_f$, if there exists $x_0 \in A$ such that $f(x_0) = f_{\min} \leq f(x)$ for all $x \in A$.

Theorem 8.10. *Assume that f is continuous on an interval.*

(i) $f(x)$ *is invertible* $\Leftrightarrow$ f *is strictly monotone.*
(ii) *If the inverse exists, then the inverse is continuous.*

Theorem 8.11. *Assume that f is a continuous function on a compact interval $[a, b]$.*

(i) (*The theorem of the largest and smallest values*) f *assumes a greatest and a smallest value in the interval $[a, b]$ (both $f_{\max}$ and $f_{\min}$).*
(ii) (*The theorem of intermediate value*) f *assumes all values between its smallest and largest values.*
(iii) (a) *The map $f([a, b]) := \{f(x) : x \in [a, b]\}$ of a compact interval $[a, b]$ is a compact interval $[f_{\min}, f_{\max}]$.*
(b) *If f is defined in an interval I, it follows that its map under f is also an interval.*
(iv) f *is uniformly continuous.*

8.6.4 *Riemann's z-function*

Definition 8.17. Riemann's ζ-function is defined to be the series

$$\zeta(s) := \sum_{n=1}^{\infty} \frac{1}{n^s}, \tag{8.54}$$

where $s = x + iy$ is a complex variable, where $x = \operatorname{Re} s$ and $y = \operatorname{Im} s$.

Theorem 8.12. *Put $s = x + iy$ (as above).*

The series (Riemann's ζ-function) is absolutely convergent for $x > 1$, and uniformly convergent on the set $\{s \in \mathbb{C} : x > 1 + \delta\}$ for each $\delta > 0$.

$\zeta(s)$ can be expressed by the Euler product

$$\zeta(s) = \prod_{p \text{ prime}} \frac{1}{1 - 1/p^s}, \quad \text{if } x = \operatorname{Re} s > 1. \tag{8.55}$$

$\zeta(s)$ can be extended to an analytic function on $\mathbb{C} \setminus \{1\}$, with a simple pole at $s = 1$ (Riemann).

Remarks. $\zeta(s)$ has no zeros for $x > 1$.

$\zeta(s)$ has no zeros for $x < 0$ except for $s = -2, -4, -6...$

It is proved that, for $x \geq 0$, all zeros of the ζ-function lie on the band $\{s = x + iy : 0 \leq x \leq 1\}$.

The Riemann's hypothesis states that all zeros lie on the line $x = 1/2$.

Hardy, 1915 proved that there are infinitely many zeros of $\zeta(s)$ on the line $x = 1/2$.

So far the only zeros that are found, by extensive calculations with computers, lie on this line.

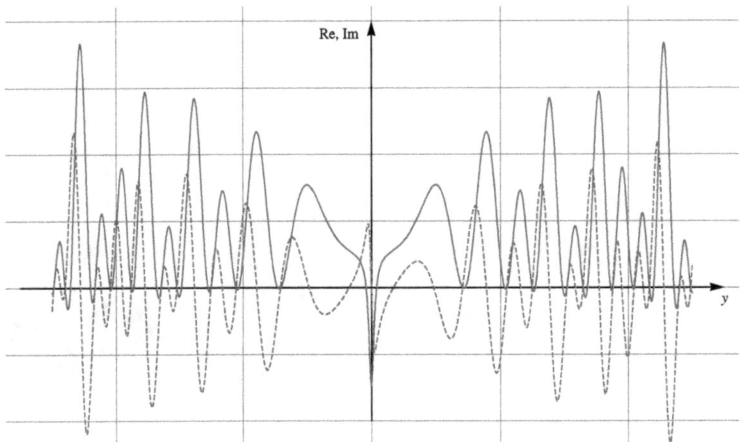

Real- and imaginary part of the Riemann z function $\zeta(1/2 + i\,y)$, solid and dashed line, respectively.

Chapter 9

Derivatives

Table with derivatives of elementary functions is found on page 172.

9.1 Directional Coefficient

Definition 9.1. Given k, $m \in \mathbb{R}$, not both equal to zero.
Equation of a line, not parallel to y-axis, is given by $\{(x, y) : y = kx + m\}$, where k is *directional coefficient* of the line and m is solely called the *m-value* (and is equal to the y-value for $x = 0$, i.e., where the line intersects the y-axis).

Remarks. For $k = 0$ the line $y = m$ is parallel to x-axis.

For $m = 0$, $y = kx$, represents all straight lines through the origin: $(0, 0)$.

The number k in the equation of a line $y = kx + m$ is a measure of the slope of the line.

If we denote the angle between the line and the positive x-axis by α, then we get $k = \tan\alpha$.

A line on the form $x = a$ has no directional coefficient. Alternatively, such a line is said to have directional coefficient $\pm\infty$.

For a line on the form $y = kx + m$, y can be considered as a function of the variable x: $f(x) = kx + m$. The directional coefficient k is a special case of the *derivative*.

9.1.1 *The one- and two-point formulas*

Theorem 9.1. *Given a line of the form* $y = kx + m$. *where* $x \in \mathbb{R}$, k *and* m *are constants (see LHS in Figure* 9.1*).*

(i) *The two-point formula gives the directional coefficient* k :

$$k = \frac{y_2 - y_1}{x_2 - x_1}. \tag{9.1}$$

(ii) *The one-point formula is*

$$k = \frac{y - y_1}{x - x_1}, \quad \textit{or equivalently} \quad y - y_1 = k(x - x_1). \tag{9.2}$$

Definition 9.2. The expression

$$\frac{\Delta f}{\Delta x} := \frac{f(x + \Delta x) - f(x)}{\Delta x} \tag{9.3}$$

is called *difference quotient* (also referred to as *Newton quotient*), and is equal to the directional coefficient for the secant through the points $(x, f(x))$ and $(x + \Delta x, f(x + \Delta x))$ (right Figure 9.1). If the limit

$$\begin{aligned} f'(x) &:= \lim_{\Delta x \to 0} \frac{f(x + \Delta x) - f(x)}{\Delta x} \\ &= \{\text{or}\} = \lim_{h \to 0} \frac{f(x + h) - f(x)}{h} \end{aligned} \tag{9.4}$$

exists, then f is differentiable at x. The limit $f'(x)$ is the *derivative* of f at the point x.

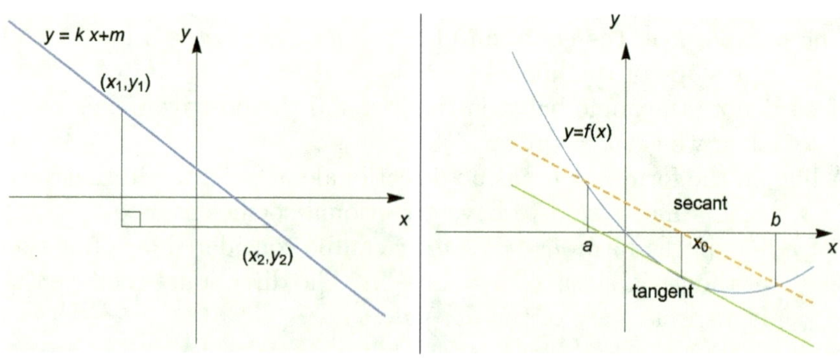

Figure 9.1: LHS: Line of the form $y = kx + m$. RHS: Secant- and tangent-lines for the curve $y = f(x)$.

Equation (9.4) can be represented as a *differential quotient* (differential coefficient), i.e., a ratio between two infinitely small numbers denoted by df and dx. In the following, a number of ways to present the derivative are indicated:

$$Df(x) = f'(x) = \frac{df}{dx} = \frac{d}{dx}f. \tag{9.5}$$

The derivative at the point $x = a$ is

$$f'(a) = \lim_{x \to a} \frac{f(x) - f(a)}{x - a}. \tag{9.6}$$

The following interpretations are crucial in the concept of derivatives:

- Geometrically, the derivative of $f(x)$ at a point x is the directional coefficient of the tangent to the graph of f at $(x, f(x))$.
- Analytically, the derivative of f at x is a measure for the instantaneous change of $f(x)$ with respect to x.
- The relationship between the instantaneous displacement and instantaneous velocity, s and v, respectively, at the time t, is $\frac{ds}{dt} = v$, a *time derivative.*
- If the limit

$$f'_R(x) := \lim_{\Delta x \to 0_+} \frac{f(x + \Delta x) - f(x)}{\Delta x}, \tag{9.7}$$

  exists, it is called right derivative of f. The left derivative, $f'_L(x)$, is defined similarly.
- $\lim_{x \to a} f'(x)$ need not exist although $f'(a)$ exists.
- To denote the derivative of $f(x)$ for a specific x−value, say $x = 2$, one writes $f'(2)$ or $\left.\frac{df}{dx}\right|_{x=2}$.

Definition 9.3. A function $f(x)$ is *differentiable* at x if there is a function $\rho(h)$, for which $\rho(h) \to 0$, as $h \to 0$ and

$$f(x + h) - f(x) = h[A + \rho(h)]. \tag{9.8}$$

Theorem 9.2. *That a function is differentiable at the point x is equivalent to that f has a derivative at x and $f'(x) = A$.*

9.1.2 *Continuity and differentability*

Differentiability is a *sufficient (but not necessary)* condition for continuity.

Theorem 9.3. *Suppose that the function $f(x)$ is differentiable at $x = a$. Then the function is continuous at this point.*

A note on infinitesimals

The infinitesimal calculus is based on the "differential quotient" concept. The arithmetic with infinitely small and large numbers has long been well established in the physical community, while it is considered with skepticism by the mathematicians. The infinitesimals include, e.g., expressions as dx, dy, which are known as differentials, a reason for calling $\frac{dy}{dx}$ a differential quotient. The notion of differentials was introduced by Newton and Leibniz[1] As for the *infinite small/large numbers*, the contemporary George Berkeley[2] wrote: "And what are these same evanescent increments? They are neither finite quantities, nor quantities infinitely small, nor yet nothing. May we not call them ghosts of departed quantities?" A.E. Hurd, P.A. Loeb, An introduction to nonstandard real analysis. It took a long time for the mathematicians to provide a satisfactory explanation/theory to integrate infinite small/large quantities in the real number system. It was around 1960 that a mathematical theory, the so-called Nonstandard analysis, emerged to bridge the gap. Equipped with this nonstandard theory, the real numbers are extended to include infinitely small and large numbers. The extension is the so-called set of hyper-real numbers.

9.1.3 *Tangent, normal, and asymptote*

Definition 9.4.

(i) For a function $y = f(x)$, which is differentiable at $x = a$, the equation of the tangent to its graph at the point $(a, f(a))$ is given by

[1]Isaac Newton, English physicist and mathematician, 1643–1727. Gottfried Wilhelm von Leibniz German mathematician, 1646–1716. Both are considered inventors of the infinitesimal calculus, independently. The integral sign $\int$ was introduced by Leibniz and is a stylized form of the german *Summe*.

[2]George Berkeley, 1685–1753.

$$y = f(a) + f'(a)(x - a). \tag{9.9}$$

(ii) If the function $f(x)$ is defined at the point $x = a$ and

$$\frac{f(x) - f(a)}{x - a} \to \infty \quad \text{or} \quad -\infty \text{ as } x \to a_- \quad \text{or} \quad x \to a_+,$$

then the tangent to the curve is a vertical line, and its equation is given by $x = a$.

(iii) A line l_1 is called normal to a line l_2 if l_1 intersects l_2 at right angle.

(iv) Asymptotes:

(a) An oblique *asymptote* of a function $f(x)$ is a line $y = kx+m$ such that

$$f(x) - (kx + m) \to 0 \quad \text{as} \quad x \to -\infty \text{ or } x \to +\infty. \tag{9.10}$$

(b) A vertical asymptote is defined as a line of the form $x = a$ such that $f(x) \to -\infty$ or $f(x) \to \infty$ as $x \to a_-$ or $x \to a_+$.

Theorem 9.4. *The necessary and sufficient conditions for $y = kx + m$ to be an oblique asymptote for the curve/function $y = f(x)$ as $x \to \infty$ is that the following limits hold:*

$$\lim_{x\to\infty} \frac{f(x)}{x} = k \quad \text{and} \quad \lim_{x\to\infty} f(x) - kx = m. \tag{9.11}$$

A corresponding assertion applies for $x \to -\infty$.

Theorem 9.5. *If the coordinate axes have the same scale, the following equivalence holds.*

A line l_1 has directional coefficient k for $k \neq 0$.

$$\Leftrightarrow$$

Every normal line l_2 has directional coefficient $-1/k$.

9.2 The Differentiation Rules

Theorem 9.6. $D = \frac{d}{dx}$. *If the functions* $f(x)$ *and* $g(x)$ *are differentiable, then*

$$\begin{aligned} &D(af(x) + bg(x)) = aDf(x) + bDg(x) && \text{(Linearity properties)} \\ &Df(g(x)) = f'(g(x)) \cdot g'(x) && \text{(The chain rule)} \\ &D(f(x)g(x)) = f'(x)g(x) + f(x)g'(x) && \text{(Product rule)} \\ &D\left(\frac{f(x)}{g(x)}\right) = \frac{f'(x)g(x) - f(x)g'(x)}{[g(x)]^2} && \text{(Quotient rule),} \end{aligned} \tag{9.12}$$

where a, b *are constants and* $g(x) \neq 0$, *and the composition* $f(g(x))$ *exists.*

Theorem 9.7. *The derivative of the polynomial*

$$f(x) = a_n x^n + a_{n-1}x^{n-1} + \cdots + a_1 x + a_0$$

is

$$f'(x) = na_n x^{n-1} + (n-1)a_{n-1}x^{n-2} + \cdots + 2a_2\, x + a_1. \tag{9.13}$$

Remarks. By $f'(g(x))$ is meant the derivative of f with respect to $z = g(x)$. This derivative is called exterior derivative, and $g'(x)$ is the derivative with respect to x and is called the interior derivative.

An easy way to remember the formula and at the same time an example of differential calculus is through writing

$$f'(g(x)) = f'(z) = \frac{df}{dz} \quad \text{and} \quad g'(x) = \frac{dz}{dx},$$

wherein the derivative of the composite function $f(g(x))$ can be written as

$$\frac{df}{dx} = \frac{df}{dz} \cdot \frac{dz}{dx} \text{ (Chain rule).} \tag{9.14}$$

Definition 9.5. The second derivative, if it exists, is defined insofar as

$$\frac{d^2 f}{dx^2} := \frac{d}{dx}\left(\frac{df}{dx}\right). \tag{9.15}$$

Higher order derivatives, if they exist, are defined inductively, *viz.*

$$\frac{d^{n+1} f}{dx^{n+1}} := \frac{d}{dx}\left(\frac{d^n f}{dx^n}\right), \quad n = 0, 1, 2, \dots \tag{9.16}$$

The set of functions f, such that $\frac{d^n f}{dx^n}$, with n being a positive integer, are continuous in the interval $I = (a, b)$, is denoted by $\mathcal{C}^n(I)$.

Theorem 9.8.

$$\mathcal{C}^{n+1}(I) \subset \mathcal{C}^n(I), \quad n = 0, 1, 2, \dots$$

Theorem 9.9. *If f is differentiable and $f(x) \neq 0$, then*

$$D \ln|f(x)| = \frac{f'(x)}{f(x)} \text{ or equivalently } f'(x) = f(x) \cdot D \ln|f(x)|. \tag{9.17}$$

To calculate $f'(x)$ by means of the second identity is called logarithmic differentiation.

Assume that $g(x) > 0$, and $g(x)$ and $h(x)$ are differentiable. Then

$$f(x) = g(x)^{h(x)} \Longrightarrow f'(x) = g(x)^{h(x)} \cdot \left(\frac{g'(x)}{g(x)} \cdot h(x) + \ln(g(x)) \cdot h'(x)\right). \tag{9.18}$$

Theorem 9.10. *If $f^{(n)}(x)$ and $g^{(n)}(x)$ exist, then the derivative $\frac{d^n}{dx^n}(f(x)g(x))$ exists and*

$$\frac{d^n}{dx^n}(f(x)\, g(x)) = \sum_{k=0}^{n} \binom{n}{k} f^{(k)}(x)\, g^{(n-k)}(x). \tag{9.19}$$

9.3 Applications of Derivatives

9.3.1 *Newton–Raphson iteration method*

The method is used to (numerically) find roots of the equation $f(x) = 0$.

The recursion sequence $(x_1, x_2, \ldots)$ is given by

$$x_{n+1} = x_n - \frac{f(x_n)}{f'(x_n)}, \quad (9.20)$$

which yields a convergent sequence $x_n \to x$ such that $f(x) = 0$. The curve $y = f(x)$, passing through the points $(x_1, f(x_1))$ and $(x_2, f(x_2))$ in Newton–Raphson's iteration method, where $n = 1$.

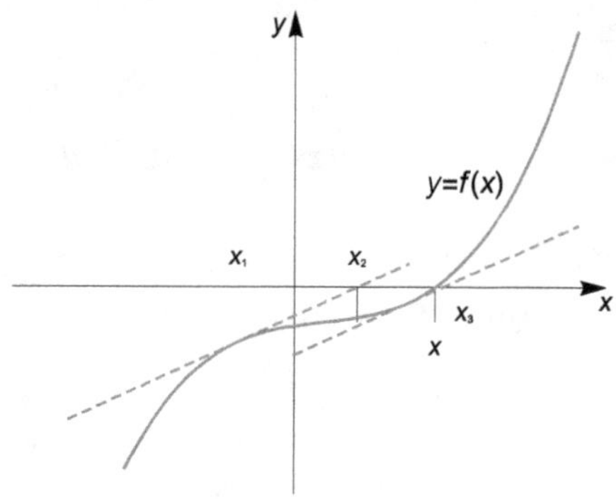

9.3.2 *L'Hôspital's rule*

Theorem 9.11 (L'Hospital's rule). *If f and g are differentable in a punctured neighborhood of $x = x_0$ and if $\lim_{x\to x_0} \frac{f(x)}{g(x)}$ is of type "$\frac{0}{0}$" or "$\frac{\infty}{\infty}$", then*

$$\lim_{x\to x_0} \frac{f(x)}{g(x)} = \lim_{x\to x_0} \frac{f'(x)}{g'(x)}, \quad (9.21)$$

if the latter limit exists. Here punctured neighborhood of $x_0 = \infty$ and $x_0 = -\infty$ are interpreted as intervals of type $[a, \infty)$ and $(-\infty, b]$, respectively.

9.3.3 *Lagrange's mean value theorem*

Local maximum- and minimum points

Definition 9.6.

(i) A function is increasing (decreasing) in an interval if for all x_1, x_2 in the interval

$$x_1 < x_2 \Longrightarrow f(x_1) \leq f(x_2), \quad (f(x_1) \geq f(x_2)). \tag{9.22}$$

If strong inequality, that is "<" or ">", holds in the respective RHS, then the function is *strongly* increasing or decreasing, respectively.

(ii) A decreasing or increasing function is called a monotonous function. Strongly monotonous is defined in the same way.

(iii) Assume that f is a real function with domain $D_f \subseteq \mathbb{R}$. Assume further there is a $x_0 \in D_f$ such that $f(x_0) \geq f(x)$, for all $x \in D_f \cap I$, for some neighborhood $I = (x_0 - \delta, x_0 - \delta)$, $(\delta > 0)$ of x_0.

Then the point $(x_0, f(x_0))$ is called a local maximum point, and $f(x_0)$ a local maximum.

(iv) If $(x_0, -f(x_0))$ is a local maximum point, the point $(x_0, f(x_0))$ is called a local minimum point, and $f(x_0)$, a local minimum.

(v) A point x_0, such that $f'(x_0) = 0$, is called stationary point, or critical point.

(vi) If f is increasing/decreasing in a neighborhood of a stationary point, then the point is called a terrace point.

Remarks. For a continuously differentiable function, a stationary point is a local max/min (extreme points), or a terrace point. The interior stationary points have horizontal tangents (see Figure 9.2).

Sometimes only the $x-$ coordinate is mentioned:

"$f(x)$ has (local) maximum at the point x_0."
Abbreviations are generally used as "loc. max" and "loc. min" for local maximum and local minimum, respectively.
A point which is a local max- or min point is a local extreme point. The corresponding function value is called an extreme value.

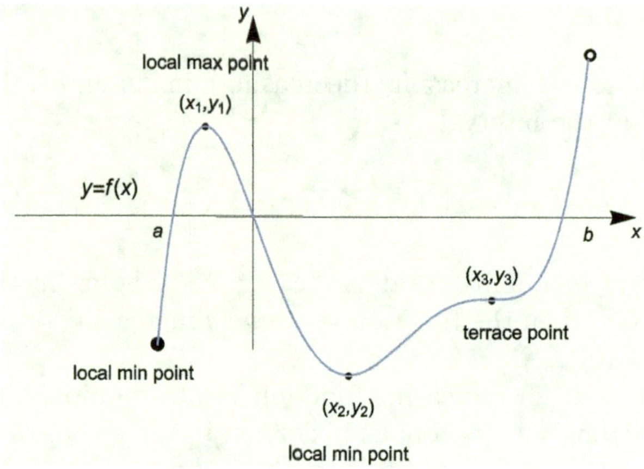

Figure 9.2: Curve with local max- and min. In the figure the right endpoint is not a local maximum, since $b \notin D_f$ and f has no largest value, but has a smallest value y_2 (also global minimum).

Theorem 9.12 (Lagrange's mean value theorem).

If the function $y = f(x)$ is differentiable in the interval (a, b) and continuous in the closed interval $I := [a, b]$, then there exists $x_0 \in (a, b)$ such that

$$f'(x_0) = \frac{f(b) - f(a)}{b - a}. \quad (9.23)$$

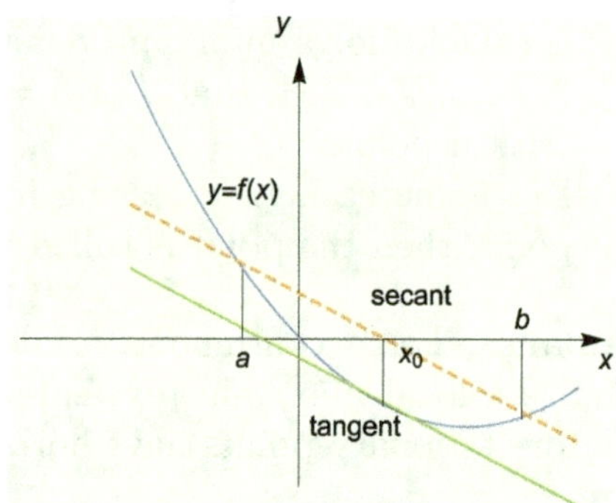

Illustration of Lagrange's mean value theorem.

Theorem 9.13. *For a function, with the same conditions as in the mean value theorem, the following hold true:*

$$\begin{aligned} f'(x) \geq 0 \quad x \in I &\implies f \text{ is increasing in } I \\ f'(x) \leq 0 \quad x \in I &\implies f \text{ is decreasing in } I. \end{aligned} \quad (9.24)$$

Remarks. From the above theorem, it follows that $f'(x) > 0$ implies that f is strongly increasing. Likewise, $f'(x) < 0$ implies that f is strongly decreasing. The theorem and the above two properties are referred to as monotonic and strictly monotonic, respectively.

If $f'(x) > 0$, except on an isolated point x_0 with $f'(x_0) = 0$, then f is still strictly increasing.

9.3.4 *Derivative of inverse function and implicit derivation*

Differentiating an inverse function

Theorem 9.14. *Assume that $f'(x) \neq 0$ in an open interval $I = (a, b)$. Then the following holds:*

(i) *f has inverse f^{-1}.*
(ii) *f^{-1} is differentiable with derivative $\frac{d}{dy} f^{-1}(y) = \frac{1}{f'(x)}$, where $f^{-1}(y) = x$.*
(iii) *Either $f' > 0$ or $f' < 0$ and hence f and f^{-1} are either strongly increasing or strongly decreasing.*

Remarks.

One can write

$$f'(x) = \frac{1}{(f^{-1})'(y)} \quad \text{(assumptions as above).} \tag{9.25}$$

Expressed by differentials, (9.25) means that

$$\frac{dy}{dx} = \frac{1}{\frac{dx}{dy}} \text{ or equivalently } \frac{dx}{dy} \cdot \frac{dy}{dx} = 1. \tag{9.26}$$

9.3.5 *Second derivative of inverse function*

Theorem 9.15. *Let $y = y(x)$ and $x = x(y)$ be well-defined functions. Assume that all following derivatives exist and $x'(y) = \frac{dx}{dy} \neq 0$. Then*

$$\frac{d^2y}{dx^2} = -\frac{\frac{d^2x}{dy^2}}{\left(\frac{dx}{dy}\right)^3}. \tag{9.27}$$

9.3.6 *Implicit differentiation*

Theorem 9.16. *If* $F(x,y) = 0$ *(implicitly) defines a function* $y = f(x)$*, then*

$$\frac{dF}{dx} = \frac{dF}{dy} \cdot \frac{dy}{dx}, \quad \text{and if} \quad \frac{dF}{dy} \neq 0, \quad \text{then} \ \frac{dy}{dx} = \frac{\frac{dF}{dx}}{\frac{dF}{dy}}. \tag{9.28}$$

9.3.7 *Convex and concave functions*

Definition 9.7. A function f is convex in an interval I if for all $x_1, x_2 \in I$ and every $\lambda : 0 \leq \lambda \leq 1$

$$f(\lambda x_1 + (1-\lambda)x_2) \leq \lambda f(x_1) + (1-\lambda) f(x_2). \tag{9.29}$$

With $<$ instead of $\leq$ in (9.29), *strictly* convex applies.

A concave function is defined analogously. (If f is concave, then $-f$ is convex.)

x_0 is called an inflexion point of $f(x)$, if x_0 is an interior point in an interval I and $f(x)$ is convex in the interval $I \cap \{x : x < x_0\}$ and concave in $I \cap \{x : x > x_0\}$, or vice versa.

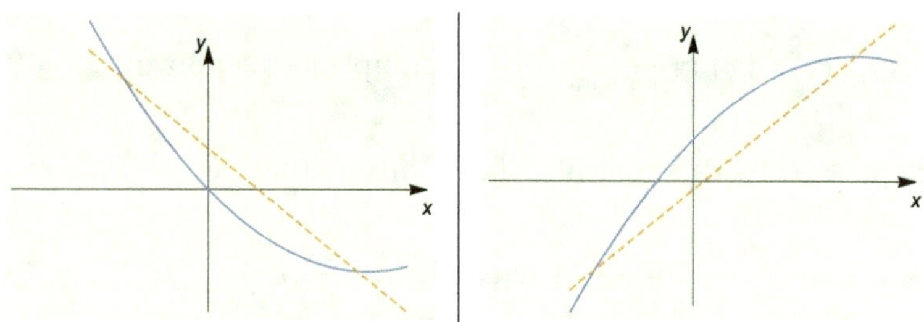

Convex curve *Concave curve*

Remarks. Convexity (9.29) for a function $f(x)$ defined on an interval, can alternatively be expressed as, for the triple x coordinates a, b, c in I, as follows:

$$a < b < c \Longrightarrow \frac{f(b) - f(a)}{b - a} \leq \frac{f(c) - f(b)}{c - b}. \tag{9.30}$$

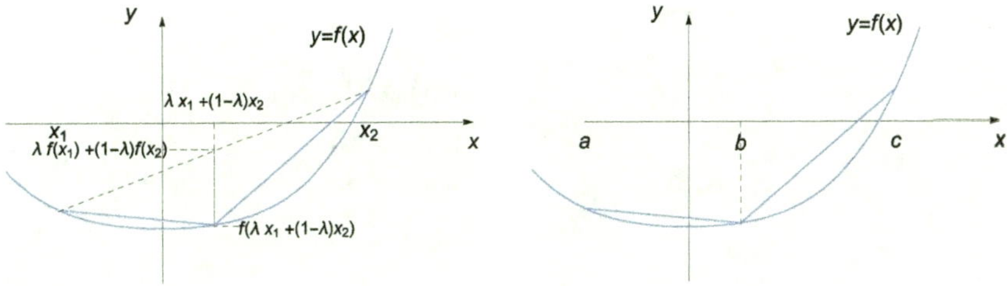

Theorem 9.17. *A convex/concave function f in an open interval $I = (a, b)$ is continuous.*

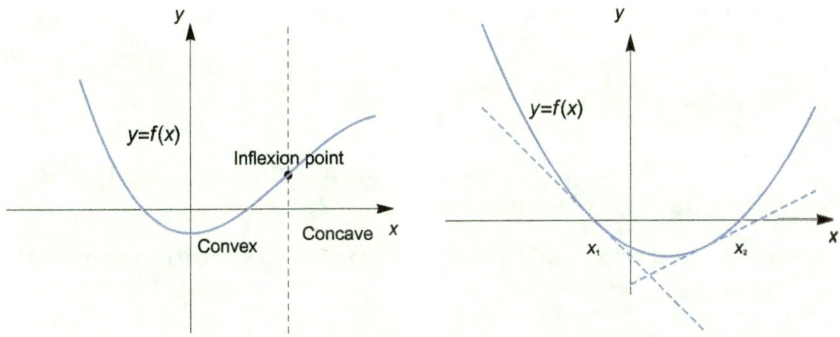

Theorem 9.18. *Assume that the function $f(x)$ is twice continuously differentiable in an interval, i.e., $f''(x)$ is continuous. Then*

(i)
$$f''(x) > 0 \Longrightarrow f'(x) \text{ increasing}$$
$$\Longrightarrow \text{the curve } y = f(x) \text{ is convex} \tag{9.31}$$

and

$$f''(x) < 0 \Longrightarrow f'(x) \text{ decreasing}$$
$$\Longrightarrow \text{the curve } y = f(x) \text{ is concave.} \tag{9.32}$$

(ii) *If the condition (a) is satisfied, then f has a local minimum at $x = x_0$. If (b) is satisfied, then f has a local maximum at*

$x = x_0$.

$$\text{(a)} \quad \begin{cases} f'(x_0) = 0, \\ f''(x_0) > 0. \end{cases} \qquad \text{(b)} \quad \begin{cases} f'(x_0) = 0, \\ f''(x_0) < 0. \end{cases} \tag{9.33}$$

(iii) *A point* $(x, f(x))$ *is an inflexion point* $\Longrightarrow f''(x) = 0$.
(iv) *A point* $(x, f(x))$ *is a terrace point* $\Longrightarrow$ *the point is an inflexion point* $\Longrightarrow f''(x) = 0$.

Theorem 9.19. *If* $x_0 \in I$, I *is an open interval,* $f'(x_0) = f''(x_0) = \cdots = f^{2k-1}(x_0) = 0$ *and* $f^{(2k)}(x_0) > 0\,(< 0)$, *then* $(x_0, f(x_0))$ *is a local minimum point (local maximum point).*

9.4 Tables

The derivative of products and composite functions. Derivative of elementary functions is found on page 172.

Table I: Derivatives of some products and composite functions

Function	Derivative
$e^{ax}\sin bx$	$e^{ax}(a\sin bx + b\cos bx)$
$e^{ax}\cos bx$	$e^{ax}(a\cos bx - b\sin bx)$
$\ln\left\lvert x + \sqrt{x^2 + a}\right\rvert$	$\dfrac{1}{\sqrt{x^2 + a}}$
$\ln\lvert\tan(x/2)\rvert$	$\dfrac{1}{\sin x}$
$\ln\lvert\cot(x/2)\rvert$	$-\dfrac{1}{\sin x}$
$\ln\lvert\sin x\rvert$	$\cot x$
$\ln\lvert\cos x\rvert$	$-\tan x$
$\ln\left\lvert\dfrac{x-1}{x+1}\right\rvert$	$\dfrac{2}{x^2 - 1}$

(9.34)

Table II: Derivatives of some special functions

Function	Derivative
$x^n e^{kx}$	$x^{n-1}(kx+n)e^{kx}$
$\tan^n x$	$n(1+\tan^2 x)\tan^{n-1} x$
$x^m \sin(\ln x)$	$x^{m-1}(m\sin(\ln x)+\cos(\ln x))$
$x^m(\ln\lvert x\rvert)^n$	$x^{m-1}(\ln\lvert x\rvert)^{n-1}(m\ln\lvert x\rvert+n)$

(9.35)

Table III: Some nth order derivatives

Function	The nth derivative
$f(x)=\ln\lvert x+a\rvert$	$f^{(n)}(x)=(-1)^{n-1}\dfrac{(n-1)!}{(x+a)^n}$
$f(x)=\sqrt{x}$	$f^{(n)}(x)=(-1)^{n-1}\dfrac{(2n-3)!!\,\sqrt{x}}{2^n\,x^n}$
$f(x)=e^{ax}\cos bx$	$f^{(n)}(x)=\sum_{k=0}^{n}\binom{n}{k}a^{n-k}b^k\cos(bx+k\pi/2)$
$f(x)=e^{ax}\sin bx$	$f^{(n)}(x)=\sum_{k=0}^{n}\binom{n}{k}a^{n-k}b^k\sin(bx+k\pi/2)$

(9.36)

Chapter 10

Integral

Integral of elementary functions are also introduced on page 172. Integral of functions of several variables can be found in Section 17.6, page 427.

10.1 Definitions and Theorems

10.1.1 *Lower and upper sums*

Definition 10.1 (Definition of lower and upper sums). Assume that the function f is bounded on an interval $[a, b]$.

Partitioning the interval in a finite number of sub-intervals

$$a = x_0 < x_1 < x_2 < \cdots < x_n = b,$$

and choosing real numbers $l_k \leq f(x)$ and $u_k \geq f(x)$ in each interval $[x_{k-1}, x_k]$, a *lower sum* L and an *upper sum* U are obtained setting

$$L = \sum_{k=1}^{n} l_k(x_k - x_{k-1}), \tag{10.1}$$

and

$$U := \sum_{k=1}^{n} u_k(x_k - x_{k-1}), \tag{10.2}$$

(as illustrated in Figure 10.1).

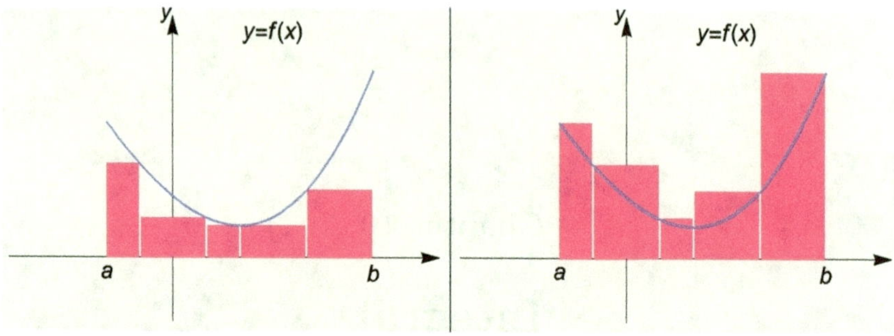

Figure 10.1: Lower sum (to the left) and upper sum (to the right).

Remarks. Using the notation $\Delta x_k := x_k - x_{k-1}$,

$$L = \sum_{k=1}^{n} l_k \Delta x_k \quad \text{and} \quad U = \sum_{k=1}^{n} u_k \Delta x_k, \text{ respectively.} \tag{10.3}$$

The definition does not require that the lower and upper sums have the same partitioning as sub-intervals.

Theorem 10.1. *All lower and upper sums for a given function satisfy*

$$L \leq U. \tag{10.4}$$

Definition 10.2. A function is integrable (in the meaning of Riemann integration) if there is only one number, I, between all lower and upper sums.

This number is called **the definite integral** of $f(x)$ over the interval $[a, b]$ and is denoted by

$$I := \int_a^b f(x)dx.$$

$$\text{Integral sign} \rightarrow \int_{a,\text{lower limit}}^{b,\text{upper limit}} \underbrace{f(x)}_{\text{integrand}} \underbrace{dx}_{\text{differential}}. \tag{10.5}$$

Interchanging limits of integration switches sign of the integral:

$$\int_b^a f(x)dx = -\int_a^b f(x)dx. \tag{10.6}$$

Remarks. We have that

$$L \le I \le U \quad \text{for all } L \text{ and all } U.$$

The definition of the definite integral (10.5) can be expressed as

$$\forall\, \varepsilon > 0, \quad \exists L \quad \text{and } U, \text{ such that} \quad U - L < \varepsilon.$$

In particular,

$$\int_a^a f(x)dx = 0.$$

dx is a differential, which occurs in dy/dx. $f(x)dx$ should be considered as a product in $f(x) \cdot dx$.

x is called the integration variable and can be replaced, for instance, by t (or any other symbol) without affecting the value of the integral.

The definition of upper and lower sums, in principal, containing the terms $f(x) \cdot \Delta x$, would have negative (positive) value in the part of the interval where $f(x) < 0(> 0)$. To interpret the integral as an area, the negative contributions would change the sign rendering them positive. In this way, they preserve the notion of integral as an area.

Integral gives "area with sign".

A sum $\sum_{k=1}^{n} f(\xi_k)\Delta x_k$ of rectangle areas, where ξ_k is an arbitrary point in the interval $[x_{k-1}, x_k]$, approximates the integral and is referred to as *Riemann sum*.

Monotonicity for integrals:

$$a \le b \quad \text{and} \quad h(x) \le g(x) \implies \int_a^b h(x)dx \le \int_a^b g(x)dx. \tag{10.7}$$

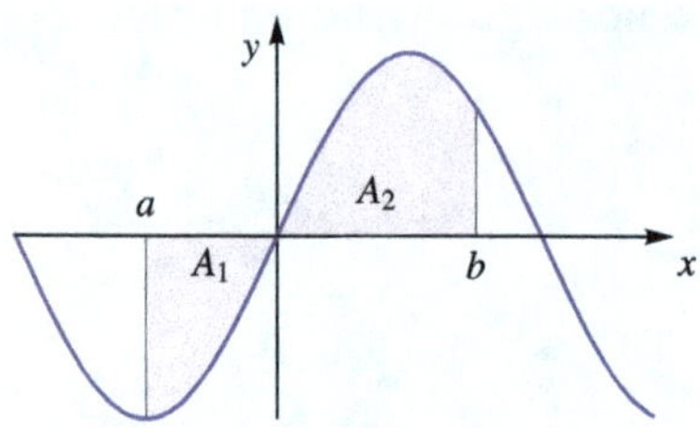

The integral $\int_a^b f(x)dx = -A_1 + A_2$, i.e., the area with respect to sign.

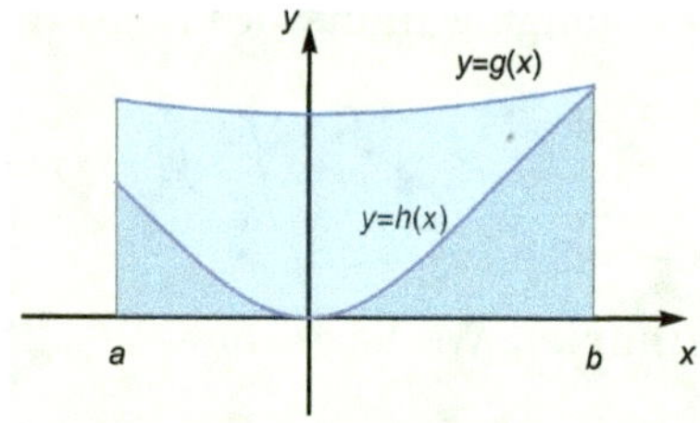

Monotonity (10.7): $h(x) \le g(x) \Longrightarrow \int_a^b h(x)dx \le \int_a^b g(x))dx$.

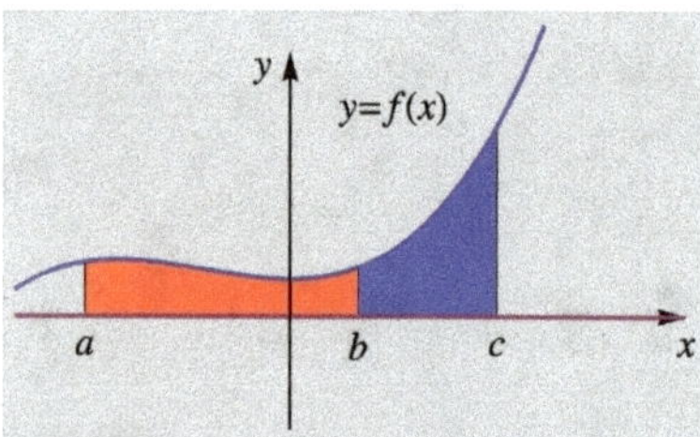

Illustration of (10.8).

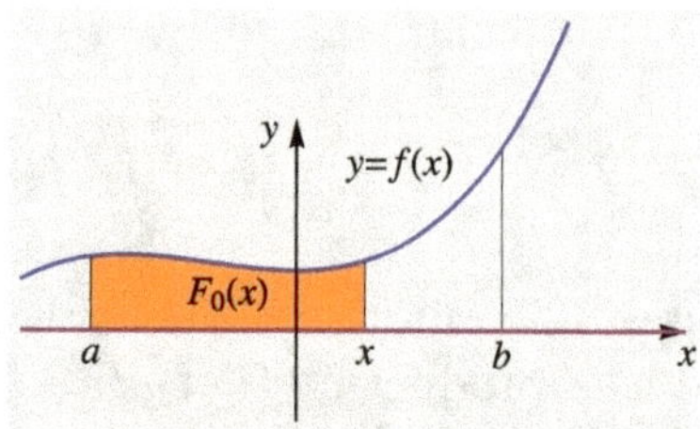

Illustration of (10.9).

Theorem 10.2. *If $f(x)$ is integrable on the interval $[a, c]$ and $b \in [a, c]$, then*

$$\int_a^b f(x)dx + \int_b^c f(x)dx = \int_a^c f(x)dx. \tag{10.8}$$

10.2 Primitive Function

Definition 10.3. A function $F(x)$, whose derivative is $f(x)$, is called a *primitive* function of $f(x)$, i.e., $F'(x) = f(x)$.

Definition 10.4. For x such that $a \le x \le b$, we define the function $F(x)$ as

$$F(x) := \int_a^x f(t)dt, \tag{10.9}$$

where $f(x)$ is assumed to be a continuous function on $[a, b]$.

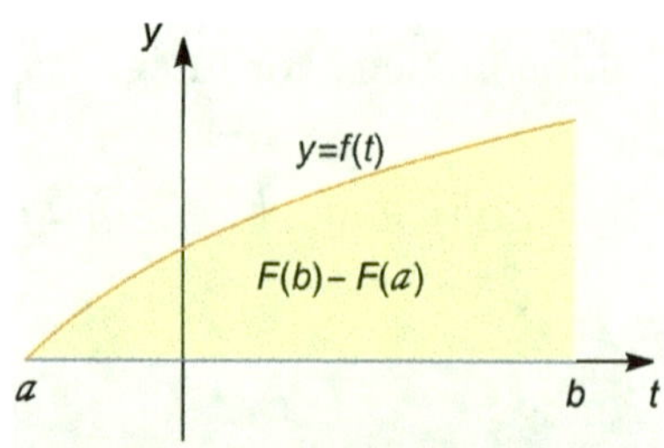

Remarks. One has to take the integration variable t (or any other symbol except x) to be able to use x as the upper limit of the integration.

In particular, it follows that for $x = b$,

$$F(b) = \int_a^b f(x)dx, \quad \text{and} \quad F(a) = 0.$$

Theorem 10.3. (The fundamental theorem of calculus (I). *With $F(x)$ defined as* (10.9)

$$F'(x) = f(x). \tag{10.10}$$

Theorem 10.4. *Assume that F_1 and F_2 are primitive functions of the same function f defined on a common interval. Then they satisfy*

$$\begin{aligned} F_1(x) &= F_2(x) + C \quad \text{for some constant } C. \\ f(x) &= F_1'(x) = F_2'(x). \end{aligned} \tag{10.11}$$

In particular, the difference between two primitive functions F_1 and F_2, of the same function f, is a constant (C).

Theorem 10.5 (The fundamental theorem of calculus (II)). *If $f(x)$ is a continuous function on the interval $[a, b]$, then for a primitive function $F(x)$, of f, the following holds true:*

$$\int_a^b f(x)dx = F(b) - F(a). \tag{10.12}$$

Definition 10.5. The right-hand side in (10.12) is also written as

$$[F(x)]_a^b := F(b) - F(a). \tag{10.13}$$

Remarks. The fundamental theorem of calculus and the formula (10.13) are both valid even if the upper limit is smaller than the lower limit, see (10.5).

Definition 10.6.

$$\int_a^b f(x)dx \quad \text{is called } \textbf{definite} \text{ integral,} \tag{10.14}$$

while

$$\int f(x)dx = F(x) + C \quad \text{is called } \textbf{indefinite} \text{ integral,} \tag{10.15}$$

where C is an arbitrary constant.

(i) Indefinite integral means *all* primitive functions of $f(x)$.
(ii) A definite integral is a *number* while indefinite integral is a function, or rather, a function determined as an additive constant.

Remarks.

(i) For instance, the integral $\int \sin x dx = -\cos x + C$.
To integrate $\sin(kx+m)$, $k \neq 0$, with $kx+m$ an inner function, we have

$$\int \sin(kx+m)dx = -\frac{1}{k} \cdot \cos(kx+m) + C.$$

In this way, to "compensate" for the derivative k of the inner function $kx+m$, we multiply by $1/k$.
(ii) In general,

$$\int f(kx+m)dx = \frac{1}{k} \cdot F(kx+m) + C, \tag{10.16}$$

where F is a primitive function of f.

10.3 Rules of Integral Calculus

Theorem 10.6.

$$\left.\begin{aligned} D\int_a^x f(t)dt = f(x) \\ D\left(\int f(x)dx\right) = f(x) \end{aligned}\right\} \quad \text{if } f(x) \text{ is continuous.} \tag{10.17}$$

$$\frac{d}{dx}\int_a^x f(x,t)dt = f(x,x) + \int_a^x \frac{\partial}{\partial x} f(x,t)dt. \tag{10.18}$$

$$\frac{d}{dx}\int_{u(x)}^{v(x)} f(t)dt = f(u(x))u'(x) - f(v(x))v'(x). \tag{10.19}$$

$$\int f(x)dx = F(x) + C \;\Leftrightarrow\; f(x) = F'(x). \tag{10.20}$$

$$\int D(f(x))dx = f(x) + C \quad \text{if } f'(x) \text{ is continuous.} \tag{10.21}$$

10.3.1 *Linearity of integral*

Theorem 10.7. *If $f(x)$ and $g(x)$ are continuous functions on the interval $[a,\ b]$ and k is a constant, then the following equalities hold:*

$$k\int_a^b f(x)dx = \int_a^b kf(x)dx. \tag{10.22}$$

$$\int_a^b f(x)dx + \int_a^b g(x)dx = \int_a^b [f(x)+g(x)]dx. \tag{10.23}$$

Theorem 10.8. *Corresponding relations for indefinite integrals are*

$$k\int f(x)dx = \int k\cdot f(x)dx. \tag{10.24}$$

$$\int f(x)dx + \int g(x)dx = \int (f(x)+g(x))dx. \tag{10.25}$$

10.3.2 *Area between function curves*

Definition 10.7. For two functions, with $f(x) \geq g(x)$ in the interval $[a, b]$, the area A between their function curves is given by

$$A = \int_a^b (f(x)-g(x))dx. \tag{10.26}$$

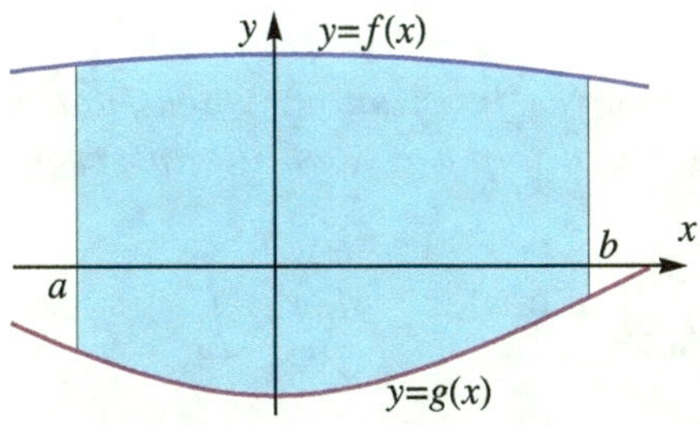

The formula holds even if some of the curves are below the x-axis.

Formally, the area A between two function curves is

$$A = \int_a^b |f(x)-g(x)|dx. \tag{10.27}$$

10.3.3 The integral mean value theorem (I)

Theorem 10.9. *If the function $f(x)$ is continuous in the interval $[a,b]$, then there exists an x_0 in the interval such that*

$$\int_a^b f(x)dx = (b-a)\, f(x_0). \tag{10.28}$$

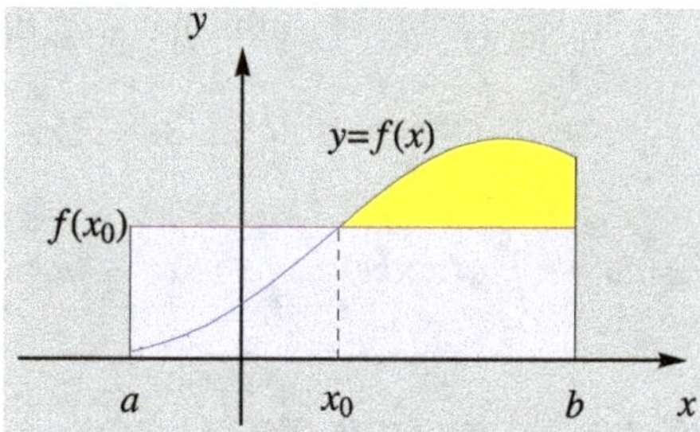

The mean value theorem I of the integral calculus.
$\int_a^b f(x)dx =$ The area of the rectangle in the figure.

10.3.4 The integral mean theorem (II)

Theorem 10.10. *If the functions $f(x)$ and $g(x)$ are continuous in the interval $[a,b]$ and the function $g(x)$ does not change sign in the interval, then there is a number x_0 in the interval such that*

$$\int_a^b f(x)g(x)dx = f(x_0)\int_a^b g(x)dx. \tag{10.29}$$

10.3.5 Some common inequalities for integrals

Theorem 10.11. The triangle inequality *for an integral. With $a \le b$, the inequality holds:*

$$\left|\int_a^b f(x)dx\right| \le \int_a^b |f(x)|\, dx. \tag{10.30}$$

Theorem 10.12 (Wirtinger inequality). *Assume that $f'(x)$ is continuous on $[0,\pi]$ and $f(0)=f(\pi)=0$. Then*

$$\int_0^{\pi}[f'(x)]^2\,dx \geq \int_0^{\pi}[f(x)]^2\,dx. \tag{10.31}$$

With equality if and only if $f(x)=A\sin x$.

Theorem 10.13 (Jensen's inequality). *Let f be a real-valued function defined on $[0,1]$ and $\varphi(x)$, a convex function on $f([0,1])$. Then*

$$\varphi\left(\int_0^1 f(x)dx\right) \leq \int_0^1 \varphi(f(x))\,dx. \tag{10.32}$$

(*See also* (20.25) *on page* 460.)

Remark. In this elementary form of Jensen's inequality, the interval $[0,1]$ can be replaced by any interval of length 1.

10.4 Methods of Integration

10.4.1 *Symmetry; even and odd functions (II)*

If there is some kind of symmetry between function and the integration interval, then the calculation of the integral can be substantially simplified.

Theorem 10.14. *If $f(x)$ is continuous in the interval $[-a,a]$, then we have that*

$$f(x)\text{ odd} \Rightarrow \int_{-a}^{a} f(x)dx = 0. \tag{10.33}$$

$$f(x)\text{ even} \Rightarrow \int_{-a}^{a} f(x)dx = 2\int_0^a f(x)dx. \tag{10.34}$$

In the figures below, $A=\displaystyle\int_0^a f(x)dx$.

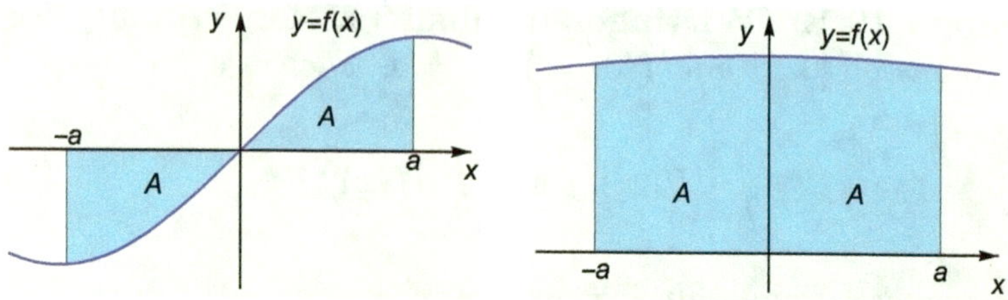

Further, if $F(x)$ is a primitive function of $f(x)$, then

$$\begin{aligned} f(x) \text{ even} &\iff F(x) - F(0) \text{ odd.} \\ f(x) \text{ odd} &\iff F(x) \text{ even.} \end{aligned} \tag{10.35}$$

10.4.2 *Integration by parts*

Theorem 10.15. *If f is a continuous function, F a primitive function of f, and g a continuously differentiable function (in the interval $[a, b]$ in (ii)), then*

$$\begin{aligned} &\text{(i)} \quad \int f(x)g(x)dx = F(x)g(x) - \int F(x)g'(x)dx. \\ &\text{(ii)} \quad \int_a^b f(x)g(x)dx = [F(x)g(x)]_a^b - \int_a^b F(x)g'(x)dx. \end{aligned} \tag{10.36}$$

Remarks. Integration by parts constitutes the identities (10.36).

The terms $F(x)g(x)$ and $[F(x)g(x)]_a^b$ in (10.36) are called outintegrated terms. In (ii), this term equals $F(b)g(b) - F(a)g(a)$.

If in (10.36) $f(x) = p(x)$ is a polynomial: To perform integration by parts it is convenient to choose:

- $f(x) = p(x)$, and

$$g(x) = \begin{cases} \ln q(x), \\ \arcsin q(x), \\ \arctan q(x), \end{cases}$$

where $q(x)$ is a polynomial.

- Or $p(x) = g(x)$ and

$$f(x) = \begin{cases} e^{kx+m}, \\ \sin(kx+m), \\ \cos(kx+m), \end{cases}$$

adjusting the notations as in (10.36).

Integration of product of polynomial $p(x)$ and exponential function $e^{k\,x}$, where $k \neq 0$, can be performed by *substituting.*

$$\int p(x)\, e^{kx} dx = q(x) e^{kx} + C,$$

where $\operatorname{degree} p = \operatorname{degree} q$. The polynomial $q(x)$ then satisfies

$$q'(x) + k\, q(x) = p(x).$$

Integration of the product of polynomial $p(x)$ and $\cos k\, x$ can be performed using the ansatz

$$\int p(x) \cos k\, x\, dx = q_1(x) \cos k\, x + q_2(x) \sin k\, x + C,$$

where $\deg p = \deg q_2 = 1 + \deg q_1$. Integration of the product of polynomial $p(x)$ and $\sin k\, x$ can be performed using the ansatz

$$\int p(x)\, \sin k\, x\, dx = q_1(x) \sin k\, x + q_2(x) \cos k\, x + C,$$

where $\deg p = \deg q_2 = 1 + \deg q_1$.

10.4.3 *Variable substitution*

Theorem 10.16. *If $x = x(t)$ is a continuously differentiable function of t and f is a continuous function, then*

$$\int f(x) dx = \int f(x(t)) \frac{dx}{dt} dt. \tag{10.37}$$

Remarks. It is sufficient that the "old" variable (x) is a function of the "new" variable (t), and $x'(t) = \frac{dx}{dt}$ is continuous.

If for instance $t = \ln x$, then the *differentiation* $dt = \frac{1}{x}dx$ is equivalent to the following derivation:

$$\frac{dt}{dx} = \frac{1}{x}.$$

Theorem 10.17. *If the function $f(x)$ is continuous in the interval $[a, b]$, $x(t)$ is a continuously differentiable function of t on $[\alpha, \beta]$ or $[\beta, \alpha]$ and $x(\alpha) = a$, $x(\beta) = b$, then*

$$\int_a^b f(x)dx = \int_\alpha^\beta f(x(t))\frac{dx}{dt}dt. \tag{10.38}$$

Remarks. In the above theorem, let $t_1 = \alpha_1$ and $t_2 = \alpha_2$ be two distinct points such that $x(\alpha_1) = x(\alpha_2) = a$. Then, one can easily verify that the value of the integral is independent of the choice of the integration bound α_1 or α_2.

Note further that $x(t)$ should be defined in an interval $[\alpha, \beta]$ with $x([\alpha, \beta]) = [a, b]$.[1]

Suppose that we have the reverse order variable substitution from x to t. Then, it is important to have $x = x(t)$ for $t \in [\alpha, \beta]$, i.e., that x is a function of t.

If the substitution from x to t is presented in the form $t = t(x)$, then it is important that the relation has an inverse $(x(t))$ (in the integration interval).

Theorem 10.18. *If the integrand contains an inner derivative as a multiplicative factor, i.e., it is of the form $f(t(x))t'(x)$, then*

$$\int f(t(x))t'(x)dx = \int f(t)dt. \tag{10.39}$$

In particular,

$$\int \frac{f'(x)}{f(x)}dx = \ln |f(x)| + C.$$

[1]Note that the image of a compact interval is a compact interval, since we assume that $x(t)$ is continuous.

Theorem 10.19. *If a differentiable function $y = f(x)$ is invertible, then the following formula holds true:*

$$\int y dx = xy - \int x dy \quad (\text{where } y = f(x), x = f^{-1}(y)). \quad (10.40)$$

10.4.4 The $\tan \dfrac{x}{2}$—substitution

Integration of functions of type $f(\cos x, \sin x)$:
For this type of functions the appropriate variable substitution is $t = \tan \frac{x}{2}$, which yields

$$\begin{aligned} \cos x &= \frac{1-t^2}{1+t^2}, \quad \sin x = \frac{2t}{1+t^2}, \\ \tan x &= \frac{2t}{1-t^2}, \quad dx = \frac{2dt}{1+t^2}. \end{aligned} \quad (10.41)$$

This substitution gives rise to the equality

$$\int f(\cos x, \sin x) dx = \int f\left(\frac{1-t^2}{1+t^2}, \frac{2t}{1+t^2}\right) \frac{2}{1+t^2} dt. \quad (10.42)$$

10.5 Improper Integral

Definition 10.8.

(i) Assume that $f(x)$ is continuous in $[a, \infty)$

$$\lim_{b \to \infty} \int_a^b f(x) dx = \int_a^\infty f(x) dx. \quad (10.43)$$

The integral on the RHS of (10.43) is called an improper integral. More specifically, (10.43) is improper in ∞. An improper integral at $-\infty$ is defined analogously.

(ii) Assume that $|f| \to \infty$, as $x = b \to c_-$. Then

$$\lim_{b \to c_-} \int_a^b f(x) dx = \int_a^c f(x) dx \quad (10.44)$$

is an improper integral with respect to its upper limit $x = c$. Similarly, an improper integral with respect to the lower limit $x = a$ is defined.

(iii) The integrals in i and ii are said to be convergent if their corresponding limit exists. Otherwise, the integral is divergent.

(iv) A *conditionally convergent* improper integral of f is convergent, while the improper integral of $|f|$ is divergent.

(v) If the integral in (10.43) or in (10.44) has integrand $f(x) = |g(x)|$ and is convergent,

$$\int_a^\infty f(x)dx \quad \text{and} \quad \int_a^c f(x)dx,$$

are *absolutely convergent.*

Remarks. Examples of improper integrals can be found on page 224 and subsequent pages.

The above definitions (over unbounded interval or an unbounded range) both are referred to as improper integral, and together they represent integration over an unbounded surface.[2]

A definite integral over whole real line; $(-\infty, \infty)$ or over the positive part of the real line $(0, \infty)$ can be written as

$$\int_{\mathbb{R}} \cdot \quad \text{and} \quad \int_{\mathbb{R}_+} \cdot, \text{ respectively.}$$

Theorem 10.20. *Let $p(x)$ be a polynomial (in variable x). Then*

$$\int_0^\infty p(x)e^{-kx}dx \text{ is convergent if and only if } k > 0.$$

$$\int_0^1 p(\ln x)dx \text{ is convergent.}$$

Theorem 10.21. *In the following, $f(x)$ and $g(x)$ are assumed to be continuous on the interval (a, b) with $-\infty \leq a < b \leq \infty$, and we write an improper integral just like*

$$\int_a^b f(x)\,dx.$$

[2]One can extend the definition, but this formulation is appropriate.

(i) *An absolutely convergent integral is convergent, i.e.,*

$$\int_a^b |f(x)|\,dx \textit{ convergent} \implies \int_a^b f(x)\,dx \textit{ convergent}.$$

Theorem 10.22.

(i) *Assume that* $f(x) \geq 0$*. Then*

$$\int_a^b f(x)dx \textit{ divergent} \iff \int_a^b f(x)dx = \infty.$$

(ii) *Assume that* $f(x) \geq g(x) \geq 0$*. Then*

$$\int_a^b f(x)dx \text{ convergent} \implies \int_a^b g(x)dx \text{ convergent} \tag{10.45}$$

(the comparison criterion).

(iii) *Suppose that the functions* $f(x)$ *and* $g(x)$ *are continuous and non-negative on* $[a, b)$ *with lower limit* a *a real number and with* $\lim_{x\to b} \frac{f(x)}{g(x)} = C$.

If $0 < C < \infty$*, then*

$$\int_a^b f(x)dx \textit{ convergent} \iff \int_a^b g(x)dx \textit{ convergent}.$$

If $C = 0$*, then*

$$\int_a^b f(x)dx \textit{ convergent} \impliedby \int_a^b g(x)dx \textit{ convergent}.$$

If $C = \infty$*, then*

$$\int_a^b f(x)dx \textit{ convergent} \implies \int_a^b g(x)dx \textit{ convergent}.$$

Theorem 10.23 (The integral criterion). *Assume that* $f(x)$ *is a non-negative, decreasing function on* $[1, \infty)$*. Then,*

$$\int_1^\infty f(x)dx \text{ convergent} \iff \sum_{k=1}^{\infty} f(k) \text{ convergent}. \tag{10.46}$$

10.6 Tables

Primitive functions of elementary functions can be found on page 172. The functions in the right column are primitive functions (indefinite integrals) to corresponding functions in the left column. The case $n = 1$ is found in (8.21) page 172.

10.6.1 *Common indefinite integrals with algebraic integrand*

$$\int \frac{ax+b}{cx+d}\,dx = \frac{ax}{c} + \frac{(bc-ad)\ln(cx+d)}{c^2} + C,\ c \neq 0$$

$$\int \frac{dx}{(ax+b)(cx+d)} = \frac{1}{ad-bc}\ln\left|\frac{ax+b}{cx+d}\right| + C,\ ad-bc \neq 0$$

$$\int (ax+b)\sqrt{cx+d}dx$$

$$= \sqrt{cx+d}\left(\frac{2d\,(5bc-2ad)}{15c^2} + \frac{2\,(5bc+ad)\,x}{15c} + \frac{2ax^2}{5}\right) + C$$

$$\int \frac{dx}{(1+x^2)^2} = \frac{x}{2(1+x^2)} + \frac{1}{2}\arctan x + C$$

$$\int \frac{dx}{x^3+a^3} = \frac{1}{a^2\sqrt{3}}\arctan\left(\frac{2x-a}{a\sqrt{3}}\right)$$

$$+ \frac{1}{6a^2}\left(2\ln(x+a) - \ln(x^2-ax+a^2)\right) + C$$

$$\int \frac{dx}{x^4+a^4} = \frac{1}{2a^3\sqrt{2}}\left[\arctan\left(\frac{x\sqrt{2}}{a}-1\right) + \arctan\left(\frac{x\sqrt{2}}{a}+1\right)\right]$$

$$+ \frac{1}{4a^3\sqrt{2}}\left[\ln\left(x^2+ax\sqrt{2}+a^2\right) - \ln\left(x^2-ax\sqrt{2}+a^2\right)\right] + C$$

$$\int \frac{ax+b}{\sqrt{cx+d}}dx = \frac{2\sqrt{cx+d}\,(acx+3bc-2ad)}{3c^2} + C$$

$$\int \frac{\sqrt{a^2-x^2}dx}{x} = \sqrt{a^2-x^2} - a\ln|\frac{a+\sqrt{a^2-x^2}}{x}| + C$$

$$\int (ax+b)\sqrt{c-x^2}dx = \sqrt{c-x^2}\left(\frac{ax^2}{3}+\frac{bx}{2}-\frac{ac}{3}\right) + \frac{1}{2}bc\arctan\left(\frac{x}{\sqrt{c-x^2}}\right)+C$$

$$\int \frac{ax+b}{x^2+px+q}dx = \frac{(2b-ap)\arctan\left(\frac{2x+p}{\sqrt{4q-p^2}}\right)}{\sqrt{4q-p^2}} + \frac{a\ln(x^2+px+q)}{2}+C, \text{ if } 4q-p^2>0$$

$$\int \frac{ax+b}{x^2+px+q}dx = \frac{a}{2}\ln|x^2+px+q| + \frac{(2b-ap)}{4c}\ln\left|\frac{2x+p-2c}{2x+p+2c}\right|+C, \quad \text{if } c^2=p^2-4q>0. \tag{10.47}$$

10.6.2 *Common indefinite integrals with non-algebraic integrands*

$$\int \ln(ax)dx = x\ln(ax)-x+C$$

$$\int \frac{1}{x}(\ln|x|)^n dx \begin{cases} \ln|\ln x|+C, & \text{if } n=-1 \\ \frac{1}{n+1}(\ln x)^{n+1}+C, & \text{if } n\neq -1 \end{cases}$$

$$\int x^n \ln x dx = \frac{x^{n+1}}{n+1}\left(\ln x - \frac{1}{n+1}\right)+C, \quad n\neq -1$$

$$\int \arctan\sqrt{x}dx = (x+1)\arctan\sqrt{x}-\sqrt{x}+C$$

$$\int \arcsin x dx = \sqrt{1-x^2}+x\arcsin x+C$$

$$\int \frac{dx}{\cosh x} = 2\arctan(\tanh(x/2))+C$$

$$\int \frac{dx}{\sinh x} = \ln\left|\tanh\left(\frac{x}{2}\right)\right|+C$$

$$\int \frac{1}{\sqrt{e^x - 1}}\, dx = 2\arctan\left(\sqrt{e^x - 1}\right) + C$$

$$\int e^{ax} \sin bx dx = \frac{e^{ax}}{a^2 + b^2}(a \sin bx - b \cos bx) + C$$

$$\int e^{ax} \cos bx dx = \frac{e^{ax}}{a^2 + b^2}(b \sin bx + a \cos bx) + C$$

$$\int \sin(\ln x) = \frac{x}{2}\left(\sin(\ln x) - \cos(\ln x)\right) + C$$

$$\int \cos(\ln x) = \frac{x}{2}\left(\sin(\ln x) + \cos(\ln x)\right) + C$$

$$\int \tan^{2n} x\, dx = \sum_{k=1}^{n} (-1)^{n-k} \frac{\tan^{2k-1} x}{2k - 1} + (-1)^n x + C, \quad n = 1, 2, \ldots$$

$$\int \tan^{2n+1} x\, dx = \sum_{k=1}^{n} (-1)^{n-k} \frac{\tan^{2k} x}{2k} + (-1)^{n-1} \ln|\cos x| + C,$$

$$n = 0, 1, \ldots \tag{10.48}$$

For $\alpha \neq 0$, we have

$$\int \frac{\sqrt{x^\alpha + 1}}{x}\, dx = \frac{1}{\alpha}\left(2\sqrt{x^a + 1} + \ln\left|1 - \sqrt{x^\alpha + 1}\right| - \ln\left|1 + \sqrt{x^\alpha + 1}\right|\right) + C,$$

$$\int \frac{dx}{x\sqrt{x^\alpha + 1}} = \frac{1}{\alpha}\left(\ln\left|1 - \sqrt{x^\alpha + 1}\right| - \ln\left|1 + \sqrt{x^\alpha + 1}\right|\right) + C. \tag{10.49}$$

10.6.3 *Some integrals with trigonometric integrands*

$f(x)$	$F(x)+C$
$\dfrac{1}{\cos^2 x} = 1+\tan^2 x$	$\tan x + C$
$\dfrac{1}{\sin^2 x} = 1+\cot^2 x$	$-\cot x + C$
$\dfrac{1}{\sin x}$	$\ln\left\lvert\tan\dfrac{x}{2}\right\rvert + C$
$\dfrac{1}{\cos x}$	$\ln\left\lvert\tan\left(\dfrac{x}{2}+\dfrac{\pi}{4}\right)\right\rvert + C$
$\sin^2 x$	$\dfrac{x}{2} - \dfrac{1}{4}\sin 2x + C$
$\cos^2 x$	$\dfrac{x}{2} + \dfrac{1}{4}\sin 2x + C$
$\sin^3 x$	$\dfrac{\cos^3 x}{3} - \cos x + C$
$\cos^3 x$	$\sin x - \dfrac{\sin^3 x}{3} + C$
$\sin^4 x$	$\dfrac{3x}{8} - \dfrac{1}{4}\sin 2x + \dfrac{1}{32}\sin 4x + C$
$\cos^4 x$	$\dfrac{3x}{8} + \dfrac{1}{4}\sin 2x + \dfrac{1}{32}\sin 4x + C$

10.6.4 *Recursion formulas*

$$\int (x^2+px+q)^n\,dx = \frac{1}{(2n+1)}\left[(x+p/2)(x^2+px+q)^n \right.$$

$$\left. + 2n(q-(p/2)^2)\int (x^2+px+q)^{n-1}\,dx\right],$$

$$n = 1, 2, \ldots$$

$$\int x^m (\ln x)^n\,dx = \begin{cases} \dfrac{x^{m+1}}{m+1}(\ln x)^n - \dfrac{n}{m+1}\displaystyle\int x^m (\ln x)^{n-1}\,dx, \\ \quad \text{if } \neq -1. \\ \displaystyle\int \dfrac{(\ln x)^n}{x}\,dx = \dfrac{(\ln x)^{n+1}}{n+1} + C, \\ \quad \text{if } m = -1. \end{cases}$$

$$\int x^n e^{ax^2}dx = \frac{x^{n-1}}{2a}\cdot e^{ax^2} - \frac{n-1}{2a}\int x^{n-2}e^{ax^2}dx,$$
$$n = 2, 3, \dots$$

$$\int_0^\infty \cos x\left(\frac{\sin x}{x}\right)^n dx = \frac{n}{n+1}\int_0^\infty \left(\frac{\sin x}{x}\right)^{n+1} dx, \quad n = 1, 2, \dots$$

$$\int \tan^n x\,dx = \frac{\tan^{n-1} x}{n-1} - \int \tan^{n-2} x\,dx, \quad n = 2, 3, \dots$$

$$\int_0^\pi x\,f(\sin x)dx = \frac{\pi}{2}\int_0^\pi f(\sin x)dx,$$
$$\text{if } f \text{ iscontinuous in } [0,1]. \tag{10.50}$$

10.6.5 *Tables of some definite integrals*

Powers of sine and cosine

$$\int_0^{\pi/2} \sin^{2n-1} xdx = \int_0^{\pi/2} \cos^{2n-1} xdx = \frac{(2n-2)!!}{(2n-1)!!}, \quad n = 1, 2, \dots$$
$$\int_0^{\pi/2} \sin^{2n} xdx = \int_0^{\pi/2} \cos^{2n} xdx = \frac{(2n-1)!!}{(2n)!!}\cdot\frac{\pi}{2}, \quad n = 1, 2, \dots \tag{10.51}$$

10.6.6 *Tables of improper integrals*

Elementary integrals

$$\int_1^\infty x^{-n}\ln x\,dx = \frac{1}{(n-1)^2},\ n = 2, 3, \dots$$
$$\int_0^\infty x^n e^{-kx}dx = \frac{n!}{k^{n+1}}, \quad k > 0,\ n = 0, 1, \dots \tag{10.52}$$

In the following table, most integrals are elementary.

Theorem 10.24. *Let α be a real number, then the following integrals are convergent under the corresponding conditions on α.*

$$\int_1^\infty \frac{dx}{x^\alpha} = \frac{1}{\alpha-1}, \quad \alpha > 1.$$

$$\int_0^1 \frac{dx}{x^\alpha} = \frac{1}{1-\alpha}, \quad \alpha < 1.$$

$$\int_1^\infty \frac{(\ln x)^n\,dx}{x^\alpha} = \frac{n!}{(\alpha-1)^{n+1}}, \quad \alpha > 1,\ n = 0, 1, 2, \ldots$$

$$\int_0^1 \frac{(\ln x)^n\,dx}{x^\alpha} = -\frac{n!}{(1-\alpha)^{n+1}}, \quad \alpha < 1,\ n = 0, 1, 2, \ldots$$

$$\int_0^{\pi/2} \tan^\alpha x\,dx = \frac{\pi}{2\cos(\alpha\,\pi/2)}, \quad -1 < \alpha < 1. \tag{10.53}$$

$$\int_1^\infty \frac{dx}{x\sqrt{x^\alpha+1}} = \frac{2\ln\left(1+\sqrt{2}\right)}{\alpha}, \quad \alpha > 0.$$

$$\int_0^\infty \frac{dx}{(x^2+1)^\alpha} = \frac{\Gamma\left(\alpha-\frac{1}{2}\right)\sqrt{\pi}}{2\Gamma(\alpha)}, \quad \alpha > 1/2.$$

$$\int_1^\infty \frac{dx}{x^\alpha\sqrt{\ln x}} = \sqrt{\frac{\pi}{\alpha-1}}, \quad \alpha > 1.$$

$$\int_0^\infty \frac{\arctan x\,dx}{x^\alpha} = \frac{\pi}{2(\alpha-1)\,\sin(\alpha\pi/2)}, \quad 1 < \alpha < 2.$$

10.6.7 *Tables of some non-elementary integrals*

Definition 10.9. By a non-elementary integral we mean an integral whose primitive function cannot be expressed in terms of elementary functions.

$$\int_0^\infty \frac{xdx}{e^x-1} = 2\int_0^\infty \frac{xdx}{e^x+1} = \frac{1}{4}\int_{-\infty}^\infty \frac{x^2e^xdx}{(e^x-1)^2} = \frac{\pi^2}{6}$$

$$\int_0^\infty \frac{x^3dx}{e^x-1} = \frac{\pi^4}{15}$$

$$\int_0^\infty \frac{\sin x}{x}dx = \int_0^\infty \left(\frac{\sin x}{x}\right)^2 dx = \frac{\pi}{2} \tag{10.54}$$

$$\int_{-\infty}^\infty x^2e^{-x^2/2}dx = \int_{-\infty}^\infty e^{-x^2/2}dx = \sqrt{2\pi} \quad (*)$$

$$\int_0^{\pi/2} \ln(\sin x)dx = \int_0^{\pi/2} \ln(\cos x)dx = -\frac{\pi}{2}\ln 2$$

$$\int_0^\pi x\ln(\sin x)dx = -\frac{\pi^2\ln 2}{2}$$

$$\int_0^{\pi/4} \ln(1+\tan x)dx = \frac{\pi\ln 2}{8}$$

$$\int_0^\infty \frac{e^{-bx^c}-e^{-ax^c}}{x}dx = \frac{1}{c}\ln\frac{b}{a}, \quad \text{if } a>0,\ b>0,\ c>0.$$

$$\int_0^\infty \cos(x^2)dx = \int_0^\infty \sin(x^2)dx = \sqrt{\frac{\pi}{8}}. \tag{10.55}$$

$$\int_0^\infty \frac{x^\alpha dx}{e^x-1} = \Gamma(\alpha+1)\cdot\zeta(\alpha+1), \quad \alpha>0. \tag{10.56}$$

$$\int_{-\infty}^\infty x^{2n}e^{-x^2/2}dx = (2n-1)!!\cdot\sqrt{2\pi}, \quad n=0,1,2,\dots \tag{10.57}$$

The first two integrals in (10.55) are special cases of the integrals in (10.56).

(∗) in (10.54) is the special case of (10.57). Note that $(-1)!! = 1$.

$$\int_0^\infty \frac{\sin x}{\sqrt{x}}\,dx = \int_0^\infty \frac{\cos x}{\sqrt{x}}\,dx = \sqrt{\frac{\pi}{2}}$$
$$\int_0^\infty \frac{\sin^3 x}{x^3}\,dx = \frac{3\pi}{8}, \quad \int_0^\infty \frac{\sin^4 x}{x^4}\,dx = \frac{\pi}{3} \tag{10.58}$$
$$\int_0^\infty \frac{\sin^5 x}{x^5}\,dx = \frac{115\pi}{384}.$$

$$\begin{array}{c|c}
\int_0^1 \frac{\ln x}{x-1}\,dx = \frac{\pi^2}{6} & \int_0^1 \frac{\ln x}{x+1}\,dx = -\frac{\pi^2}{12} \\
\int_0^1 \frac{\ln x}{\sqrt{1-x^2}}\,dx = -\frac{1}{2}\pi\ln 2 & \int_0^\infty \frac{\ln x}{x^2+1}\,dx = 0 \\
\int_0^\infty \frac{\ln x}{(x^2+1)^2}\,dx = -\frac{\pi}{4} & \int_0^\infty \frac{\ln x}{(x^2+1)^3}\,dx = -\frac{\pi}{4} \\
\int_1^\infty \frac{\ln x}{x(x^2+1)}\,dx = \frac{\pi^2}{48} & \\
\int_0^\infty \frac{x^{\alpha-1}}{x+1}\,dx = \frac{\pi}{\sin(\pi\,\alpha)} & \int_0^1 \frac{x^{\alpha-1}}{(1-x)^\alpha}\,dx = \frac{\pi}{\sin(\pi\,\alpha)} \\
0 < \alpha < 1 & 0 < \alpha < 1
\end{array} \tag{10.59}$$

Definition 10.10 (Elliptic integrals). *Incomplete elliptic integrals*

$$\begin{array}{ll}
\int_0^\varphi (1 - m\sin^2\theta)^{-1/2} d\theta & \text{Elliptic integral of order 1} \\
\int_0^\varphi (1 - m\sin^2\theta)^{1/2} d\theta & \text{Elliptic integral of order 2} \\
\int_0^\varphi (1 - n\sin^2\theta)^{-1}(1 - m\sin^2\theta)^{-1/2} d\theta & \text{Elliptic integral of order 3.}
\end{array} \tag{10.60}$$

With $\varphi = \pi/2$ the corresponding **complete elliptic integrals are obtained.**

Definition 10.11. The convolution of two functions f and g is defined as

$$(f * g)(x) := \int_{-\infty}^{\infty} f(x-y)g(y)dy, \tag{10.61}$$

when the integral exists.

Theorem 10.25. *Convolution satisfies*

$$\begin{aligned}(f * g) * h(x) &= f * (g * h)(x), \quad &\text{(associative)}.\\ (f * g)(x) &= (g * f)(x), \quad &\text{(commutative)}.\end{aligned}$$

Remarks.

(i) $\Gamma(x)$, introduced in (10.56), is the gamma function for complex x, where $\operatorname{Re} x > 0$. Especially $\Gamma(x) = (x-1)!$ for integers $x = 1, 2, 3, \ldots$.
The gamma function satisfies

$$\Gamma(x+1) = x\,\Gamma(x) \text{ and } \Gamma(x) = \int_0^1 (-\ln t)^{x-1} dt.$$

(ii) By *defining* the logarithm $\ln x = \int_1^x \frac{dt}{t}$, one may prove logarithm laws, define a power a^y, and derive the corresponding power laws.

(a) For instance, by a variable substitution, one can show that $\ln(ab) = \ln a + \ln b$, viz.

$$\int_1^{ab} \frac{dt}{t} = \int_1^a \frac{dt}{t} + \int_a^{ab} \frac{dt}{t}.$$

In the last integral, let $t = as$. Then for $b \geq 1$, $a \leq t = as \leq ab$ (if $b < 1$, the result follows in a similar way) and $a\,ds = dt$, hence,

$$\int_a^{ab} \frac{dt}{t} = \int_1^b \frac{a\,ds}{as} = \ln b.$$

(b) If $a = 1$, then $\ln a = 0$. Assume that $a > 0$ and $a \neq 1$. Define the "a-logarithm" $\log_a x := \dfrac{\ln x}{\ln a}$, giving $D \log_a x = \frac{1}{x \ln a} \neq$

0, i.e., $\log_a x = y$ is invertible. Now *define* the power a^y as the inverse of this function.

$$\log_a x = y \Leftrightarrow x = a^y.$$

Especially it follows that

$$\log_a x = \frac{\ln x}{\ln a} = \frac{\ln a^y}{\ln a} = y \text{ i.e., } \ln a^y = y \ln a.$$

10.6.8 *The Dirac function*

Dirac's delta function $\delta(x)$ is actually not a function in the usual sense. It belongs to a larger class, the so-called generalized functions or distributions. Nevertheless, it can be intuitively defined by

$$f_\varepsilon(x) = \begin{cases} 0, & \text{if } x < 0, \\ 1/\varepsilon, & \text{if } 0 \le x \le \varepsilon, \\ 0, & \text{if } x > \varepsilon. \end{cases}$$

Then

$$\int_0^\infty f_\varepsilon(x)dx = 1.$$

Definition 10.12. $\delta(x)$ is defined as

$$\lim_{\varepsilon \to 0} f_\varepsilon(x)dx =: \int_0^\infty \delta(x)dx. \tag{10.62}$$

Theorem 10.26. $\delta(x)$ *has the following evaluation property.*

$$\int_0^\infty \delta(x-a)g(x)dx = g(a), \text{ if } g \text{ is continuous.} \tag{10.63}$$

10.7 Numerical Integration

Definition 10.13 (The rectangle method (Midpoint rule)). Assume that $f(x)$ is continuous on $[a, b]$. Let

$$\Delta x = \frac{b-a}{n}, \quad a = x_0 \text{ and } x_k = x_0 + k\Delta x,\ k = 0, 1, 2, \ldots, n,$$

so that $x_n = b$. Then, the rectangular method is given by

$$R_n := \Delta x \sum_{k=1}^{n} f\left(\frac{x_{k-1} + x_k}{2}\right). \tag{10.64}$$

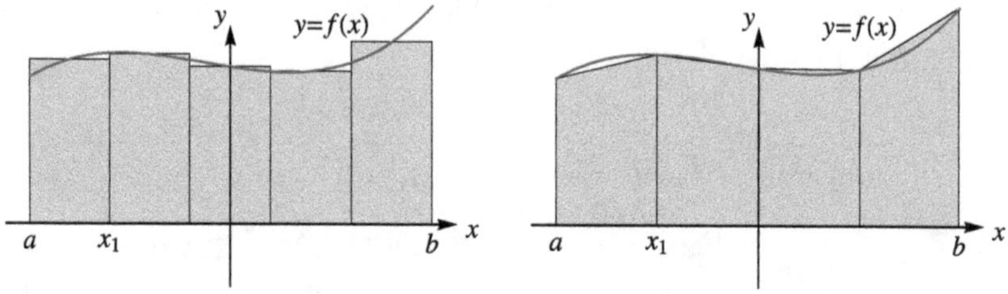

Definition 10.14 The secant method (The trapezoidal rule). Assume that $f(x)$ is continuous and $[a, b]$. Let

$$\Delta x = \frac{b-a}{n}, \ a = x_0, \quad \text{and} \quad x_k = x_0 + k\Delta x, \ k = 0, 1, 2, \ldots, n,$$

so that $x_n = b$. Then the secant method is given by

$$T_n := \Delta x \left[\frac{f(a) + f(b)}{2} + \sum_{k=1}^{n-1} f(x_k) \right]. \tag{10.65}$$

Simpson's formula

Assume that f is four times continuously differentiable, and $a < b$. Let

$$\Delta x = \frac{b-a}{2n}, \ a = x_0, \quad \text{and} \quad x_k = x_0 + k\Delta x, \ k = 0, 1, 2, \ldots, 2n,$$

so that $x_{2n} = b$. Then the following expression is Simpson's formula:

$$S_n = \frac{b-a}{6n} \sum_{k=1}^{n} \left(f(x_{2k-2}) + 4f(x_{2k-1}) + f(x_{2k}) \right).$$

It yields

$$\int_a^b f(x)dx = \frac{b-a}{6n} \sum_{k=1}^{n} \left(f(x_{2k-2}) + 4f(x_{2k-1}) + f(x_{2k}) \right) + R, \tag{10.66}$$

where $R = \dfrac{b-a}{180}(\Delta x)^4 f^{(4)}(\xi)$ is the remainder at some point $\xi \in (a, b)$.

Remarks. The rectangular method means a Riemann sum.

An open version of Simpson's formula is (10.66)

$$\frac{b-a}{n}(f(a)+f(b)+4(f(x_1)+f(x_3)+\cdots+f(x_{2n-1}))$$
$$+2(f(x_2)+f(x_4)+\cdots+f(x_{2n-2})))\,.$$

A connection between these three numerical methods of integration.

Let the interval $[a,b]$ be partitioned into $2n$ subintervals of equal size:

$$a =: x_0 < x_1 < \cdots < x_{2n-1} < x_{2n} := b \quad \text{and} \quad x_k - x_{k-1} = \frac{b-a}{2n}.$$

Then $x_{2k} - x_{2k-2} = \dfrac{b-a}{n}$. Let

$$R_n = \frac{b-a}{n}\sum_{k=1}^{n} f(x_{2k-1}), \quad \text{and}$$

$$T_n = \frac{b-a}{2n}\sum_{k=1}^{n} f(x_{2k-2}) + f(x_{2k}).$$

With S_n given by (10.66), the following equality holds true:

$$\frac{2}{3}R_n + \frac{1}{3}T_n = S_n. \tag{10.67}$$

Chapter 11

Differential Equations

An ordinary differential equation (in the following, abbreviated DE or ODE) is an equation containing derivatives of the function $y(x)$, with $x \in \mathbb{R}$. A differential equation with independent variable $\boldsymbol{x} \in \mathbb{R}^n$, $n = 2, 3, \ldots$, and derivatives with respect to more than one variable is called partial differential equation (PDE). The highest derivative in a DE is the order of the DE.

11.1 ODEs of Order 1 and 2

The equation

$$f(y)g(x) = y' \tag{11.1}$$

is solved by *separation of variables* for those y such that $f(y) \neq 0$:

$$\underbrace{g(x)dx = \frac{dy}{f(y)}}_{\text{separation of variables}} \quad \Longleftrightarrow \int g(x)dx = \int \frac{dy}{f(y)}. \tag{11.2}$$

If $f(0) = 0$, then $y = y(x) \equiv 0$, is its *singular solution*. The equation

$$f(y)g(y') = y'' \tag{11.3}$$

is solved by setting $p(y) = \dfrac{dy}{dx}$ whereby the equation can be written as

$$f(y)g(p) = p\frac{dp}{dy}, \tag{11.4}$$

and then can be integrated as in (11.2).

11.2 Linear ODE

A *differential operator* of the form

$$L(y) := \sum_{k=0}^{n} g_k(x)\, y^{(k)}(x) \tag{11.5}$$

is referred as *linear differential operator.*
L is linear in the sense that for two, sufficiently regular functions y_1 and y_2 and a constant C,

$$\begin{aligned} L(y_1 + y_2) &= L(y_1) + L(y_2) \\ L(Cy_1) &= CL(y_1). \end{aligned} \tag{11.6}$$

A linear differential equation is of the form

$$L(y) = \sum_{k=0}^{n} g_k(x)\, y^{(k)}(x) = f(x). \tag{11.7}$$

11.2.1 *Linear ODE of first order*

Definition 11.1. A linear ODE of first order can be written as

$$y' + f(x)y = g(x). \tag{11.8}$$

Theorem 11.1. *The solution of* (11.8) *is given by*

$$y = y_p + y_h = e^{-F(x)} \int e^{F(x)} g(x)dx + \underbrace{Ce^{-F(x)}}_{= y_h}, \tag{11.9}$$

where F is a primitive function of f and C, an arbitrary constant.

Remark. $e^{F(x)}$ is called *integrating factor*, abbreviated IF.

Note that the integral on the right-hand side of (11.9) is an indefinite integral where F means all primitive functions of f. Hence, the integral itself contains a constant C. One inserts yet another constant C in the "non-integral" part in (11.9). This is to keep in mind the homogeneous term $y_h = Ce^{-F(x)}$.

Sometimes one writes $y(x)$ to emphasize that y is a function of x. When this is obvious from the context, one only writes y.

A differential equation containing y' as its highest order derivative is of the first order.

A differential equation containing y'' as its highest order derivative is of the second order (and so on).

A differential equation of type $y' + f(x)y = g(x)$ is called linear DE of first order, likewise the DE $y' + ay = 0$, where a is a constant. This DE is called homogeneous (since its RHS $= 0$), with constant coefficients (here 1 and a).

The solution of (11.9) consists of two terms, y_p, corresponding to the particular RHS in the DE, that is $g(x)$ and y_h, the homogeneous solution.

11.3 Linear DE with Constant Coefficients

Definition 11.2. Let a_i be (complex) constants and $a_n \neq 0$. The differential equation

$$\text{DE} \quad a_n y^{(n)} + a_{n-1} y^{(n-1)} + \cdots + a_1 y^{prime} + a_0 y = g(x) \tag{11.10}$$

is linear of order n (in the variable x) with constant coefficients a_j, $j = 0,\ 1, \ldots,\ n$.

With the differential operator $D := \dfrac{d}{dx}$, generally $D^k := \dfrac{d^k}{dx^k}$, $k = 0, 1, 2, \ldots$, the DE can be written as

$$P(D)y = g(x), \quad \text{where the corresponding differential operator is}$$

$$P(D) = a_n D^n + a_{n-1} D^{(n-1)} + \cdots + a_1 D + a_0. \tag{11.11}$$

Let λ be a complex number. The characteristic polynomial of (11.11) is then

$$P(\lambda) = a_n \lambda^n + a_{n-1} \lambda^{(n-1)} + \cdots + a_1 \lambda + a_0$$

and the corresponding characteristic equation is (11.12)

$$P(\lambda) = 0.$$

For linear DE of degree 2, with constant coefficients, see page 240.

The boundary conditions for a differential equation are conditions on y and its derivatives at boundary points $x_i \in \partial\Omega$ (or a point on the boundary). The number of boundary conditions equals the order of the DE (they determine the integration constants).

$$y^{(0)}(x_0) = y_0,\, y^{(1)}(x_0) = y_1, \ldots, y^{(n-1)}(x_0) = y_{n-1}. \tag{11.13}$$

Heaviside's displacement rule
The following reformulation of $P(D)$ in (11.11) is known as Heaviside's displacement rule

$$P(D)(y \cdot e^{\alpha x}) = e^{\alpha x} P(D + \alpha) y. \tag{11.14}$$

11.3.1 *Solution of linear DE*

Theorem 11.2. *The solution y of* (11.10) *is the sum of y_h and y_p:*

$$y = y_h + y_p,$$

where

(i) y_h *is the solution of* (11.10) *with* $g(x) \equiv 0$.
Consider the polynomial $P(\lambda)$ (11.12), *let* $\lambda_r, r = 1, 2, \ldots, k$, *be its k different complex zeros of multiplicity n_r i.e.,*

$$P(\lambda) = a_n \prod_{r=1}^{k} (\lambda - \lambda_r)^{n_r}, \quad (n_1 + n_2 + \cdots + n_r = n). \tag{11.15}$$

Make the anstaz

$$y_h = \sum_{r=1}^{k} p_r(x) e^{\lambda_r x}, \tag{11.16}$$

where $p_r(x) = b_{n_r-1}x^{n_r-1} + b_{n_r-2}x^{n_r-2} + \cdots + b_1 x^1 + b_0$.
(ii) y_p *is a solution which solves* (11.10).

Due to the factorization of $P(\lambda)$ *in* (11.11), *the differential operator* $P(D)$ *can be written as*

$$P(D) = a_n \prod_{r=1}^{k} (D - \lambda_r)^{n_r}. \tag{11.17}$$

11.3.2 *Ansatz to determine* y_p

(i) If $g(x) = p(x)e^{\alpha x}$ in (11.10), where p is a polynomial of degree n, set $y_p(x) = q(x)e^{\alpha x}$ where $q(x)$ is a polynomial as follows:

(a) If α *is not* a root of $P(\lambda) = 0$, one puts $q(x)$ as a polynomial of the same degree as p.

(b) If $\alpha = \lambda = \lambda_r$ such that $P(\lambda_r) = 0$, $q(x)$ is chosen so that $\text{degree}\, q = n_r + \text{degree}\, p$, where n_r is the multiplicity of λ_r.

One can (technically) eliminate $e^{\alpha x}$ using displacement rule: $y = ze^{\alpha x}$. Then (11.14) gives

$$P(D)(y) = e^{\alpha x} P(D + \alpha)z = p(x)e^{\alpha x},$$

which is equivalent to

$$P(D + \alpha)z = p(x).$$

(ii) If $g(x) = p(x)\cos\beta x$ or $p(x)\sin\beta x$, one can change the RHS to $p(x)e^{i\beta x}$ and replace y on the LHS by w, and finally set $ze^{i\beta x} = w$.

This case can be reduced to case 1 above. Now one may use (11.14).

11.4 Linear DE with Continuous Coefficients

Definition 11.3.

$$L = L[y] = \sum_{k=0}^{n} a_k(x)\,\frac{d^k}{dx^k}, \quad a_n(x) \neq 0,$$

is called differential operator of degree n, where $a_0(x)$, $a_1(x), \ldots, a_n(x)$ are continuous functions defined on an interval I.

$$L[y] := a_n(x)y^{(n)} + a_{n-1}(x)y^{(n-1)} + \cdots + a_1(x)y' + a_0(x)y = g(x) \tag{11.18}$$

is a linear differential equation of order n.

With the boundary conditions on $x_0 \in I$ as

$$y(x_0) = y_0, \quad y'(x_0) = y_1, \ldots, y^{(n-1)}(x_0) = y_{n-1},$$

for some (complex) numbers $y_0, y_1, \ldots, y_{n-1}$, $y = y(x)$ is uniquely determined.

In particular, if all a_k are constant (as in (11.10)), then the n constants in the solution can be uniquely determined.

Theorem 11.3. *Euler's differential equation is given by*

$$a_n x^n y^{(n)}(x) + a_{n-1}x^{n-1}y^{(n-1)}(x) + \cdots + a_1 x y'(x) + a_0 y(x) = g(x), \tag{11.19}$$

where a_k are constants. By the substitutions: $x = e^t$ for $x > 0$ (or $x = -e^t$, for $x < 0$), (11.19) transforms to a linear differential equation with constant coefficients. The operators $D^k := \frac{d^k}{dx^k}$ and $T^k := \frac{d^k}{dt^k}$ fulfill

$$x^k \cdot D^k = T(T-1)\ldots(T-k+1) = \prod_{j=0}^{k-1}(T-j), \quad k = 1, 2, \ldots$$

The DE (11.19) is then equivalent to

$$a_n \prod_{j=0}^{n-1}(T-j)y + a_{n-1}\prod_{j=0}^{n-2}(T-j)y + \cdots + a_1 T y + a_0 y$$
$$= \begin{cases} g(e^t), & \text{if } x > 0, \\ g(-e^t), & \text{if } x < 0. \end{cases} \tag{11.20}$$

11.4.1 *Linear ODE of second order*

Definition 11.4. The operator

$$L[y] := p_0(x)y''(x) + p_1(x)y'(x) + p_2(x)y(x), \tag{11.21}$$

with $p_0(x) \neq 0$ being a linear differential operator of second order.

The equation

$$L[y] := p_0(x)y''(x) + p_1(x)y'(x) + p_2(x)y(x) = p_3(x), \tag{11.22}$$

with $p_0(x) \neq 0$ being a linear differential equation (linear ODE or DE) of second order. If $p_3(x) \equiv 0$, the ODE (11.22) is homogeneous.

A differential operator of type (11.21) is exact if the following equality holds:

$$p_0(x)y''(x) + p_1(x)y'(x) + p_2(x)y(x) = \frac{d}{dx}\left(A(x)y'(x) + B(x)y(x)\right), \tag{11.23}$$

for some functions $A, B \in \mathcal{C}^1$.

An integrating factor $v = v(x)$ is a function such that $v(x)\,L[y]$ is exact.

A function $v \in \mathcal{C}^2$ is an integrating factor if and only if v solves the *adjoint equation* of (11.21):

$$M[y] := \frac{d^2}{dx^2}(p_0(x)v(x)) - \frac{d}{dx}(p_1(x)v(x)) + p_2(x)v(x) = 0. \tag{11.24}$$

A differential equation for which $L(y) \equiv M(y)$ is called self-adjoint. $L(y)$ in (11.21) is self-adjoint if and only if

$$\frac{d}{dx}\left[p(x)\frac{dy}{dx}\right] + q(x)y(x) = 0. \tag{11.25}$$

Let f and g be two solutions for a homogeneous version of (11.21), (i.e., $p_3(x) \equiv 0$). Then every solution can be written as $af(x)+bg(x)$, where a and b are scalars (real or complex constants).

(i) The Wronskian, $W(f, g; x)$, for two differentiable functions f and g is defined as

$$W(f, g; x) := f(x)g'(x) - f'(x)g(x). \tag{11.26}$$

(ii) If $W(f, g; x)$ = (for short) = $W(x)$ is given from two linearly independent solutions, then $W(x) > 0$ or $W(x) < 0$ for all x.

(iii) If $W(f, g; x)$ is obtained by two linearly dependent solutions, f and g, then $W(x) \equiv 0$ for all x.

(a) Dividing (11.22) by $p_0 \neq 0$ yields the *normal* form

$$\begin{aligned} y'' + p(x)y' + q(x)y &= r(x), \\ \text{where} \quad p = p_1/p_0, \quad q = p_2/p_0, \quad r &= p_3/p_0. \end{aligned} \tag{11.27}$$

Let $f(x)$ and $g(x)$ be two solutions of (11.27). Then the Wronskian $W(x) = W(f, g; x)$ fulfills

$$W'(x) + p(x)\,W(x) = 0, \text{i.e., } W(f, g; x) = W(f, g; a)e^{-\int_a^x p(t)dt}. \tag{11.28}$$

(b) If $f(x)$ is a non-trivial solution to the homogenous DE (11.27), that is with $r(x) \equiv 0$ in (11.27), a linearly independent solution $g(x)$ of $f(x)$ is

$$g(x) = f(x) \int \frac{dx}{[f(x)]^2\, e^{\int p(x)dx}}.$$

(c) If p and q are constants, then the homogenous differential equation

$$L[y] = y'' + py' + qy = 0 \tag{11.29}$$

has the *characteristic equation*

$$\lambda^2 + p\lambda + q = 0.$$

(i) Suppose that an ODE has real constant coefficients. Then the roots of the characteristic equation are complex conjugated. In this case, if the roots of the characteristic equation in (11.29) are $\lambda = \alpha \pm i\beta$ with $\beta \neq 0$, α and β reals, then the solution is of the form

$$y(x) = e^{\alpha x}(A\cos\beta x + B\sin\beta x).$$

If the characteristic equation has a real double root λ, then the solution becomes

$$y(x) = e^{\lambda x}(Ax + B).$$

(ii) If in (11.27) $p = p(x)$, and $q = q(x)$ are continuous functions of x, $r(x) = 0$, and $y(x) \neq 0$, then the differential equation can be reduced to a first-order ODE putting $v(x) = y'(x)/y(x)$, *viz.*

$$v' + v^2 + p(x)v + q(x) = 0 \text{ (Riccati's equation)} \qquad (11.30)$$

and

$$y = y(x) = Ce^{\int v(x)dx}.$$

(iii) Let f and g be two linearly independent homogeneous solutions of the differential equation (11.27), with initial conditions $y(a) = y'(a) = 0$. Then the general solution is given by

$$y(x) = \int_a^x \frac{f(x)\,g(t) - f(t)\,g(x)}{g(t)\,f'(t) - f(t)\,g'(t)} \cdot r(t)\,dt. \qquad (11.31)$$

11.4.2 *Some special ODEs of second order*

In the following, m and n are positive integers, and α and β, real numbers.

Definition 11.5.

$y'' - 2xy' + 2ny = 0$ Hermite's DE

$\frac{d}{dx}\left[(1-x^2)\frac{dy}{dx}\right] + \lambda y = 0$ Legendre's DE

$(1-x^2)\,y'' - 2\,x\,y' + \left(n\,(n+1) - \frac{m^2}{1-x^2}\right)\,y = 0$ Associated Legendre's DE

$xy'' + (1-x)y' + \alpha y = 0$ Laguerre's DE

$(1-x^2)\,y'' - x\,y' + \lambda\,y = 0$ Chebyshev's DE

$y'' + \frac{y'}{x} + \left(1 - \frac{n^2}{x^2}\right)\,y = 0$ Bessel's DE

$n = 0, 1, 2, 3\ldots$

(11.32)

Definition 11.6.

$$\varphi'' + \frac{2m}{\hbar^2}\left[E - V(x)\right]\varphi = 0$$ Schrödinger's DE (One-dimensional and time indep.)

$$xy'' + (k+1-x)y' + (n-k)y = 0$$ Associated Laguerre's DE

$$(1-x^2)y'' + [a-b-(a+b+2)x]y' + n(n+a+b+1)y = 0$$ Jacobi's DE

$$x(1-x)y'' + [\gamma - (\alpha+\beta+1)x]y' - \alpha\beta y = 0$$ Hypergeometric DE

(11.33)

In Schrödinger's DE, the unknown function denoted by φ. E is an energy parameter, $2\pi\hbar$ is Planck's constant, and $V(x) = E_p(x)$ is potential energy.

$\varphi \cdot \overline{\varphi} dx = |\varphi|^2 dx$ is the probability that the particle with mass m is in the interval $[x, x+dx]$.

Theorem 11.4.

(i) *Hermite's DE is solved using Hermite polynomials* (14.53) *page* 329 *for* $n = 0, 1, 2, \ldots$
(ii) *Legendre's DE is solved using Legendre polynomials* (14.59) *page* 332, *if* $\lambda = n(n+1), \quad n = 0, 1, 2, \ldots$
(iii) *Laguerre's DE is solved using Laguerre's polynomials* (14.52) *when* $\alpha = n$ *is a positive integer.*
(iv) *Chebyshev's DE is solved using Chebyshev polynomials* $T_n(\cos\theta) = \cos n\theta$ *with* $\lambda = n^2$.
(v) *Bessel's DE is solved by* $A\,J_n(x) + B\,Y_n(x)$ (*see* 14.55 *and* 14.57).

(*The polynomials and the Bessel functions can be found on page* 329).

11.4.3 *Linear system of differential equations*

Definition 11.7. A linear system of differential equations is of the form

$$\begin{cases} \dfrac{dy_1}{dx} = a_{11}(x)y_1(x) + a_{12}(x)y_2(x) + \cdots + a_{1n}(x)y_n(x), \\ \quad\vdots \qquad\qquad\qquad \ddots \\ \dfrac{dy_n}{dx} = a_{n1}(x)y_1(x) + a_{n2}(x)y_2(x) + \cdots + a_{nn}(x)y_n(x), \end{cases} \tag{11.34}$$

or in matrix form, with $\boldsymbol{A} = (a_{jk})_{n\times n}$ and $\boldsymbol{y} = [x_1, y_2, \ldots, y_n]^T$:

$$\boldsymbol{y}' = \boldsymbol{A} \cdot \boldsymbol{y}.$$

The norm of $\boldsymbol{y}$ is defined as

$$|\boldsymbol{y}| = \sqrt{\boldsymbol{y}^T \cdot \mathbf{y}} = \left(\sum_{k=1}^{n} y_k^2\right)^{1/2}.$$

The norm of the matrix $\boldsymbol{A}$ is defined as

$$\|\boldsymbol{A}\| = \sup_{\boldsymbol{y}\neq 0} \frac{|\boldsymbol{Ay}|}{|\boldsymbol{y}|}.$$

The function $e^{x\,\boldsymbol{A}}$ is defined as

$$e^{x\cdot \boldsymbol{A}} := \mathbf{I} + x\mathbf{A} + \frac{x^2}{2!}\mathbf{A}^2 + \cdots = \sum_{k=0}^{\infty} \frac{x^k}{k!}\mathbf{A}^k, \tag{11.35}$$

where $\mathbf{I}$ is the identity matrix of order n.

Any system of linear differential equations can be reduced to this form. For instance, for

$$a_n y^{(n)} + a_{n-1} y^{(n-1)} + \cdots + a_1 y' + a_0 y = 0,$$

one can write $y^{(k)} =: y_k$ and the equation system becomes

$$\begin{cases} \quad y^{(n)} = -\left(y_n + \dfrac{a_{n-1}}{a-n} y_{n-1} + \cdots + \dfrac{a_1}{a_n} y_1 + \dfrac{a_0}{a_n} y_0\right), \\ y^{(n-1)} = y_{n-1}, \\ \qquad \vdots \;\vdots\; \vdots \\ y^{(0)} \equiv y = y_0. \end{cases}$$

11.5 Existence and Uniqueness of the Solution

Here we consider a differential equation of the variable x written in an implicit form as

$$\boldsymbol{F}(\boldsymbol{y}(x), x) = \boldsymbol{y}'(x). \tag{11.36}$$

Definition 11.8. Let $\boldsymbol{y} \in M \subseteq \mathbb{R}^{n+1}$ and $\boldsymbol{F}(\mathbf{y}, x) \in \mathbb{R}^n$ be a function defined on M. Then F satisfies a Lipschitz condition in the set M in the variable $\boldsymbol{y}$, if there is a constant K such that

$$\left|\boldsymbol{F}(\boldsymbol{y}, x) - \boldsymbol{F}(\boldsymbol{y}', x)\right| \leq K \left|\boldsymbol{y} - \boldsymbol{y}'\right|, \tag{11.37}$$

for all $(\boldsymbol{y}, x)$ and $(\boldsymbol{y}', x)$ in M.

Theorem 11.5.

(i) **Uniqueness:** *If the function* $\boldsymbol{F}(\boldsymbol{y}, x)$ *in* (11.36) *satisfies* (11.37), *then the differential equation in* (11.36), *with the solution passing through a given point* $(\boldsymbol{y}_0, x_0) \in M$ *has at most one solution.*

(ii) **Existence:** *Assume that* $\boldsymbol{F}(\boldsymbol{y}, x)$ *is continuous on* M *and satisfies* (11.37) *in an interval* $I_\delta := (x_0 - \delta, x_0 + \delta)$ *for all* $\boldsymbol{y}$ *and that* $(\boldsymbol{y}_0, x_0) \in M$. *Then, for all* $x \in I_\delta$ *there exists a solution* $\boldsymbol{y}(x)$ *to* (11.36), *with* $\boldsymbol{y}(x_0) = \boldsymbol{y}_0$.

11.6 Partial Differential Equations (PDEs)

Definition 11.9. A partial differential equation (PDE) is an equation containing partial derivatives of a function in two or more independent variables. The highest partial derivative in the equation is the order of the equation.

Note that in the following, the differentiation is not denoted by the *primes*, e.g., u_{xx} is used to denote $\frac{\partial^2 u}{\partial x^2} = u''_{xx}$, etc.

The solution process for the first-order linear PDE
For simplicity, the following layout is restricted to two-dimensional cases (two independent variables). Generalizations to higher dimensions are straightforward.

Consider

$$a(x,y)u_x + b(x,y)u_y = f(x,y,u), \qquad u = u(x,y). \tag{11.38}$$

Because of the u-dependence in f, the PDE (11.38) is called *quasi-linear.*

For the general solution, the following steps are performed:

(i) Find characteristic curves, $\frac{dy}{dx} = \frac{b(x,y)}{a(x,y)}$ with the general solution $\xi(x,y) = C$.

(ii) Perform the coordinate transformation

$$\begin{cases} \xi = \xi(x,y), \\ \eta = \text{ a suitable function of } x, y \text{ (e.g., } \eta = x, \quad \text{or } \eta = y). \end{cases}$$

(iii) The equation (11.38) is transformed into an ordinary differential equation

$$(a\eta_x + b\eta_y)\frac{\partial u}{\partial \eta} = f.$$

This last PDE can now be solved for u.

Remark. The general solution contains an arbitrary function of ξ.

Example 11.1.

$$xu_x + yu_y = u.$$

Consider

$$\frac{dy}{dx} = \frac{y}{x} \implies \int \frac{dy}{y} = \int \frac{dx}{x} \implies \frac{y}{x} = C.$$

Let $\xi = y/x$, and $\eta = x$, then the above PDE is transformed to

$$\eta \frac{\partial u}{\partial \eta} = u.$$

By separating the variables, one easily gets $u = \eta f(\xi) = xf(y/x)$.

Second-order linear PDE

$$a(x,y)u_{xx} + 2b(x,y)u_{xy} + c(x,y)u_{yy} = f(x,y,u,u_x,u_y). \quad (11.39)$$

Here $ac - b^2$ is called the *discriminant* of (11.39).

Classification of second-order PDE (Trinities):
There are three types of partial second-order differential equations:

(i) Elliptic if $ac - b^2 > 0$ (e.g., $\Delta u = u_{xx} + u_{yy} = 0$, Laplace's equation).
(ii) Parabolic if $ac - b^2 = 0$ (e.g., $u_t = ku_{xx}$, One-dim. heat conduction equation).
(iii) Hyperbolic if $ac - b^2 < 0$ (e.g., $u_{tt} = c^2u_{xx}$, One-dim. wave equation).

Remark. The classifications above are local. For example, the Tricomi equation for gas dynamics

$$yu_{xx} + u_{yy} = f(x,y)$$

is elliptic for $y > 0$, parabolic for $y = 0$, and hyperbolic for $y < 0$.

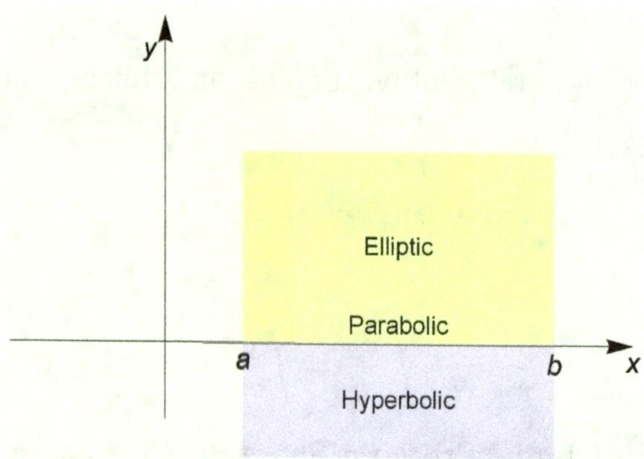

Characteristic curves: Note that

$$a\left(\frac{dy}{dx}\right)^2 - 2b\frac{dy}{dx} + c = 0 \quad \Longrightarrow \frac{dy}{dx} = \frac{1}{a}(b \pm \sqrt{b^2 - ac}).$$

Hence, if (11.39) is elliptic (case 1), then there are no real characteristics. In the parabolic case (case 2), there is a family of characteristic curves, and in the hyperbolic cases (case 3), there are two families of characteristic curves.

11.6.1 *The most common initial and boundary value problems*

The wave equation:

$$u_{tt} - c^2 u_{xx} = 0, \quad c = \text{constant}.$$

The coordinate transformation $\xi = x + ct, \quad \eta = x - ct$ from Cartesian to asymptotes gives $u_{\xi\eta} = 0$ with the general solution $u = \varphi(x + ct) + \psi(x - ct)$.

The initial–boundary value problem:

$$\begin{cases} u_{tt} - c^2 u_{xx} = 0, & t > 0, \qquad\qquad -\infty < x < \infty, \\ u(x,0) = f(x), & u_t(x,0) = g(x), \quad -\infty < x < \infty, \end{cases}$$

has the solution

$$u(x,t) = \frac{1}{2}[f(x+ct) + f(x-ct)] + \frac{1}{2c}\int_{x-ct}^{x+ct} g(y)\, dy,$$

which is also known as *d'Almembert's formula.*

The Dirichlet problem. Assume that $u \in C^2(\Omega)$ and $\Omega \subset \mathbb{R}^2$ is an open set. The problem

$$\begin{cases} \Delta u = u_{xx} + u_{yy} = 0, & \text{in } \Omega, \\ u = f, & \text{on } \partial\Omega \ (f \text{ continuous}), \end{cases} \tag{11.40}$$

has a unique solution.

Poisson's integral formula

(i) The equation (11.40), with $\Omega = \{(x,y) : x^2 + y^2 \leq 1\}$, has the solution

$$u = u(r,\theta) = \frac{1}{2\pi}\int_0^{2\pi} \frac{(1-r^2)f(\varphi)}{1 - 2r\cos(\theta - \varphi) + r^2}\, d\varphi.$$

(ii) The equation (11.40) with Ω as the upper half plane $\{(x, y) \in \mathbb{R}^2 : y \geq 0\}$ has the solution

$$u(r, \theta) = \frac{1}{\pi} \int_{-\infty}^{\infty} \frac{y f(s)}{y^2 + (x - s)^2} \, ds.$$

(iii) The equation (11.40) with Ω, an arbitrary domain in $\mathbb{R}^2$ which, by means of conformal mappings, can be mapped to geometrically regular domain where the problem is analytically solvable, and then maps back to Ω.

The Neumann problem

$$\begin{cases} \Delta u = 0, & \text{in } \Omega, \\ \frac{\partial u}{\partial n} = g, & \text{on } \partial\Omega. \end{cases} \tag{11.41}$$

For the problem (11.41), to have a unique solution, up to an additive constant, it is necessary that

$$\oint_{\partial\Omega} g(s) \, ds = 0.$$

Poisson's integral formula

(i) $\Omega := \{(x, y) : x^2 + y^2 \leq 1\}$, the unit disk.
Solution:

$$u = u(r, \theta) = -\frac{1}{2\pi} \int_0^{2\pi} \ln(1 - 2r\cos(\theta - \varphi) + r^2) g(\varphi) \, d\varphi + C.$$

(ii) $\Omega := \{(x,\, y) : y \geq 0\}$, the upper half plane.
Solution:

$$u(r, \theta) = \frac{1}{2\pi} \int_{-\infty}^{\infty} \ln(y^2 + (x - s)^2) g(s) \, ds.$$

(iii) Ω is an arbitrary domain in $\mathbb{R}^2$ which can be described by mathematical expressions. Then one can solve the problem making use of conformal mappings.

11.6.2 *Representation with orthogonal series*

I. (Heat conduction on a rod)

$$\begin{cases} \text{(PDE)} & u_t = \beta^2 u_{xx}, & t > 0, \quad 0 < x < L, \\ \text{(BC)} & u(0,t) = u(L,t) = 0, & t > 0, \\ \text{(IC)} & u(x,0) = f(x), & \qquad\quad 0 < x < L. \end{cases}$$

BC:= Boundary conditions, IC:= Initial condition.

Solution by separation of variable: The Fourier method

(i) Separation of variable: $u(x,t) = X(x)T(t)(\neq 0) \implies$ [(Inserting in PDE)] yields

$$\frac{T'(t)}{\beta^2 T(t)} = \frac{X''(x)}{X(x)} = \lambda \quad (\lambda = \text{ the separation constant}). \tag{11.42}$$

(ii) $$X'' - \lambda X = 0. \qquad \text{(BC)} \implies X(0) = X(L) = 0 \implies$$

$$\begin{cases} X_n(x) = \sin \frac{n\pi x}{L}, \ n = 1, 2, 3, \ldots \text{(eigenfunctions)}, \\ \lambda_n = -\frac{n^2\pi^2}{L^2}, \ n = 1, 2, 3, \ldots \text{(eigenvalues)}. \end{cases}$$

(iii) Then, the equation for T becomes (11.42)

$$T' + \frac{\beta^2 n^2 \pi^2}{L^2} T = 0 \implies T_n(t) = b_n e^{-\beta^2 n^2 \pi^2 t / L^2}.$$

(iv) The super position means that we have the general solution as

$$u(x,t) = \sum_{n=1}^{\infty} T_n(t) X_n(x) = \sum_{n=1}^{\infty} b_n e^{-\beta^2 n^2 \pi^2 t / L^2} \sin \frac{n\pi x}{L}.$$

(v) By using (IC), one may write

$$f(x) = \sum_{n=1}^{\infty} b_n \sin \frac{n\pi x}{L} \quad \implies \quad b_n = \frac{2}{L} \int_0^L f(x) \sin \frac{n\pi x}{L} dx.$$

II. (Generalization of I)

$$\begin{cases} \text{(PDE)} & u_t - \beta^2 u_{xx} = g(x,t), \quad t > 0, \quad 0 < x < L, \\ \text{(BC)} & u(0,t) = u(L,t) = 0, \quad t > 0, \\ \text{(IC)} & u(x,0) = f(x), \qquad\qquad\qquad 0 < x < L. \end{cases}$$

By following the steps in problem **I**, since (BC) is the same in both examples, we have that

$$u(x,t) = \sum_{n=1}^{\infty} u_n(t) \sin \frac{n\pi x}{L}.$$

In this setting the Fourier series expansions for $f(x)$ and $g(x,t)$ have the forms

$$f(x) = \sum_{n=1}^{\infty} f_n \sin \frac{n\pi x}{L} \quad \text{and} \quad g(x,t) = \sum_{n=1}^{\infty} g_n(t) \sin \frac{n\pi x}{L}.$$

By inserting in (PDE) and using (IC), the following ODE is obtained for the coefficient u_n:

$$\begin{cases} u_n'(t) + \dfrac{n^2\pi^2}{L^2} u_n(t) = g_n(t), \quad n = 1, 2, 3, \ldots \\ u_n(0) = f_n. \end{cases} \tag{11.43}$$

III. (The Dirichlet problem for a sphere)

Assuming that u is independent of φ, one gets $u = u(r, \theta, \varphi) = u(r, \theta)$.

Set $\xi = \cos\theta$, then we obtain the so-called *Laplace–Beltrami operator* which is the same as the Laplace operator on the sphere:

$$\text{(PDE)}\ \Delta u = \frac{1}{r^2}\frac{\partial}{\partial r}\left(r^2 \frac{\partial u}{\partial r}\right) + \frac{1}{r^2}\frac{\partial}{\partial \xi}\left((1-\xi^2)\frac{\partial u}{\partial \xi}\right) = 0, \qquad 0 < r < R,$$

$$\text{(RV)}\ u(R,\xi) = g(\xi), \qquad -1 < \xi < 1.$$

The general solution to the above PDE is given by

$$u(r,\xi) = \sum_{n=0}^{\infty} (A_n r^n + B_n r^{-n-1}) P_n(\xi),$$

where $P_n(\xi)$ is the Legendre polynomial of order n. Further,

$$\begin{cases} A_n = \dfrac{2n+1}{2R^n} \displaystyle\int_{-1}^{1} f(\xi) P_n(\xi)\, d\xi, \\ B_n \quad (B_n = 0 \text{ if } u \text{ is bounded for } r = 0) \\ \text{is determined from (BC) as Legendre–Fourier coefficients.} \end{cases}$$

IV. (Oscillations of a circular membrane)
Polar coordinates: $u = u(r, \theta, t) = u(r, t)$ with the assumption that u is independent of θ.

$$\begin{aligned}
&\text{(PDE)} \quad \Delta u = u_{rr} \frac{1}{r} u_r = \frac{1}{c^2} u_{tt}, && 0 < r < R, \quad t > 0, \\
&\text{(BC)} \quad u(R, \xi) = 0, && \qquad\qquad\quad t > 0, \\
&\text{(IC 1)} \quad u(r, 0) = g(r), && 0 \le r \le R, \\
&\text{(IC 2)} \quad u_t(r, 0) = 0, && 0 \le r \le R.
\end{aligned}$$

With the technique of variable separation, one gets

$$u(r,t) = \sum_{n=1}^{\infty} \left(A_n \cos \frac{c\alpha_n t}{R} + B_n \sin \frac{c\alpha_n t}{R} \right) J_0\left(\frac{\alpha_n r}{R} \right),$$

where α_n are zeros of the Bessel function J_0 and where

$$A_n = \frac{2}{R^2 J_1(\alpha_n)^2} \int_0^R J_0\left(\frac{\alpha_n r}{R} \right) dr \quad \text{and} \quad (B_n = 0),$$

are Bessel–Fourier coefficients which are determined using (IC 1) and (IC 2).

V. ($\hat{u}(\xi)$ is the Fourier transform for $u(r)$)

$$\begin{aligned}
&\text{(PDE)} \quad \Delta u = u_{xx} + u_{yy} = 0, && -\infty < x < \infty, \quad 0 < y < 1, \\
&\text{(RV1)} \quad u(x, 0) = g(x), && -\infty < x < \infty, \\
&\text{(RV2)} \quad u(x, 1) = 0, && -\infty < x < \infty.
\end{aligned}$$

Fourier's inversion formula in x-direction gives

$$u(x,y) = \frac{1}{2\pi} \int_{-\infty}^{\infty} \hat{u}(\xi, y) e^{i\xi x} \, d\xi.$$

Inserting in the (PDE) yields

$$\hat{u}_{yy} - \xi^2 \hat{u} = 0 \quad \Longrightarrow \quad \hat{u}(\xi, y) = A(\xi) \cosh \xi y + B(\xi) \sinh \xi y,$$

i.e.,

$$u(x,y) = \frac{1}{2\pi} \int_{-\infty}^{\infty} (A(\xi) \cosh \xi y + B(\xi) \sinh \xi y) \, e^{i\xi x} \, d\xi. \tag{11.44}$$

$$\begin{cases} \text{(RV1)} \implies A(\xi) = \hat{g}(\xi), \\ \text{(RV2)} \implies B(\xi) = -A(\xi) \dfrac{\cosh \xi}{\sinh \xi} = -\hat{g}(\xi) \dfrac{\cosh \xi}{\sinh \xi}. \end{cases} \tag{11.45}$$

Hence,

$$u(x,y) = \frac{1}{2\pi}\int_{-\infty}^{\infty} \hat{g}(\xi)\frac{\sinh(\xi(1-y))}{\sinh\xi}\, e^{i\xi x}\, d\xi.$$

VI. ($U(s)$ is the Laplace transform of $u(t)$)

$$\begin{array}{lllr}
\text{(PDE)} & u_{xx} = u_t, & x>0, & t>0,\\
\text{(BC1)} & u_x(0,t) = g(t), & & t>0,\\
\text{(BC2)} & \lim_{x\to\infty} u(x,t) = 0, & & t>0,\\
\text{(IC)} & u(x,0) = 0, & x>0. &
\end{array}$$

The Laplace transform of (PDE) $\Longrightarrow$

$$U_{xx}(x,s) = sU(x,s) \quad \Longrightarrow \quad U(x,s) = A(s)e^{x\sqrt{s}} + B(s)e^{-x\sqrt{s}}.$$

$$\text{(BC2)} \quad \Longrightarrow \quad A(s) = 0, \quad U_x = -B(s)\sqrt{s}e^{-x\sqrt{s}},$$

$$B(s) = -\frac{1}{\sqrt{s}}G(s).$$

Thus,

$$U(x,s) = -\frac{G(s)}{\sqrt{s}}e^{-x\sqrt{s}} \quad \Longrightarrow \quad u(x,t) = -\int_0^t \frac{1}{\sqrt{\pi\tau}}e^{-x^2/4\tau} g(t-\tau)\, d\tau.$$

11.6.3 *Green's functions*

Case I: Ordinary differential equations/Boundary value problems

Consider the boundary value problem

$$\begin{cases} L[f](x) = \displaystyle\sum_{j=1}^{n} a_j(x) f^{(k)}(x) = g(x), & a<x<b, \\ B_j f = \displaystyle\sum_{i=1}^{n-1} [\alpha_{ij} f^{(i)}(a) + \beta_{ij} f^{(i)}(b)] = c_j, & j = 1,\dots,n, \end{cases} \tag{11.46}$$

where $\alpha_{ij}, \beta_{ij}, c_j$, are constants and $g(x), a_j(x) \in C[a,b]$, $a(x) \neq 0$.

The problem (11.46) has a unique solution $\iff$

$$\det(B_j f_k) \neq 0. \tag{11.47}$$

Green's function $G(x,y)$ is continuous for $(x,y) \in [a,b]^2$ (if $n=2$), and is defined as *the fundamental solution* to a differential equation with differential operator L:

$$\begin{cases} L[G](x,y) = \delta(x-y), & a < y < b, \\ B_j G(x,y) = 0, & a < y < b, \quad j = 1,2,\ldots,n. \end{cases}$$

Theorem 11.6. *Assume* (11.47), *then the solution to* (11.46) *with* $c_j = 0$ *can be written as*

$$f(x) = \int_a^b G(x,y) g(y)\, dy.$$

Example 11.2. Find the Green function for the boundary value problem

$$\begin{cases} (3+x)f'' + f' = g(x), 0 < x < 1, \\ f'(0) = f(1) = 0. \end{cases}$$

Solution:
To find the Green function, we consider the equation

$$(3+x)f'' + f' = \delta(x-y) \quad \iff \quad ((3+x)f')' = \delta(x-y).$$

By taking primitive function, one gets

$$(3+x)f' = H(x-y) + C, \quad f(0) = 0 \Longrightarrow C = 0.$$

$$f' = \frac{H(x-y)}{3+x} \quad \iff \quad f = [\ln(3+x) - \ln(3+y)]H(x-y) + D.$$

$$f(1) = 0 \quad \iff \quad D = \ln(3+y) - \ln 4.$$

Thus,

$$G(x,y) = \begin{cases} \ln(3+x) - \ln 4, & 0 \le y \le x \le 1, \\ \ln(3+y) - \ln 4, & 0 \le x \le y \le 1. \end{cases}$$

Case II: Partial differential equations $\mathcal{L}$ is a linear differential operator with *regular* coefficients and $\mathcal{L}_{\boldsymbol{x}}$ is the operator $\mathcal{L}$ with respect to $\boldsymbol{x}$.

The fundamental solution
Consider the differential equation

$$\mathcal{L}u(\boldsymbol{x}) = f(\boldsymbol{x}), \qquad \boldsymbol{x} \in \mathbb{R}^n. \tag{11.48}$$

$Q(\boldsymbol{x})$ is called a fundamental solution to the differential operator $\mathcal{L}$ if

$$\mathcal{L}Q(\boldsymbol{x}) = \delta(\boldsymbol{x}), \qquad \boldsymbol{x} \in \mathbb{R}^n.$$

Then,

$$u(\boldsymbol{x}) = Q(\boldsymbol{x}) * f(\boldsymbol{x}) = \int_{\mathbb{R}^n} Q(\boldsymbol{x} - \boldsymbol{y}) f(\boldsymbol{y})\, d\boldsymbol{y}$$

is a solution of (11.48).

Example 11.3. The fundamental solution of the Laplace-operator $-\Delta$ in 2 and 3 dimensions is as follows:

$$Q(\boldsymbol{x}) = Q(x, y) = -\frac{1}{2\pi} \ln|\boldsymbol{x}| = \frac{1}{4\pi} \ln|x^2 + y^2|, \qquad \boldsymbol{x} \in \mathbb{R}^2, \quad \text{and}$$

$$Q(\boldsymbol{x}) = Q(x, y, z) = -\frac{1}{4\pi|\boldsymbol{x}|} = \frac{1}{4\pi\sqrt{x^2 + y^2 + z^2}}, \qquad \boldsymbol{x} \in \mathbb{R}^3,$$

respectively.

Definition 11.10. Let $\Omega \subset \mathbb{R}^n$, and consider the boundary value problem

$$\mathcal{L}u(\boldsymbol{x}) = f(\boldsymbol{x}), \quad \boldsymbol{x} \in \Omega \qquad \mathcal{B}u(\boldsymbol{x}) = 0, \quad \boldsymbol{x} \in \partial\Omega. \tag{11.49}$$

$G(\boldsymbol{x}, \boldsymbol{y})$ is called the *Green function* for $\mathcal{L}$, with respect to $\boldsymbol{x}$, if

$$\begin{cases} \mathcal{L}_{\boldsymbol{x}} G(\boldsymbol{x},\, \boldsymbol{y}) = \delta(\boldsymbol{x} - \boldsymbol{y}), & \\ \mathcal{B}_{\boldsymbol{x}} G(\boldsymbol{x}, \boldsymbol{y}) = 0, & \boldsymbol{x} \in \partial\Omega, \ \ \boldsymbol{y} \in \Omega. \end{cases}$$

$$u(\boldsymbol{x}) = \int_{\Omega} G(\boldsymbol{x}\,,\, \boldsymbol{y})\, f(\boldsymbol{y})\, d\boldsymbol{y},$$

is a solution of (11.49).

Dirichlet problem for Laplace operator
If $G(\boldsymbol{x}, \boldsymbol{y})$ is the Green function of the problem

$$-\Delta u = f, \qquad \text{i } \Omega, \qquad u = 0 \quad \text{on } \partial\Omega,$$

i.e., if

$$\begin{cases} -\Delta_{\boldsymbol{x}} G(\boldsymbol{x}\,, \boldsymbol{y}) = \delta(\boldsymbol{x} - \boldsymbol{y}), & \boldsymbol{x}\,, \boldsymbol{y} \in \Omega, \\ \mathcal{B}_{\boldsymbol{x}} G(\boldsymbol{x}\,, \boldsymbol{y}) = 0, & \boldsymbol{x} \in \partial\Omega, \ \boldsymbol{y} \in \Omega. \end{cases}$$

Then,

$$u(\boldsymbol{x}) = \int_{\Omega} G(\boldsymbol{x}\,, \boldsymbol{y})\, f(\boldsymbol{y})\, d\boldsymbol{y} - \int_{\partial\Omega} \frac{\partial G}{\partial n_{\boldsymbol{y}}}(\boldsymbol{x}\,, \boldsymbol{y})\, b(\boldsymbol{y})\, d\sigma_{\boldsymbol{y}},$$

is a solution of the problem

$$-\Delta u = f, \qquad \text{in } \Omega, \qquad u = b \quad \text{on } \partial\Omega.$$

Example 11.4. Green function for Laplace operators in 2 and 3 dimensions:

Consider the problem

$$\begin{cases} -\Delta_x G(\boldsymbol{x}, \boldsymbol{y}) = \delta(\boldsymbol{x} - \boldsymbol{y}), & \boldsymbol{x}, \boldsymbol{y} \in \Omega, \\ G(\boldsymbol{x}, \boldsymbol{y}) = 0, & \boldsymbol{x} \in \partial\Omega, \ \boldsymbol{y} \in \Omega. \end{cases}$$

(i) in upper half-plane $\Omega = \{\boldsymbol{x} = (x_1, x_2) : \ x_2 > 0\}$: Then so is

$$G(\boldsymbol{x}, \boldsymbol{y}) = -\frac{1}{2\pi}(\ln|\boldsymbol{x} - \boldsymbol{y}| - \ln|\boldsymbol{x} - \bar{\boldsymbol{y}}|),$$
$$\boldsymbol{y} = (y_1, y_2), \quad \bar{\boldsymbol{y}} = (y_1, -y_2).$$

(ii) in half-space $\Omega = \{\boldsymbol{x} = (x_1, x_2, x_3) : \ x_3 > 0\} \subset \mathbb{R}^3$: Then

$$G(\boldsymbol{x}\,, \boldsymbol{y}) = \frac{1}{4\pi}\Big(\frac{1}{|\boldsymbol{x} - \boldsymbol{y}|} - \frac{1}{|\boldsymbol{x} - \bar{\boldsymbol{y}}|}\Big),$$
$$\boldsymbol{y} = (y_1, y_2, y_3), \quad \bar{\boldsymbol{y}} = (y_1, y_2, -y_3).$$

(iii) In the disk $\Omega = \{\boldsymbol{x} : |\boldsymbol{x}| = \sqrt{(x_1^2 + x_2^2} < r\} \subset \mathbb{R}^2$: Then

$$G(\boldsymbol{x}\,, \boldsymbol{y}) = -\frac{1}{2\pi}\Big(\ln|\boldsymbol{x} - \boldsymbol{y}| - \ln|\boldsymbol{x} - \bar{\boldsymbol{y}}| - \ln\frac{|\bar{\mathbf{y}}|}{r}\Big), \qquad \bar{\mathbf{y}} = \frac{r^2}{|\boldsymbol{y}|^2}\boldsymbol{y}.$$

(iv) On the sphere $\Omega = \{\boldsymbol{x} : |\boldsymbol{x}| == \sqrt{(x_1^2 + x_2^2 + x_3^2} < r\} \subset \mathbb{R}^3$: Then

$$G(\boldsymbol{x}\,, \boldsymbol{y}) = -\frac{1}{4\pi}\Big(\frac{1}{|\boldsymbol{x} - \boldsymbol{y}|} - \frac{r}{|\bar{\mathbf{y}}|}\frac{1}{|\boldsymbol{x} - \bar{\boldsymbol{y}}|}\Big), \qquad \bar{\mathbf{y}} = \frac{r^2}{|\boldsymbol{y}|^2}\boldsymbol{y}.$$

Chapter 12

Numerical Analysis

12.1 Computer Language Approach

Some basic concepts

In the following, x is an exact real number and x^* is a real number approximating x.

Absolute error: $|x - x^*|$.

Absolute error bound: $\delta(x^*) \geq |x - x^*|$.

Relative error: $\dfrac{|x - x^*|}{|x|}$.

Relative error bound: $\mathrm{Rel}(x^*) \geq \dfrac{|x - x^*|}{|x|}$.

A floating point representation of a real number: x,

$$x = f \cdot \beta^E; \quad f = \text{fraction}, \quad \beta = \text{base},$$
$$E = \text{exponent, where } \beta^{-1} \leq f < 1.$$

A computational environment has the property that a real numbers x, called *word*, has a computable (limited) range.

$|\text{word}|$:= number of positions occupied by floating point representation of a number (word).

Example 12.1. Let β =10, then

$$x = f \cdot 10^E, \quad \frac{1}{10} \leq f < 1.$$

Hypothetical computer: Consider a computational environment that stores only digits: $0, 1, \ldots, 9$.

Floating point representations in this computer follow the arrangement

sign of nr.	fract. part		sign of exp.	exp part	
(d_1)	$d_2\, d_3$	$d_4\, d_5$	(d_6)	d_7	d_8

where a real number a is denoted as $a = (0.d_1 d_2 d_3 d_4) \times 10^E$.

The following is the range $[\mathbf{x}^*_{\text{min}}, \mathbf{x}^*_{\text{max}}]$ of the floating point representation of positive numbers in this hypothetical computer:

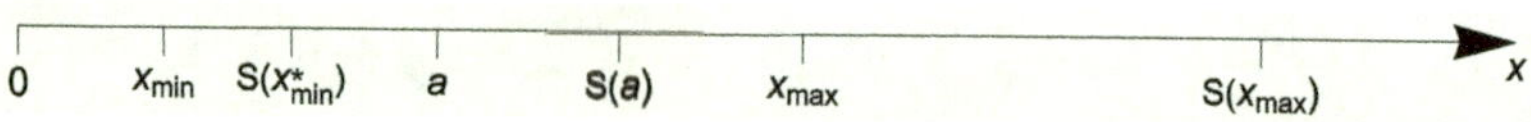

$$0, \qquad \mathbf{x}^*_{\text{min}} = (0.1000) \cdot 10^{-99}, \qquad S(\mathbf{x}^*_{\text{min}}) = (0.1001) \cdot 10^{-99},$$

$$\mathbf{a} = (0.d_1 d_2 d_3 d_4) \cdot 10^E, \quad S(\mathbf{a}) = (0.d_1 d_2 d_3 (d_4 + 1)) \cdot 10^E, \quad \mathbf{x}^*_{\text{max}} = (0.9999) \cdot 10^{99},$$

where $\mathbf{a}$ and $S(\mathbf{a})$ are two successive numbers.

The whole range for both positive and negative numbers is

$$[-\mathbf{x}^*_{\text{max}}, -\mathbf{x}^*_{\text{min}}] \cup [\mathbf{x}^*_{\text{min}}, \mathbf{x}^*_{\text{max}}].$$

Obviously,

$$|\mathbf{a} - S(\mathbf{a})| = 10^{E-4}.$$

A quantity which is large for large E and small for small E.

This means that gaps are not uniformly distributed. In other words, we have larger gaps between large machine numbers than those gaps between smaller machine numbers.

Machine epsilon

$$\varepsilon := \frac{1}{2} \cdot 10^{-s+1}.$$

$s =$ number of significant digits that the fraction f has in a decimal machine.

Property: $1 + \varepsilon$ is the smallest number greater than 1 that the machine in question will distinguish from 1.

All arithmetic with real numbers in this hypothetic computer needs to fall in the realm of the above representation form.

12.2 Numerical Differentiation and Integration

Main Idea: Given a function $f(x)$ to differentiate or integrate.

(i) Approximate $f(x)$ by an interpolating polynomial $P(x)$.
(ii) Differentiate or integrate $P(x)$.

12.2.1 *Numerical differentiation*

Forward difference: Use Taylor's formula

$$f(x_0 + h) = f(x_0) + f'(x_0)h + \frac{1}{2}f''(\xi)h^2, \quad (x_0 < \xi < x_0 + h),$$

to obtain

$$\underbrace{f'(x_0)}_{\text{Exact derivative}} = \underbrace{\frac{f(x_0 + h) - f(x_0)}{h}}_{\text{Approx. derivative}} - \underbrace{\frac{1}{2}f''(\xi)h^1}_{\text{Truncation error}}.$$

Here, $f'(x_0)$ = slope of tangent line $T \approx$ slope of $P_1(x)$, where $P_1(x)$ is the linear *interpolant* of f.

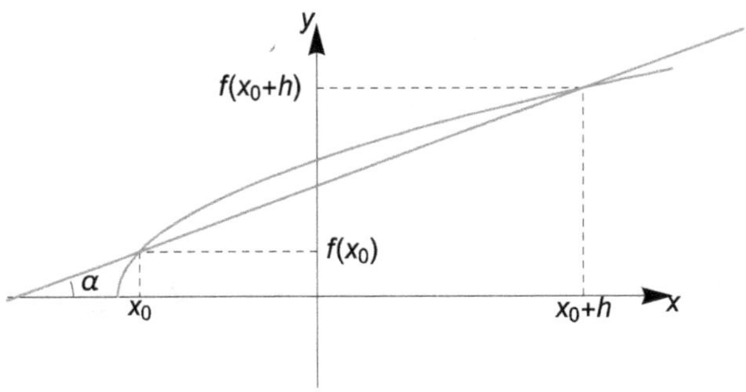

Backward difference:

$$f'(x_0) = \frac{f(x_0) - f(x_0 - h)}{h} + \mathcal{O}(h^1).$$

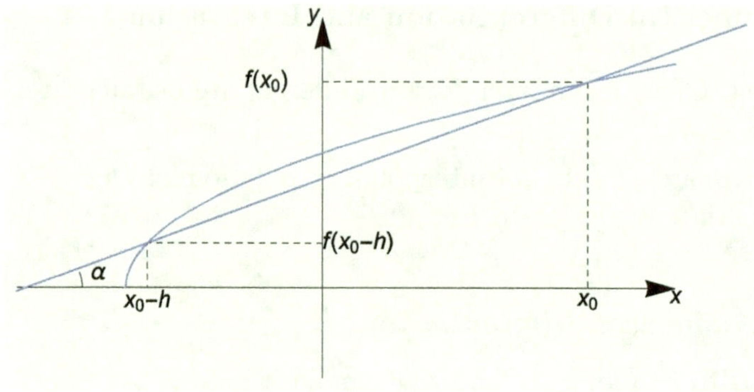

Central difference:

$$f'(x_0) = \frac{f(x_0 + h) - f(x_0 - h)}{2h} + \mathcal{O}(h^2).$$

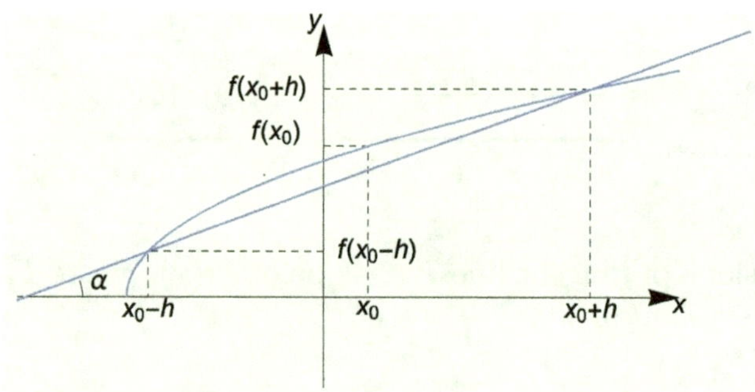

Sensitivity:

$$f'(x_0) = \frac{f(x_0 + h) - f(x_0)}{h} \equiv \frac{f_1 - f_0}{h}.$$

Let f_0^* and f_1^* be known approximations of f_0 and f_1, respectively, such that

$$|f_0^* - f_0| \leq \delta, \quad |f_1^* - f_1| \leq \delta.$$

Then

$$f'(x_0) \approx \frac{f_1^* - f_0^*}{h} \quad \text{is the actual approximation,}$$

and the computed error is

$$\left|f'(x_0) - \frac{f_1^* - f_0^*}{h}\right| \le \underbrace{\left|f'(x_0) - \frac{f_1 - f_0}{h}\right|}_{\mathcal{O}(h)} + \underbrace{\left|\frac{f_1 - f_0}{h} - \frac{f_1^* - f_0^*}{h}\right|}_{\mathcal{O}(1/h)}$$
$$\le \underbrace{\frac{M_2\, h}{2}}_{\text{Truncation error}} + \underbrace{\frac{2\,\delta}{h}}_{\text{Roundoff error}}, \tag{12.1}$$

where truncation error dominates for $h \gg 1$ and roundoff error dominates for $h \ll 1$, and $M_2 \approx f''(\xi)$.

To the right: The total error equals the sum of truncation error and the roundoff error.

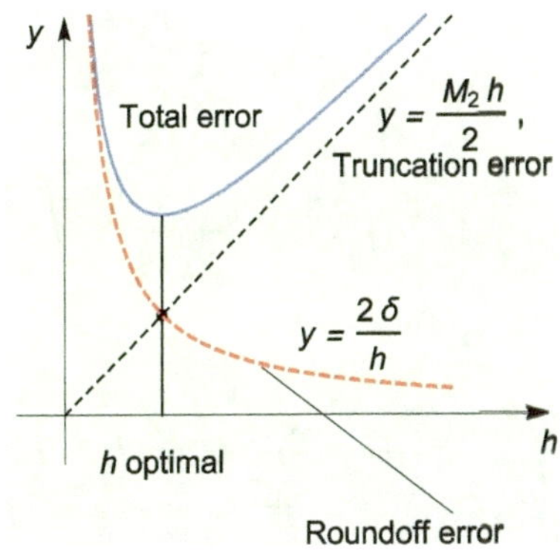

12.2.2 *Numerical integration*

General Idea:

(i) Divide the domain of integration $[a, b]$ into n subintervals of arbitrary lengths as follows:

$$a = x_0 < x_1 < \cdots < x_{n-1} < x_n = b.$$

(ii) Write

$$\int_a^b f(x)\, dx = \int_{x_0}^{x_1} f(x)\, dx + \int_{x_1}^{x_2} f(x)\, dx + \cdots + \int_{x_{n-1}}^{x_n} f(x)\, dx$$
$$= \sum_{k=0}^{n-1} \int_{x_k}^{x_{k+1}} f(x)\, dx.$$

(iii) On each interval $I_k := [x_k, x_{k+1}]$, $k = 0, 1, \ldots, n-1$, apply the same given integration rule.

To begin with, we consider *uniform partition:*

$$x_{k+1} - x_k = \frac{b-a}{n} = h. \tag{12.2}$$

Midpoint formula: (n points).
Here the interval $[a, b]$ is divided in n subintervals $[x_k, x_{k+1}]$ of equal lengths, $k = 0, 1, 2, \ldots, n$ and $x_0 = a$, $x_n = b$.

$$\bar{x}_{k+1} = \frac{x_k + x_{k+1}}{2}, \quad k = 0, 1, \ldots, n-1.$$

$$\int_{x_k}^{x_{k+1}} f(x)\,dx \approx \int_{x_k}^{x_{k+1}} f(\bar{x}_{k+1})\,dx = hf(\bar{x}_{k+1}),$$

and so

$$\int_a^b f(x)\,dx \approx \sum_{k=0}^{n} f(\bar{x}_k)\,h = h[f(\bar{x}_1) + f(\bar{x}_2) + \cdots + f(\bar{x}_n)] =: m_n.$$

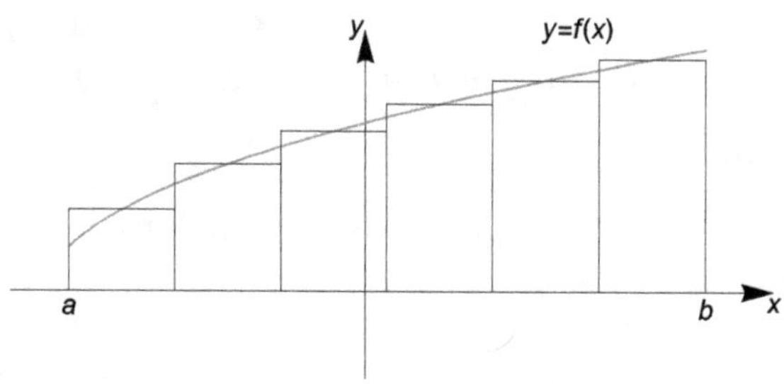

Midpoint formula approximates the integral $\int_a^b f(x)dx$, by a sum of area of rectangles with base $h = x_{k+1} - x_k$ and heights $f(\overline{x}_k)$, where $\overline{x}_k = \dfrac{x_k + x_{k+1}}{2}$, the midpoint of x_k and x_{k+1}.

Trapezoidal rule: $(n+1$ points, and hence n subintervals)

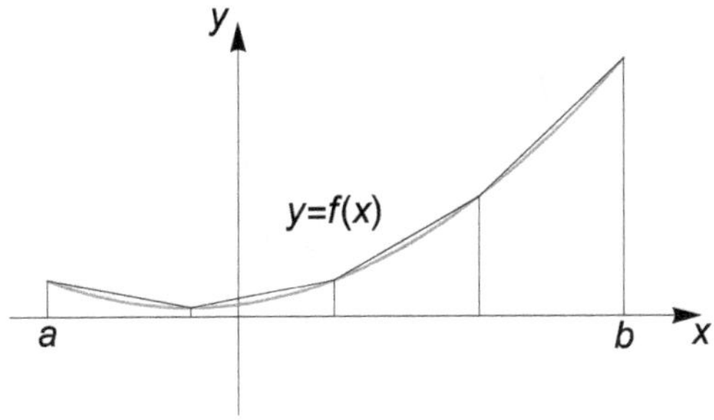

$$\int_a^b f(x)\,dx \approx \sum_{k=0}^{n-1} \frac{h}{2}[f(x_k) + f(x_{k+1})]$$
$$= h\left[\frac{1}{2}f(a) + f(x_1) + f(x_2) + \cdots \right.$$
$$\left. + \; f(x_{n-1}) + \frac{1}{2}f(b)\right] =: t_n.$$

Definition 12.1. Let $P_n(x)$ be Legendre polynomial of degree $\leq n$ and define the Lagrange Cardinal functions of degree n as

$$\ell_k(x) = \prod_{j=0, j\neq k}^{n} \frac{x - x_j}{x_k - x_j},$$

where x_k, $k = 0, 1, \ldots n$ are the roots of nth Legendre polynomial. Note that

$$\ell_k(x_i) = \begin{cases} 1, & i = k, \\ 0, & i \neq k. \end{cases}$$

The Lagrange interpolation polynomial of degree n for the function $f(x)$, here denoted by $\Pi_n f(x)$, is defined as

$$\prod_n f(x) := \sum_{k=0}^{n} f(x_k)\ell_k(x).$$

Using Lagrange interpolation polynomials we may apply the following approximation for integrals:

$$\int_a^b f(x)\,dx \approx \int_a^b \prod_n f(x)\,dx = \int_a^b \sum_{k=1}^n f(x_k)\ell_k(x)\,dx.$$

The error of this approximation is then

$$E(x) := f(x) - \prod_n f(x) = \frac{f^{n+1}(\eta)}{(n+1)!} \prod_{i=0}^n (x - x_i), \quad \eta \in (a, b),$$

with the obvious bound for the error of integration, *viz.*

$$\int_a^b \left| f(x) - \prod_n f(x) \right| dx \le \frac{1}{(n+1)!} \max_x \left| f^{n+1}(x) \right| \int_a^b \prod_{i=1}^n |x - x_i|\,dx.$$

Quadrature formula. The above approximation may be generalized and interpreted as

$$\int_a^b f(x)\,dx \approx \sum_{k=1}^n f(x_k) A_k\,dx.$$

where x_k are called quadrature points and

$$A_k = \int_a^b \ell_k(x)\,dx$$

are called quadrature weights.

Example 12.2. For $[a, b] = [-1, 1]$ and $n = 3$, we have that

$$P_3(x) = 0 \Longrightarrow \begin{cases} x_1 = -1, & x_2 = 0, & x_3 = 1, \\ A_1 = \dfrac{1}{3}, & A_2 = \dfrac{4}{3}, & A_3 = \dfrac{1}{3}. \end{cases}$$

Hence, in this case

$$\int_{-1}^1 f(x)\,dx \approx \frac{1}{3} f(-1) + \frac{4}{3} f(0) + \frac{1}{3} f(1). \tag{12.3}$$

Note that this approximation is exact for all polynomials "f" of degree $\leq$ 5 (such a polynomial is uniquely determined by 6 coefficients, and we have 6 degrees of freedom: $(x_i,\, A_i)$, $i = 1, 2, 3$, in above).

Simpson's rule: The interval $[a, b]$ is divided in $2n$ subintervals of the same length, that is with $2n + 1$ points.

(i) Use quadrature polynomials with data points; x_k, $\frac{x_k+x_{k+1}}{2}$, x_{k+1}.
(ii)

$$\int_{x_k}^{x_{k+1}} f(x)\,dx \approx \frac{h}{3}\Big[f(x_k) + 4f\Big(\frac{x_k + x_{k+1}}{2}\Big) + f(x_{k+1})\Big]. \quad (12.3)$$

Summing over k, $k = 0, 1, \ldots, 2n$:

$$\int_a^b f(x)\,dx \approx \frac{h}{3}\Big[f(a) + 4f(x_1) + 2f(x_2) + 4f(x_3) + 2f(x_4) + \cdots$$
$$+\ 2f(x_{2n-2}) + 4f(x_{2n-1}) + f(b)\Big] =: S_{2n}.$$

Richardson's extrapolation and corrected formulas

Correcting trapezoidal rule:

$$\int_a^b f(x)\,dx \approx h\left[\frac{1}{2}f(a) + f(x_1) + f(x_2) + \cdots\right.$$
$$\left. + f(x_{n-1}) + \frac{1}{2}f(b)\right] =: t_n, \quad (12.4)$$

$$\text{where } h = \frac{b-a}{n}, \quad x_k = a + hk, \quad k = 0, 1, \ldots, n.$$

Using Taylor expansion $\Longrightarrow$ there are constants C_2, $C_4, \ldots$ (depending on f, not on n and h) such that

$$I_n\colon\ e_n = \int_a^b f(x)\,dx - t_n = C_2h^2 + C_4h^4 + C_6h^6 + \cdots, \quad h = \frac{b-a}{n}.$$

If n is doubled, then $\bar{h} = \frac{b-a}{2n} = \frac{h}{2}$ is the new step six and

$$I_{2n} : e_{2n} = \int_a^b f(x)\,dx - t_{2n} = C_2\frac{h^2}{4} + C_4\frac{h^4}{16} + C_6\frac{h^6}{64} + \cdots$$

Now

$$\begin{aligned} I_n - 4I_{2n} &\Longrightarrow -3\int_a^b f(x)\,dx + 4t_{2n} - t_n \\ &= C_4\left(1 - \frac{1}{4}\right)h^4 + C_6\left(1 - \frac{1}{16}h^6\right) + \cdots, \end{aligned}$$

and we get the corrected trapezoidal formula

$$[n \to 2n] \quad J_M : \int_a^b f(x)\,dx - \underbrace{\frac{1}{3}(4t_{2n} - t_n)}_{S_{2n}} = d_4h^4 + d_6h^6 + \cdots$$

Hence, the error in corrected trapezoidal formula is of order $\mathcal{O}(h^4)$ instead of $\mathcal{O}(h^2)$.

We start with $J_M \Longleftrightarrow S_{2n}$ which is a Simpson rule, i.e.,

$$\int_a^b f(x)\,dx - S_{2n} = d_4h^4 + d_6h^6 + \cdots, \quad h = \frac{b-a}{2n} = \frac{b-a}{M}.$$

Now doubling $M \Longrightarrow$ New $h = \frac{b-a}{2M} =$ old $\frac{h}{2}$. Thus,

$$J_{2M} : \int_a^b f(x)\,dx - S_{2M} = d_4\frac{h^4}{16} + d_6\frac{h^6}{64} + \cdots$$

Then, $S_{2n} = S_M$ and

$$J_M - 16J_{2M} \Longrightarrow -15\int_a^b f(x)\,dx - \left(S_M - 16S_{2M}\right) = \mathcal{O}(h^6),$$

i.e., we have **the corrected Simpson rule**

$$\int_a^b f(x)\,dx - \frac{1}{15}\Big(16S_{2M} - S_M\Big) = \mathcal{O}(h^6).$$

Rule	Formula	Error
Simple midpoint	$\int_{x_0}^{x_1} f(x)\,dx \approx h\,f\left(\frac{x_0+x_1}{2}\right)$, $h = x_1 - x_0$,	$-\frac{h^3}{24}f''(\xi), \quad \xi \in [x_0, x_1]$
Simple trapezoidal	$\int_{x_0}^{x_1} f(x)\,dx \approx \frac{1}{2}(f(x_0) + f(x_1))$	$-\frac{h^3}{12}f''(\xi), \quad \xi \in [x_0, x_1]$
Simple Simpson's	$\int_{x_0}^{x_1} f(x)\,dx \approx \frac{1}{3}(f(x_0) + 4f(x_1) + f(x_2))$	$-\frac{h^5}{90}f^{(4)}(\xi), \quad \xi \in [x_0, x_1]$
Midpoint	$\int_a^b f(x)\,dx \approx h\sum_{k=1}^{n} f\left(\frac{x_{k-1}+x_k}{2}\right)$	$-\frac{(b-a)h^2}{24}f''(\xi), \quad \xi \in [a, b]$
Trapezoidal	$\int_a^b f(x)\,dx \approx h\Big(\frac{1}{2}f(a) + f(x_1) + \cdots + f(x_{n-1}) + \frac{1}{2}f(b)\Big) := t_N$	$-\frac{(b-a)h^2}{12}f''(\xi), \quad \xi \in [a, b]$
Simpson's	$\int_a^b f(x)\,dx \approx \frac{h}{3}\left(f_0 + f_{2N} + 4\sum_{k=1}^{N} f_{2k-1} + 2\sum_{k=1}^{N-1} f_{2k}\right) := S_M$ $(M = 2N)$	$-\frac{(b-a)h^4}{180}f^{(4)}(\xi), \quad \xi \in [a, b]$ $h = \frac{b-a}{M}$
Modified trapezoidal	$\int_a^b f(x)\,dx \approx t_N + \frac{h}{24}\Big(f_{-1} + f_1 + f_{n-1} + f_{n+1}\Big)$	$-\frac{(b-a)h^4}{720}f^{(4)}(\xi), \quad \xi \in [a, b]$

12.3 Solving $f(x) = 0$

Any real or complex equation can, equivalently, be written in the form of $f(x) = 0$.

12.3.1 *Linear case*

$x \in \mathbb{R}$:

$$a \neq 0: \quad ax + b = 0, \quad \Longrightarrow x = -b/a.$$

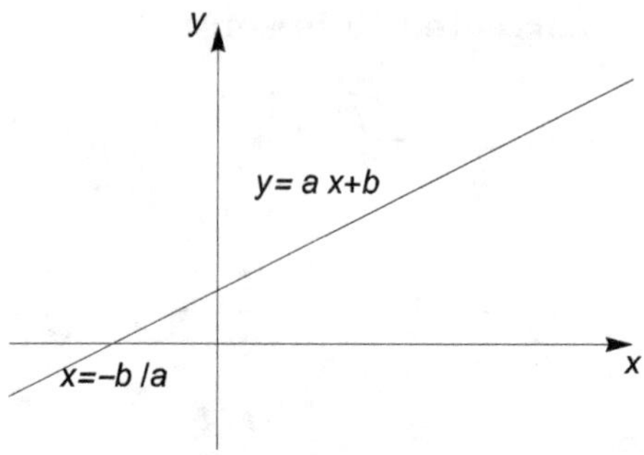

The line intersects the x-axis at $x = -b/a$.

$\boldsymbol{x} \in \mathbb{R}^n$, $n = 2, 3, \ldots$, $\boldsymbol{A}$ matrix of type $n \times n$, with $\det \boldsymbol{A} \neq 0$.

$$\boldsymbol{Ax} + \boldsymbol{b} = \boldsymbol{0} \Longleftrightarrow \boldsymbol{x} = -\boldsymbol{A}^{-1}\boldsymbol{b}.$$

12.3.2 *Numerical solution of nonlinear equations*

- **Iteration methods:** These are based on an *initial guess.*
- **Convergence** of an iteration method depends on a clever choice of initial guess.

For a polynomial

$$P(x) = a_n x^n + a_{n-1}x^{n-1} + \cdots a_1 x + a_0 = 0$$

all (real or complex) roots x^* satisfy

$$|x^*| \leq 1 + \frac{1}{|a_n|}\max\{|a_0|, |a_1|, \ldots, |a_{n-1}|\}.$$

Example 12.3. If $f(x) = 2x^3 - 3x^2 - 3x + 4 = 0$, then

$$|x^*| \leq 1 + \frac{1}{2}\max\{3, 4\} = 3, \quad \text{i.e.,} \ x^* \in [-3, 3].$$

Actually the roots are $x^* = \pm 1.686$ and $x^* = 1.0$.

12.3.3 *Common methods of iterations*

Bisection method [finding a root of $f(x) = 0$]. Assume $f(x)$ is continuous on $[a, b]$, and $f(a)$ and $f(b)$ do not have the same sign, that is $f(a)f(b) < 0$. Then there exists at least one root between a and b.

Description:

- Given $f(x)$, find an interval $[a, b]$ such that $f(a) < 0 < f(b)$ (if not possible, consider $-f(x)$).
- Make a guess that the root is $r = \frac{a+b}{2}$.
- There are now three possibilities for $f\left(\frac{a+b}{2}\right)$.

Case 1. $f\left(\frac{a+b}{2}\right) = 0$, then you are done.
Case 2. $f\left(\frac{a+b}{2}\right) < 0$.
Case 3. $f\left(\frac{a+b}{2}\right) > 0$.
Now we repeat the process using the proper interval in place of $[a, b]$.

Secant method: x_0, $x_1 \in \mathbb{R}$.

(Does not insist $x_0 < x_1$, has no condition of the sign of $f(x_0) \cdot f(x_1)$.)

Approximate $f(x)$ by a linear polynomial $P_1(x)$ with data x_0, x_1: [Secant line for $f(x)$ at x_0, x_1]; i.e., a linear Lagrange interpolation polynomial

$$f(x) \approx P_1(x) = f(x_0)\frac{x - x_1}{x_0 - x_1} + f(x_1)\frac{x - x_0}{x_1 - x_0}.$$

Now

$$f(x) = 0 \Longrightarrow \{x_0 \neq x_1\} \quad x \approx \frac{f(x_1)x_0 - f(x_0)x_1}{f(x_1) - f(x_0)} =: \frac{N}{T}.$$

Then, the value of the new approximation of the root depends on the previous two approximations and corresponding functional values, *viz.*

$$N_2 = x_1[f(x_1) - f(x_0)] - f(x_1)(x_1 - x_0)$$
$$\Longrightarrow x_2 \approx \frac{N_2}{T} = x_1 - f(x_1)\frac{x_1 - x_0}{f(x_1) - f(x_0)}.$$

General algorithm:

$$x_{k+1} = x_k - f(x_k)\frac{x_k - x_{k-1}}{f(x_k) - f(x_{k-1})}. \tag{12.5}$$

Comparing bisection and secant methods:
In a neighborhood of the root, secant method converges faster than the bisection method in the sense that much fewer steps were used to obtain the same accuracy.

Bisection method always converges (is fault free) provided that one can find points on either side of a root.

The secant method may diverge unless the starting points are sufficiently close to the root.

Example 12.4 (A horrible example).

$$f(x) = \frac{1}{x} - 1, \quad (x > 0), \quad \text{where} \quad x \to +\infty \Longrightarrow y \to -1,$$
$$x \to 0 \Longrightarrow y \to +\infty,$$

while the exact solution of $f(x) = 0$ is $x = 1$. By choosing $x_0 = 2.8$ and $x_1 = 2.0$ and making use of (12.5), one gets

$$x_2 = x_1 - f(x_1)\frac{x_1 - x_0}{f(x_1) - f(x_0)} = -0.8.$$

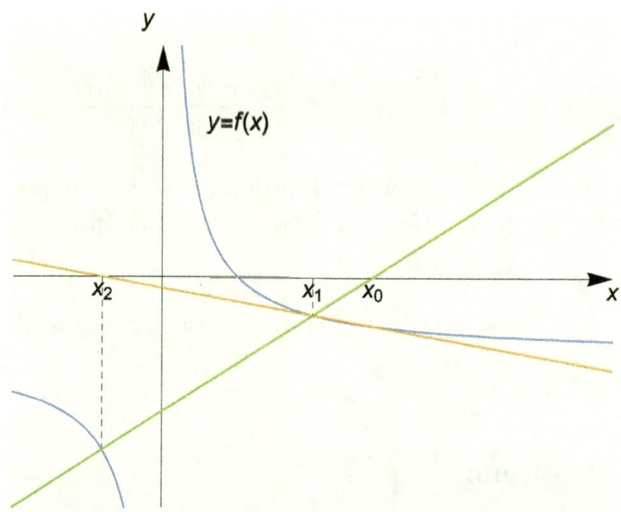

The choices $x_0 = 2.8$ and $x_1 = 2.0$ give $x_2 = -0.8$. The next steps give $x_3 = 2.8$, $x_4 = 4.24$ and $x_5 = 38.4653\ldots$ It can be shown that the sequence $(x_k)_{k=0}^{\infty}$ diverges.

Newton's method (Newton–Raphson method):
We start with an initial guess x_0 (note that secant method requires two points x_0, x_1).

$$f(x) \approx P_1(x) = f(x_0) + f'(x_0)(x - x_0).$$

$$P_1(x) = 0 \Longrightarrow x_1 - x_0 = -\frac{f(x_0)}{f'(x_0)}, \quad (f'(x_0) \neq 0)$$

$$x_1 = x_0 - \frac{f(x_0)}{f'(x_0)}.$$

General Newton:

$$x_{k+1} = x_k - \frac{f(x_k)}{f'(x_k)}, \quad f'(x_k) \neq 0. \tag{12.6}$$

Properties of Newton Method:

- Effective if $f'(x_k)$ is available.
- Requires two functional calls $f(x_k)$ and $f'(x_k)$ for each iteration.

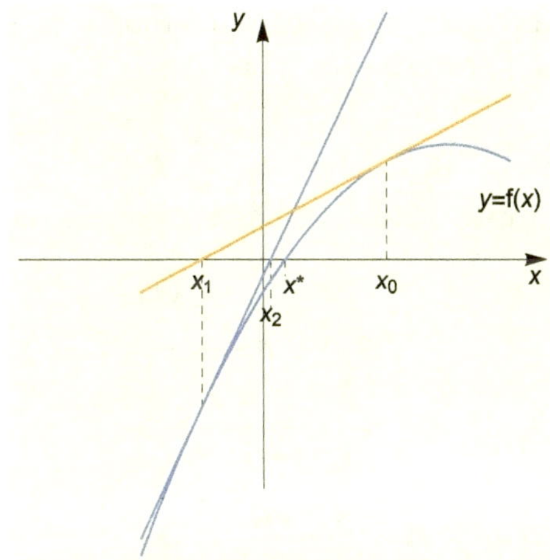

The starting point is x_0. The tangent at $(x_0, f(x_0))$ intersects $y = 0$ at $x = x_1$. In turn, the tangent at $(x_1, f(x_1))$ intersects $y = 0$ at $x = x_2$ and so on. In most cases, the sequence $(x_1, x_2, \ldots)$ converges to x^* a zero of $f(x)$.

Comparisons:
Newton method converges faster than both bisection and secant method.

Newton method requires both $f(x_k)$ and $f'(x_k)$ at each step.

By making the step size h (in the formula for the derivative) proportional to $|f(x_k)|$, we can approximate the derivative appearing in Newton's method by an approximating formula and still preserve the rapid convergence property of Newton's method, i.e.,

$$f'(x_k) \approx \frac{f(x_k) - f(x_{k-1})}{x_k - x_{k-1}}.$$

Then

$$\text{Newton's Method} \Longrightarrow x_{k+1} \approx x_k - f(x_k)\frac{x_k - x_{k-1}}{f(x_k) - f(x_{k-1})},$$

which is actually

$$\text{The Secant Method} \Longrightarrow x_{k+1} = x_k - f(x_k)\frac{x_k - x_{k-1}}{f(x_k) - f(x_{k-1})}.$$

However, in the secant method the step size is proportional to $f(x_{k-1})$ rather than $f(x_k)$.

Applications

Example 12.5. Finding the square root of a number:

$$f(x) = x^2 - a.$$
$$f(x) = 0 \Longrightarrow x^2 - a = 0 \Longrightarrow x = \pm\sqrt{a}.$$

Newton:

$$x_{k+1} = x_k - \frac{f(x_k)}{f'(x_k)} = x_k - \frac{x_k^2 - a}{2x_k} \Longrightarrow x_{k+1} = \frac{1}{2}\left(x_k + \frac{a}{x_k}\right).$$

For $a = 2$ and $x_0 = 2$, one gets

$$x_0 = 2.0 \quad x_1 = 1.5, \quad x_2 = 1.41667, \quad x_3 = 1.41422, \text{ etc.},$$

giving even better approximations of $\sqrt{2}$.

Generally: To obtain $\sqrt[n]{a}$, we write

$$f(x) = x^n - a, \qquad \text{then} \quad f(x) = 0 \Longrightarrow x = \sqrt[n]{a}.$$

Newton's Method: $\Longrightarrow$:

$$x_{k+1} = x_k - \frac{x_k^n - a}{n x_k^{n-1}} = \frac{1}{n}\left[(n-1)x_k + \frac{a}{x_k^{n-1}}\right].$$

Remarks.
Failure of Newton's Method: Let x^* be the root of $f(x) = 0$. Then, for

$$f(x) = 1 - 2e^{-|x|} \qquad \text{an even function}$$

at a distance from the root the slope is close to 0, i.e., no convergence for x_0 far from x^* and rapid convergence for x_0 close to x^*.

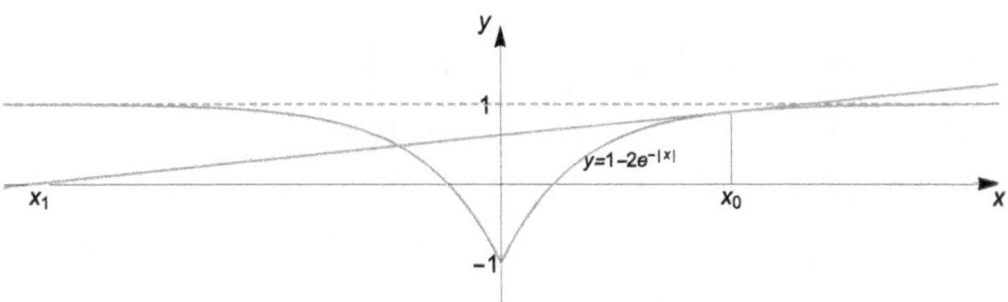

By using x_0 sufficiently far from x^*, the number x_1 becomes even more distant from x^* of the function. Thus, the sequence $x_0, x_1, x_2, \ldots$ does not converge to a zero. More precisely, $|x_n| \longrightarrow \infty$.

For general, $f(x)$, k-Bisection steps $\Longrightarrow$ x^* is located inside an interval of length $(\frac{b-a}{2^k})$:

$$|x^* - \text{endpoint}| \leq \frac{b-a}{2^k}.$$

Secant and Newton: If $f''(x)$ exists and is bounded and $f'(x) \neq 0$ in a neighborhood of x^*, then the successive Secant- and/or Newton-Method converges with a *convergence rate* r as follows:

$$|x_{k+1} - x^*| \leq C\,|x_k - x^*|^r \Longrightarrow \begin{cases} \text{Secant;} & r = \frac{1+\sqrt{5}}{2}, \\ \text{Newton;} & r = 2. \end{cases}$$

Polynomial roots: $i^2 = -1,\ x \in \mathbb{C}$.

$$f(x) = a_n x^n + a_{n-1}x^{n-1} + \cdots + a_1 x + a_0 = 0.$$

Bisection is no good to find complex roots, because ordering of the values of $f(x)$ is required with the method.

Newton's and Secant Methods Converge, if the roots $r_1, r_2, \ldots, r_n$ are simple.

Newton is more suitable, since derivative of polynomials are available.

12.4 Ordinary Differential Equations (ODEs)

Here follow some algorithms finding solutions of ODE.

12.4.1 *The initial value problems*

First-order ODE:

$$\begin{cases} y'(x) = f(x, y(x)), \\ y(x_0) = y_0. \end{cases}$$

Taylor series method.

Motivation: The derivatives of the solution are easily found from the differential equation itself.

Example 12.6.

$$\begin{cases} y'(x) = x^2 + y^2, & \text{(DE)} \\ y(0) = 0, & \text{(IV)} \end{cases} \qquad \text{OBS!} \quad y'(0) = 0.$$

Solution:

$$\begin{aligned} \text{(DE)}' \Longrightarrow \quad & y'' = 2x + 2yy' \Longrightarrow & y''(0) = 0 \\ & y''' = 2 + 2(y')^2 + 2y\prime\prime \Longrightarrow & y'''(0) = 2 \\ & y^{(4)} = (2yy'')'0 = 0 = y^{(5)} = \ldots & \end{aligned}$$

$$\text{Taylor} \Longrightarrow y(x) \approx y(0) + xy'(0) + \frac{x^2}{2!}y''(0) + \frac{x^3}{3!}y'''(0) + \cdots$$
$$\Longrightarrow y(x) \approx \frac{x^3}{3!}y'''(0) = \frac{x^3}{3}.$$

Generally:

$$\begin{cases} y(x) \approx y(x_0) + (x - x_0)\dfrac{y'(x_0)}{1!} + (x - x_0)^2\dfrac{y''(x_0)}{2!} + \cdots \\ \quad + (x - x_0)^m \dfrac{y^{(m)}(x_0)}{m!}, \\ \quad \text{Error} = \mathcal{O}\Big(|x - x_0|^{m+1} \Big) \quad \text{is large for } |x - x_0| \text{ large.} \end{cases}$$

Define the grid points $x_j = x_0 + jh$ $(j = 0, 1, , 2, \ldots, N)$.
For $x \in [x_j, x_{j+1}]$:

$$(j) \quad \begin{cases} y(x) \approx y(x_j) + \dfrac{y'(x_j)}{1!}(x - x_j) + \dfrac{y''(x_j)}{2!}(x - x_j)^2 + \cdots \\ \quad + \dfrac{y^{(m)}(x_j)}{m!}(x - x_j)^m, \\ \quad \text{Error} \le |x_{j+1} - x_j|^m = h^m. \end{cases}$$

Now for $x = x_1$, let

$$y_1 = y_0 + \frac{y_0'}{1!}h + \cdots + \frac{y_0^{(m)}}{m!}h^m,$$

and successively define

$$y_2 = y_1 + \frac{y_1'}{1!}h + \cdots + \frac{y_1^{(m)}}{m!}h^m.$$

Generally: The iteration at the subinterval $[x_j, x_{j+1}]$ is given by

$$y_{j+1} = y_j + \frac{y_j'}{1!}h + \cdots + \frac{y_j^{(m)}}{m!}h^m. \tag{12.7}$$

Note that

$$\begin{cases} y_j = y(x_j), \\ y_j^{(k)} = \dfrac{d^k}{dx^k}y(x)\Big|_{x=x_j}. \end{cases}$$

12.4.2 *Some common methods*

Euler Method:

$$y_{j+1} = y_j + hy'_j,$$

for solving

$$\begin{cases} y'(x) = f(x, y(x)), \\ y(x_0) = y_0. \end{cases}$$

Thus, we have

$$y_{j+1} = y_j + hf(x_j, y_j),$$

giving recursively $y_1, y_2, y_3, \ldots,$ approximating $y_1(x), y_2(x), y_3(x), \ldots$.

Observe that here no differentiation is required.

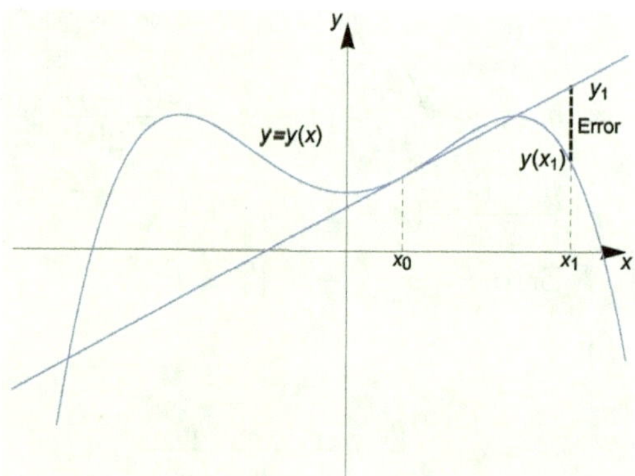

The error equals $|y_1 - y(x_1)| = |\,(x_1 - x_0)\, y'\,(x_0) + y\,(x_0) - y(x_1)|$.

Example 12.7.

$$\begin{cases} y' = x^2 + y^2, \quad x \in [0, 1], \\ y(0) = 0, \qquad (x_0 = 0) \quad N = 10 \Longrightarrow h = \dfrac{b - x_0}{N} = \dfrac{1}{10}. \end{cases}$$

This yields

$$\begin{aligned} y_1 &= y_0 + f(x_0, y_0) = 0 + \frac{1}{10}(x_0^2 + y_0^2) = y_0 = 0 \\ y_2 &= y_1 + f(x_1, y_1) = \{x_1 = x_0 + h \cdot 1 = h = 0.1\} \\ &= 0 + \frac{1}{10} f(0.1, 0) = \frac{1}{10} \cdot \frac{1}{10}^2 = 0.001. \end{aligned}$$

Euler Method

Advantages:
No differentiation required.
Simple in computer implementations.
Disadvantages: Large truncation error.

12.4.3 *More accurate methods*

Consider the Taylor series method: (12.7) with $m = 2$:

$$y_{j+1} = y_j + \frac{y_j'}{1!}h + \frac{y_j''}{2!}h^2. \qquad (12.8)$$
$$y_j' \approx y'(x_j) = f(x_j, y(x_j)) \approx f(x_j, y_j).$$

Let h^* be a small step (not necessarily $= h$).

Approximation I:

$$\begin{aligned} y_j'' = y''(x_j) &\approx \frac{y'(x_j + h^*) - y'(x_j)}{h^*} \\ &= \frac{1}{h^*}\left[f\left(x_j + h^*, \underbrace{y(x_j + h^*)}_{\approx y(x_j) + h^* y'(x_j)}\right) - f(x_j, y(x_j))\right] \qquad (12.9) \\ &\approx \frac{1}{h^*}\Big[f(x_j + h^*, y_j + h^* f(x_j, y_j)) - f(x_j, y(x_j))\Big]. \end{aligned}$$

With $h^* = \lambda h (\lambda \text{ constant})$ (12.8) $\Longrightarrow$

$$y_{j+1} = y_j + h\Big[\Big(1 - \frac{1}{2\lambda}\Big)f(x_j, y_j) + \frac{1}{2\lambda}f(x_j + \lambda h, y_j + \lambda h f(x_j, y_j))\Big].$$

Now letting $\alpha_1 = 1 - \frac{1}{2\lambda}$, $K_1 = f(x_j, y(x_j))$, $\alpha_2 = \frac{1}{2\lambda}$,

$$K_2 = f\Big(x_j + \lambda h, y_j + \lambda h f(x_j, y(x_j))\Big),$$

we get explicitly for Approximation I:

$$y_{j+1} = y_j + h[\alpha_1 K_1 + \alpha_2 K_2].$$

Generalizing Approximation I $\Longrightarrow$ Following Runge–Kutta method

Approximation II:

Backward Euler. $[x_{j-1} = x_j - h \cdot 1]$.

$$\begin{aligned} y_j'' &= \frac{y'(x_j) - y'(x_j - h)}{h} = \frac{y'(x_j) - y'(x_{j-1})}{h} \\ &= \frac{1}{h}[f(x_j, y(x_j)) - f(x_{j-1}, y(x_{j-1}))] \\ &\approx \frac{1}{h}[f(x_j, y_j) - f(x_{j-1}, y_{j-1})]. \end{aligned} \tag{12.10}$$

Now, recall (12.8): $y_{j+1} = y_j + \dfrac{y_j'}{1!}h + \dfrac{y_j''}{2!}h^2$,
then

$$y_{j+1} = y_j + h\left[\frac{3}{2}f(x_j, y_j) - \frac{1}{2}f(x_{j-1}, y_{j-1})\right].$$

Then, the final form of **Approximation II** reads

$$y_{j+1} = y_j + h[\beta_1 f_j + \beta_2 f_{j-1}].$$

Generalizing Approximation II yields **Adam's method.**

Runge–Kutta Method:
Approximates Taylor's without taking derivatives.

The Implicit Runge–Kutta: Start at $y_0 = y(x_0)$. Compute

$$\begin{array}{ccc} y_1 & \text{approximating} & y(x_0 + h) \\ y_2 & \text{approximating} & y(x_0 + 2h) \\ \vdots & \vdots & \vdots \end{array}$$

t = number of (stages)

$$y_{j+1} = y_j + h\left(\sum_{i=1}^{t} \alpha_i K_i\right),$$

where

$$\begin{cases} K_1 &= f(x_j, y_j), \\ K_i &= f\Big(x_j + h\mu_j,\, y_j + h\Big(\sum_{m=1}^{i-1} \lambda_{im} K_m\Big)\Big), \quad i = 2, 3, \ldots, t \end{cases}$$

and α_i, μ_i, λ_{im}, $1 \leq m \leq i-1$, $1 \leq i \leq t$, are parameters to be chosen to make the method as accurate as possible.

Example 12.8. "1-stage" Runge–Kutta, i.e., $(t = 1)$ in

$$y_{j+1} = y_j + h\left(\sum_{i=1}^{t} \alpha_i K_i\right) \Longrightarrow y_{j+1} = y_j + h\alpha_1 f(x_j, y_j).$$

The truncation error is minimal for $\alpha_1 = 1 \implies$
"1-stage" Runge–Kutta coincides with Euler's method.

Example 12.9. "2-stage" Runge–Kutta, (corrected Euler) i.e., $(t = 2) : \mu_i = \mu_2 = 1/2,\ \lambda_{im} = \lambda_{21} = 1/2$

$$\begin{cases} K_1 &= f(x_j, y_j), \\ K_2 &= f\left(x_j + \frac{h}{2}, y_j + \frac{h}{2}K_1\right), \quad i = 2, 3, \ldots, t. \end{cases}$$

Then,

$$y_{j+1} = y_j + hK_2, \qquad (*).$$

Comparing with Approximation I:

Example 12.10. $(*)$ yields

$$y_{j+1} = y_j + h[\alpha_1 K_1 + \alpha_2 K_2] \Longrightarrow (\alpha_1 = 0,\ \alpha_2 = 1).$$

Applying to

$$\begin{cases} y' = x^2 + y^2, \\ y(0) = 0, \qquad \text{with } h = \frac{1}{10}, \end{cases}$$

where

$$j = 0 \Longrightarrow \begin{cases} K_1 = x_0^2 + y_0^2, \\ K_2 = \left(\left(0 + \frac{0.1}{2}\right)^2 + \left(0 + \frac{0.1}{2} \cdot 0\right)^2\right) = 0.0025, \end{cases}$$

gives

$$y_1 = y_0 + hK_2 = 0 + \frac{1}{10} \cdot (0.0025) = 0.00025.$$

Heun Method: "2-Stage" Runge–Kutta ($t = 2$), $\alpha_1 = \frac{1}{4}$, $\alpha_2 = \frac{3}{4}$, $\mu_2 = \lambda_{21} = \frac{2}{3}$ yields Heun method:

$$\begin{cases} K_1 = f(x_j, y_j), \\ K_2 = f\left(x_j + \dfrac{2}{3}h,\ y_j + \dfrac{2}{3}hK_1\right), \\ y_{j+1} = y_j + h\left(\dfrac{1}{4}K_1 + \dfrac{3}{4}K_2\right). \end{cases}$$

Applied to our canonical example:

Example 12.11.

$$\begin{cases} y' = x^2 + y^2, \\ y(0) = 0, \qquad \text{with } h = \dfrac{1}{10}, \end{cases}$$

we have $K_1 = 0$, $K_2 = (0 + \frac{2}{3} \cdot 0.1)^2 + (0 + \frac{2}{3} \cdot 0.1 \cdot (0))^2 = 0.004444$ and hence

$$y_1 = y_0 + h\left[\frac{1}{4}K_1 + \frac{3}{4}K_2\right] = 0 + \frac{1}{10} \cdot \frac{3}{4}(0.004444) = 0.000333.$$

[The most popular case]: "4-Stage" Runge–Kutta Method ($t = 4$).

Or the so-called Classical Runge–Kutta formula:

$$\begin{cases} K_1 = f(x_j,\ y_j), \\ K_2 = f\left(x_j + \dfrac{1}{2}h,\ y_j + \dfrac{1}{2}hK_1\right), \\ K_3 = f\left(x_j + \dfrac{1}{2}h,\ y_j + \dfrac{1}{2}hK_2\right), \\ K_4 = f(x_j + h,\ y_j + hK_3), \\ \cdots \\ y_{j+1} = y_j + \dfrac{h}{6}\Big(K_1 + 2K_2 + 2K_3 + K_4\Big). \end{cases}$$

Example 12.12. Applied to

$$\begin{cases} y' = x^2 + y^2, \\ y(0) = 0, \qquad \text{with } h = \dfrac{1}{10} \text{ and } j = 0, \end{cases}$$

implies

$$\begin{aligned} K_1 &= 0^2 + 0^2 = 0 \\ K_2 &= \left(0 + \frac{0.1}{2}\right)^2 + \left(0 + \frac{0.1}{2} \cdot 0\right)^2 = 0.0025 \\ K_3 &= \left(0 + \frac{0.1}{2}\right)^2 + \left(0 + \frac{0.1}{2} \cdot 0.0025\right)^2 \approx 0.0025 \\ K_4 &= (0 + (0.1))^2 + (0 + (0.1) \cdot 0.0025)^2 \approx 0.01, \end{aligned}$$

and

$$y_1 = 0 + \frac{0.1}{6}(0 + 0.0050 + 0.0050 + 0.01) \approx 0.000333.$$

12.5 Finite Element Method (FEM)

A model problem. We consider a convection–diffusion–absorption, boundary-value, problem with positive, constant coefficients and homogeneous Dirichlet data, *viz.*

$$\begin{cases} -du''(x) + cu'(x) + au(x) = f(x), \quad 0 < x < 1, \\ u(0) = u(1) = 0. \end{cases} \tag{12.11}$$

Here, in general,

$d = d(x)$ is the diffusion coefficient and the first term is diffusion term.

$c = c(x)$ is the convection coefficient and the second term is the convection term.

$a = a(x)$ is the absorption coefficient and the third term is the absorption term.

Finally, $f(x)$ is the load.

The idea is to follow the Fourier analysis technique where we multiply the function by a test function (an ON-basis) and integrate over the domain to get an information (Fourier transform at a point/Fourier coefficient) about the function at a point in, e.g., $\mathbb{R}$.

Now, to get an approximate solution for the differential equations, we need a finite number (infinite number of points/data are not computable) of approximate values of the solution at certain points and adopt a polynomial of a certain degree to these approximate values. Then, the Fourier technique for (12.11) is in the form of the so-called *variational formulation* where now the multipliers, instead of being orthonormal basis as, e.g., $e^{-ix\xi}$, are *almost orthogonal* test functions consisting of piecewise polynomial basis of certain degree at the discrete nodes (being 1 at one node and 0 at the others).

To proceed we let V be the space of all continuous, piecewise differentiable functions $v(x)$ with $v(0) = v(1) = 0$:

$$V := H_0^1[0,1] = \left\{ v : \int_0^1 v'(x)^2\,dx < \infty, \quad v(0) = v(1) = 0 \right\}.$$

Now we multiply the equation (12.11) by test functions $v \in V$ and integrate over $[0,1]$, where using the notation

$$(u, v) = \int_0^1 u(x)v(x)\,dx,$$

after partial integration, we have (for simplicity we let $d = c = a = 1$) the following:

Variational formulation: Find $u \in V$ such that

$$(VF) \qquad (u', v') + (u', v) + (u, v) = (f, v) \quad \forall v \in V. \tag{12.12}$$

Now, let $0 = x_0 < x_1 < \cdots < x_n = 1$ be a subdivision $[0,1]$, and set

$$\begin{aligned} &I_k := [x_{k-1}, x_k], && h_k := x_k - x_{k-1}, \\ &h = \max_k h_k = \max_k |I_k|\,; && k = 1, 2, \ldots, n, \end{aligned}$$

and let V_h be the space of all continuous piecewise polynomial functions $v_h(x)$ of degree $\leq k$ with $v_h(0) = v_h(1) = 0$, and such that the derivatives of v_h of degree $\leq k-1$ are continuous splines.

Now, for simplicity, let $k = 1$. Then

$$V_h := \Big\{ v : v \text{ is piecewise linear, continuous, and} \\ v_h(0) = v_h(1) = 0 \Big\}.$$

Then, V_h has the basis φ_i, $i = 1, 2, \ldots n$ consisting of *hat-functions*: piecewise linear continuous such that

$$\varphi_i(x_j) = \begin{cases} 1, & i = j, \\ 0, & i \neq j. \end{cases} \tag{12.13}$$

More specifically,

$$\varphi_i(x) = \begin{cases} \dfrac{x - x_{i-1}}{x_i - x_{i-1}} = \dfrac{x - x_{i-1}}{h_i} & x_{i-1} \le x \le x_i, \\ \dfrac{x - x_{i+1}}{x_i - x_{i+1}} = \dfrac{x_{i+1} - x}{h_{i+1}} & x_i \le x \le x_{i+1}. \end{cases} \tag{12.14}$$

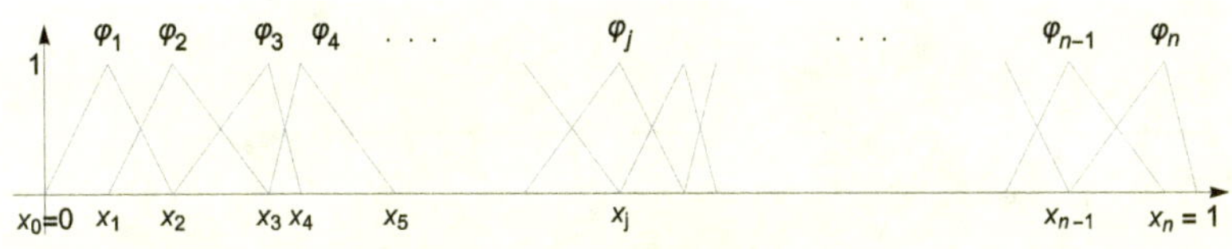

The hat-functions in (12.14).

- Continuous Galerkin finite element basis functions of degree 2: cG(2):
- cG(2) basis in single interval $[x_{j-1}, x_j]$ and midpoint $\overline{x}_j = \frac{x_j - x_{j-1}}{2}$:

$$\lambda_{j-1}(x) = \begin{cases} \dfrac{(x - \overline{x}_j)(x - x_j)}{(x_j - \overline{x}_j)(x_{j-1} - x_j)}, \\ x_{j-1} \le x \le x_j \\ 0, \quad \text{otherwise.} \end{cases} \qquad \overline{\lambda}_j(x) = \begin{cases} \dfrac{(x - x_{j-1})(x - x_j)}{(\overline{x}_j - x_j)(\overline{x}_j - x_j)}, \\ x_{j-1} \le x \le x_j \\ 0, \quad \text{otherwise.} \end{cases}$$

$$\overline{\lambda}_j(x) = \begin{cases} \dfrac{(x - x_{j-1})(x - x_j)}{(x_j - \overline{x}_j)(x_{j-1} - x_j)}, \\ x_{j-1} \le x \le x_j \\ 0, \quad \text{otherwise.} \end{cases}$$

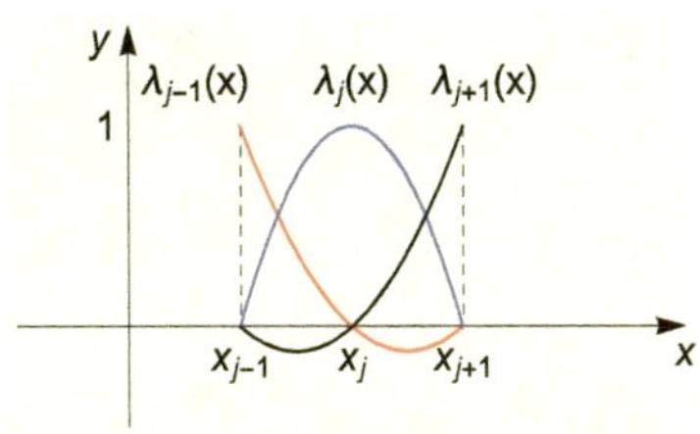

Remarks. The distances $x_j - x_{j-1}$ need not be equal.

Instead of two points x_{j-1}, and x_j, three points , x_{j-1}, x_j and x_{j+1} can be used (see figure to the right).

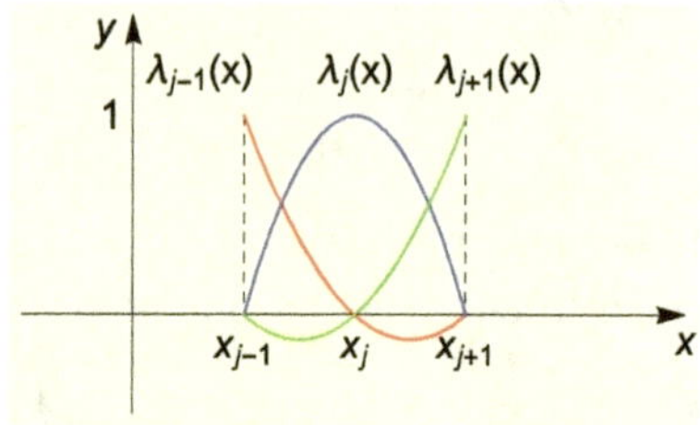

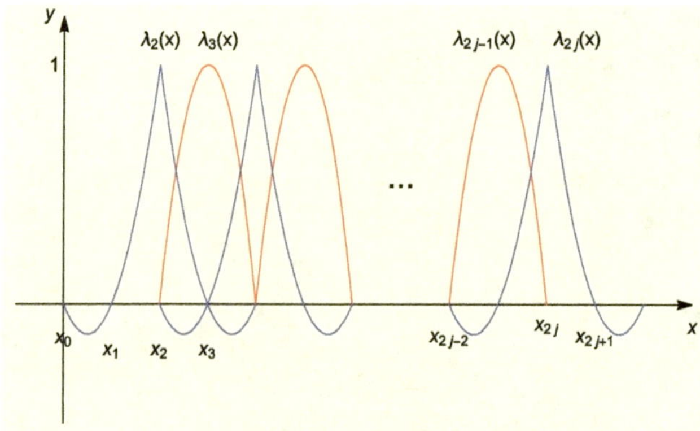

Then $u_h \in V_h$ is given by

$$u_h(x) = \sum_{j=1}^{n} \xi_j \lambda_j(x),$$

where ξ_j are the approximate values of $u(x)$ at $x = x_j$.

Then, the corresponding discrete analogue of the variational formulation (12.12) reads as follows:

Find $u_h \in V_h$ such that

$$(FEM) \qquad (u_h', v_h') + (u_h', v_h) + (u_h, v_h) = (f, v_h) \quad \forall v_h \in V_h. \tag{12.15}$$

Which is equivalent to finding the constants ξ_j such that:

$$\sum_{j=1}^{n} \xi_j(\varphi_j', \varphi_i') + \sum_{j=1}^{n} \xi_j(\varphi_j', \varphi_i) + \sum_{j=1}^{n} \xi_j(\varphi_j, \varphi_i) = (f, \varphi_i),$$

$$i = 1, 2, \ldots, n. \tag{12.16}$$

Or in compact (matrix) form to find the constants ξ_j such that

$$(S + C + A)\xi = F, \tag{12.17}$$

where

$$\begin{cases} S = (\varphi_j', \varphi_i')_{i,j=1}^{n} & \text{is the stiffness matrix,} \\ C = (\varphi_j', \varphi_i)_{i,j=1}^{n} & \text{is the convection matrix,} \\ M = (\varphi_j, \varphi_i)_{i,j=1}^{n} & \text{is the mass matrix,} \\ F = (f, \varphi_i)_{i,j=1}^{n} & \text{is the load vector.} \end{cases} \tag{12.18}$$

Now, using (12.14) we can compute the elements of the coefficient matrices.

The following are elements of stiffness and mass matrices

$$\begin{cases} S_{ii} = \dfrac{1}{h_i} + \dfrac{1}{h_{i+1}}, & S_{i-1,i} = S_{i,i-1} = -\dfrac{1}{h_i}, & S_{ij} = 0, \ |i-j| > 1, \\ M_{ii} = \dfrac{1}{3}h_i + \dfrac{1}{3}h_{i+1}, & M_{i-1,i} = M_{i,i-1} = \dfrac{1}{6}h_i, & M_{ij} = 0, \ |i-j| > 1. \end{cases} \tag{12.19}$$

$$F_j = (f, \varphi_i) = \frac{1}{6}h_i f(x_{i-1}) + \frac{1}{3}(h_i + h_{i+1}) f(x_i) + \frac{1}{6}h_{i+1} f(x_{i+1}). \tag{12.20}$$

In a *uniform mesh*, i.e, a partition of the interval $I = [0, 1]$ into n equal subintervals, of length h, we have that

$$\begin{cases} S_{ii} = \dfrac{2}{h}, & S_{i-1,i} = S_{i,i-1} = -\dfrac{1}{h}, & S_{ij} = 0, & |i-j| > 1, \\ C_{ii} = 0, & C_{i-1,i} = 1/2, \ C_{i,i-1} = -1/2, & C_{ij} = 0, & |i-j| > 1, \\ M_{ii} = \dfrac{2}{3}h_i, & M_{i-1,i} = M_{i,i-1} = \dfrac{1}{6}h, & M_{ij} = 0, & |i-j| > 1, \end{cases} \tag{12.21}$$

which can be written in the matrix form as

$$S = \frac{1}{h} = \begin{bmatrix} 2 & 1 & 0 & 0 & \dots & 0 & 0 & 0 \\ -1 & 2 & 1 & 0 & \dots & 0 & 0 & 0 \\ 0 & -1 & 2 & 1 & \dots & 0 & 0 & 0 \\ \vdots & \vdots & \vdots & \vdots & \ddots & \vdots & \vdots & \vdots \\ 0 & 0 & 0 & 0 & \dots & -1 & 2 & 1 \\ 0 & 0 & 0 & 0 & \dots & 0 & -1 & 2 \end{bmatrix}.$$

$$M = \frac{h}{6}\begin{bmatrix} 4 & 1 & 0 & 0 & \dots & 0 & 0 & 0 \\ 1 & 4 & 1 & 0 & \dots & 0 & 0 & 0 \\ 0 & 1 & 4 & 1 & \dots & 0 & 0 & 0 \\ \vdots & \vdots & \vdots & \vdots & \ddots & \vdots & \vdots & \vdots \\ 0 & 0 & 0 & 0 & \dots & 1 & 4 & 1 \\ 0 & 0 & 0 & 0 & \dots & 0 & 1 & 4 \end{bmatrix}.$$

As seen both S and M are symmetric, positive, definite sparse (3-diagonal) matrices thus they are invertible, likewise the following *uniform convection matrix*,

$$C = \frac{1}{2}\begin{bmatrix} 0 & 1 & 0 & 0 & \dots & 0 & 0 & 0 \\ -1 & 0 & 1 & 0 & \dots & 0 & 0 & 0 \\ 0 & -1 & 0 & 1 & \dots & 0 & 0 & 0 \\ \vdots & \vdots & \vdots & \vdots & \ddots & \vdots & \vdots & \vdots \\ 0 & 0 & 0 & 0 & \dots & -1 & 0 & 1 \\ 0 & 0 & 0 & 0 & \dots & 0 & -1 & 0 \end{bmatrix}.$$

Thus, the system $(S + C + M)\,\xi = F$ is uniformly solvable and we have the unknown coefficient vector given as

$$\xi = (S + C + M)^{-1} \cdot F, \quad \xi = (\xi_1, \xi_1, \dots, \xi_n)^T,$$

where F is the uniform version of F which can be obtained from (12.20) as

$$F_i = (f, \varphi_i) = \frac{h}{6}\Big(f(x_{i-1}) + 4f(x_i) + f(x_{i+1})\Big). \qquad (12.22)$$

It is easy to verify that if f and all involved coefficients ($d(x)$, $c(x)$, $a(x)$ that we assumed to be $\equiv 1$) are smooth, then the equation system (12.16) has a unique solution u_h approximating the exact solution of (12.11) u and it converges to u in the L_2-sense and with an error of order $\mathcal{O}(h^2)$

$$\| u - v \| \leq Ch^2, \quad \forall v \in V_k \quad C \text{ is a constant,} \tag{12.23}$$

where $\| \cdot \|$ denotes the L_2−norm.

Remark. There is a corresponding finite difference procedure which is an alternative to the finite elements but is based, basically, on Euler and Crank–Nicolson type approaches. For instance, one gets the same matrix S in finite difference approximation as in the finite element case.

12.6 Monte Carlo Methods

12.6.1 *Monte Carlo method for DEs (indirect method)*

(i) **Monte Carlo techniques for first-order equations (ODEs).**
Consider the quadrature-type problem:

$$y' = f(x); \quad a \leq x \leq b, \quad \text{and} \quad 0 \leq f(x) \leq c. \tag{12.24}$$

The definition set $[a, b]$ is subdivided into n subintervals, not necessarily uniform, with $x_0 = a$, $x_n = b$, and $I_j := [x_j, x_{j+1}]$; $j = 0, 1, \ldots, n - 1$.

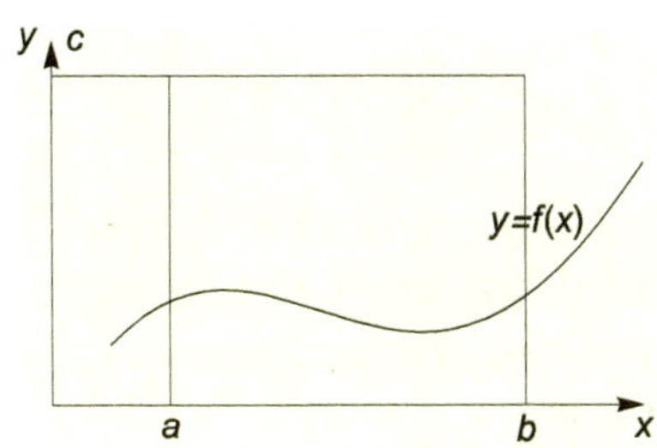

For the subinterval I_j,

Step 1. Set $N_j = 0$, and after each trial increase T_j by unity ($N_j \Longleftarrow N_j + 1$). The most recent N_j indicates the number of completed trials.

Step 2. Set $T_j = 0$, and after each trial increase T_j by unity ($T_j \Longleftarrow T_j + 1$) *only if the trial was successful*, otherwise let it retain its previous value:

$$T_j \Longleftarrow \begin{cases} T_j + 1, & \text{if the trial was successful,} \\ T_j, & \text{else.} \end{cases}$$

Step 3. Make a trial: A trial consists of selecting two random numbers r_1 and r_2, from a uniform distribution in I_j and $[0, c]$, respectively.
If $r_2 \leq f(r_1)$, then the trial is deemed to be successful. Otherwise, it is a failure.

$$\begin{cases} \dfrac{T_j}{N_j}(x_{j+1} - x_j)c \longrightarrow \displaystyle\int_{x_j}^{x_{j+1}} f(x)\,dx, & \text{as } N_j \to \infty, \\ y_{j+1} = y_j + \dfrac{T_j(x_{j+1} - x_j)c}{N_j}, & \text{as } N_j \to \infty. \end{cases}$$

Step 4. If the function f has a negative lower bound: $-d \leq f(x) \leq c$, then by an axis transformation replace $f(x)$ by $d + f(x)$ and c by $c + d$ in the above, that yields

$$y_{j+1} = y_j + \frac{T_j(x_{j+1} - x_j)(c + d)}{N_j}, \quad \text{as } N_j \to \infty.$$

Remarks. In general, the range of values for which a uniform distribution is required will not be standard, then to circumvent this difficulty a transformation is made, *viz.*

If r^* is a random number from a uniform distribution in $I = [a, b]$, then a random number r from a uniform distribution in $[p, q]$

is obtained as

$$r = \left(\frac{bp - aq}{b - a}\right) - r^*\left(\frac{p - q}{b - a}\right).$$

In particular, in $[0, 1]$ (i.e., $a = 0$, $b = 1$),

$$r = p + r^*(q - p), \quad (r^* \text{ can be replaced by } r),$$

and in $[-1, 1]$,

$$r = \frac{1}{2}(p + q) + \frac{1}{2}r^*(q - p), \qquad (r^* \text{ can be replaced by } r).$$

In all Monte Carlo approaches, the number of trials must be large; in general, several thousand.

(ii) **Monte Carlo techniques for partial differential equations**

(iia) **Elliptic Equations.** For simplicity, we shall discuss the two-dimensional problem. Extensions to higher dimensions follow by increasing the number of the directions of the random walks by two for each added dimension.

A finite difference approximation for the elliptic equation $\nabla^2 u = 0$ in $2D$:

$$\frac{\partial^2 u}{\partial x^2} = \frac{u_{i+1,j} - 2u_{i,j} + u_{i-1,j}}{h_x^2}$$
$$\frac{\partial^2 u}{\partial y^2} = \frac{u_{i,j+1} - 2u_{i,j} + u_{i,j-1}}{h_y^2},$$

yields

$$u_{i,j} = \frac{1}{2(h_x^2 + h_y^2)}(h_y^2 u_{i+1,j} + h_y^2 u_{i-1,j} + h_x^2 u_{i,j+1} + h_x^2 u_{i,j-1}).$$

Monte Carlo for elliptic equations. Define

$$p_1 = p_2 = \frac{h_y^2}{2(h_x^2 + h_y^2)}$$
$$p_3 = p_4 = \frac{h_x^2}{2(h_x^2 + h_y^2)}.$$

Then, the condition for carrying out a random walk is

$$\sum_{k=1}^{4} p_k = 1,$$

along the four directions: x increasing, x decreasing, y increasing, y decreasing. Hence, to the value of $u_{i,j} := u(x_i, y_j)$, the procedure will be as follows: Initially, set $T = N = 0$. For each walk completed add 1 to N.
A walk is completed when the boundary is reached and the value of u *at that point at that boundary* has been added to T. Then

$$\frac{T}{N} \longrightarrow u_{i,j} \quad \text{as} \quad N \to \infty.$$

As for the non-homogeneous case:

$$\nabla^2 = f,$$

the following relationship is applied.

$$u_{i,j} + \frac{h_x^2 h_y^2}{2(h_x^2 + h_y^2)} f_{i,j} = p_1 u_{i+1,j} + p_2 u_{i-1,j}$$
$$+ p_3 u_{i,j+1} + p_4 u_{i,j+1}.$$

Then as before

$$\frac{T}{N} \longrightarrow u_{i,j} + \frac{h_x^2 h_y^2}{2(h_x^2 + h_y^2)} f_{i,j} \quad \text{as} \quad N \to \infty.$$

(iib) **Parabolic Equations.** For simplicity, we consider the one-dimensional heat conduction

$$\frac{\partial u}{\partial t} = \frac{\partial^2 u}{\partial x^2},$$

in a (x, t) rectangular domain. The technique involved can easily be extended to higher spatial dimensions. For the mesh function h and k in x and t directions, respectively, starting with (x_0, t_0), we let

$$x_s = x_0 + sh,$$
$$t_r = t_0 + rk,$$
$$u_{s,r} = u(x_s, t_r).$$

12.6.2 *Examples of finite difference approximations for parabolic equations*

$$\left(\frac{\partial u}{\partial t}\right)_{s,r} = (u_{s,r+1} - u_{s,r})/k, \qquad \text{Forward Euler}$$

or

$$\left(\frac{\partial u}{\partial t}\right)_{s,r} = (u_{s,r} - u_{s,r-1})/k, \qquad \text{Backward Euler}$$

or

$$\left(\frac{\partial u}{\partial t}\right)_{s,r} = (u_{s,r+1} - u_{s,r-1})/2k, \quad \text{Central difference.}$$

As for the second-order differentiation, we have, e.g.,

$$\left(\frac{\partial^2 u}{\partial x^2}\right)_{s,r} = (u_{s+1,r} - 2u_{s,r} + u_{s-1,r})/h^2.$$

Some combinations of space time finite difference discretizarions are given below.

Examples of typical explicit formulas

Type 1.

$$u_{s,r+1} = u_{s,r} + (k/h^2)(u_{s-1,r} - 2u_{s,r} + u_{s+1,r}),$$
$$\text{stable if } 0 < k/h^2 \leq 1/2. \tag{12.25}$$
$$\text{Truncation error} = \mathcal{O}(k) + \mathcal{O}(h^2)$$

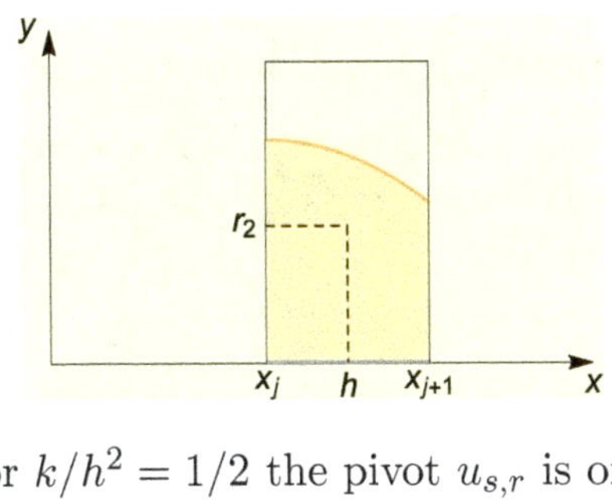

In particular, for $k/h^2 = 1/2$ the pivot $u_{s,r}$ is omitted:

$$u_{s,r+1} = \frac{1}{2}(u_{s-1,r} + u_{s+1,r}).$$

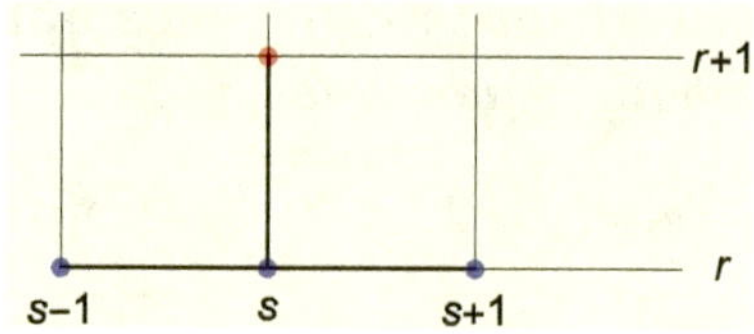

Milne method: $k/h^2 = 1/6$ yields the scheme

$$u_{s,r+1} = \frac{1}{6}(u_{s-1,r} + 4u_{s,r} + u_{s+1,r}),$$
$$\text{with improved truncation error } = \mathcal{O}(k^2) + \mathcal{O}(h^4). \tag{12.26}$$

Type 2.

$$u_{s,r+1} = \frac{(h^2 - 2k)u_{s,r-1} + 2k(u_{s-1,r} + u_{s+1,r})}{h^2 + 2k}, \tag{12.27}$$

Always stable, but $u_{s,r-1}$ is required

Truncation error $= \mathcal{O}(k) + \mathcal{O}(h^2) + \mathcal{O}(k/h)$

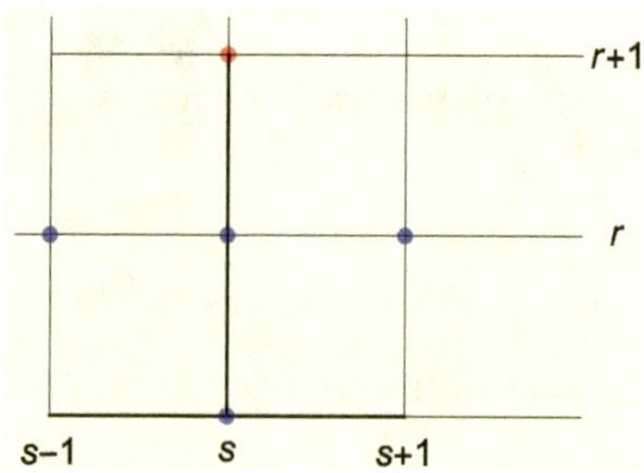

Example of typical implicit formulas

These are stable, however require simultaneous solutions for a number of equations. The remedy is using matrix methods, where only one inversion is needed.

Type 1.

$$(k/h^2)(u_{s+1,r+1} + u_{s-1,r+1} + u_{s+1,r} + u_{s-1,r}) - 2[1 + (k/h^2)]f_{s,r+1} + 2[1 - (k/h^2)]f_{s,r} = 0, \tag{12.28}$$

always stable.

Truncation error $= \mathcal{O}(k^2) + \mathcal{O}(h^2)$

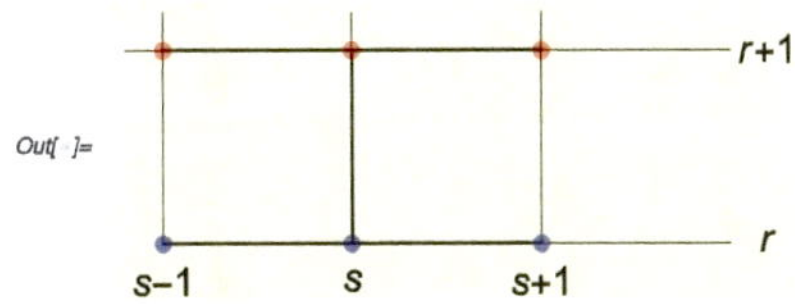

Type 2.

$$u_{s,r+1} - u_{s,r} = (k/h^2)(u_{s-1,r+1} - 2u_{s,r+1} + u_{s+1,r+1}),$$
$$\text{always stable.} \tag{12.29}$$

$$\text{Truncation error} = \mathcal{O}(k) + \mathcal{O}(h^2)$$

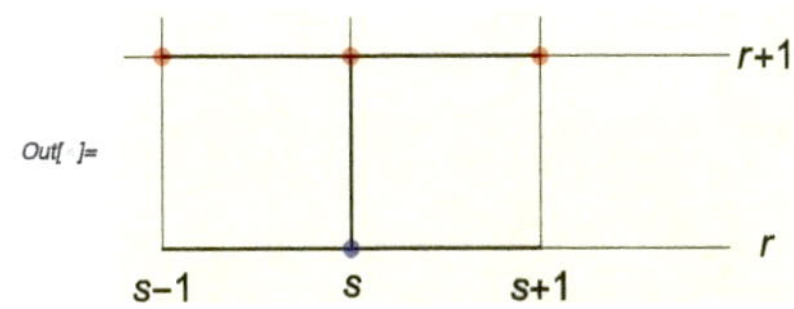

12.6.3 *Monte Carlo for elliptic equations*

Formula (12.29) yields, as a finite difference representation of

$$\frac{\partial u_{s,r}}{\partial t} = \frac{\partial^2 u_{s,r}}{\partial x^2}. \tag{12.30}$$

$$u_{s,r+1} = \left(\frac{h^2 - 2k}{h^2 + 2k}\right) u_{s,r-1} + \left(\frac{2k}{h^2 + 2k}\right) u_{s-1,r} + \left(\frac{2k}{h^2 + 2k}\right) u_{s+1,r}, \tag{12.31}$$

which can be written as

$$u_{s,r+1} = p_1 u_{s-1,r} + p_2 u_{s+1,r} + p_3 u_{s,r-1},$$

where

$$p_1 = p_2 = \left(\frac{2k}{h^2 + 2k}\right); \qquad p_3 = \left(\frac{h^2 - 2k}{h^2 + 2k}\right), \quad \text{thus} \quad \sum_{m=1}^{3} p_m = 1.$$

The three directions are x increasing, x decreasing, and t decreasing. In other words, the process is restricted to be either toward the boundaries in the x-directions, or back wards in the initial conditions along the decreasing t direction. As above N will stand for the total number of completed walks and T denotes the sum of boundary values or initial values reached.

Chapter 13

Differential Geometry

13.1 Curve

Definition 13.1. A curve is a mapping $\boldsymbol{r}$: $[a, b] \to \mathbb{R}^n$, such that $\boldsymbol{r}(t)$ is continuous and $\boldsymbol{r}'(t)$ is continuous everywhere except, possibly, at a finite number of points t_k,

$$a \le t_1 < t_2 < \cdots < t_m \le b,$$

where both left and right derivatives exist: $\boldsymbol{r}'_L(t)$ and $\boldsymbol{r}'_R(t)$.

$$t \curvearrowright \boldsymbol{r}(t) = (x_1(t), x_2(t), \ldots, x_n(t)), \quad t \in [a, b]. \tag{13.1}$$

Definition 13.2. The derivative of $\boldsymbol{r}(t) = (x_1(t),\, x_2(t),\, \ldots, x_n(t)) \in \mathbb{R}^n$, $t \in [a, b]$, is given by

$$\dot{\boldsymbol{r}}(t) := \frac{d\boldsymbol{r}}{dt}(t) = \left(\frac{dx_1}{dt}, \frac{dx_2}{dt}, \ldots, \frac{dx_n}{dt}\right), \tag{13.2}$$

and is a tangent vector to the curve, if not all $\dfrac{dx_i}{dt} = 0$.

The length $L(\boldsymbol{r})$; $a < t < b$ of a curve is

$$L(\boldsymbol{r}) := \int_a^b \sqrt{\left(\frac{dx_1}{dt}\right)^2 + \left(\frac{dx_2}{dt}\right)^2 + \cdots + \left(\frac{dx_n}{dt}\right)^2}\, dt. \tag{13.3}$$

The derivative/differentiating rules
If the function $f(t) : [a,b] \to \mathbb{R}$ is differentiable, $\alpha,\ \beta \in \mathbb{R}$, and

$$\boldsymbol{r}(t) = (x_1(t), x_2(t), \dots, x_n(t)),$$
$$\boldsymbol{s}(t) = (y_1(t), y_2(t), \dots, y_n(t)), \quad t \in [a,b],$$

both are differentiable, then

$$\begin{aligned}
\frac{d}{dt}\Big(\alpha\boldsymbol{r}(t) + \beta\boldsymbol{s}(t)\Big) &= \alpha\dot{\boldsymbol{r}}(t) + \beta\dot{\boldsymbol{s}}(t) \\
\frac{d}{dt}\Big(f(t)\boldsymbol{r}(t)\Big) &= \dot{f}(t)\boldsymbol{r}(t) + f(t)\dot{\boldsymbol{r}}(t) \\
\frac{d}{dt}\Big(\boldsymbol{r}(t)\cdot\boldsymbol{s}(t)\Big) &= \dot{\boldsymbol{r}}(t)\cdot\boldsymbol{s}(t) + \boldsymbol{r}(t)\cdot\dot{\boldsymbol{s}}(t) \\
\frac{d}{dt}\Big(\boldsymbol{r}(t)\times\boldsymbol{s}(t)\Big) &= \dot{\boldsymbol{r}}(t)\times\boldsymbol{s}(t) + \boldsymbol{r}(t)\times\dot{\boldsymbol{s}}(t) \\
\frac{d}{dt}[\boldsymbol{r},\ \boldsymbol{s},\ \boldsymbol{u}] &= [\dot{\boldsymbol{r}},\ \boldsymbol{s},\ \boldsymbol{u}] + [\boldsymbol{r},\ \dot{\boldsymbol{s}},\ \boldsymbol{u}] + [\boldsymbol{r},\ \boldsymbol{s},\ \dot{\boldsymbol{u}}] \\
\frac{d}{dt}f\Big(\boldsymbol{r}(t)\Big) &= \nabla f\Big(\boldsymbol{r}(t)\Big)\cdot\dot{\boldsymbol{r}}(t).
\end{aligned} \tag{13.4}$$

Remark.

The symbols $\cdot$ and $\times$ denote scalar- (inner) and cross-product, respectively.
The cross-product is defined in $\mathbb{R}^3$, see page 78.
The Length $L(\boldsymbol{r})$ of a curve is independent of the choice of its parametrization.

Three different parametrizations for curves and their corresponding lengths

In the following, we assume that the involved integrals exist, $t_1 \le t_2$, $\theta_1 \le \theta_2$ and $x_1 \le x_2$.

Parametrization	Length $L(\boldsymbol{r})$
$\boldsymbol{r}(t) = (x, y) = (x(t), y(t))$	$\int_{t_1}^{t_2} \sqrt{\left(\frac{dx}{dt}\right)^2 + \left(\frac{dy}{dt}\right)^2}\, dt$
$\boldsymbol{r}(\theta) = (x, y) = (r(\theta)\cos\theta, r(\theta)\sin\theta)$	$\int_{\theta_{1cf}}^{\theta_2} \sqrt{r^2 + \left(\frac{dr}{d\theta}\right)^2}\, d\theta$
$(x, y) = (x, f(x))$	$\int_{x_1}^{x_2} \sqrt{1 + (f'(x))^2}\, dx$

(13.5)

13.1.1 *Examples of curves and surfaces in* $\mathbb{R}^2$

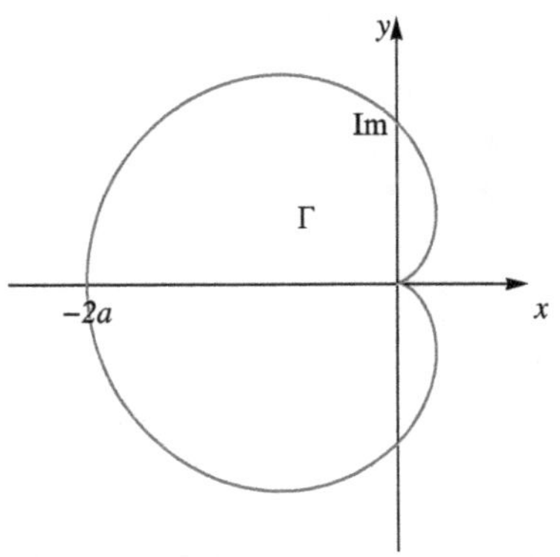

Cardioid
$\begin{cases} x(t) = a(1 - \cos t)\cos t \\ y(t) = a(1 - \cos t)\sin t \end{cases}$
Cartesian coord.
$(x^2 + y^2 + ax)^2 = a^2(x^2 + y^2)$
Polar coord.
$r = a(1 - \cos\theta)$
Length $L = 8a$
Enclosed area $A = \frac{3}{2}\pi a^2$
Arc length $8a\sin^2(t/4)$
Curvature $\kappa = \frac{3}{4a\sin(t/2)}$
Tangent's angle $\phi = 3t/2$

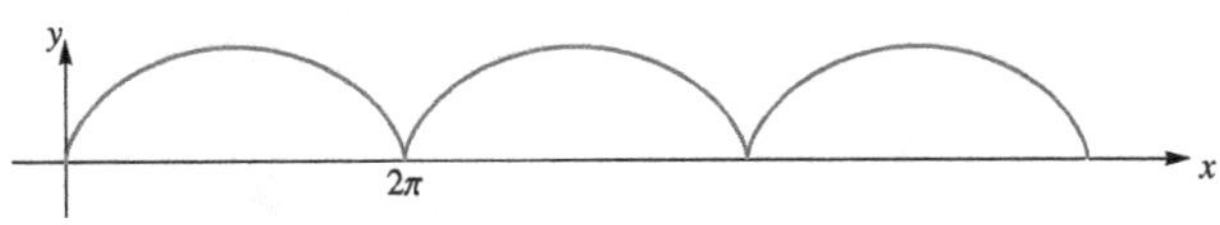

Cycloid $(x, y) = (t - \sin t, 1 - \cos t)$ for $0 \le t \le 6\pi$.

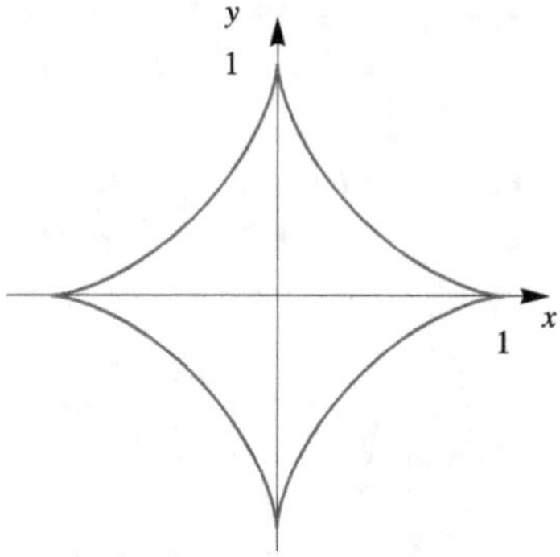

Astroid $(x, y) = a(\cos^3 t, \sin^3 t)$

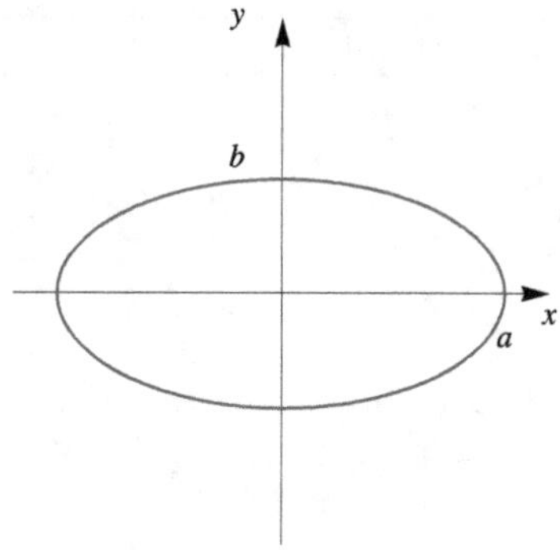

Ellipse $(x, y) = (a \cos t, b \sin t)$

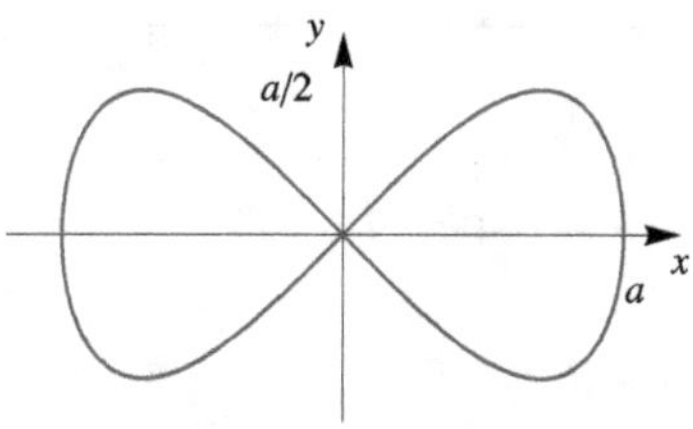

Geronos lemniscate
$(x, y) = a(\sin 3t, \sin t \, \cos t)$

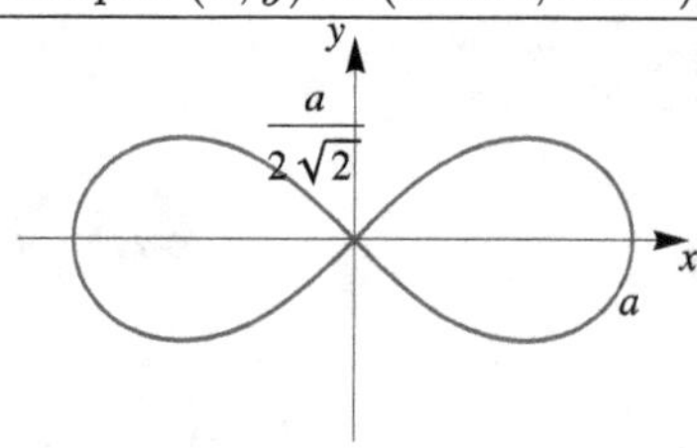

Bernoulli's lemniscate
$r^2 = 2a^2 \, \cos 2\theta$

The length of astroid: $L = 6a$. The enclosed surface area: $A = \dfrac{3\pi}{8} a^2$. The length of ellipse L can not be presented by elementary expressions. Area of the enclosed surface: $A = \pi a \, b$.

The surface area A of a region enclosed by the curve

$$\gamma(\theta) = (r \cos\theta, r \sin\theta), \qquad \theta_1 \le \theta_2, \quad r = r(\theta),$$

is given by

$$A = \int_{\theta_1}^{\theta_2} \frac{r^2}{2} \, d\theta. \tag{13.6}$$

Area of the rotation surface generated by the curve $\boldsymbol{r}(t) = (x(t), y(t))$ rotating about x-axis, with $t_1 < t_2$, is given by

$$A = \int_{t_1}^{t_2} 2\pi |y(t)| \sqrt{\left(\frac{dx}{dt}\right)^2 + \left(\frac{dy}{dt}\right)^2}\, dt. \qquad (13.7)$$

13.2 $\mathbb{R}^3$

13.2.1 *Notations in* $\mathbb{R}^3$

In the following, the notations $\boldsymbol{r} = (x_1, x_2, x_3) \in \mathbb{R}^3$, $\dot{\boldsymbol{r}} = \frac{d\boldsymbol{r}}{dt}$ and $\boldsymbol{r}' = \frac{d\boldsymbol{r}}{ds}$, are used.

Concept	Arbitrary parameter t	Curve length $s = \int_c^t \sqrt{\dot{x}^2 + \dot{y}^2 + \dot{z}^2}\, dt$
Unit tangent vector	$\boldsymbol{t} = \frac{\dot{\boldsymbol{r}}}{\lvert\dot{\boldsymbol{r}}\rvert} = \frac{\dot{\boldsymbol{r}}}{\nu}$	$\boldsymbol{t} = \boldsymbol{r}' = \frac{d\boldsymbol{r}}{ds}$
Unit normal vector	$\boldsymbol{n} = \frac{\ddot{\boldsymbol{r}} - \dot{\nu}\boldsymbol{t}}{\lvert\ddot{\boldsymbol{r}} - \dot{\nu}\boldsymbol{t}\rvert}$	$\boldsymbol{n} = \frac{\boldsymbol{r}''}{\lvert\boldsymbol{r}''\rvert}$
Unit binormal vector	$\boldsymbol{b} = \boldsymbol{t} \times \boldsymbol{n}$	$\boldsymbol{b} = \boldsymbol{t} \times \boldsymbol{n}$
Curvature	$\kappa = \frac{\lvert\dot{\boldsymbol{r}} \times \ddot{\boldsymbol{r}}\rvert}{\lvert\dot{\boldsymbol{r}}\rvert^3}$	$\kappa = \lvert\boldsymbol{r}''\rvert$
Curvature radius	$\rho_\kappa = \frac{1}{\kappa}$	$\rho_\kappa = \frac{1}{\kappa}$
Torsion	$\tau = \frac{\dot{\boldsymbol{r}} \cdot (\ddot{\boldsymbol{r}} \times \dddot{\boldsymbol{r}})}{\lvert\dot{\boldsymbol{r}} \times \ddot{\boldsymbol{r}}\rvert^2}$	$\tau = \frac{\boldsymbol{r}' \cdot (\boldsymbol{r}'' \times \boldsymbol{r}''')}{\lvert\boldsymbol{r}''\rvert^2}$
Torsion radius	$\rho_\tau = \frac{1}{\tau}$	$\rho_\tau = \frac{1}{\tau}$

Frenet's formulas: $\dot{\boldsymbol{t}} = \kappa\nu\boldsymbol{n}, \quad \dot{\boldsymbol{n}} = -\kappa\nu\boldsymbol{t} + \tau\nu\boldsymbol{b}, \quad \dot{\boldsymbol{b}} = -\tau\nu\boldsymbol{n}$

$$\begin{array}{lll} C: & \text{(Curve) is a line} \iff & \kappa = 0. \\ C: & \text{A plane curve} \quad \iff & \tau = 0. \end{array}$$

Definition 13.3. For a curve in $\mathbb{R}^3$, i.e., $\boldsymbol{r}(t) = (x(t), y(t), z(t))$ the tangent vector $\frac{d\boldsymbol{r}}{dt}(t)$ to the curve is a normal vector to the normal plane at the point $\boldsymbol{r}(t)$, provided that $\frac{d\boldsymbol{r}}{dt}(t) \neq \boldsymbol{0}$.

13.2.2 *Curve and surface in* $\mathbb{R}^3$

Some *second-degree surfaces*, given in general form

$$\boldsymbol{r}^T \cdot \boldsymbol{A} \cdot \boldsymbol{r} = d, \tag{13.8}$$

where $\boldsymbol{r} = [x\ y\ z]^T$ and $\boldsymbol{A} = \begin{bmatrix} a_{1,1} & \frac{a_{1,2}}{2} & \frac{a_{1,3}}{2} \\ \frac{a_{1,2}}{2} & a_{2,2} & \frac{a_{2,3}}{2} \\ \frac{a_{1,3}}{2} & \frac{a_{2,3}}{2} & a_{3,3} \end{bmatrix}$.

The equation (13.8) is rewritten as

$$a_{1,1}x^2 + a_{1,2}xy + a_{1,3}xz + a_{2,2}y^2 + a_{2,3}yz + a_{3,3}z^2 = d.$$

Equations of some second-degree objects

Name	Equation
Elliptical spiral	$\boldsymbol{r} = (a\cos t, b\sin t, c\,t)$
Double cone	$x^2 + y^2 = z^2$
Paraboloid	$z = \dfrac{x^2 + y^2}{a^2}$
One mantled hyperboloid	$x^2 + y^2 = z^2 - 1$
Double mantled hyperboloid	$x^2 + y^2 = z^2 + 1$
Ellipsoid	$\left(\frac{x}{a}\right)^2 + \left(\frac{y}{b}\right)^2 + \left(\frac{z}{c}\right)^2 = 1$
Elliptical cylinder	$a^2x^2 + b^2y^2 = c, \quad c > 0,\ a, b \neq 0$
Hyperbolic cylinder	$x^2 - y^2 = a, \quad a > 0$
Parabolic cylinder	$a\,y^2 = z, \quad a > 0$
Hyperbolic paraboloid	$x^2 - y^2 = z$

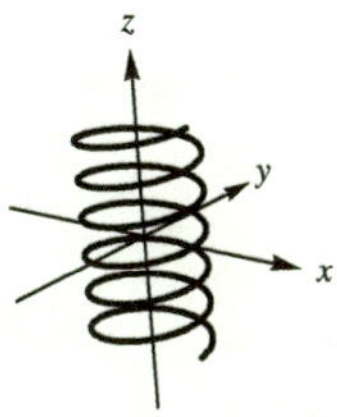

Circular spiral

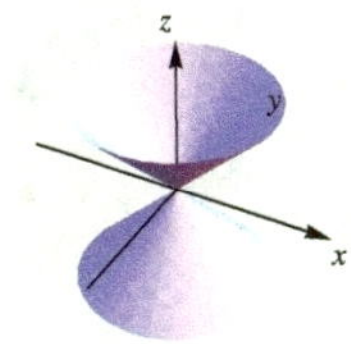

Double cone

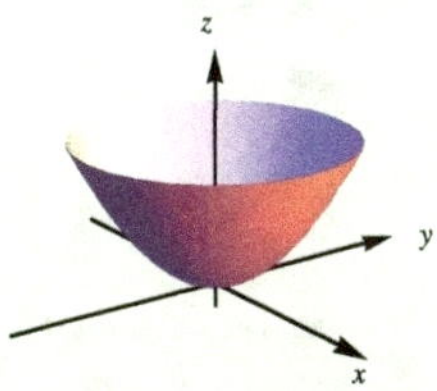

Paraboloid

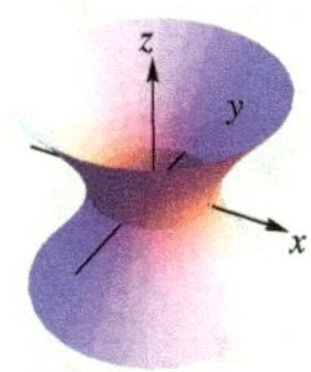

One mantled hyperboloid

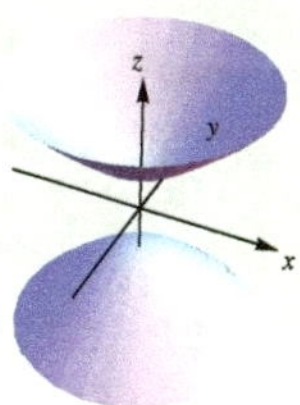

Double mantled hyperboloid

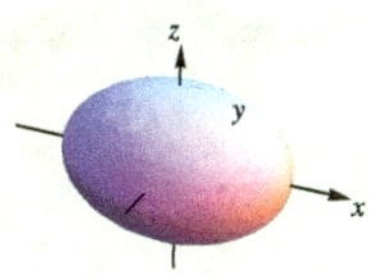

Ellipsoid

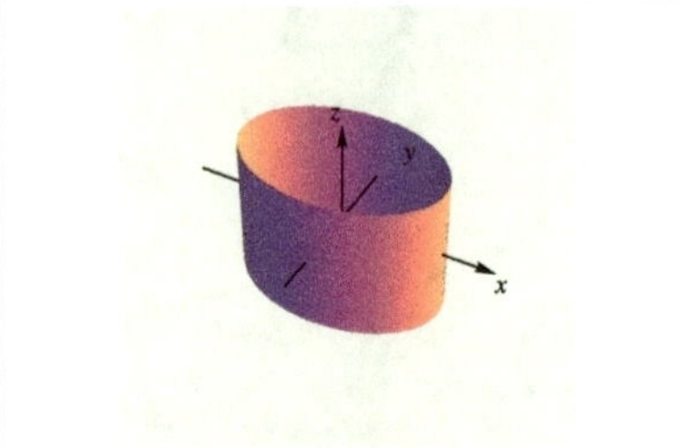

Elliptic cylinder

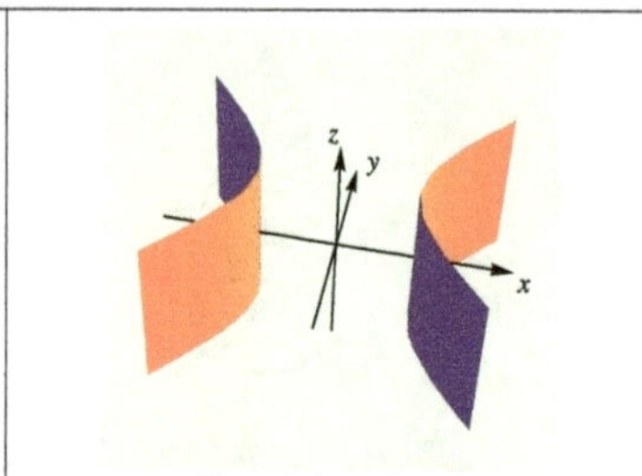

Hyperbolic cylinder

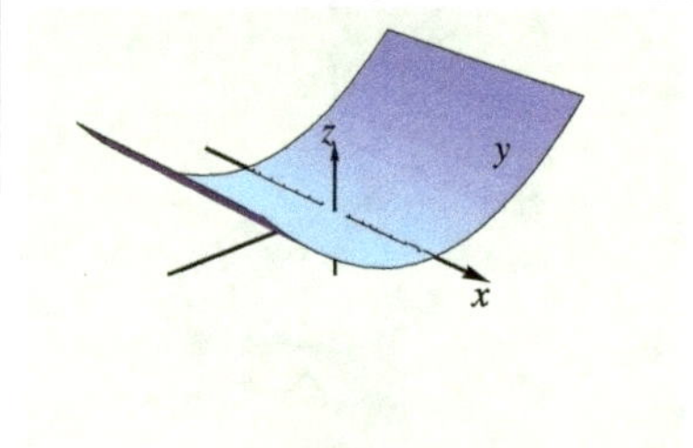

Parabolic cylinder

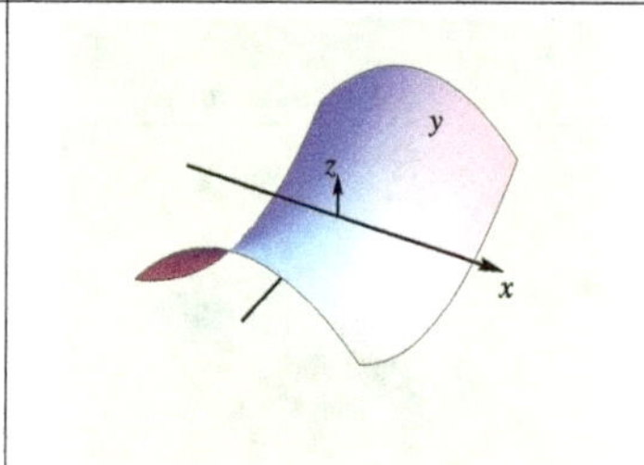

Hyperbolic paraboloid

Möbius band

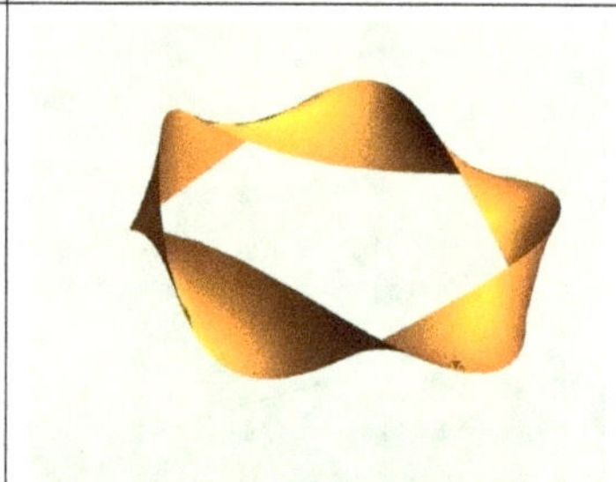
Two times twisted Möbius band

Remark. Mathematica syntax for a Möbius band i found on page 546.

13.2.3 *Slice method*

Given a compact connected set $K \subset \mathbb{R}^3$ with $a \leq x \leq b$ for all $(x, y, z) \in K$.

The Slice method
Let $\Pi_x = \{(x, y, z) : (y, z \in \mathbb{R})\}$ denote the plane $\perp x$-axis and $C_x = K \cap \Pi_x$ with area $A(x)$. The volume of K is then

$$V = \int_a^b A(x)dx. \tag{13.9}$$

13.2.4 *Volume of rotation bodies*

The Slice and Shell methods are used to express the volume of rotation bodies.

Consider the surface enclosed by $x = a$, $x = b$, $y = f(x)$ and $y = 0$.

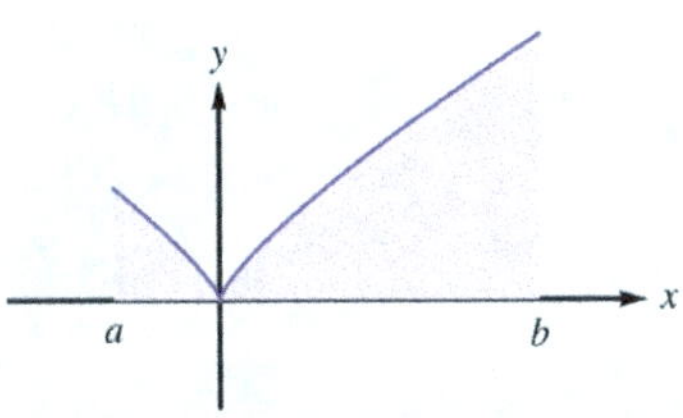

Surface between f:s curve and $x-axis$

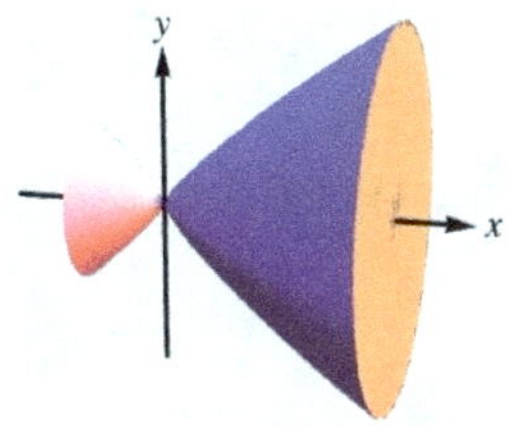

Rotation body is obtained rotating the surface on the left about the $x-axis$

Slice method

The body on the right is obtained rotating the above (left) surface about x-axis. Its volume is

$$\begin{aligned} V &= \int_a^b \pi f(x)^2 dx \\ &= (b-a)f(a)^2 + 2\int_a^b (b-x)f(x)\, f'(x)\, dx \\ &= (b-a)f(b)^2 + 2\int_a^b (a-x)f(x)\, f'(x)\, dx\,. \end{aligned} \tag{13.10}$$

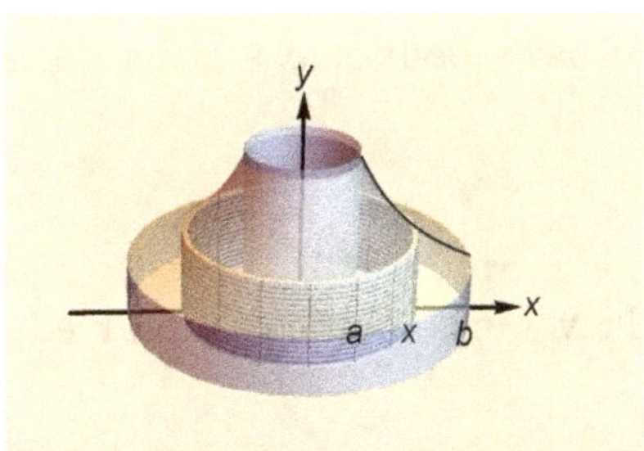

Figure 13.1: Cylindrical shell has the infinitesimal volume $dV = 2\pi x|f(x)|dx$.

Shell method
Assume that $f(x)$ is defined for $0 \le a \le x \le b$. Rotating the infinitesimal rectangle of area $|f(x)|dx$ (described in Figure 13.1) about the y-axis generates a cylinder shell of thickness dx, radius x, and height $f(x)$. The infinitesimal volume dV of the mantle surface is

$$dV = \pi \cdot 2x \cdot |f(x)| \cdot dx \text{ or } \frac{dV}{dx} = 2\pi x|f(x)|.$$

The volume V of the rotation body is

$$V = 2\,\pi \int_a^b x\,|f(x)|dx\,. \tag{13.11}$$

13.2.5 *Guldin's rules*

That following two rules are called Guldin's rules.

(i) Given a curve in $\mathbb{R}^2$, $\boldsymbol{r}(t) = (x(t), y(t))$, $\alpha \le t \le \beta$ and $y(t) \ge 0$. Let y_T be the y-coordinate of the mass center and L the curve's length.
The curve rotating about x-axis generates a surface of area A:

$$A = 2\pi\, y_T\, L. \tag{13.12}$$

(ii) Given a bounded surface $S \subset \mathbb{R} \times \mathbb{R}_+ = \{(x, y) : y > 0\}$.
Let y_T be the y-coordinate for the mass center of the surface S with area A.
Rotating the surface about x-axis generates a body around the x-axis, of volume V.

$$V = 2\pi y_T\, A. \tag{13.13}$$

Example 13.1. The volume of a torus (obtained by Guldin's second rule)

$$V = 2\pi^2 r^2 R. \tag{13.14}$$

Its area (obtained by Guldin's first rule)

$$A = 4\pi^2 r\, R. \tag{13.15}$$

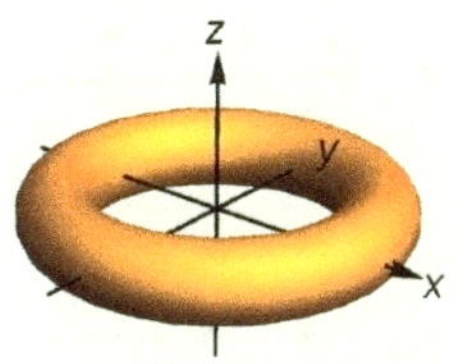

Illustration of a circular torus

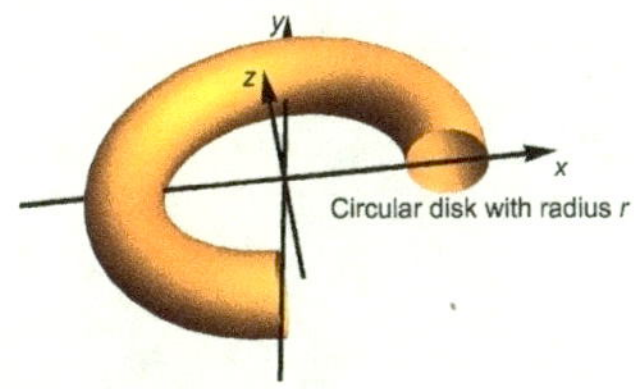

Part of torus obtained rotating a circular disk with radius r

Chapter 14

Sequence and Series

14.1 General Theory

Definition 14.1. In what follows, a_k denotes a real or complex number.

(i) A (finite) sum is given by (the lower index need not be $= 1$)

$$a_1 + a_2 + \cdots + a_n = \sum_{k=1}^{n} a_k. \tag{14.1}$$

(ii) A sequence is defined by $(a_n)_{n=1}^{\infty} = \{a_1, a_2, \ldots, a_n, \ldots\}$. The sequence is *convergent*, if $\lim_{n\to\infty} a_n$ exists as a real (or complex) number. Otherwise, it is *divergent*.

A sequence, $(a_n)_{n=1}^{\infty}$, for which all $a_n \geq 0$ is called *a positive sequence*.

(iii) A series is formally written as

$$\sum_{k=1}^{\infty} a_k = a_1 + a_2 + \cdots + a_n + \cdots \quad \text{and means} \quad \lim_{n\to\infty} \sum_{k=1}^{n} a_k, \tag{14.2}$$

if the limit exists. The series is then called convergent, otherwise, divergent.

Then nth partial sum of this series is given in (14.1).

(iv) A series, as in (14.2), for which all $a_n \geq 0$, is called *positive*.

(v) A series where $\sum_{k=1}^{\infty} a_k$ is convergent, but $\sum_{k=1}^{\infty} |a_k| = \infty$ (i.e., divergent), is called *conditionally convergent.*

(vi) Two sequences $(a_n)_{n=1}^{\infty}$ and $(b_n)_{n=1}^{\infty}$ are asymptotically equivalent if

$$\frac{a_n}{b_n} \to 1, \quad \text{as} \quad n \to \infty. \tag{14.3}$$

(vii) For a real sequence $(a_n)_{n=1}^{\infty}$, put $b_n = \sup(a_n, a_{n+1}, \ldots)$. Then $\limsup_{n\to\infty} a_n$ ("limes superior") is defined as $\lim_{n\to\infty} b_n$ (even if $b_n \to -\infty$ or $+\infty$). "Limes inferior" is defined as $\liminf a_n := -\limsup(-a_n)$.

(viii) A series $\sum_{k=1}^{\infty} a_k$, where $\sum_{k=1}^{\infty} |a_k|$ is convergent, is called *absolutely convergent.*

Theorem 14.1.

(i) *If $(a_n)_{n=1}^{\infty}$ and $(b_n)_{n=1}^{\infty}$ are two convergent sequences, A and B constants, then*

$$\lim_{n\to\infty} (Aa_n + Bb_n) = A \lim_{n\to\infty} a_n + B \lim_{n\to\infty} b_n, \tag{14.4}$$

and thus $(Aa_n + Bb_n)_{n=1}^{\infty}$ is also convergent.

(ii) *If $\sum_{k=1}^{\infty} a_k$ and $\sum_{k=1}^{\infty} b_k$ are convergent, and A and B are constants, then*

$$\sum_{k=1}^{\infty}(Aa_k + Bb_k) = A\sum_{k=1}^{\infty} a_k + B\sum_{k=1}^{\infty} b_k. \tag{14.5}$$

Hence, $\sum_{k=1}^{\infty}(Aa_k + Bb_k)$ is also convergent.

(iii) *A positive sequence either converges to a real number ≥ 0 or diverges to $+\infty$.*
Similarly, the positive series $\sum_{k=1}^{\infty} a_k$ either equals a real number $s \geq 0$ and is convergent or equals ∞ and is divergent.

Theorem 14.2. (A necessary condition for the convergence of series). *The series $\sum_{n=1}^{\infty} a_n$ is convergent $\Longrightarrow$ $a_n \longrightarrow 0$, as $n \longrightarrow \infty$.*

Theorem 14.3.

(1) *The terms in a conditionally convergent series can be rearranged so that the resulting series can assume any real value including $\pm\infty$, or diverge.*

(2) *Every rearrangement of the terms in an absolutely convergent series gives rise to a series with the same sum.*

Theorem 14.4 (Abel's partial summation formula). *Let* $A_n = \sum_{k=1}^{n} a_k$, a_k *and* b_k *being real (or complex) numbers,* $k = 1, 2, \ldots, n$.

Then

$$\sum_{k=1}^{n} a_k b_k = A_n b_n - \sum_{k=1}^{n-1} A_k (b_{k+1} - b_k). \tag{14.6}$$

Rules of double sums

$$\begin{aligned} \sum_{k=1}^{n} \left(\sum_{j=1}^{m} a_{j,k} \right) &= \sum_{j=1}^{m} \left(\sum_{k=1}^{n} a_{j,k} \right), \\ \sum_{k=1}^{n} \left(\sum_{j=1}^{k} a_{j,k} \right) &= \sum_{j=1}^{k} \left(\sum_{k=j}^{n} a_{j,k} \right). \end{aligned} \tag{14.7}$$

14.2 Positive Series

Theorem 14.5 (The principal theorem for positive series). *A positive series is convergent if and only if its partial sums build an (upper) bounded sequence.*

Theorem 14.6. *The series*

$$\sum_{n=1}^{\infty} \frac{1}{n^p} \text{ is } \begin{cases} \text{convergent,} & \text{if } p > 1, \\ \text{divergent,} & \text{if } p \leq 1. \end{cases}$$

Theorem 14.7 (Convergence criteria for series). *The series* $\sum_{k=1}^{\infty} a_k$ *is convergent if any of the following conditions are fulfilled:*

(i) $\sum_{k=1}^{\infty} |a_k|$ *is convergent. The series* $\sum_{k=1}^{\infty} a_k$ *is then said to be absolutely convergent.*

(ii) $|a_k| \leq M b_k$, $k = 1, 2, \ldots$ *for some real constant* M, *and* $\sum_{k=1}^{\infty} b_k$ *is convergent.*

(iii) $0 \leq \lim_{k\to\infty} \dfrac{|a_k|}{|b_k|} < \infty$ *and* $\sum_{k=1}^{\infty} |b_k|$ *is convergent (The comparison criterion).*

(iv) $a_k = (-1)^k b_k$, *where the* $b_k \geq 0$, $b_k \geq b_{k+1}$, $k = 1, 2, \ldots$ *and* $b_k \to 0$, *as* $k \to \infty$ *(Leibniz's criterion).*

(v) $a_k = b_k c_k$, $\sum_{k=1}^{n} c_k$ *is bounded (independent of* n*),* $|\sum_{k=1}^{n} c_k| \leq C$, $b_k \geq b_{k+1}$ *and* $b_k \to 0$, *as* $k \to \infty$ *(Dirichlet's criterion).*

(vi) $a_k = b_k c_k$, $\sum_{k=1}^{n} b_k$ *convergent and* $(c_k)_{k=1}^{\infty}$ *is monotone (increasing or decreasing) and convergent (Abel's criterion).*

(vii) $\alpha > 1$ *and* $B > 0$ *are numbers (independent of* k*) and* $|a_k| \leq \dfrac{B}{k^{\alpha}}$.

(viii) (a) *If* $\lim_{k\to\infty} \left|\dfrac{a_{k+1}}{a_k}\right| < 1$ *(The ratio test).*

(b) *If* $\limsup_{k\to\infty} \sqrt[k]{|a_k|} < 1$ *(The root test).*

(c) *If* $\int_1^{\infty} f(x)dx$ *is convergent,* $f(x) \geq 0$ *decreasing and* $|a_k| = f(k)$ *(The integral criterion).*

(ix) $\prod_{k=1}^{\infty} \ln(1 + a_k)$ *is convergent and* $a_k \geq 0$.

Remarks.

If $\lim_{k\to\infty} |\frac{a_{k+1}}{a_k}| > 1$ or $\lim_{k\to\infty} \sqrt[k]{|a_k|} > 1$, the series is divergent.
If $\lim_{k\to\infty} |\frac{a_{k+1}}{a_k}|$ exists,

$$\lim_{k\to\infty} \left|\frac{a_{k+1}}{a_k}\right| = \limsup_{k\to\infty} \sqrt[k]{|a_k|} = \lim_{k\to\infty} \sqrt[k]{|a_k|}.$$

A series which satisfies (iv) also satisfies (iii).
A series satisfying (iv), but where $\sum_{k=1}^{\infty} |a_k| = \infty$, is conditionally convergent (page 308).
The reversion to (ix) yields, if $\sum_{k=1}^{\infty} a_k$ is convergent and $a_k \geq 0$, then
$\prod_{k=1}^{\infty} \ln(1 + a_k)$ is also convergent.

Collatz conjecture

A sequence starting with a natural number, say n_0, gives rise to a sequence by using the following two rules:

(i) If n_0 is odd, it is followed by the number $n_1 = 3n_0 + 1$.
(ii) If n_0 is even, $n_1 = \frac{n_0}{2}$.

Recursively, n_1 is followed by $n_2 = 3n_1 + 1$, if n_1 is odd and by $n_2 = \frac{n_1}{2}$, if n_1 is even.

$$n_k \text{ is followed by } n_{k+1} = \begin{cases} 3n_k + 1, & \text{if } n_k \text{ is odd,} \\ & \text{and} \\ \dfrac{n_k}{2}, & \text{if } n_k \text{ is even.} \end{cases}$$

With $n_0 = 19$, one gets $n_1 = 3 \cdot 19 + 1 = 58$ and $n_2 = 29$. Continuing one gets

$$\{19,\ 58,\ 29,\ 88,\ 44,\ 22,\ 11,\ 34,\ 17,\ 52,\ 26,\ 13,\ 40,$$
$$20,\ 10,\ 5,\ 16,\ 8,\ 4,\ 2,\ 1\}.$$

Since $n_{20} = 1$, $n_{21} = 4$, so $n_{22} = 2$ and $n_{21} = 1$, hence a loop.

Collatz conjecture states that, independent of starting value n_0, the sequence eventually reaches the loop $\{4,\ 2,\ 1\}$. So far, no proof of this conjecture is available.

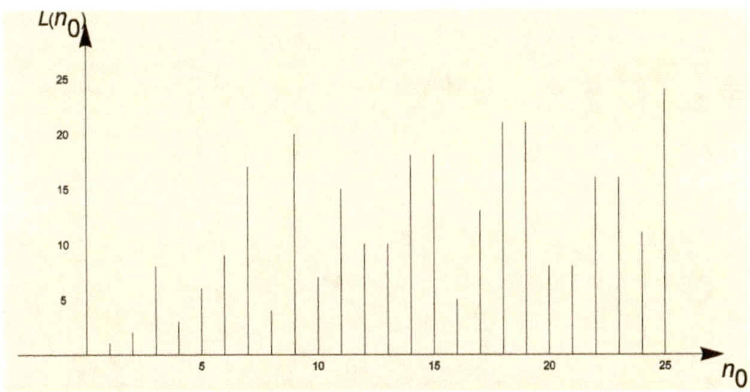

The length of the sequence $L(n_0)$, until reaching $4, 2, 1$ for $n_0 = 1, 2, \ldots, 25$. In particular, for $n_0 = 15$ the length of the line is $L(15) = 18$.

Theorem 14.8. *Assume that* $M_n > 0$, $M_n^2 \le M_{n-1}\, M_{n+1}$, $n = 1, 2, \ldots,$ *then the following equivalence holds true:*

$$\sum_{n=1}^{\infty} \left(\frac{1}{M_n}\right)^{1/n} < \infty \quad \Leftrightarrow \quad \sum_{n=1}^{\infty} \frac{M_{n-1}}{M_n} < \infty. \tag{14.8}$$

14.2.1 *Examples of sequences and series*

The Fibonacci sequence, $f_0 = 1$, $f_1 = 1$, $f_2 = 2$, $f_3 = 3$, $f_4 = 5$, $f_5 = 8, \ldots$, i.e., where the next number is the sum of the two previous. Explicitly,

$$f_n = \frac{\left(\frac{1+\sqrt{5}}{2}\right)^{n+1} - \left(\frac{1-\sqrt{5}}{2}\right)^{n+1}}{\sqrt{5}},$$
$$n = 0, 1, 2, \ldots$$

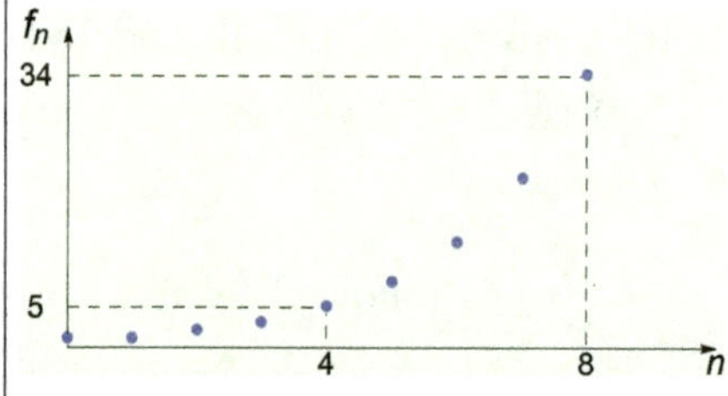

Arithmetic partial sum of an arithmetic sequence $(a_n)_{n=1}^{\infty}$:

$$a_n = a_{n-1} + d = a_1 + (n-1)d \quad (d = \text{ the difference}).$$

$$S(n) := \sum_{k=1}^{n} a_k = \sum_{k=1}^{n} [a_{n-1} + d = a_1 + (n-1)d]$$
$$= \frac{n(a_1 + a_n)}{2} = \frac{n}{2}[2a_1 + (n-1)d].$$

Geometric sum and series of geometric sequence $(a_n)_{n=1}^{\infty}$:

$$a_n = a_{n-1} \cdot q = a_0 q^n \quad (q = \text{the ratio})$$

$$S_q(n) := \sum_{k=0}^{n-1} aq^k = a + aq + aq^2 + \cdots + aq^{n-1}$$
$$= a \cdot \frac{q^n - 1}{q - 1} = a \cdot \frac{1 - q^n}{1 - q}, \quad \text{if } q \neq 1.$$

$$S_q := \sum_{k=0}^{\infty} aq^k = a + aq + aq^2 + \cdots = \frac{a}{1-q}, \quad \text{if } -1 < q < 1.$$

Some exponential sums

$$\sum_{k=1}^{n} e^{kx} = e^x \cdot \frac{e^{nx} - 1}{e^x - 1} = \frac{\sinh \frac{nx}{2}}{\sinh \frac{x}{2}} e^{(n+1)x/2}, \quad (x \neq 0)$$

$$\sum_{k=0}^{\infty} e^{-kx} = \frac{1}{1 - e^{-x}} = \frac{e^x}{e^x - 1}, \quad (x > 0)$$

$$\sum_{k=1}^{n} e^{ikx} = e^{ix} \cdot \frac{1-e^{inx}}{1-e^{ix}} = \frac{\sin\frac{nx}{2}}{\sin\frac{x}{2}} e^{i(n+1)x/2}, \quad (x \neq 2m\pi,\ m \in \mathbb{Z}).$$

Telescopic sum is a sum $\sum_{k=1}^{n} a_k$ with summand $a_k = b_k - b_{k+1}$. The sum therefore equals

$$\sum_{k=1}^{n} a_k = b_1 - b_{n+1}.$$

The last sum in the right column of (14.9), with summand $a_k = \dfrac{1}{k(k+1)} = \dfrac{1}{k} - \dfrac{1}{k+1}$ and $b_k = \dfrac{1}{k}$, gives

$$\sum_{k=1}^{n} \frac{1}{k(k+1)} = 1 - \frac{1}{n+1}.$$

Furthermore, the corresponding series is convergent:

$$\sum_{k=1}^{\infty} \frac{1}{k(k+1)} = 1.$$

Some common sums

$\sum_{k=1}^{n} k = \frac{n(n+1)}{2}$	$\sum_{k=1}^{n} k^2 = \frac{n\,(n+1)\,(2n+1)}{6}$
$\sum_{k=1}^{n} k^3 = \left[\frac{n(n+1)}{2}\right]^2$	$\sum_{k=0}^{n-1} x^k = \frac{1-x^n}{1-x}, \quad x \neq 1$ (Geometric sum)
$\sum_{k=1}^{n} k(k+1) = \frac{n(n+1)(n+2)}{6}$	$\sum_{k=2}^{n} k(k-1) = \frac{n^3-n}{3}$
$\sum_{k=1}^{n} (2k-1) = n^2$	$\sum_{k=1}^{n} \frac{1}{k\,(1+k)} = \frac{n}{n+1}$
$\sum_{k=1}^{n} \ln k = \ln n!$	$\sum_{k=1}^{n} k! \cdot k = (n+1)! - 1.$

(14.9)

(i) **The Bernoulli numbers** $(B_j)_{j=0}^{\infty}$ constitute a sequence which in recursive form is

$$\begin{cases} B_0 &= 1, \\ B_n &= -\dfrac{1}{n+1}\displaystyle\sum_{k=0}^{n-1}\binom{n+1}{k}B_k. \end{cases}$$

(ii) The Bernoulli numbers are explicitly given by

$$\begin{cases} B_1 = -\dfrac{1}{2}, \quad B_{2k+1} = 0, & k = 1, 2, \dots \\ B_{2k} = (-1)^{k+1}\dfrac{2(2k)!}{(2\pi)^{2k}}\zeta(2k), & k = 0, 1, \dots \end{cases}$$

(iii) **Faulhaber's identity**

$$\sum_{k=1}^{n} k^{\alpha} = \frac{1}{\alpha+1}\sum_{j=0}^{\alpha}(-1)^j\binom{\alpha+1}{j}B_j\, n^{\alpha+1-j}, \quad \alpha = 0, 1, 2, \dots \tag{14.10}$$

where B_j is the jth Bernoulli number.

Remarks.

For $\alpha = 0, 1, 2, \dots$, $\sum_{k=1}^{n} k^{\alpha}$ is a polynomial of degree $\alpha + 1$ in the variable n:

$$\sum_{k=1}^{n} k^{\alpha} = b_{\alpha,\alpha+1}n^{\alpha+1} + b_{\alpha,\alpha}n^{\alpha} + \cdots + b_{\alpha,1}n^1. \tag{14.11}$$

The coefficients $\boldsymbol{b}_{\alpha} := (b_{\alpha,1}, \dots, b_{\alpha,\alpha}, b_{\alpha,\alpha+1})^T$ satisfy the matrix equation

$$\boldsymbol{A}_{\alpha} \cdot \boldsymbol{b}_{\alpha} = \boldsymbol{c}_{\alpha}, \tag{14.12}$$

where

$$\boldsymbol{A}_{\alpha} = \begin{bmatrix} \binom{1}{0} & \binom{2}{0} & \dots & \binom{\alpha+1}{0} \\ 0 & \binom{2}{1} & \dots & \binom{\alpha+1}{1} \\ & & \ddots & \\ 0 & 0 & \dots & \binom{\alpha+1}{\alpha} \end{bmatrix},$$

i.e., the elements a_{ij} i $\boldsymbol{A}_\alpha$ are given by

$$a_{ij} = \begin{cases} \binom{j}{i-1}, & \text{if } i < j, \\ 0, & \text{if } i \geq j, \end{cases}$$

and

$$\boldsymbol{c}_\alpha = \left[\binom{\alpha}{0}, \binom{\alpha}{1}, \dots, \binom{\alpha}{\alpha}\right]^T.$$

In particular, $b_{\alpha,\alpha+1} = \dfrac{1}{\alpha+1}$ and $b_{\alpha,\alpha} = \dfrac{1}{2}$.

Values for some common series

$$\begin{array}{|l|l|l|} \displaystyle\sum_{k=1}^{\infty} \frac{1}{k^2} = \frac{\pi^2}{6} & \displaystyle\sum_{k=1}^{\infty} \frac{1}{k^4} = \frac{\pi^4}{90} & \displaystyle\sum_{k=1}^{\infty} \frac{1}{k^6} = \frac{\pi^6}{945} \\ \displaystyle\sum_{k=1}^{\infty} \frac{(-1)^{k-1}}{k} = \ln 2 & \displaystyle\sum_{k=0}^{\infty} \frac{1}{k!} = e & \displaystyle\sum_{k=1}^{\infty} \frac{1}{4k^2-1} = \frac{1}{2} \\ \displaystyle\sum_{k=0}^{\infty} x^k = \frac{1}{1-x}, \text{ if } |x|<1 & \displaystyle\sum_{k=1}^{\infty} \frac{1}{k^2+1} = \frac{\pi \coth \pi - 1}{2} & \\ \text{Conv. geometric series} & & \end{array} \quad (14.13)$$

Definition 14.2. An l_p–space, where $0 < p \leq \infty$, consists of sequences $\boldsymbol{x} = (x_n)_{n=1}^{\infty}$, where $x_n \in \mathbb{C}$, such that

$$\|\boldsymbol{x}\|_p = \begin{cases} \left(\displaystyle\sum_{n=1}^{\infty} |x_n|^p\right)^{1/p} = \sqrt[p]{\left(\displaystyle\sum_{n=1}^{\infty} |x_n|^p\right)} < \infty, & \text{if } 0 < p < \infty, \\ \|\boldsymbol{x}\|_\infty = \displaystyle\sup_{k=1,2,\dots} |x_k| < \infty, & \text{if } p = \infty. \end{cases} \quad (14.14)$$

Theorem 14.9. *l_p–space is a normed vector space over the field $\mathbb{C}$, for $1 \leq p \leq \infty$. The norm $\|\cdot\|_p$ satisfies*

$$\begin{aligned} &\|\boldsymbol{x}\|_p \geq 0 \text{ with equality if and only if } \boldsymbol{x} = \boldsymbol{0} = (0,0,\dots,0) \\ &\|\boldsymbol{x}+\boldsymbol{y}\|_p \leq \|\boldsymbol{x}\|_p + \|\boldsymbol{y}\|_p, \quad \|\alpha\boldsymbol{x}\|_p = |\alpha|\|\boldsymbol{x}\|_p, \end{aligned} \quad (14.15)$$

where α is a complex number.

$$l_p \subseteq l_q, \quad \text{if } p \leq q.$$

14.3 Function Sequences and Function Series

Here, sequences and series of single-variable functions are treated, loosely speaking, functions $\mathbb{R} \to \mathbb{R}$. (Some of the results hold for functions of several variables, as well.)

14.3.1 *General theory*

Definition 14.3. Assume that $f_n(x)$ are functions in the variable x, $n = 1, 2, \ldots$.

$$(f_n(x))_{n=1}^{\infty} = (f_1(x), f_2(x), \ldots, f_n(x), \ldots), \tag{14.16}$$

is called a function sequence.

$$f(x) = \sum_{k=1}^{\infty} u_k(x), \tag{14.17}$$

is called a function series.

Definition 14.4.

(i) A sequence, $(f_n(x))_{n=1}^{\infty}$, is (pointwise) convergent on M if it is convergent for each $x \in M$. The limit is a function $f(x)$ and is called the limit function.

(ii) A sequence is uniformly convergent if it is convergent on a set $M \subseteq \mathbb{R}$, to a limit f, and if for every $\varepsilon > 0$ there is an integer N such that $n > N \Longrightarrow |f(x) - f_n(x)| < \varepsilon$, for all $x \in M$.

(iii) A series, $\sum_{k=1}^{n} u_k(x)$, is pointwise or uniformly convergent if, the sequence $\sum_{k=1}^{\infty} f_k(x)$, where $\sum_{k=1}^{n} u_k(x) =: f_n(x)$, is a pointwise or uniformly convergent sequence on M, respectively.

(iv) A sequence, $(f_n(x))_{n=1}^{\infty}$, is orthogonal, with weight function $\rho(x)$ over the interval $I \subset M$, if

$$\int_I f_m(x) f_n(x) \rho(x) dx = \begin{cases} 0 & \text{if } m \neq n, \\ 1 & \text{om } m = n. \end{cases} \tag{14.18}$$

Remarks. Uniform convergence of a sequence $(f_n(x))_{n=1}^{\infty}$, to a function $f(x)$, can be expressed as $\sup_{x\in M}|f(x)-f_n(x)| \to 0$, as $n\to\infty$.

Uniform convergence is denoted by $\lim_{n\to\infty} f_n \overset{\text{unif.}}{\longrightarrow} f$.

Theorem 14.10.

(i) *If the sequence $(f_n(x))_{n=1}^{\infty}$ is uniformly convergent and f_n are continuous in a set $M \subseteq \mathbb{R}$, then the limit function is also continuous.*

(ii) *If the series $\sum_{k=1}^{\infty} u_k(x) =: f(x)$ is uniformly convergent and $u_k(x)$ are continuous on a set M, then the limit function f is continuous as well.*

(iii) *Under the above conditions on f_k and $\sum_{k=1}^{\infty} u_k(x)$, and for $M = [a,b]$, then*

$$\begin{aligned}
\lim_{k\to\infty}\int_a^b f_k(x)dx &= \int_a^b \lim_{k\to\infty} f_k(x)dx = \int_a^b f(x)dx \\
\sum_{k=1}^{\infty}\int_a^b u_k(x)dx &= \int_a^b \sum_{k=1}^{\infty} u_k(x)dx = \int_a^b f(x)dx.
\end{aligned} \tag{14.19}$$

Theorem 14.11. *The sequence $f(x) := \sum_{k=1}^{\infty} u_k(x)$ is uniformly convergent if either of the following conditions hold:*

(i) $|u_k(x)| \le a_k$ *and* $\sum_{k=1}^{\infty} a_k$ *is convergent.*

(ii) $u_k(x) = a_k(x)b_k(x)$, $a_k(x) \ge a_{k+1}(x) \ge 0$, $a_k \overset{unif.}{\longrightarrow} 0$, *and* $|\sum_{k=1}^{n} b_k(x)| \le B$, *where* $B \ge 0$ *is independent of* x *and* n.

Theorem 14.12.

(i) *If $f_n(x) \to f(x)$ pointwise and if the derivatives f_n' in the sequence $(f_n')_{k=1}^{\infty}$ are continuous and converge uniformly to, say, g, then $g(x) = f'(x)$.*

(ii) *From (i). it follows that, if* $\sum_{k=1}^{n} u_k(x) \to f(x)$ *pointwise, as* $n \longrightarrow \infty$ *and* $\sum_{k=1}^{n} u_k'(x) \overset{unif.}{\longrightarrow} g(x)$, *as* $n \longrightarrow \infty$, *then* $g = f'$.

Theorem 14.13.

(i) ***Cauchy's criterion for uniform convergence***
Let $(f_1(x), f_2(x), \ldots)$ *be a sequence of functions, such that for each* $\varepsilon > 0$, *and all integers* $k > 0$, *there is an integer* n_ε, *such that*

$$n > n_\varepsilon \Longrightarrow |f_{n+k}(x) - f_n(x)| < \varepsilon, \tag{14.20}$$

then there exists a function $f(x)$, *such that* $f_n(x) \to f(x)$ *uniformly.*

(ii) ***Abel's test for uniform convergence***
Assume that $(u_1(x,t), u_2(x,t), \ldots)$ *is a sequence of functions* $(x,t) \in \Omega \subseteq \mathbb{R}^2$. *Assume further* $u_k(x,t) = T_k(t)X_k(x)$, *where* T_k *is a bounded, monotone, sequence, i.e.,*

$$\begin{matrix} T_k(t) \le T_{k+1}(t) & \\ or & |T_k(t)| \le K \quad \text{for } k = 1, 2, \ldots, \\ T_k(t) \ge T_{k+1}(t), & \end{matrix}$$

and that $\sum_{k=1}^{\infty} X_k(x)$ *is a uniformly convergent series. Then the series*

$$\sum_{k=1}^{\infty} u_k(x,t), \tag{14.21}$$

converges uniformly in Ω.

14.3.2 *Power series*

Definition 14.5. A Power series is a function series of the form

$$\sum_{k=0}^{\infty} a_k(x - x_0)^k =: f(x), \text{ i.e., } u_k(x) = a_k(x - x_0)^k. \tag{14.22}$$

The *radius of convergence* R for the power series is defined as

$$\frac{1}{R} = \limsup_{k\to\infty} |a_k|^{1/k}, \tag{14.23}$$

or

$$\frac{1}{R} = \limsup_{k\to\infty} \left|\frac{a_{k+1}}{a_k}\right|. \tag{14.24}$$

Remarks. $0 \le R \le \infty$. If $R = \infty$, the power series converges for all $x \in \mathbb{R}$, i.e., for all $x : -\infty < x < \infty$, see the following theorem.

One can show that the limits in (14.23) and (14.24) yield the same value.

In view of the above definitions/criterion:

Theorem 14.14.

(i) *For every x, such that $|x - x_0| < R$, the power series converges pointwise to a function $f(x)$.*
(ii) *For every r such that $0 \le r < R$, the power series converges uniformly to $f(x)$ on $\{x : |x - x_0| \le r\}$.*
(iii) *The coefficients are $a_k = \dfrac{f^{(k)}(x_0)}{k!}$, $k = 0, 1, \ldots$*
(iv) *For every x such that $|x - x_0| > R$ the series is divergent.*

Theorem 14.15. *Suppose that a power series given by (14.22) has radius of convergence $R > 0$. Then*

$$\frac{d}{dx}\sum_{k=1}^{\infty} a_k(x - x_0)^k = \sum_{k=1}^{\infty} k a_k (x - x_0)^{k-1} \tag{14.25}$$

if $|x - x_0| < R$, and

$$\int_a^b \sum_{k=1}^{\infty} a_k(x - x_0)^k = \sum_{k=1}^{\infty} \int_a^b a_k(x - x_0)^k dx \tag{14.26}$$

if $x_0 - R < a \le b < x_0 + R$.

Theorem 14.16. *If the coefficients a_k satisfy the difference equation*

$$a_{k+2} + \alpha a_{k+1} + \beta a_k = 0, \quad k = 0, 1, 2, \ldots$$

then

$$\sum_{k=0}^{\infty} a_k x^k = \frac{a_0 + (a_1 + \alpha a_0)x}{1 + \alpha x + \beta x^2}. \tag{14.27}$$

14.3.3 *Taylor expansions*

Theorem 14.17. *If a real function f is continuously differentiable of order $n+1$ in a neighborhood (a, b) of x_0, then the function can be Taylor expanded about $x = x_0$. Its Taylor expansion is given by*

$$f(x) = \underbrace{\sum_{k=0}^{n} \frac{(x - x_0)^k}{k!} f^{(k)}(x_0)}_{\text{Taylor polynomial}} + \underbrace{R_n(x)}_{\text{Rest term}} \tag{14.28}$$

where the rest term can be written, on Lagrange's form, as the following RHS,

$$R_n(x) = (-1)^n \int_{x_0}^{x} \frac{(t - x)^n}{n!} f^{(n+1)}(t)dt = f^{(n+1)}(\xi) \frac{(x - x_0)^{n+1}}{(n+1)!}, \tag{14.29}$$

for some ξ between x_0 and x.

Remarks. The rest term is usually written on *Ordo-form.* With $\mathcal{O}((x - x_0)^n)$ ("Big ordo $(x - x_0)^n$") means the class of functions $x \curvearrowright g(x)$ such that

$$\frac{g(x)}{(x - x_0)^n},$$

is bounded in a neighborhood $(x_0 - \delta, x_0 + \delta)$ of $x = x_0$.

Alternatively, one writes $\mathcal{O}((x - x_0)^n) := (x - x_0)^n \, B_n(x)$, where $B_n(x)$ is bounded in a neighborhood of x_0.

The rest term is usually denoted by R_n (rather than R_{n+1}).

In the Taylor expansion, $R_n(x) = \mathcal{O}((x - x_0)^{n+1})$.

The Taylor expansion can be rewritten as

$$f(x) = \sum_{k=1}^{n} \frac{f^{(k)}(x_0)(x - x_0)^k}{k!} + \mathcal{O}((x - x_0)^{n+1}). \tag{14.30}$$

The Taylor expansion yields a corresponding Taylor series for $f(x)$:

$$f(x) = \sum_{k=0}^{\infty} \frac{f^{(k)}(x_0)(x - x_0)^k}{k!}, \tag{14.31}$$

where the equality holds for all x, for which the series is convergent.

MacLaurin expansion for some common functions with rest terms

MacLaurin expansion is a special case of Taylor expansion (14.28) with $x_0 = 0$.

Function	MacLaurin expansion

$$\begin{aligned}
e^x &= 1 + x + \frac{x^2}{2!} + \frac{x^3}{3!} + \frac{x^4}{4!} + \dots \frac{x^n}{n!} + \mathcal{O}(x^{n+1}) \\
\sin x &= x - \frac{x^3}{3!} + \frac{x^5}{5!} + \cdots + (-1)^{n-1}\frac{x^{2n-1}}{(2n-1)!} + \mathcal{O}(x^{2n+1}) \\
\cos x &= 1 - \frac{x^2}{2!} + \frac{x^4}{4!} - \frac{x^6}{6!} + \cdots + (-1)^n \frac{x^{2n}}{(2n)!} + O(x^{2n+2}) \\
\ln(x+1) &= x - \frac{x^2}{2} + \frac{x^3}{3} - \frac{x^4}{4} + \cdots + (-1)^{n-1}\frac{x^n}{n} + \mathcal{O}(x^{n+1}) \\
\sinh x &= x + \frac{x^3}{3!} + \frac{x^5}{5!} + \cdots + \frac{x^{2n-1}}{(2n-1)!} + \mathcal{O}(x^{2n+1}) \\
\cosh x &= 1 + \frac{x^2}{2!} + \frac{x^4}{4!} + \frac{x^6}{6!} + \cdots + \frac{x^{2n}}{(2n)!} + \mathcal{O}(x^{2n+2}) \\
\arctan x &= x - \frac{x^3}{3} + \frac{x^5}{5} + \cdots + (-1)^{n-1}\frac{x^{2n-1}}{2n-1} + \mathcal{O}(x^{2n+1}) \\
(1+x)^\alpha &= 1 + x + \binom{\alpha}{2}x^2 + \cdots + \binom{\alpha}{n}x^n + \mathcal{O}(x^{n+1}) \\
&\text{where } \binom{\alpha}{n} = \frac{\alpha(\alpha-1)\cdot \dots \cdot(\alpha-n+1)}{n!}
\end{aligned}$$

The rest terms in different forms: ξ is a number between $x_0 = 0$ and x.

Function	Lagrange's form	Ordo-form
e^x	$R_n(x) = e^{\xi} \dfrac{x^{n+1}}{(n+1)!}$	$= \mathcal{O}(x^{n+1})$
$\sin x$	$R_{2n-1}(x) = \cos\xi\,(-1)^n \dfrac{x^{2n+1}}{(2n+1)!}$	$= \mathcal{O}(x^{2n+1})$
$\cos x$	$R_{2n}(x) = \cos\xi \dfrac{x^{2n+2}}{(2n+2)!}$	$= O(x^{2n+2})$
$\ln(x+1)$	$R_n(x) = \dfrac{1}{(\xi+1)^{n+1}}(-1)^n \dfrac{x^{n+1}}{n+1}$	$= \mathcal{O}(x^{n+1})$
$\sinh x$	$R_{2n-1}(x) = \cosh\xi \dfrac{x^{2n+1}}{(2n+1)!}$	$= \mathcal{O}(x^{2n+1})$
$\cosh x$	$R_{2n}(x) = \cosh\xi \dfrac{x^{2n+2}}{(2n+2)!}$	$= \mathcal{O}(x^{2n+2})$
$\arctan x$	$R_{2n-1}(x) = (-1)^n \dfrac{x^{2n+1}}{(2n+1)(1+\xi^2)}$	$= \mathcal{O}(x^{2n+1})$

$$(1+x)^{\alpha} R_n(x) = (1+\xi)^{\alpha-n-1}\binom{\alpha}{n+1} x^{n+1} = \mathcal{O}(x^{n+1}). \quad (14.32)$$

MacLaurin series for some functions with specified range of convergence

Function	MacLaurin series	Range of convergence
$e^x =$	$\sum_{n=0}^{\infty} \dfrac{x^n}{n!}$	$\{x : -\infty < x < \infty\} = \mathbb{R}$
$\sin x =$	$\sum_{n=1}^{\infty} \dfrac{(-1)^{n-1} x^{2n-1}}{(2n-1)!}$	$\{x : -\infty < x < \infty\} = \mathbb{R}$
$\cos x =$	$\sum_{n=0}^{\infty} \dfrac{(-1)^n x^{2n}}{(2n)!}$	$\{x : -\infty < x < \infty\} = \mathbb{R}$
$\ln(x+1) =$	$\sum_{n=1}^{\infty} \dfrac{(-1)^{n-1} x^n}{n}$	$\{x : -1 < x < 1\} = (-1,1)$
$\sinh x =$	$\sum_{n=1}^{\infty} \dfrac{x^{2n-1}}{(2n-1)!}$	$\{x : -\infty < x < \infty\} = \mathbb{R}$
$\cosh x =$	$\sum_{n=0}^{\infty} \dfrac{x^{2n}}{(2n)!}$	$\{x : -\infty < x < \infty\} = \mathbb{R}$
$\arctan x =$	$\sum_{n=1}^{\infty} (-1)^{n-1} \dfrac{x^{2n-1}}{2n-1}$	$\{x : -1 \le x \le 1\} = [-1,1]$
$(1+x)^{\alpha} =$	$\sum_{n=1}^{\infty} \binom{\alpha}{n} x^n$	$\{x : -1 < x < 1\} = (-1,1)^{(*)}$

(14.33)

$^{(*)}$ With $\alpha = 0, 1, 2, \ldots$, $(1+x)^{\alpha}$ is a polynomial with domain of convergence $\{x : -\infty < x < \infty\} = \mathbb{R}$.

14.3.4 *Fourier series*

Definition 14.6. For a periodic function $f : \mathbb{R} \to \mathbb{R}$, there is a smallest number $T > 0$, such that $f(t+T) = f(t)$ for all $t \in \mathbb{R}$. T is called the period of f.

The basic angular frequency is defined as $\Omega = \dfrac{2\pi}{T}$.

The Fourier coefficients of a T-periodic function are defined as

$$a_n = \frac{2}{T}\int_0^T f(t)\cos(n\Omega t)dt, \quad n = 0, 1, 2, \ldots$$
$$b_n = \frac{2}{T}\int_0^T f(t)\sin(n\Omega t)dt, \quad n = 1, 2, \ldots \tag{14.34}$$

Remarks. Integration over any interval of length T, e.g., $[a, a+T]$ gives rise to the same result. Here, for simplicity $[0, T]$ is chosen, where $\Omega = \frac{2\pi}{T}$ and hence is the basic angular frequency.

The Fourier series of a T-periodic function f is defined as

$$f(t) \sim \frac{a_0}{2} + \sum_{n=1}^{\infty} a_n \sin(n\Omega t) + b_n \cos(n\Omega t). \tag{14.35}$$

Equality holds at the points of continuity, otherwise the right-hand side is the mean value of the left and right limits of f in the discontinuity point t (see the following equation).

We define the left and right limits as $f_L(t_0) = \lim_{s\to 0+} f(t-s)$ and $f_R(t_0) = \lim_{s\to 0+} f(t+s)$, respectively.

Left and right continuity for the function f at $t = t_0$ are defined as

$$f_L(t_0) := \lim_{t\to t_0-} f(t) \quad \text{and} \quad f_R(t_0) := \lim_{t\to t_0+} f(t).$$

Left- and right derivative of a function f are defined as

$$f'_L(t_0) = \lim_{\Delta t\to 0, \Delta t<0} \frac{f(t_0+\Delta t) - f_L(t_0)}{\Delta t}$$
$$f'_R(t_0) = \lim_{\Delta t\to 0, \Delta t>0} \frac{f(t_0+\Delta t) - f_R(t_0)}{\Delta t}$$

as far as the limits exist.

Theorem 14.18. *If both left and right limits of f at $t = t_0$ exist, then the Fourier series of f at t_0 converges to $\frac{1}{2}[f_L(t_0) + f_R(t_0)]$.*
Especially if f is continuous at $t = t_0$, then

$$f(t_0) = \frac{a_0}{2} + \sum_{n=1}^{\infty} a_n \cos(n\Omega t_0) + b_n \sin(n\Omega t_0). \tag{14.36}$$

Theorem 14.19. *The Fourier series* (14.35) *can also be written as*

$$\begin{aligned} f(x) &\sim A_0 + \sum_{n=1}^{\infty} A_n \sin(n\Omega t + \alpha_n) \quad \text{(amplitude-phase angle form)}, \\ f(x) &\sim \sum_{n=-\infty}^{\infty} c_n e^{in\Omega t} \quad \text{(complex form)}, \end{aligned} \tag{14.37}$$

where

$$\begin{aligned} A_0 &= \tfrac{a_0}{2} = \frac{1}{T}\int_0^T f(t)\,dt, && \text{(the mean value of } f), \\ A_n &= \sqrt{a_n^2 + b_n^2}, && \alpha_n = \arg(a_n - ib_n), \\ \sin\alpha_n &= -\frac{b_n}{A_n}, && \cos\alpha_n = \frac{a_n}{A_n}. \end{aligned}$$

$c_n = a - ib_n$ *(i is the imaginary unit) and* $c_{-n} = \overline{c_n}$.

Orthogonality of $(\sin n\Omega t, \cos n\Omega t)$, $n = 1, 2, \ldots$

Theorem 14.20. *Let* $T = \frac{2\pi}{\Omega}$ *och* $m, n = 1, 2, \ldots$
The class of functions

$$\{\sin n\Omega t, \cos n\Omega t\}_{n=1}^{\infty}$$

is orthogonal in the following sense:

$$\begin{aligned} &\frac{2}{T}\int_0^T \cos m\Omega t \sin n\Omega t dt = 0, \quad \textit{and} \\ &\frac{2}{T}\int_0^T \cos m\Omega t \cos n\Omega t dt = \tfrac{2}{T}\int_0^T \sin m\Omega t \sin n\Omega t dt \\ &\qquad = \begin{cases} 1, & \textit{if } m = n, \\ 0, & \textit{if } m \neq n. \end{cases} \end{aligned} \tag{14.38}$$

Fourier series of even and odd functions

Theorem 14.21. *If f is even, then $b_n = 0$ for all $n = 1, 2, \ldots$ and*

$$a_n = \frac{4}{T}\int_0^{T/2} f(t)\cos n\Omega t dt, \ n = 0, 1, \ldots \tag{14.39}$$

If f is odd, then $a_n = 0$ for all $n = 0, 1, 2, \ldots$ and

$$b_n = \frac{4}{T}\int_0^{T/2} f(t)\sin n\Omega t dt, \ n = 1, 2 \ldots \tag{14.40}$$

Theorem 14.22 (Parseval's formulas). *Let f and g be two periodic functions with the same period T and with the complex Fourier series as*

$$\sum_{n=-\infty}^{\infty} c_n(f)e^{in\Omega t} \quad \text{and} \quad \sum_{n=-\infty}^{\infty} c_n(g)e^{in\Omega t}. \tag{14.41}$$

Then,

$$\frac{1}{T}\int_0^T f(t)g(t)dt = \sum_{n=-\infty}^{\infty} c_n(f)c_n(g) = \langle f, g\rangle. \tag{14.42}$$

In particular, if $f = g$ with $c_n(f) = c_n(g) = c_n$, then

$$\langle f, f\rangle =: \|f\|_2^2 = \frac{1}{T}\int_0^T (f(t))^2 dt = |c_0|^2 + 2\sum_{n=1}^{\infty} |c_n|^2. \tag{14.43}$$

Remarks. In electrical engineering literature, often, inner product $\langle f , g\rangle$ is written as $\overline{f \cdot g}$.

$\|f\|_2$ is the L_2-norm of the function f, here defined on an interval of length T, for instance $[-T/2, T/2]$.

Fourier series of some T-periodic functions The functions $f(t)$ on the left are given in a symmetric interval $[-T/2, T/2]$ and are assumed to have period T, hence with frequency $\Omega = \frac{2\pi}{T}$.

Function	Fourier series
$f(t) = t$	$\sum_{k=1}^{\infty} \frac{2\pi(-1)^{n-1}}{n\Omega} \sin(n\Omega t)$
$f(t) = \|t\|$	$\frac{1}{2} - \frac{4}{\Omega} \sum_{m=1}^{\infty} \frac{1}{(2m-1)^2 \pi} \cos(\Omega(2m-1)t)$
$f(t) = t\left(t^2 - \frac{T^2}{4}\right)$	$\frac{12}{\Omega^3} \sum_{n=1}^{\infty} \frac{(-1)^n}{n^3} \sin(n\Omega t)$
$f(t) = \begin{cases} 0, & -T/2 \le t < 0 \\ t, & 0 \le t < T/2 \end{cases}$	$\frac{\pi}{4\Omega} - \frac{2}{\Omega\pi} \sum_{m=1}^{\infty} \frac{1}{(2m-1)^2} \cos((2m-1)\Omega t)$ $+\frac{1}{\Omega} \sum_{n=1}^{\infty} \frac{(-1)^{n-1}}{n} \sin(n\Omega t)$

(14.44)

Remarks. The periodic function $f(t) = t$, for $t \in [-T/2, T/2]$ has discontinuities at $t = T/2 + lT,\ l \in \mathbb{Z}$. This is due to the fact that the Fourier coefficients are of the order $1/n$.

The Fourier series converges pointwise to $f(t)$ except at the points of discontinuity, where the series converges to $\frac{f(T/2 + lT_+) + f(T/2 + lT_-)}{2}$.

The function $f(t) = |t|$ is continuous but not differentiable at all points. The Fourier coefficients are of order $1/n^2$. The convergence of the series to f is uniform. Thus, the limit function is continuous.

The function $f(t) = t\left(t^2 - \frac{T^2}{4}\right)$ is differentiable over $\mathbb{R}$. The Fourier coefficients are of order $1/n^3$. Then its Fourier series converges uniformly to f. Furthermore, termwise differentiation is permitted at all points.

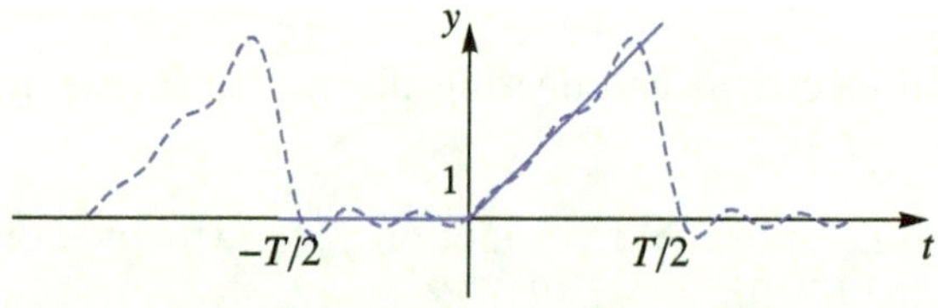

The graph of the last function in (14.44) including its partial sum with 5 cosine- and 5 sine terms (dashed) over the interval $[-T/2, T/2]$.

Fourier series of some 2π-periodic functions

The following table is for most common, simple, $T = 2\pi$ periodic, $\Omega = 1$, functions.

Function	Fourier series
$f(t) = t, \quad (-\pi < t < \pi)$	$2\sum_{n=1}^{\infty} \frac{(-1)^{n+1}}{n} \sin(nt)$
$f(t) = \|t\|, \quad (-\pi < t < \pi)$	$\frac{\pi}{2} - \frac{4}{\pi}\sum_{n=1}^{\infty} \frac{\cos(2n-1)t}{(2n-1)^2}$
$f(t) = \pi - t, \quad (-\pi < t < \pi)$	$2\sum_{n=1}^{\infty} \frac{\sin nt}{t}$
$f(t) = \begin{cases} 0, & -\pi \le t < 0 \\ t, & 0 \le t < \pi \end{cases}$	$\frac{\pi}{4} - \frac{2}{\pi}\sum_{n=1}^{\infty} \frac{\cos(2n-1)t}{(2n-1)^2} + \sum_{n=1}^{\infty} \frac{(-1)^{n+1}}{n} \sin(nt)$
$f(t) = \sin^2 t,$	$\frac{1}{2} - \frac{1}{2}\cos 2t$
$f(t) = \begin{cases} -1, & -\pi \le t < 0 \\ 1, & 0 \le t < \pi \end{cases}$	$\frac{4}{\pi}\sum_{n=1}^{\infty} \frac{\sin(2n-1)t}{2n-1}$
$f(t) = \begin{cases} 0, & -\pi \le t < 0 \\ 1, & 0 \le t < \pi \end{cases}$	$\frac{1}{2} + \frac{2}{\pi}\sum_{n=1}^{\infty} \frac{\sin(2n-1)t}{2n-1}$
$f(t) = \|\sin t\|$	$\frac{2}{\pi} - \frac{4}{\pi}\sum_{n=1}^{\infty} \frac{\cos 2nt}{4n^2-1}$
$f(t) = \|\cos t\|$	$\frac{2}{\pi} - \frac{4}{\pi}\sum_{n=1}^{\infty} \frac{(-1)^n \cos 2nt}{4n^2-1}$
$f(t) = \begin{cases} 0, & -\pi \le t < 0 \\ \sin t, & 0 \le t < \pi \end{cases}$	$\frac{1}{\pi} - \frac{2}{\pi}\sum_{n=1}^{\infty} \frac{\cos 2nt}{4n^2-1} + \frac{1}{2}\sin t$

(14.45)

Continuation of Table 1

Function	Fourier series
$f(t) = \begin{cases} a^{-2}(a - \|t\|), & \|t\| < a \\ 0, & a < \|t\| < \pi \end{cases}$	$\frac{1}{2\pi} + \frac{2}{\pi}\sum_{n=1}^{\infty} \frac{1-\cos na}{n^2a^2} \cos nt$
$f(t) = t^2, \quad -\pi < t < \pi$	$\frac{\pi^2}{3} + 4\sum_{n=1}^{\infty} \frac{(-1)^n}{n^2} \cos nt$
$f(t) = t(\pi - \|t\|), \quad -\pi < t < \pi$	$\frac{8}{\pi}\sum_{n=1}^{\infty} \frac{\sin(2n-1)t}{(2n-1)^3}$
$f(t) = e^{bt}, \quad -\pi < t < \pi$	$\frac{\sinh b\pi}{\pi} \sum_{n=-\infty}^{\infty} \frac{(-1)^n}{b - in} e^{int}$
$f(t) = e^{bt}, \quad -0 < t < 2\pi$	$\frac{e^{2\pi b} - 1}{2\pi} \sum_{n=-\infty}^{\infty} \frac{e^{int}}{b - in}$
$f(t) = \sinh t, \quad -\pi < t < \pi$	$\frac{2\sinh \pi}{\pi} \sum_{n=1}^{\infty} \frac{(-1)^{n+1} n}{n^2+1} \sin nt$

(14.46)

14.3.5 *Some sums, series, and inequalities*

$$
\begin{aligned}
\sum_{k=1}^{n} \sin kx &= \frac{\sin \frac{nx}{2} \sin \frac{(1+n)x}{2}}{\sin \frac{x}{2}} \\
&= \sin \frac{nx}{2} \left(\cos \frac{nx}{2} + \cot \frac{x}{2} \sin \frac{nx}{2} \right) \\
\frac{1}{2} + \sum_{k=1}^{n} \cos kx &= \frac{\sin(n+1/2)\,x}{2\sin(x/2)}, \quad x/(2\pi) \notin \mathbb{Z} \text{ (Dirichlet kernel)} \\
\sum_{k=0}^{\infty} z^k &= \frac{1}{1-z}, \quad |z| < 1 \text{ and } z \text{ complex} \\
\sum_{k=1}^{\infty} r^k \cos(k\theta) &= \frac{1 - r\cos\theta}{1 - 2r\cos\theta + r^2}, \quad |r| < 1 \\
\sum_{k=1}^{\infty} r^k \sin(k\theta) &= \frac{r\sin\theta}{1 - 2r\cos\theta + r^2}, \quad |r| < 1 \\
\sum_{k=1}^{\infty} \cos(k\theta) &\le \frac{2}{\sqrt{2 - 2\cos\theta}} \\
\sum_{k=1}^{\infty} \sin(k\theta) &\le \frac{2}{\sqrt{2 - 2\cos\theta}}.
\end{aligned}
\tag{14.47}
$$

14.4 Some Important Orthogonal Functions

In the following, we present the polynomial classes that are orthogonal on a bounded interval (a, b) or $(-1, 1)$ and other polynomial classes that are orthogonal on $\mathbb{R}^+$, or $\mathbb{R}$. We start with different types of orthogonal polynomials on $(-1, 1)$.

Definition 14.7.

(i) Legendre polynomials are given by

$$
p_n(x) = 2^{-n} \sum_{k=0}^{[n/2]} (-1)^k \binom{n}{k} \binom{2(n-k)}{n} x^{n-2k}, \quad n = 0, 1, 2, \ldots
\tag{14.48}
$$

(ii) The associated Legendre functions are defined as

$$P_l^m(x) := (-1)^m \left(1-x^2\right)^{\frac{m}{2}} \frac{d^m}{dx^m} P_l(x). \tag{14.49}$$

(iii) Chebyshev polynomials of first and second order are defined as

$$T_n(\cos\theta) = \cos n\theta \text{ and } U_n(\cos\theta) = \frac{\sin(n+1)\theta}{\sin n\theta}, \text{ respectively.} \tag{14.50}$$

(iv) The Jacobi polynomials are given by

$$P_n^{a,b}(x) = \frac{(-1)^n}{2^n n!}(1-x)^{-a}(1+x)^{-b}\frac{d^n}{dx^n} \times[(1-x)^{a+n}(1+x)^{b+n}]. \tag{14.51}$$

(v) Laguerre polynomials are defined as

$$L_n(x) = \sum_{k=0}^{n}(-1)^k \binom{n}{k}\frac{x^k}{k!}. \tag{14.52}$$

(vi) Hermite polynomials are

$$H_n(x) = (-1)^n e^{x^2}\frac{d^n}{dx^n}(e^{-x^2}). \tag{14.53}$$

(vii) Gegenbauer polynomials $G_n(x)$ are given by using the corresponding Chebyshev polynomials of the first kind:

$$G_n(x) := \frac{2}{n}T_n(x), \quad n = 0, 1, 2, \ldots \tag{14.54}$$

(viii) The Bessel functions (of the first kind) constitute

$$J_n(x) = \sum_{k=0}^{\infty}\frac{(-1)^k (x/2)^{2k+n}}{2^k k!(n+k)!}, \quad J_{-n}(x) = (-1)^n J_n(x) \quad n = 0, 1, 2, \ldots \tag{14.55}$$

(ix) Spherical surface functions are defined as

$$Y_l^m(\theta,\phi) = \frac{e^{im\phi}\sqrt{1+2l}\sqrt{\frac{(l-m)!}{(l+m)!}}\,P_l^m(\cos\theta)}{2\sqrt{\pi}}, \quad \text{for integers } |m| \le l, \tag{14.56}$$

where $P_l^m(x)$ are the associated Legendre functions.

(x) The Neumann functions or Bessel functions of the second kind are defined as

$$Y_\nu(x) = \frac{J_\nu(x)\cos\nu x - J_{-\nu}(x)}{\sin\nu x}, \qquad \text{non-integer } \nu$$

and with $\nu = n$ integer:

$$Y_n(x) = -\frac{1}{\pi}\left(\frac{2}{x}\right)^n \sum_{k=0}^{n-1} \frac{(n-k-1)!}{k!}\left(\frac{x}{2}\right)^{2k} + \frac{2}{\pi}\ln(x/2)J_n(x)$$
$$-\frac{1}{\pi}\left(\frac{x}{2}\right)^n \sum_{k=0}^{\infty}[\psi(k+1)+\psi(n+k+1)]$$
$$\frac{1}{k!(n+k)!}\cdot\left(-\frac{x}{2}\right)^{2k}, \tag{14.57}$$

where $\psi(x)$ is the digamma function defined as

$$\psi(x) := \frac{d}{dx}\ln\Gamma(x) = \frac{\Gamma'(x)}{\Gamma(x)} \text{ and } \Gamma(x) = \int_0^\infty t^{x-1}e^{-t}dt. \tag{14.58}$$

Table of the first 10 Bernoulli polynomials

n	$B_n(x)$
0	1
1	$x - \frac{1}{2}$
2	$x^2 - x + \frac{1}{6}$
3	$x^3 - \frac{3x^2}{2} + \frac{x}{2}$
4	$x^4 - 2x^3 + x^2 - \frac{1}{30}$
5	$x^5 - \frac{5x^4}{2} + \frac{5x^3}{3} - \frac{x}{6}$
6	$x^6 - 3x^5 + \frac{5x^4}{2} - \frac{x^2}{2} + \frac{1}{42}$
7	$x^7 - \frac{7x^6}{2} + \frac{7x^5}{2} - \frac{7x^3}{6} + \frac{x}{6}$
8	$x^8 - 4x^7 + \frac{14x^6}{3} - \frac{7x^4}{3} + \frac{2x^2}{3} - \frac{1}{30}$
9	$x^9 - \frac{9x^8}{2} + 6x^7 - \frac{21x^5}{5} + 2x^3 - \frac{3x}{10}$

Table of the first 10 Euler polynomials

n	$E_n(x)$
0	1
1	$x - \frac{1}{2}$
2	$x^2 - x$
3	$x^3 - \frac{3x^2}{2} + \frac{1}{4}$
4	$x^4 - 2x^3 + x$
5	$x^5 - \frac{5x^4}{2} + \frac{5x^2}{2} - \frac{1}{2}$
6	$x^6 - 3x^5 + 5x^3 - 3x$
7	$x^7 - \frac{7x^6}{2} + \frac{35x^4}{4} - \frac{21x^2}{2} + \frac{17}{8}$
8	$x^8 - 4x^7 + 14x^5 - 28x^3 + 17x$
9	$x^9 - \frac{9x^8}{2} + 21x^6 - 63x^4 + \frac{153x^2}{2} - \frac{31}{2}$

Relations for Bernoulli and Euler polynomials:

$$B_n(x) = n\,B'_{n-1}(x), \quad E_n(x) = n\,E'_{n-1}(x), \quad n = 1, 2, \ldots$$

where the prime "$'$" denotes derivative.

Properties of some common function classes
(I) Legendre polynomials $P_n(x)$:

The set $\{P_n(x)\}_{n=0}^{\infty}$ is an orthogonal polynomial class in the interval $I = (-1, 1)$ with the following properties:

$$|P_n(x)| \leq 1, \ -1 \leq x \leq 1; \quad P_n(-x) = (-1)^n P_n(x);$$

$$P_n(1) = 1; \quad P_n(0) = \begin{cases} 0, & n \text{ odd}, \\ (-1)^n \dfrac{(n-1)!!}{n!!}, & n \text{ even}. \end{cases}$$

Explicit forms:

$$P_n(x) = \frac{1}{2^n} \sum_{m=0}^{[n/2]} (-1)^m \binom{n}{m} \binom{2n-2m}{n} x^{n-2m}$$

$$= \sum_{k=0}^{n} \binom{n}{k} \binom{-n-1}{k} \left[\frac{1-x}{2}\right]^k$$

$$= \sum_{k=0}^{n} \binom{n}{k}^2 (x-1)^{n-k}(x+1)^k = 2^n \sum_{k=0}^{n} \binom{n}{k} \binom{\frac{n+k-1}{2}}{n} x^k.$$

Rodrigues' formula:

$$P_n(x) = \frac{1}{n!2^n} \frac{d^n}{dx^n} \Big((x^2-1) \Big).$$

Weight function: $w(x) = 1$

Orthogonality:

$$\int_{-1}^{1} P_m(x) P_n(x)\, dx = \begin{cases} \dfrac{2}{2n+1}, & m = n, \\ 0, & m \neq n. \end{cases}$$

Orthogonal series:

$$f(x) = \sum_{n=0}^{\infty} C_n P_n(x), \quad C_n = \frac{2n+1}{2} \int_{-1}^{1} f(x) P_n(x)\, dx.$$

Differential equation: $f(x) := P_n(x)$ satisfies

$$(1-x^2) f''(x) - 2x f'(x) + n(n+1) f(x) = 0.$$

Recursive formulas:

$$\begin{cases} (n+1)P_{n+1}(x) = (2n+1)xP_n(x) - nP_{n-1}(x), \\ P'_{n+1}(x) - P'_{n-1}(x) = (2n+1)P_n(x), \\ (x^2-1)P'_n(x) = nxP_n(x) - nP_{n-1}(x), \\ \int P_n(x)\, dx = \frac{1}{2n+1}\Big(P_{n+1}(x) - P_{n-1}(x)\Big) + C. \end{cases}$$

Generating function:

$$\frac{1}{\sqrt{1-2xz+z^2}} = \sum_{n=0}^{\infty} P_n(x) z^n; \quad |z| < 1, \quad |x| \leq 1.$$

(14.59)

Legendre polynomial in an arbitrary bounded interval $I = (a, b)$:
Let $\rho = (b - a)/2$, $\eta = (b + a)/2$. The polynomials $P_n(\frac{x-\eta}{\rho})$, $n = 0, 1, 2, \ldots$, define an orthogonal system over the interval (a, b), with orthogonality relation

$$\int_a^b P_m\Big(\frac{x-\eta}{\rho}\Big)\, P_n\Big(\frac{x-\eta}{\rho}\Big)\, dx = \begin{cases} \dfrac{b-a}{2n+1}, & m = n, \\ 0, & m \neq n. \end{cases} \tag{14.60}$$

$$\begin{aligned} P_1(x) &= x, \\ P_2(x) &= \frac{1}{4}\,(6x^2 - 2), \\ P_3(x) &= \tfrac{1}{8}\,(20x^3 - 12x), \\ P_4(x) &= \frac{1}{16}\,(70x^4 - 60x^2 + 6). \end{aligned}$$

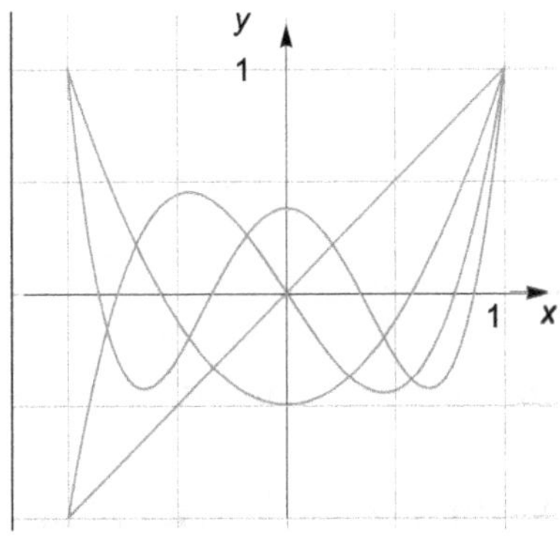

Associated Legendre polynomials $P_n^k(x)$, $0 \le k \le n$:

Orthogonality:

$$\int_{-1}^{1} P_m^k(x) P_n^k(x)\, dx = \begin{cases} \dfrac{(n+k)!}{(n-k)!}\dfrac{2}{2n+1}, & m = n, \\ 0, & m \neq n. \end{cases}$$

Orthogonal series:

$$f(x) = \sum_{n=k}^{\infty} C_n P_n^k(x), \quad C_n = \frac{2n+1}{2} \cdot \frac{(n-k)!}{(n+k)!} \int_{-1}^{1} f(x) P_n^k(x)\, dx.$$

Differential equation:
$f(x) = P_n^k(x) = (1 - x^2)^{k/2} D^k P_n(x)$, $0 \le k \le n$ satisfies

$$(1 - x^2) f''(x) - 2x f'(x) + \Big[n(n+1) - \frac{k^2}{1 - x^2}\Big] f(x) = 0.$$

Recursive formulas:

$$(n-k+1)P_{n+1}^k(x) = (2n+1)xP_n^k(x) - (n+k)P_{n-1}^k(x)$$

$$P_n^{k+1}(x) = 2kx(1-x^2)^{-1/2}P_n^k(x) - (n-k+1)(n+k)P_n^{k-1}(x).$$

Generating function:

$$\frac{(2k-1)!!(1-x^2)^{k/2}z^k}{(1-2xz+z^2)^{k+1/2}} = \sum_{n=k}^{\infty} P_n^k(x)z^n, \quad |z|<1,\ |x|\le 1. \quad (14.61)$$

Spherical harmonics:
The functions $f = \cos(k\varphi)\,P_n^k(\cos\theta)$ and $f = \sin(k\varphi)\,P_n^k(\cos\theta)$ satisfy the partial differential equation

$$\frac{1}{\sin\theta}\frac{\partial}{\partial\theta}\Big(\sin\theta\frac{\partial f}{\partial\theta}\Big) + \frac{1}{\sin^2\theta}\frac{\partial^2 f}{\partial\varphi^2} + n(n+1) = 0.$$

Remarks. There are other orthogonal polynomials in $(-1,1)$, different from the Legendre polynomials, with weight functions $w(x)\neq 1$.

(II) Chebyshev polynomials of the first kind $T_n(x)$:

$T_n(x)$ have the following properties:

$$T_n(1)=1, \quad T_n(-x) = (-1)^n T_n(x), \quad |T_n(x)|\le 1, \quad -1\le x\le 1.$$

Explicit form:

$$T_n(x) := \cos(n\arccos x) = \frac{n}{2}\sum_{m=0}^{[n/2]}(-1)^m\frac{(n-m-1)!}{m!(n-2m)!}(2x)^{n-2m}.$$

Rodrigues' formula:

$$T_n(x) = \frac{(-1)^n(1-x^2)^{1/2}\sqrt{\pi}}{2^n\Gamma(n+\frac{1}{2})}\frac{d^n}{dx^n}\Big((1-x^2)^{n-1/2}\Big).$$

Weight function: $w(x) = (1-x^2)^{-1/2}$.
Orthogonality:

$$\int_{-1}^{1}(1-x^2)^{-1/2}T_m(x)T_n(x)\,dx = \begin{cases} 0, & m\neq n, \\ \pi/2, & m=n\neq 0, \\ \pi, & m=n=0. \end{cases}$$

Orthogonal series:

$$f(x) = \frac{1}{2}C_0 + \sum_{n=1}^{\infty} C_n T_n(x), \quad C_n = \frac{2}{\pi}\int_{-1}^{1} \frac{f(x)}{\sqrt{1-x^2}} T_n(x)\, dx.$$

The *differential equation*: $f(x) := T_n(x)$ satisfies

$$(1-x^2)f''(x) - xf'(x) + n^2 f(x) = 0.$$

Recursive formula:

$$T_{n+1}(x) = 2xT_n(x) - T_{n-1}(x).$$

Generating function:

$$\frac{1-xz}{1-2xz+z^2} = \sum_{n=0}^{\infty} T_n(x) z^n, \quad |z| < 1, \quad |x| < 1.$$

(III) Chebyshev polynomials of the second kind $U_n(x)$:

$U_n(x)$ have the following properties: $U_n(1) = n+1$ and

$$U_n(-x) = (-1)^n U_n(x), \quad |U_n(x)| \le n+1, \quad -1 \le x \le 1.$$

Explicit form:

$$\begin{aligned} U_n(x) &:= \frac{\sin((n+1)\arccos x)}{\sqrt{1-x^2}} \\ &= \sum_{m=0}^{[n/2]} (-1)^m \frac{(m-n)!}{m!(n-2m)!} (2x)^{n-2m}. \end{aligned}$$

Rodrigues' formula:

$$U_n(x) = \frac{(-1)^n (n+1)\sqrt{\pi}}{(1-x^2)^{1/2} 2^{n+1} \Gamma(n+\frac{1}{2})} \frac{d^n}{dx^n}\left((1-x^2)^{n+1/2}\right).$$

Weight function: $w(x) = (1-x^2)^{1/2}$.
Orthogonality:

$$\int_{-1}^{1} (1-x^2)^{1/2} U_m(x) U_n(x)\, dx = \begin{cases} \pi/2, & m = n, \\ 0, & m \ne n. \end{cases}$$

Orthogonal series:

$$f(x) = \sum_{n=0}^{\infty} C_n U_n(x), \quad C_n = \frac{2}{\pi} \int_{-1}^{1} f(x) U_n(x) \sqrt{1-x^2}\, dx.$$

Differential equation: $f(x) := U_n(x)$ satisfies

$$(1-x^2) f''(x) - 3x f'(x) + n(n+2) f(x) = 0.$$

Recursive formula:

$$U_{n+1}(x) = 2x U_n(x) - U_{n-1}(x).$$

Generating function:

$$\frac{1}{1-2xz+z^2} = \sum_{n=0}^{\infty} U_n(x) z^n, \quad |z| < 1, \quad |x| < 1.$$

Shifted Chebyshev polynomials $\widetilde{T}_n(x)$:

$$\widetilde{T}_n(x) = T_n(2x-1) = T_{2n}(\sqrt{x}), \quad 0 \le x \le 1.$$

Orthogonality:

$$\int_0^1 \widetilde{T}_k(x) \widetilde{T}_n(x) (x-x^2)^{-1/2}\, dx = \begin{cases} 0, & k \neq n, \\ \pi, & k = n = 0, \\ \pi/2, & k = n \neq 0. \end{cases}$$

Differential equation: $f(x) := \widetilde{T}_n(x)$ satisfies

$$(x-x^2) f''(x) - (x-1/2) f'(x) + n^2 f(x) = 0.$$

Recursive formula:

$$\widetilde{T}_{n+1}(x) = (4x-2) \widetilde{T}_n(x) - \widetilde{T}_{n-1}(x).$$

Shifted Chebyshev polynomials $\widetilde{U}_n(x)$:

$$\widetilde{U}_n(x) = U_n(2x-1), \quad 0 \le x \le 1.$$

Orthogonality:

$$\int_0^1 \widetilde{U}_k(x) = \widetilde{U}_n(x)(x-x^2)^{1/2}\,dx = \begin{cases} 0, & k \neq n, \\ \pi/8, & k = n. \end{cases}$$

Differential equation: $f(x) := \widetilde{U}_n(x)$ satisfies

$$(x-x^2)f''(x) - 3(x-1/2)f'(x) + n(n+2)f(x) = 0.$$

Recursive formula:

$$\widetilde{U}_{n+1}(x) = (4x-2)\widetilde{U}_n(x) - \widetilde{T}_{n-1}(x).$$

(IV) Jacobi polynomials $P_n^{(\alpha,\beta)}(x)$:

$P_n^{(\alpha,\beta)}(x)$ have the following properties: $P_n^{(\alpha,\beta)}(1) = \binom{n+\alpha}{n}$.

Explicit form:

$$P_n^{(\alpha,\beta)}(x) = \frac{1}{2^n}\sum_{m=0}^{n}\binom{n+\alpha}{m}\binom{n+\beta}{n-m}.(x-1)^{n-m}(x+1)^m.$$

Rodrigues' formula:

$$P_n^{(\alpha,\beta)}(x) = \frac{(-1)^n}{2^n n!(1-x)^\alpha(1+x)^\beta}\frac{d^n}{dx^n}\Big((1-x)^{n+\alpha}(1+x)^{n+\beta}\Big).$$

Weight function: $w(x) = (1-x)^\alpha(1+x)^\beta;\quad \alpha,\beta > 1.$

Orthogonality:

$$\int_{-1}^{1}(1-x)^\alpha(1+x)^\beta P_m^{(\alpha,\beta)}(x)P_n^{(\alpha,\beta)}(x)\,dx$$

$$= \begin{cases} \dfrac{2^{\alpha+\beta+1}\Gamma(n+\alpha+1)\Gamma(n+\beta+1)}{(2n+\alpha+\beta+1)\Gamma(n+\alpha+\beta+1)}, & \text{if } m = n, \\ 0, & \text{if } m \neq n. \end{cases}$$

Differential equation: $f(x) := P_n^{(\alpha,\beta)}(x)$ satisfies

$$(1-x^2)f'' + \Big(\beta - \alpha - (\alpha+\beta+2)x\Big)f' + n(n+\alpha+\beta+1)f = 0.$$

Generating function:

$$v^{-1}(1-z+v)^{-\alpha}(1+z+v)^{-\beta} = \sum_{n=0}^{\infty} 2^{-\alpha-\beta}P_n^{(\alpha,\beta)}(x)z^n, \quad |x| < 1$$

$$v = \sqrt{1-2xz+z^2}, \quad |z| < 1.$$

(V) Laguerre polynomials $L_n^{(\alpha)}(x)$ and $L_n(x) = L_n^{(0)}(x),\ 0 \le x < \infty$:

Explicit form:

$$L_n^{(\alpha)}(x) = \sum_{m=0}^{n} (-1)^m \binom{n+\alpha}{n-m} \frac{1}{m!} x^m.$$

$L_n^{(\alpha)}(x)$ satisfy the following inequality

$$|L_n^{(\alpha)}(x)| \le e^{x/2} \times \begin{cases} \dfrac{\Gamma(n+\alpha+1)}{n!\Gamma(\alpha+1)}, & x \ge 0,\ \alpha > 0, \\ 2 - \dfrac{\Gamma(n+\alpha+1)}{n!\Gamma(\alpha+1)}, & x \ge 0,\ 0 < \alpha < 1. \end{cases}$$

Rodrigues' formula:

$$L_n^{(\alpha)}(x) = \frac{x^{-\alpha}e^x}{n!} \frac{d^n}{dx^n}\left(x^{n+\alpha}e^{-x}\right), \quad L_n(x) = L_n^{(0)}(x),\ n = 0, 1, 2, \ldots$$

Laguerre function: $l_n(x) = e^{-x/2} L_n(x)$.
Weight function: $w(x) = x^{\alpha} e^{-x},\ \alpha > -1$.
Orthogonality:

$$\int_0^{\infty} L_m^{(\alpha)}(x) L_n^{(\alpha)}(x) x^{\alpha} e^{-x}\, dx = \frac{\Gamma(n+\alpha+1)}{n!} \delta_{mn}$$

$$\int_0^{\infty} L_m(x) L_n(x) e^{-x}\, dx = \int_0^{\infty} l_m(x) l_n(x) dx = \delta_{mn}, \quad \alpha > -1.$$

Orthogonal series:

$$f(x) = \sum_{n=0}^{\infty} C_n L_n^{(\alpha)}(x),$$

$$C_n = \frac{n!}{\Gamma(1+\alpha-n)} \int_0^{\infty} f(x) L_n^{(\alpha)}(x) x^{\alpha} e^{-x}\, dx,$$

$$f(x) = \sum_{n=0}^{\infty} C_n L_n(x), \quad C_n = \int_0^{\infty} f(x) L_n(x) e^{-x}\, dx.$$

Differential equation: $f(x) := L_n^{(\alpha)}(x)$ satisfies

$$xf''(x) + (1 + \alpha - x)f'(x) + nf(x) = 0.$$

Recursive formula:

$$(n+1)L_{n+1}^{(\alpha)}(x) = (2n + \alpha + 1)L_n^{(\alpha)}(x) - (n + \alpha)L_{n-1}^{(\alpha)}(x)$$

$$\frac{d}{dx} L_n^{(\alpha)}(x) = -L_{n-1}^{(\alpha+1)}(x).$$

Generating function:

$$(1 - z)^{-\alpha-1} \exp\left(\frac{xz}{x-1}\right) = \sum_{n=0}^{\infty} L_n^{(\alpha)}(x)z^n, \quad |z| < 1.$$

(VI) Hermite polynomials $H_n(x)$, $-\infty < x < \infty$:

Explicit form:

$$H_n(x) = \sum_{m=0}^{[n/2]} (-1)^m \frac{n!}{m!(n-2m)!}(2x)^{n-2m}.$$

Rodrigues' formula:

$$H_n(x) = (-1)^n e^{x^2} \frac{d^n}{dx^n}\left(e^{-x^2}\right), \quad n = 0, 1, \ldots$$

Hermite function: $h_n(x) = e^{-x^2/2}H_n(x)$.

Weight function: $w(x) = e^{-x^2}$.

Orthogonality:

$$\int_{-\infty}^{\infty} H_m(x)H_n(x)e^{-x^2}\,dx = \int_{-\infty}^{\infty} h_m(x)h_n(x)\,dx = n!2^n\sqrt{\pi}\delta_{mn}.$$

Orthogonal series:

$$f(x) = \sum_{n=0}^{\infty} C_n H_n(x), \quad C_n = \frac{1}{n!2^n\sqrt{\pi}} \int_{-\infty}^{\infty} f(x)H_n(x)e^{-x^2}\,dx.$$

Differential equation: $f(x) := H_n(x)$ och $g(x) = h_n(x)$ satisfies

$$f''(x) - 2xf'(x) + 2nf(x) = 0, \quad g''(x) - (2n + 1 - x^2)g(x) = 0.$$

Recursive formula:

$$H_{n+1}(x) = 2xH_n(x) - 2nH_{n-1}(x).$$

$$H_n'(x) = 2nH_{n-1}(x), \quad \left(e^{-x^2} H_n(x)\right)' = -e^{-x^2} H_n(x).$$

Generating function:

$$e^{2xz-z^2} = \sum_{n=0}^{\infty} H_n(x)\frac{z^n}{n!}, \quad -\infty < z < \infty, \quad -\infty < x < \infty.$$

Some properties of the function classes

Theorem 14.23.

(i) *Legendre polynomials are orthogonal with weight function* 1 *in the interval* $[-1, 1]$. *More precisely*

$$\int_{-1}^{1} P_m(x)P_n(x)\frac{\sqrt{(2m+1)(2n+1)}}{2}dx = \delta_{mn}.$$

(ii) *Chebyshev polynomials satisfy*

$$\int_{-1}^{1} T_m(x)T_n(x) \cdot \frac{1}{\sqrt{1-x^2}}dx = \frac{\pi}{2} \cdot \delta_{mn}$$

and

$$\int_{-1}^{1} U_m(x)U_n(x)\sqrt{1-x^2}dx = \frac{\pi}{2} \cdot \delta_{mn}.$$

(iii) *Laguerre polynomials* $L_n(x)$ *satisfy*

$$\int_{0}^{\infty} L_m(x)L_n(x)e^{-x}dx = \delta_{mn}.$$

(iv) *Bessel functions* $J_n(x)$ *satisfy*

$$J_{n-1}(x) + J_{n+1}(x) = \frac{2n}{x} J_n(x), \quad J_{n-1}(x) - J_{n+1}(x) = 2J_n'(x). \tag{14.62}$$

(v) *Jakobi polynomials can be rewritten as*

$$P_n^{(a,b)}(x) = \frac{1}{2^n} \sum_{k=0}^{n} \binom{n+a}{k} \binom{n+b}{n-k} (x-1)^{n-k} (x+1)^k \tag{14.63}$$

and have orthogonality property

$$\int_{-1}^{1} P_m^{(a,b)} P_n^{(a,b)} (1-x)^a (1+x)^b dx = 0, \quad m \neq n, \ a, b > -1. \tag{14.64}$$

(vi) *Generating function:*

$$\frac{1}{\sqrt{1 - 2xz + z^2}} = \sum_{n=0}^{\infty} P_n(x) z^n; \quad |z| < 1, \ |x| \leq 1.$$

Remarks. Legendre, Gegenbauer, and Chebyshev polynomials are special cases of Jacobian polynomials. By $a = b$, the ultra-spherical or Gegenbauer polynomials are obtained by normalization

$$P_n^{(a)}(x) := \frac{\Gamma(2a+n+1)}{\Gamma(a+1/2+n+1)} P_n^{(a-1/2, a-1/2)}(x). \tag{14.65}$$

$P_n^{(0)}(x)$ are the Chebyshev polynomials and $P_n^{(1/2)}(x)$ are the Legendre polynomials of degree n.

14.4.1 *Generation of the most common polynomial classes*

Let p be a polynomial that satisfies the differential equation

$$p_{n+1}(x) = (A_n x + B_n) p_n(x) + C_n p_{n-1}(x). \tag{14.66}$$

Polynomial	A_n	B_n	C_n
Legendre	$\dfrac{2n+1}{n+1}$	0	$-\dfrac{n}{n+1}$
Chebyshev	2	0	-1
Gegenbauer	$\dfrac{2n+\lambda}{n+1}$	0	$\dfrac{1-n-2\lambda}{n+1}$
Hermite	2	0	$-2n$
Laguerre	$-\dfrac{1}{n+1}$	$\dfrac{2n+1}{n+1}$	$-\dfrac{n}{n+1}$

(14.67)

The integral representation of Neumann functions is given by

$$\begin{aligned} Y_\nu(x) &= \frac{1}{\pi}\int_0^\pi \sin(x\sin\theta - \nu\theta)d\theta \\ &\quad - \frac{1}{\pi}\int_0^\infty [e^{\nu t} + e^{-\nu t}(-1)^\nu]e^{-x\sinh t}dt \\ &= -\frac{2(2/x)^\nu}{\sqrt{\pi}\,\Gamma(1/2-\nu)}\int_1^\infty \frac{\cos xt\,dt}{(t^2-1)^{\nu+1/2}}. \end{aligned} \tag{14.68}$$

14.4.2 *Hypergeometric functions*

There are two classes of functions which in their general form have only one series representation.

Hypergeometric functions of the first kind

$${}_1F_1(\alpha,\beta;x) = 1+\frac{\alpha}{\beta}x+\frac{\alpha(\alpha+1)}{\beta(\beta+1)}\frac{x^2}{2!}+\cdots = \sum_{n=0}^{\infty}\frac{\mathcal{P}(\alpha,n)}{\mathcal{P}(\beta,n)}\frac{x^n}{n!} \tag{14.69}$$

Hypergeometric functions of the second kind

$$\begin{aligned} {}_2F_1(\alpha,\beta,\gamma;x) &:= 1+\frac{\alpha\beta}{\gamma}\frac{x}{1!}+\frac{\alpha(\alpha+1)\beta(\beta+1)}{\gamma(\gamma+1)}\frac{x^2}{2!}+\cdots \\ &= \sum_{n=0}^{\infty}\prod_{k=1}^{n}\left[\frac{(\alpha+k-1)(\beta+k-1)}{(\gamma+k-1)}\right]\frac{x^n}{n!} \\ &= \sum_{n=0}^{\infty}\left[\frac{\mathcal{P}(\alpha,n)\mathcal{P}(\beta,n)}{\mathcal{P}(\gamma,n)}\right]\frac{x^n}{n!}. \end{aligned} \tag{14.70}$$

$\mathcal{P}(\alpha, n)$ is the Pochhammer symbol, defined on page 176. A general hypergeometric function is given by

$$
\begin{aligned}
&{}_pF_q(\{a_1, a_2, \ldots, a_p\}, \{b_1, b_2, \ldots, p_q\}; x) \\
&\quad = \sum_{n=0}^{\infty} \frac{P(a_1, n)P(a_2, n) \cdot \ldots \cdot P(a_p, n)}{P(b_1, n)P(b_2, n) \cdot \ldots \cdot P(b_q, n)} \frac{x^n}{n!}.
\end{aligned} \tag{14.71}
$$

Remarks. The indices 2 and 1 in ${}_2F_1$ refer to the number of parameters in the numerator and the denominator, respectively (so as for p, q).

The notions "of the first" and "second kind" are not generally used.

Some correlations between Hypergeometric and elementary functions

$$
\begin{aligned}
e^x &= {}_1F_1(\alpha, \alpha; x), \text{ if } \alpha > -1 \\
1 - \frac{x}{\alpha} &= {}_1F_1(-1, \alpha; x) \\
e^{-x} L_n(x) &= {}_1F_1(n+1, 1; -x) \text{ where } L \text{ is the Laguerre polynomial of order } n.
\end{aligned} \tag{14.72}
$$

14.5 Products

14.5.1 *Basic examples*

$$
\begin{aligned}
n! &\approx \sqrt{2\pi\, n} \left(\frac{n}{e}\right)^n \quad \text{(Stirling's formula)}, \\
(2n)! &\approx \sqrt{2\pi\, n} \left(\frac{2n}{e}\right)^n, \\
(2n-1)!! &\approx \sqrt{2} \left(\frac{2n}{e}\right)^n,
\end{aligned} \tag{14.73}
$$

with asymptotic equivalence, more precisely: $\dfrac{\text{LHS}}{\text{RHS}} \longrightarrow 1$, as $n \longrightarrow \infty$.

14.5.2 *Infinite products*

Definition 14.8. An infinite product of complex numbers $u_1, u_2, \ldots$ means the limit (as far as it exists)

$$\prod_{n=1}^{\infty} u_n := \lim_{m\to\infty} \prod_{n=1}^{m} u_n. \tag{14.74}$$

Theorem 14.24. *Given the sequence* $(u_n)_{n=1}^{\infty}$ *of real numbers* $u_n > 0 (u_n \in \mathbb{R}_+)$. *Then*

$$\prod_{n=1}^{\infty} u_n \text{ convergent} \Longleftrightarrow \sum_{n=1}^{\infty} \ln(u_n) \text{ convergent.}$$

$$\prod_{n=1}^{\infty} (u_n + 1) \text{ convergent} \Longleftrightarrow \sum_{n=1}^{\infty} u_n \text{ convergent.}$$

The values of some infinite products

$$\prod_{n=1}^{\infty} \frac{(1+1/n)^2}{1+2/n} = 2$$

$$\prod_{n=1}^{\infty} \frac{(1+1/n)^3}{1+3/n} = 6$$

and in general

$$\prod_{n=1}^{\infty} \frac{(1+1/n)^k}{1+k/n} = k!$$

$$\prod_{n=3}^{\infty} \left[1 - \left(\frac{2}{n}\right)^2\right] = 6$$

$$\prod_{n=2}^{\infty} \frac{n^2-1}{n^2+1} = \frac{2\pi}{e^{\pi} - e^{-\pi}} = \frac{\pi}{\sinh \pi}$$

$$\prod_{n=1}^{\infty} \left[1 + \frac{1}{n^2}\right] = \frac{e^{\pi} - e^{-\pi}}{2\pi} = \frac{\sinh \pi}{\pi} \tag{14.75}$$

$$\prod_{n=2}^{\infty} \frac{n^3-1}{n^3+1} = \frac{2}{3}.$$

Some elementary functions expressed as infinite products

$$\begin{aligned}
\sin x &= x\prod_{n=1}^{\infty}\left[1-\left(\frac{x}{n\,\pi}\right)^2\right] = x\prod_{n=1}^{\infty}\cos\left[\frac{x}{2^n}\right] \\
\cos x &= \prod_{n=1}^{\infty}\left[1-\left(\frac{2x}{(2n-1)\,\pi}\right)^2\right] \\
\sinh x &= x\prod_{n=1}^{\infty}\left[1+\left(\frac{x}{n\,\pi}\right)^2\right] \\
\cosh x &= \prod_{n=1}^{\infty}\left[1+\left(\frac{2x}{(2n-1)\,\pi}\right)^2\right].
\end{aligned} \tag{14.76}$$

Remarks. For instance, the following infinite products are convergent

$$\prod_{n=1}^{\infty}\left(1+\frac{1}{n!}\right) \quad \text{and} \quad \prod_{n=2}^{\infty}\left(1-\frac{1}{n!}\right),$$

which also holds true if $n!$ is substituted by semi factorial.

Chapter 15

Transform Theory

15.1 Fourier Transform

The Fourier transform of a function $f : \mathbb{R} \to \mathbb{C}$ is defined as

$$\hat{f}(\omega) := \int_{-\infty}^{\infty} f(t)e^{-i\omega t}dt, \tag{15.1}$$

insofar as the integral exists. Given $\hat{f}$ one gets *the Fourier inversion formula* as

$$f(t) = \frac{1}{2\pi}\int_{-\infty}^{\infty} \hat{f}(\omega)e^{i\omega t}d\omega.$$

The Fourier transform of f is denoted by $\hat{f} = \mathcal{F}(f)$, or $\hat{f}(\omega) \subset f(t)$.

Theorem 15.1 (Linearity of the Fourier transform).

$$af(t) + bg(t) \textit{ has Fourier transform } \quad a\hat{f}(\omega) + b\hat{g}(\omega)$$

or alternatively (15.2)

$$\mathcal{F}(af(t) + bg(t)) = a\mathcal{F}(f(t)) + b\mathcal{F}(g(t)), \qquad \forall a, b \in \mathbb{R}.$$

Name	Function $g(t)$	$\mathcal{F}$ – Transform $\hat{g}(\omega)$
Frequency translation	$f(t)e^{i\alpha t}$	$\hat{f}(\omega-\alpha)$
Time translation	$f(t-\alpha)$	$\hat{f}(\omega)e^{-i\alpha\omega}$
Reflection	$f(-t)$	$\hat{f}(-\omega)$
Conjunction	$\overline{f(-t)}$	$\overline{\hat{f}(\omega)}$
Scaling	$f(t/\lambda)\quad \lambda>0$	$\lambda\hat{f}(\lambda\omega)$
Derivation	$f'(t)$	$(i\omega)\hat{f}(\omega)$
Higher derivatives	$f^{(n)}(t)$	$(i\omega)^n\hat{f}(\omega)$
Mult. by variable	$-it\,f(t)$	$\hat{f}'(\omega)$
Mult. by n-monomial	$(-it)^n\,f(t)$	$\hat{f}^{(n)}(\omega)$
Integral	$\int_{-\infty}^{t} f(\tau)\,d\tau$	$\frac{\hat{f}(\omega)}{i\omega}+\pi\hat{f}(0)\delta(\omega)$
Convolution	$(f*h)(t)$	$\hat{f}(\omega)\cdot\hat{h}(\omega)$
Product	$f(t)h(t)$	$\frac{1}{2\pi}(\hat{f}*\hat{h})(\omega)$
Inversion	$\hat{f}(t)$	$2\pi f(-\omega)$

(15.3)

Definition 15.1. Convolution of two functions

$$(f*g)(t)=\int_{-\infty}^{\infty} f(t-x)g(x)\,dx. \tag{15.4}$$

Plancherel's identity

$$\int_{-\infty}^{\infty} f(t)\overline{g(t)}dt=\frac{1}{2\pi}\int_{-\infty}^{\infty} \hat{f(\omega)}\overline{\hat{g}(\omega)}d\omega.$$

Parseval's formula (15.5)

$$\int_{-\infty}^{\infty} |f(t)|^2dt=\frac{1}{2\pi}\int_{-\infty}^{\infty} |\hat{f}(\omega)|^2d\omega.$$

The first identity holds if both integrals are absolutely convergent.

The second identity follows from the first by setting $g \equiv f$.

Some important relations

(I) **Symmetry:** $f(t) \supset g(\omega) \iff g(t) \supset 2\pi f(-\omega)$.

Example 15.1. $e^{-|t|} \supset \dfrac{2}{1+\omega^2} \iff \dfrac{2}{1+t^2} \supset 2\pi e^{-|\omega|}$.

(II) **Differentiation with respect to a parameter:**

$$f(t,\alpha) \supset \hat{f}(\omega,\alpha) \implies \frac{\partial}{\partial\alpha} f(t,\alpha) \supset \frac{\partial}{\partial\alpha} \hat{f}(\omega,\alpha)$$

(III) $f(t)$ even $\iff$ $\hat{f}(\omega)$ even, $f(t)$ odd $\iff$ $\hat{f}(\omega)$ odd.

(IV) $f^{(n)}(t) \supset \hat{g}(\omega) \implies f(t) \supset \dfrac{\hat{g}(\omega)}{(i\omega)^n} + C_1\delta(\omega) + C_2\delta'(\omega) + \cdots + C_n\delta^{(n-1)}(\omega)$

$$\hat{f}^{(n)}(\omega) \subset g(t) \implies \hat{f}(\omega) \subset \frac{g(t)}{(-it)^n} + C_1\delta(t) + C_2\delta'(t) + \cdots + C_n\delta^{(n-1)}(t).$$

(V) **Poisson's summation formula:**

$$\sum_{k=-\infty}^{\infty} f(ak) = \frac{1}{a}\sum_{n=-\infty}^{\infty} \hat{f}(2\pi n/a), \qquad a > 0.$$

(VI) **The Sampling theorem:** Assume that $f(t)$ is continuous with Fourier transform $\hat{f}(\omega) = 0$ for $|\omega| \geq \alpha$, (band-limited signal). If the signal is sampled with the frequency $\frac{1}{T} \geq \frac{\alpha}{\pi}$ (angular frequency $\Omega = \frac{2\pi}{T} \geq 2\alpha$), then $f(t)$ is recovered from the sampled signal by a low-pass filter with the chopping angle frequency α (LP_α–filtration and multiplication by T):

$$f(t) = \sum_{n=-\infty}^{\infty} f(nT)\frac{T\sin(\alpha(t-nT))}{\pi(t-nT)}.$$

$$\text{if} \quad T = \pi/\alpha, \quad f(t) = \sum_{n=-\infty}^{\infty} f\left(\frac{n\pi}{\alpha}\right)\frac{\sin(\alpha t - n\pi)}{\alpha t - n\pi}.$$

Some common Fourier transforms

Function $f(t)$	$\hat{f}(\omega) = \int_{-\infty}^{\infty} f(t)e^{-i\omega t}\,dt$
$\theta(t)$	$\pi\delta(\omega) + \dfrac{1}{i\omega}$
$e^{-at}\theta(t)$	$\dfrac{1}{a+i\omega}, \quad a > 0$
$e^{at}\theta(-t) = e^{at}(1-\theta(t))$	$\dfrac{1}{a-i\omega}, \quad a > 0$
$t^n\theta(t), \quad n = 1, 2, \ldots$	$\pi\delta(\omega) + \dfrac{n!}{(i\omega)^{n+1}} + \pi i\delta^{(n)}(\omega)$
$t^n e^{-\alpha t}\theta(t), \ \alpha > 0,$ $n = 1, 2, \ldots$	$\dfrac{n!}{(\alpha + 2\pi i\omega)^{n+1}}$
$\theta(t+\alpha) - \theta(t-\alpha)$	$\dfrac{2\sin\alpha\omega}{\omega}$
$(\theta(t+\alpha) - \theta(t-\alpha))\text{sgn}\ t$	$\dfrac{4\sin^2\frac{\alpha\omega}{2}}{i\omega}$
$(\theta(t+\alpha) - \theta(t-\alpha))e^{i\Omega t}$	$\dfrac{2\sin\alpha(\Omega-\omega)}{\Omega-\omega}$
$\text{sgn}t = \begin{cases} 1 & \text{if} \quad t > 0 \\ -1 & \text{if} \quad t < 0 \end{cases}$	$\dfrac{2}{i\omega}$

(15.6)

Some common Fourier transforms, continuation

Function	Fourier transform
$\chi_a(t) = \begin{cases} 1 & \text{if} \quad \lvert t\rvert \le a \\ 0 & \text{if} \quad \lvert t\rvert > a \end{cases}$	$\dfrac{2\sin a\omega}{\omega}$
$f(t) = \begin{cases} t & \text{if} \quad \lvert t\rvert \le 1 \\ 0 & \text{if} \quad \lvert t\rvert > 1 \end{cases}$	$\dfrac{2\,i\,\cos(\omega)}{\omega} - \dfrac{2\,i\,\sin(\omega)}{\omega^2}$
$f(t) = \begin{cases} 1 & \text{if} \quad \lvert t\rvert < 1/2 \\ 0 & \text{if} \quad \lvert t\rvert > 1/2 \end{cases}$	$\dfrac{2\sin(\omega/2)}{\omega}$
e^{-at^2}	$\sqrt{\dfrac{\pi}{a}}e^{-\omega^2/(4a)}$
$\dfrac{1}{\sqrt{4\pi a}}e^{-t^2/(4a)}$	$e^{-a\omega^2}, \quad a > 0$
$\dfrac{1}{t^2+a^2}$	$\left(\dfrac{\pi}{a}\right)e^{-a\lvert\omega\rvert}$
$\dfrac{t}{t^2+a^2}$	$-i\pi e^{-a\lvert\omega\rvert}\text{sgn }(\omega)$
$e^{i\Omega t}$	$2\pi\delta(\omega-\Omega)$
$\sin\Omega t$	$\pi[\delta(\omega+\Omega)-\delta(\omega-\Omega)]$

(15.7)

Some common Fourier transforms, continuation

Function	Fourier transform (Common shape)
$\cos \Omega t$	$\pi[\delta(\omega+\Omega)+\delta(\omega-\Omega)]$
$\dfrac{\sin \Omega t}{t}$	$\pi[\theta(\omega+\Omega)-\theta(\omega-\Omega)]$
$\sin at^2$	$\sqrt{\dfrac{\pi}{a}}\cos\left(\dfrac{\omega^2}{4a}+\dfrac{\pi}{4}\right)$
$\cos at^2$	$\sqrt{\dfrac{\pi}{a}}\cos\left(\dfrac{\omega^2}{4a}-\dfrac{\pi}{4}\right)$
$\dfrac{1}{\sinh t}$	$-i\pi \tanh \dfrac{\pi\omega}{2}$
$\dfrac{1}{\cosh t}$	$\dfrac{\pi}{\cosh \frac{\pi\omega}{2}}$
1	$2\pi\delta(\omega)$
$t^n, \quad n=1,2,\ldots$	$2\pi i^n \delta^{(n)}(\omega)$
$t^{-n}, \quad n=1,2,\ldots$	$\frac{\pi(-i)^n}{(n-1)!}\omega^{n-1}\text{sgn}\ (\omega)$
$\delta(t)$	1
$\delta^{(n)}(t)$	$(i\omega)^n$
$\delta^{(n)}(t-T)$	$(i\omega)^n e^{-i\omega T}, \quad n=0,1,\ldots$
$\text{III}\ (t) := \sum_{n=-\infty}^{\infty} \delta(t-n)$	$\text{III}\ (\omega) = \sum_{n=-\infty}^{\infty} \delta(\omega-n)$

(15.8)

Some common Fourier transforms, continuation

Function	Fourier transform II
$e^{-a\|t\|}$	$\dfrac{2a}{a^2+\omega^2}, \quad a>0$
$e^{-a\|t\|}\operatorname{sgn} t$	$-\dfrac{2i\omega}{a^2+\omega^2}, \quad a>0$
$te^{-a\|t\|}$	$-\dfrac{4ia\omega}{(a^2+\omega^2)^2}, \quad a>0$
$\|t\|e^{-a\|t\|}$	$\dfrac{2(a^2-\omega^2)}{(a^2+\omega^2)^2}, \quad a>0$
$e^{-(1+i\beta)\|t\|}, \quad -\infty<\beta<\infty$	$\dfrac{2(1+i\beta)}{(1+i\beta)^2+4\pi^2\omega^2}$
$e^{-\pi(\alpha+i\beta)^2t^2}, \ \alpha\geq\|\beta\|, \ \alpha+i\beta\neq 0$	$\dfrac{1}{\alpha+i\beta}e^{-i\pi\omega^2/(\alpha+i\beta)^2}$
$\begin{cases}(a^2-t^2)^{-1/2}, & \|t\|<a\\ 0, & \|t\|>a\end{cases}$	$\pi J_0(a\omega)$
$\begin{cases}t(a^2-t^2)^{-1/2}, & \|t\|<a\\ 0, & \|t\|>a\end{cases}$	$-ia\pi J_1(a\omega)$
$t^n \operatorname{sgn} t$	$\dfrac{2n!}{(i\omega)^{n+1}}$
$\|t\| = t\operatorname{sgn} t$	$-\dfrac{2}{\omega^2}$

(15.9)

Some common Fourier transforms, continuation

Function	Fourier transform III
$\lvert t\rvert^{2n-1}$	$2(-1)^n \frac{(2n-1)!}{\omega^{2n}}, \; n = 1, 2, \ldots$
$\lvert t\rvert^{2n} = t^{2n}$	$2\pi(-1)^n \delta^{(2n)}(\omega), \; n = 1, 2, \ldots$
$\lvert t\rvert^{r-1}$	$\dfrac{2\,\Gamma(r)\cos\frac{\pi r}{2}}{\lvert\omega\rvert^r}, \; r \notin \mathbb{Z}$
$\lvert t\rvert^{r-1}\text{sgn}\, t$	$\dfrac{-2i\,\Gamma(r)\sin\frac{\pi r}{2}\ \text{sgn}\ \omega}{\lvert\omega\rvert^r}, \; r \notin \mathbb{Z}$

(15.10)

15.1.1 *Cosine and sine transforms*

$$\hat{f}_c(\alpha) = \int_0^\infty f(x)\cos\alpha x\,dx \qquad f(x) = \frac{2}{\pi}\int_0^\infty \hat{f}_c(\alpha)\cos\alpha x\,d\alpha. \tag{15.11}$$

$$\hat{f}_s(\alpha) = \int_0^\infty f(x)\sin\alpha x\,dx \qquad f(x) = \frac{2}{\pi}\int_0^\infty \hat{f}_s(\alpha)\sin\alpha x\,d\alpha. \tag{15.12}$$

15.1.2 *Relations between Fourier transforms*

If $\hat{f}$ is the Fourier transform of $f(x)$, $-\infty < x < \infty$, then

$$\begin{aligned} f(x) \text{ even} &\implies \hat{f}(\alpha) = 2\hat{f}_c(\alpha) \\ f(x) \text{ odd} &\implies \hat{f}(\alpha) = -2i\hat{f}_s(\alpha). \end{aligned}$$

Table of some cosine transforms

$f(x), \quad x > 0$		$\hat{f}_c(\alpha), \quad \alpha > 0$
$\begin{cases} 1, & x < c \\ 0, & x > c \end{cases}$	$c > 0$	$\dfrac{\sin c\alpha}{\alpha}$
e^{-cx},	$c > 0$	$\dfrac{c}{c^2 + \alpha^2}$
e^{-cx^2},	$c > 0$	$\dfrac{1}{2}\sqrt{\dfrac{\pi}{c}}e^{-\alpha^2/4c}$
x^{c-1},	$0 < c < 1$	$\Gamma(c)\alpha^{-c}\cos\dfrac{c\pi}{2}$
$\cos cx^2$		$\dfrac{1}{2}\sqrt{\dfrac{\pi}{c}}\cos\left(\dfrac{\alpha^2}{4c} - \dfrac{\pi}{4}\right)$
$\sin cx^2$		$\dfrac{1}{2}\sqrt{\dfrac{\pi}{c}}\cos\left(\dfrac{\alpha^2}{4c} + \dfrac{\pi}{4}\right)$

(15.13)

Table of some sine transforms

$f(x), \quad x > 0$		$\hat{f}_s(\alpha), \quad \alpha > 0$
$\begin{cases} 1, & x < c \\ 0, & x > c \end{cases}$	$c > 0$	$\dfrac{1 - \cos c\alpha}{\alpha}$
e^{-cx},	$c > 0$	$\dfrac{\alpha}{c^2 + \alpha^2}$
xe^{-cx^2},	$c > 0$	$\sqrt{\dfrac{\pi}{c}}\dfrac{\alpha}{4c}e^{-\alpha^2/4c}$
x^{c-1},	$-1 < c < 1$	$\Gamma(c)\alpha^{-c}\sin\dfrac{c\pi}{2}$
$\cos cx^2$		$\sqrt{\dfrac{\pi}{2c}}\Big[\sin\dfrac{\alpha^2}{4c}C\Big(\dfrac{\alpha}{\sqrt{2\pi c}}\Big) - \cos\dfrac{\alpha^2}{4c}S\Big(\dfrac{\alpha}{\sqrt{2\pi c}}\Big)\Big]$
$\sin cx^2$		$\sqrt{\dfrac{\pi}{2c}}\Big[\cos\dfrac{\alpha^2}{4c}C\Big(\dfrac{\alpha}{\sqrt{2\pi c}}\Big) + \sin\dfrac{\alpha^2}{4c}S\Big(\dfrac{\alpha}{\sqrt{2\pi c}}\Big)\Big]$

(15.14)

C and S denote Fresnel's cosine and sine functions, respectively (page 178)

$$C(x)=\int_0^x \cos\left(\frac{\pi}{2}.\tau^2\right)d\tau, \qquad S(x)=\int_0^x \sin\left(\frac{\pi}{2}.\tau^2\right)d\tau.$$

Fourier transforms in $\mathbb{R}^n$

$n=2$	
Fourier transform	$\hat{f}(\xi,\eta)=\iint_{\mathbb{R}^2} f(x,y)e^{-i(\xi x+\eta y)}\,dxdy$
Inversion formula	$f(x,y)=\frac{1}{(2\pi)^2}\iint_{\mathbb{R}^2}\hat{f}(\xi,\eta)e^{i(\xi x+\eta y)}\,d\xi d\eta$
Plancherel	$\iint_{\mathbb{R}^2} f(x,y)\overline{g(x,y)}dxdy$
	$=\frac{1}{(2\pi)^2}\iint_{\mathbb{R}^2}\hat{f}(\xi,\eta)\overline{\hat{g}(\xi,\eta)}\,d\xi d\eta$
Parseval	$\iint_{\mathbb{R}^2}\lvert f(x,y)\rvert^2\,dxdy=\frac{1}{(2\pi)^2}\iint_{\mathbb{R}^2}\lvert\hat{f}(\xi,\eta)\rvert^2\,d\xi d\eta$
Convolution	$(f*g)(x,y)=\iint_{\mathbb{R}^2} f(u,v)g(x-u,y-v)\,dudv$

(15.15)

Fourier transforms in $\mathbb{R}^n$, $n\geq 2$

$n=2,3,\ldots$	$\xi=(\xi_1,\xi_2,\ldots,\xi_n)$
Fourier transform	$\hat{f}(\xi)=\int_{\mathbb{R}^n} f(\mathbf{x})e^{-i\mathbf{x}\cdot\xi}\,d\mathbf{x}$
Inversion formula	$f(\mathbf{x})=\frac{1}{(2\pi)^n}\int_{\mathbb{R}^n}\hat{f}(\xi)e^{i\mathbf{x}\cdot\xi}\,d\xi$
Plancherel	$\int_{\mathbb{R}^n} f(\mathbf{x})\overline{g(\mathbf{x})}\,d\mathbf{x}=\frac{1}{(2\pi)^n}\int_{\mathbb{R}^n}\hat{f}(\xi)\overline{\hat{g}(\xi)}\,d\xi$
Parseval	$\int_{\mathbb{R}^n}\lvert f(\mathbf{x})\rvert^2\,d\mathbf{x}=\frac{1}{(2\pi)^n}\int_{\mathbb{R}^n}\lvert\hat{f}(\xi)\rvert^2\,d\xi$
Convolution	$(f*g)(\mathbf{x})=\int_{\mathbb{R}^n} f(\mathbf{u})g(\mathbf{x}-\mathbf{u})\,d\mathbf{u}=\hat{f}(\xi)\hat{g}(\xi)$

(15.16)

Table of two-dimensional Fourier transform

$f(x,y)$	$\hat{f}(\xi,\eta)$
$f(ax,by)$ $(a,\ b$ real)	$\frac{1}{\lvert ab\rvert}\hat{f}\left(\frac{\xi}{a},\frac{\eta}{b}\right)$
$f(x-a,y-b)$ $(a,\ b$ real)	$e^{-i(a\xi+b\eta)}\hat{f}(\xi,\eta)$
$e^{iax}e^{iby}f(x,y)$	$\hat{f}(\xi-a,\eta-b)$
$D_x^m D_y^n f(x,y)$	$(i\xi)^m(i\eta)^n\hat{f}(\xi,\eta)$
$(-ix)^m(-iy)^n f(x,y)$	$D_\xi^m D_\eta^n \hat{f}(\xi,\eta)$
$(f\star g)(x,y)$	$\hat{f}(\xi,\eta)\hat{g}(\xi,\eta)$
$\hat{f}(x,y)$	$(2\pi)^2 f(-\xi,-\eta)$
$\delta(x-a,y-b)=\delta(x-a)\delta(y-b)$	$e^{-i(a\xi+b\eta)}$
$e^{-\frac{x^2}{4a}-\frac{y^2}{4b}},\quad (a,b>0)$	$4\pi\sqrt{ab}\,e^{-(a\xi^2+b\eta^2)}$
$\begin{cases}1, & \lvert x\rvert<a, \quad \text{(band)}\\ 0, & \text{otherwise}\end{cases}$	$4\pi\frac{\sin a\xi}{\xi}\delta(\eta)$
$\begin{cases}1, & \lvert x\rvert<a,\ \lvert y\rvert<b \quad \text{(rectangle)}\\ 0, & \text{otherwise}\end{cases}$	$\frac{4\sin a\xi\sin b\eta}{\xi\eta}$
$\begin{cases}1, & x^2+y^2<a^2 \quad \text{(circle)}\\ 0, & \text{otherwise}\end{cases}$	$\frac{2\pi a}{\xi^2+\eta^2}J_1(a(\xi^2+\eta^2))$

(15.17)

Table of n-dimensional Fourier transform

$f(\mathbf{x})$	$\hat{f}(\xi)$
$f(a\mathbf{x})$ $(a$, real)	$\frac{1}{\lvert a\rvert^n}\hat{f}\left(\frac{\xi}{a}\right)$
$f(\mathbf{x}-\mathbf{a})$	$e^{-i\mathbf{a}\cdot\xi}\hat{f}(\xi)$
$e^{i\mathbf{a}\cdot\mathbf{x}}f(\mathbf{x})$	$\hat{f}(\xi-a)$
$D^\alpha f(\mathbf{x}) = D_1^{\alpha_1}\dots D_n^{\alpha_n} f(\mathbf{x})$	$(i\xi)^\alpha \hat{f}(\xi) = (i\xi_1)^{\alpha_1}\dots$ $(i\xi_n)^{\alpha_n}\hat{f}(\xi)$
$(-i\mathbf{x})^\alpha f(\mathbf{x})$	$D^\alpha \hat{f}(\xi)$
$(f*g)(\mathbf{x})$	$\hat{f}(\xi)\hat{g}(\xi)$
$\hat{f}(\mathbf{x})$	$(2\pi)^n f(-\xi)$
$\delta(\mathbf{x}-\xi) = \delta(x_1-\xi_1)\dots$ $\delta(x_n-\xi_n)$	$e^{-i\mathbf{x}\cdot\xi}, \quad \xi = (\xi_1,\dots,\xi_n)$
$e^{-\frac{\mathbf{x}^2}{4\mathbf{a}}} = e^{-\frac{1}{4}(x_1^2/a_1+\dots+x_n^2/a_n)}$	$2^n\pi^{n/2}\sqrt{a_1\dots a_n}\, e^{-\mathbf{a}\cdot\xi^2}$ $\mathbf{a} = (a_1,\dots,a_n),$ $\boldsymbol{\xi}^2 := (\xi_1^2,\dots,\xi_n^2)$

(15.18)

15.1.3 *Special symbols*

Function	Analytical expression
Rectangle	$\Pi(x) = \begin{cases} 1 & \text{if } \|x\| < 1/2 \\ 0 & \text{if } \|x\| > 1/2 \end{cases}$
Triangle	$\wedge(x) = \begin{cases} 1 - \|x\| & \text{if } \|x\| < 1 \\ 0 & \text{if } \|x\| > 1 \end{cases}$
Heaviside	$H(x) = \begin{cases} 1 & \text{if } x > 0 \\ 0 & \text{if } x < 0 \end{cases}$
Sign	$\text{sign}(x) = \begin{cases} -1 & \text{if } \|x\| < 0 \\ 1 & \text{if } \|x\| > 0 \end{cases}$
Impulse (dirac delta)	$\delta(x)$
Sampling and copying	$\Psi(x) = \sum_{a=-\infty}^{\infty} \delta(x-a)$
Filter or interpolation	$\text{sinc}(x) = \frac{\sin \pi x}{\pi x}$, $\text{jinc}(x) = \frac{J_1(x)}{x}$
Convolution	$(f * g)(x) = \int_{-\infty}^{\infty} f(y)g(x-y)dy$
Auto correlation	$(f \star g)(x) = \int_{-\infty}^{\infty} \overline{f(-y)}g(x-y)dy$

(15.19)

Special symbols, continuation

Two-dimensional functions

$$\begin{array}{l|l} \Pi(x,y)=\Pi(x)\Pi(y) & \delta(x,y)=\delta(x)\delta(y) \\ \mathrm{III}(x,y)=\mathrm{III}(x)\,\mathrm{III}(y) & \mathrm{sinc}\,(x,y)=\mathrm{sinc}\,(x)\,\mathrm{sinc}\,(y). \end{array} \tag{15.20}$$

15.1.4 *Fourier transform in signal and system*

Definition 15.2. $f \in L_2(\mathbb{R})$ $(L_2(\mathbb{R}^n))$. The Fourier transform of f is

$$\begin{cases} \text{in } \ \mathbb{R}: \ \hat{f}(\xi) = \displaystyle\int_{-\infty}^{\infty} e^{-2\pi i x\xi} f(x)\,dx, \ \ f(x) = \int_{-\infty}^{\infty} e^{2\pi i x\xi} \hat{f}(\xi)\,d\xi \\ \text{in } \ \mathbb{R}^n: \hat{f}(\boldsymbol{\xi}) = \displaystyle\int_{\mathbb{R}^n} e^{-2\pi i\,\boldsymbol{x}\cdot\boldsymbol{\xi}} f(\boldsymbol{x})\,d\boldsymbol{x}, \ f(\boldsymbol{x}) = \int_{\mathbb{R}^n} e^{2\pi i\,\boldsymbol{x}\cdot\boldsymbol{\xi}} \hat{f}(\boldsymbol{\xi})\,d\boldsymbol{\xi}. \end{cases}$$

Note that the lack of a coefficient $\left(\frac{1}{2\pi}\right)^n$ in the transforms here is compensated in the exponent.

The Fourier transform is denoted (in the same way for $\mathbb{R}^n$) by

$$\hat{f}(\xi) = (\mathcal{F}f)(\xi) \quad \text{or} \quad f(x) \supset \hat{f}(\xi). \tag{15.21}$$

For $\quad \hat{f}(\xi) := \displaystyle\int_{-\infty}^{\infty} e^{-2\pi i x\xi} f(x)\,dx \quad$ yields

Function	Fourier transform
$\Pi(x) := \begin{cases} 1 \text{ if } & \lvert x\rvert < 1/2 \\ 0 \text{ if } & \lvert x\rvert > 1/2 \end{cases}$	$\mathrm{sinc}\ \xi := \dfrac{\sin(\pi\xi)}{\pi\xi}$
$\mathrm{sinc}\ x$	$\Pi(\xi)$
$\Lambda(x) := \begin{cases} 1 - \lvert x\rvert \text{ if } & \lvert x\rvert \le 1 \\ 0 \qquad\ \text{ if } & \lvert x\rvert > 1 \end{cases}$	$\mathrm{sinc}^2(\xi)$
$\mathrm{sinc}^2 x$	$\Lambda(\xi)$

(15.22)

Basic properties (essentially as in (15.3)).

- Linearity:

$$\begin{aligned} f+g &\supset \hat{f}+\hat{g} \\ \alpha f &\supset \alpha \hat{f}, \qquad (\alpha \in \mathbb{R},\ \text{or}\ \mathbb{C}). \end{aligned}$$

- Scaling:

$$f(x) \supset \hat{f}(\xi) \quad \Longleftrightarrow \quad \frac{1}{a} f(\frac{x}{a}) \supset \hat{f}(a\xi), \qquad (a>0).$$

- Derivation:

$$D\,f(x) \supset (2\pi i\xi)\hat{f}(\xi), \qquad\qquad -2\pi i(\cdot) f \supset (D\,\hat{f})(\xi).$$

- Translation:

$$\tau_a f := f(x-a) \supset e^{-2\pi a\xi}\hat{f}(\xi), \qquad\qquad e^{2\pi a x} f(x) \supset \tau_a \hat{f}(\xi).$$

- Convolution:

$$(f * g)(x) \supset \hat{f}(\xi)\hat{g}(\xi).$$

If f is sufficiently regular/smooth, then $\hat{f}$ is also regular. The notions mentioned above, will be specified in the definition of the Schwartz class $\mathcal{S}$, page 378.

Fourier transform in L_2:

$$L_2(\mathbb{R}) := \Big\{ f : f \text{ measurable and } \int_{\mathbb{R}} |f(x)|^2\,dx < \infty \Big\}.$$

Scalar product:

$$\langle f, g\rangle = \int_{\mathbb{R}} f(x)\overline{g(x)}\,dx.$$

Parseval's:

$$\int_{\mathbb{R}} |f(x)|^2\,dx = \int_{\mathbb{R}} |\hat{f}(\xi)|^2\,d\xi.$$

Properties of convolution

$$\begin{array}{ll} \text{Commutative} & f * g = g * f, \\ \text{Associative} & f * (g * h) = (f * g) * h, \\ \text{Distributive} & f * (g + h) = f * g + f * h. \end{array}$$

Fourier's inversion formula

$$\hat{f}(\xi) = \int_{-\infty}^{\infty} e^{-2\pi i x \xi} f(x)\, dx \quad \Longleftrightarrow \quad f(x) = \int_{-\infty}^{\infty} e^{-2\pi i x \xi} \hat{f}(\xi)\, d\xi. \tag{15.23}$$

Fixed point

$$\mathcal{F}\left(e^{-\pi x^2}\right) = e^{-\pi \xi^2}.$$

$$\text{Scaling by } a > 0 \quad \Longrightarrow \quad e^{-\pi (x/a)^2} \supset a e^{-\pi (a\xi)^2}.$$

$$f \in L_p, \qquad (1 \leq p \leq 2) \quad \Longrightarrow \quad \hat{f} \in L_q, \quad \text{for } q: \quad \frac{1}{p} + \frac{1}{q} = 1.$$

Some function classes having Fourier transform

- $L_2(\mathbb{R}) = \left\{ f : \left(\int_{\mathbb{R}} |f(x)|^2\, dx \right)^{1/2} < \infty \right\}$,
- $\mathcal{S}$, "smooth" rapidly decreasing functions,
- $L_1(\mathbb{R}) = \left\{ f : \int_{\mathbb{R}} |f(x)|\, dx < \infty \right\}$.

Discrete Fourier transform (DFT) Periodic sequence:
Let S^N be the set of N-periodic complex-valued sequences $\{x(n)\}_{n \in \mathbb{Z}}$:

$$x(n + N) = x(n), \quad n \in \mathbb{Z}.$$

The impulse on $k(\text{mod } N)$: $\quad e_k(n) = \begin{cases} 1, n = k + mN, & m \in \mathbb{Z}, \\ 0, \text{otherwise.} \end{cases}$

For $x \in S^N$ the discrete Fourier transform for the function x is defined as

$$X(\mu) = \frac{1}{N}\sum_{n=0}^{N-1} x(n)W^{-\mu n},\ W = e^{2i\pi/N},\ \mu \in \mathbb{Z},$$

with the inverse discrete Fourier transform as

$$x(n) = \sum_{\mu=0}^{N-1} X(\mu)W^{\mu n}, \quad n \in \mathbb{Z}.$$

Fast Fourier transform (FFT)
FFT is an algorithm which reduces the number of operations in DFT for 2^m. The number of operations in DFT $\approx N^2$ and in FFT $\approx N \cdot \log_2 N$.

Let $W = e^{2\pi i/N}$, if $N = 2^m$, then $W^2 = e^{2\pi i/2^{m-1}}$. The idea of FFT is to divide DFT over odd and even indices n as follows:

$$\begin{aligned} X(\mu) &= \frac{1}{N}\sum_{n=0}^{2^m-1} x(n)W^{-\mu n} = \sum_{\text{odd}\, n} + \sum_{\text{even}\, n} \\ &= \frac{1}{N}W^{-\mu}\sum_{k=0}^{2^{m-1}-1} x(2k+1)(W^2)^{-\mu k} + \frac{1}{N}\sum_{k=0}^{2^{m-1}-1} x(2k)(W^2)^{-\mu k}. \end{aligned} \tag{15.24}$$

This procedure (of taking half) continues until it reaches the final step having sum of DFT with $N = 2$.

15.1.5 Table of discrete Fourier transform

$x(n), \quad x \in S^N$	$X(\mu), \quad \text{DFT for } x(n)$
$x(n), \quad x \in S^N$	$X(\mu) = \frac{1}{N}\sum_{n=0}^{N-1} x(n)W^{-\mu n}, \quad \mu \in \mathbb{Z}$
$\sum_{\mu=0}^{N-1} X(\mu)W^{\mu n}$	$X(\mu)$
$ax(n) + by(n)$	$aX(\mu) + bY(\mu)$
$\frac{1}{N}\sum_{k=0}^{N-1} x(n-k)y(k)$	$X(\mu)Y(\mu)$
$x(n-c)$	$W^{-\mu c}X(\mu)$
$W^{\nu n}x(n)$	$X(\nu - n)$
$X(n)$	$\frac{1}{N}x(-\mu)$
$e_k(n)$	$\frac{1}{N}W^{-\mu k} = \frac{1}{N}e^{-2i\pi k\mu/N}$
$e_0(n)$	$\frac{1}{N}$
$W^{\nu n} = e^{2i\pi\nu n/N}$	$e_\nu(\mu)$
1	$e_0(\mu)$
$\sin\frac{2\pi\nu n}{N}$	$\frac{1}{2i}(e_\nu(\mu) - e_{-\nu}(\mu))$
$\cos\frac{2\pi\nu n}{N}$	$\frac{1}{2i}(e_\nu(\mu) + e_{-\nu}(\mu))$
$\sin\frac{\pi n}{N}, \quad n = 0, 1, \ldots, N-1$	$\frac{1}{N}\frac{\sin\pi/N}{\cos(2\mu\pi/N) - \sin(\pi/N)}$
Inverse formula	$x(n) = \sum_{n=0}^{N-1} X(\mu)W^{\mu n}, \quad n \in \mathbb{Z}$
Plancherel's formula	$\frac{1}{N}\sum_{n=0}^{N-1} x(n)\overline{y(n)} = \sum_{n=0}^{N-1} X(\mu)\overline{Y(\mu)}$
Parseval's formula	$\frac{1}{N}\sum_{n=0}^{N-1} \lvert x(n)\rvert^2 = \sum_{n=0}^{N-1} \lvert X(\mu)\rvert^2$

(15.25)

Definition 15.3. The discrete Fourier transform of $(u_1, u_2, \ldots, u_n)$ is given by

$$v_m = n^{a(b-2)/2} \sum_{k=1}^{n} u_k e^{2\pi i(k-1)(m-1)/n}, \quad m = 1, \ldots, m,$$

and the inverse Fourier transform is given by

$$u_k = n^{b(a-2)/2} \sum_{m=1}^{n} v_m e^{-2\pi i(k-1)(m-1)/n}, \tag{15.26}$$

where $(a, b) = (0, 1), (1, 0)$, or $(1, 1)$, i.e., the factors $n^{a(b-2)/2}$ and $n^{b(a-2)/2}$ in front of the respective sums in (15.26) are given by

	$(a,b) = (0,1)$	$(a,b) = (1,0)$	$(a,b) = (1,1)$
Fourier transform	$n^0 = 1$	$n^{-1} = 1/n$	$n^{-1/2} = 1/\sqrt{n}$
Inverse Fourier transform	$n^{-1} = 1/n$	$n^0 = 1$	$n^{-1/2} = 1/\sqrt{n}$

The discrete cosine transform of $a_1, a_2, \ldots, a_{n+1}$ is given by

$$b_m = \sqrt{\frac{2}{n}} \left[\frac{a_1}{2} + \frac{(-1)^{m-1}}{2} a_{n+1} + \sum_{k=2}^{n} \cos\left(\frac{(k-1)(m-1)\pi}{n} \right) a_k \right]$$
$$m = 1, 2, \ldots, n+1. \tag{15.27}$$

The discrete sine transforms of $a_1, a_2, \ldots, a_{n-1}$ are given by

$$b_m = \sqrt{\frac{2}{n}} \sum_{k=1}^{n-1} \sin\left(\frac{km\pi}{n} \right) a_k, \quad m = 1, 2, \ldots, n-1. \tag{15.28}$$

Remarks. The basic values of (a, b) are $(1, 1)$, i.e., $n^{a(b-2)/2} = n^{b(a-2)/2} = n^{-1/2}$. In computations, the factor $1/n$ is used. For transform- and signal-treatment, the factor 1 is used.

15.2 The $j\omega$-Method

In alternating current (AC) in electrical engineering, the imaginary unit commonly is denoted by j and not i, since i is reserved for AC.

Definition 15.4.

$$u(t) = C\,\sin(\omega\,t + \alpha), \tag{15.29}$$

has the complex pointer

$$U = Ce^{j\,\alpha}.$$

This is usually denoted as

$$u(t) \longleftrightarrow U,$$

and reads "corresponds".

Table of $j\,\omega$

$$\begin{aligned}
C\,\sin(\omega\,t + \alpha) &\longleftrightarrow Ce^{j\,\alpha} \text{ by definition.}\\
u(t) &\longleftrightarrow \operatorname{Im}\left(U \cdot e^{j\omega\,t}\right)\\
A\,u(t) + B\,v(t) &\longleftrightarrow A\,U + B\,V \text{ (Linearity)}\\
\sin\omega t &\longleftrightarrow 1\\
\cos\omega t &\longleftrightarrow j\\
A\sin\omega t + B\cos\omega t &\longleftrightarrow A + B\,j\\
\frac{d}{dt}\,u(t) &\longleftrightarrow j\omega U\\
\textstyle\int u(t)\,dt &\longleftrightarrow \frac{1}{j\omega}\,U.
\end{aligned} \tag{15.30}$$

15.3 The z-Transform

Definition 15.5. Let $(x_0, x_1, x_2, \ldots) = (x_k)_{k=0}^{\infty}$ be a real sequence. The z-transform of this sequence is given by

$$X(z) = x_0 + x_1 z^{-1} + x_2 z^{-2} + \cdots = \sum_{k=0}^{\infty} x_k z^{-k}. \tag{15.31}$$

One denotes the sequence $(x_k)_{k=0}^{\infty}$ by $(\ldots, x_{-2}, x_{-1}, \underline{x_0}, x_1, x_2, \ldots)$, where $x_{-1} = x_{-2} = \ldots = 0$. The underlined x_0 means that x_0 is in "position 0". For $x(n) \neq 0$ for some $n < 0$, the following notation is used

$$\{x(n)\}_{n=-\infty}^{\infty} \Longrightarrow \quad X^{\star}(z) = \sum_{n=-\infty}^{\infty} x(n)z^{-n}.$$

The inverse is given by

$$x(n) = \frac{1}{2\pi i}\int_{|z|=r} X(z)z^{n-1}\,dz, \quad \text{resp.}$$

$$x(n) = \frac{1}{2\pi i}\int_{|z|=r} X^{\star}(z)z^{n-1}\,dz,$$

where r is assumed to be large enough.

In short, one writes $(x_k)_{k=0}^{\infty}$ as (x_k).

Definition 15.6. Three important sequences

$$(\ldots,0,0,0,\underline{1},1,1,\ldots) = (\theta_k) \text{ (the unit step)},$$

$$(\ldots,0,0,0,\underline{1},0,0,\ldots) = (\delta_k) \text{ (the unit pulse)}, \tag{15.32}$$

$$(\ldots,0,0,0,\underline{0},1,2,3,\ldots) = (r_k) \text{ (the ramp function)}.$$

Definition 15.7.

(i) Consider the sequence $(x_k)_{k=0}^{\infty}$. The sequence $\left(x_k a^k\right)_{k=0}^{\infty}$ is called damped with damping a, if $|a| < 1$.

(ii) The convolution of two sequences $(x_k)_{k=0}^{\infty}$ and $(y_k)_{k=0}^{\infty}$ is again a sequence with element in position k as

$$(x_k) * (y_k)\,(m) = \left(\sum_{k=0}^{m} x_{m-k}\cdot y_k\right). \tag{15.33}$$

For n$\in \mathbb{Z}$, the discrete Heaviside function is defined as

$$\theta(n) = \begin{cases} 1, & n \geq 0, \\ 0, & n < 0. \end{cases} \tag{15.34}$$

It yields

$$x(n-k)\theta(n-k) = \begin{cases} x(n-k), & n \geq k, \\ 0, & n \leq k-1. \end{cases}$$

Table of z-transform

(x_k)	$X(z)$
$a\,(x_k) + b\,(y_k)$	$a\,X\,(z) + b\,Y\,(z)$
$\left(a^k x_k\right)$	$X(z/a)$
(kx_k)	$-zX'(z)$
$(x_k) * (y_k)$	$X(z) \cdot Y(z)$
$(x_k \sigma_{k-m})$	$z^{-m} X(z), \quad m \geq 0$
$(x_k \sigma_{k+m})$	$z^m X(z) - \sum_{r=0}^{m-1} x_r z^{m-r}$
(θ_k)	$\frac{z}{z-1}$
(δ_k)	1
(r_k)	$\frac{z}{(z-1)^2}$
(a^k)	$\frac{z}{z-a}$
$(a^k \sin k\theta)$	$\frac{za \sin\theta}{z^2 - 2za\cos\theta + a^2}$
$(a^k \cos k\theta)$	$\frac{z(z - a\cos\theta)}{z^2 - 2za\cos\theta + a^2}$

(15.35)

Table of z-transform

$x(n), \quad n \geq 0$	$X(z)$
$x(n)$	$X(x) = \sum_{n=0}^{\infty} x(n) z^{-n}$
$ax(n) + by(n)$	$aX(x) + bY(x)$
$x(n-k)\theta(n-k)$	$z^{-k} X(z)$
$x(n+k), \quad k > 0$	$z^k X(z) - z^k x(0)$ $-z^{k-1} x(1) - \ldots - zx(k-1)$
$a^n x(n)$	$X\left(\frac{z}{a}\right)$
$(-1)^k \frac{(n-1)!}{(n-k-1)!} \theta(n)$	$\frac{d^k}{dz^k} X(z)$
$nx(n)$	$-zX'(z)$
$\sum_{k=0}^{n} x(n-k) y(k) = \sum_{k=0}^{n} x(k) y(n-k)$	$X(z)Y(z)$
$\delta_k(n) = \begin{cases} 1, & n = k \\ 0, & n \neq k \end{cases}$	$\frac{1}{z^k}$
a^n	$\frac{z}{z-a}$ (a is a complex number $\neq 0$)
$x(n), \quad -\infty < n < \infty$	$X^\star(z) = \sum_{n=-\infty}^{\infty} x(n) z^{-n}$
$x(n+k), \quad k$ arbitrary	$x^k X^\star(z)$

(15.36)

Table of z-transform, continuation

$(x*y)(n) = \sum_{k=-\infty}^{\infty} x(n-k)y(k)$	$X^{\star}(z)Y^{\star}(z)$
$x(n), \quad n \geq 0$	$X(z)$
$a^{n-k}\theta(n-k)$	$\frac{z^{1-k}}{z-a}, \quad k=0,1,\ldots$
na^n	$\frac{az}{(z-a)^2}$
n^2a^n	$\frac{az(z+a)}{(z-a)^3}$
$\frac{a^n}{n!}$	$e^{a/z}$
$\frac{a^n}{n}\theta(n-1)$	$\ln \frac{z}{z-a}$
$\binom{n}{m}a^{n-m}\theta(n-m)$	$\frac{z}{(z-a)^{m+1}} \quad (m \geq 0, \text{ integer})$
$\binom{n+k}{m}a^{n+k-m}\theta(n+k-m)$	$\frac{z^{k+1}}{(z-a)^{m+1}} \quad (m \geq 0, \ k \leq m)$
$\frac{a^{n+k}-b^{n+k}}{a-b}\theta(n+k-1)$	$\frac{z^{k+1}}{(z-a)(z-b)}, \quad (k=1,0,-1,\ldots)$
$a^{n-1}\sin\frac{n\pi}{2}$	$\frac{z}{z^2+a^2}$
$a^{n+k-1}\sin\frac{(n+k)\pi}{2}\theta(n+k-1)$	$\frac{z^{k+1}}{z^2+a^2}, \quad (k=1,0,-1,\ldots)$

(15.37)

Table of z-transform, continuation

$x(n), \quad n \geq 0$	$X(z)$
$b > 0,\ r = \sqrt{a^2 + b^2}$	$\varphi = \arctan(b/a), \quad a > 0$ $\varphi = \pi + \arctan(b/a), \quad a < 0$
$\frac{1}{b} r^n \sin n\varphi$	$\frac{z}{(z-a)^2 + b^2}$
$\frac{1}{b} r^{n+k} \sin(n+k)\varphi\theta(n+k-1)$	$\frac{z^{k+1}}{(z-a)^2 + b^2}, \ k = 1, 0, -1, \ldots$
$a^n \sin n\varphi$	$\frac{z(z - \cos\varphi)}{z^2 - 2az\cos\varphi + a^2}$
$a^n \cos n\varphi$	$\frac{az\sin\varphi}{z^2 - 2az\cos\varphi + a^2}$

(15.38)

15.4 The Laplace Transform

Definition 15.8. Assume that $s \in \mathbb{C}$. The one-sided Laplace transform $\mathcal{L}(f) = F$ of a function f is given by

$$F(s) = \mathcal{L}(f)(s) = \int_0^\infty e^{-st} f(t)dt, \tag{15.39}$$

as far as the improper integral exists.

Remarks. For generalized functions (distributions), it is necessary that the lower bound is replaced by 0_-, i.e.,

$$\mathcal{L}(f)(s) = \int_{0_-}^\infty e^{-st} f(t)dt.$$

This is pointed out only when it is significant.

Theorem 15.2.

$$\mathcal{L}(af(t) + bg(t)) = a\mathcal{L}(f(t)) + b\mathcal{L}(g(t)) \ (Linearity)$$

$$\mathcal{L}(e^{-at}f(t)) = F(s+a) \ (damping)$$

$$\mathcal{L}(f(t/a)) = aF(as) \ (time\ scaling)$$

$$\mathcal{L}(f^{(n)}(t)) = s^n F(s) - \sum_{k=0}^{n-1} f^{(k)}(0)$$
$$n = 1, 2, \ldots$$

$$\mathcal{L}\left(\int_0^t f(x)dx\right) = \frac{F(s)}{s}$$

$$\mathcal{L}\left(\int_0^t f(x)g(t-x)dx\right) \equiv \mathcal{L}((f * g)(t)) = F(s)G(s). \tag{15.40}$$

Theorem 15.3.

$$\frac{d^n}{ds^n}F(s) = \int_{0-}^{\infty} e^{-st}(-t)^n f(t)dt$$
$$i.e.,$$
$$\mathcal{L}(t^n f(t)) = (-1)^n \frac{d^n}{ds^n}F(s), \quad n = 0, 1, 2, \ldots$$

$$\mathcal{L}(f(t)/t) = \int_s^{\infty} F(s)ds \tag{15.41}$$

$$\mathcal{L}f(t) = \frac{1}{1 - e^{-iT}} \int_0^T e^{-st} f(t)dt.$$
$$(Periodic\ function)$$

The Laplace transform of some elementary functions

$f(t)$	$F(s) = L(f(t))$
1	$\frac{1}{s}, \quad s > 0$
$\frac{t^n}{n!}$	$s^{-(n+1)}, \quad s > 0, \quad n = 0, 1, \ldots$
e^{at}	$\frac{1}{s-a}, \quad s > a$
$\sin bt$	$\frac{b}{s^2+b^2}, \quad s > 0$
$\cos bt$	$\frac{s}{s^2+b^2}, \quad s > 0$
$\cosh bt$	$\frac{s}{s^2-b^2}, \quad s > \lvert b \rvert$
$\sinh bt$	$\frac{b}{s^2-b^2}, \quad s > \lvert b \rvert$
$\delta^{(n)}(t)$	$s^n, \quad s > 0$
$\delta(t-a)$	$e^{-as}\, H(a)$
$t^n \sqrt{t}$	$\frac{(2n+1)!!\,\sqrt{\pi}}{2^{n+1}\, s^{(2n+3)/2}}, \quad s > 0, \quad n = 0, 1, \ldots$

(15.42)

Some useful Laplace transforms

$f(t)$	$F(s) = L(f(t))$
$\theta(t-T)$	$\dfrac{e^{-Ts}}{s}$
$f(t-T)\theta(t-T)$	$e^{-Ts}F(s)$
$f'(t)$	$sF(s) - f(0-)$
$f''(t)$	$s^2F(s) - f(0-) - f'(0-)$
$t^a, \quad (Re\, a > -1)$	$\dfrac{\Gamma(a+1)}{s^{a+1}}$
$\sqrt{t}$	$\dfrac{1}{2s}\sqrt{\dfrac{\pi}{s}}$
$\dfrac{1}{\sqrt{t}}$	$\sqrt{\dfrac{\pi}{s}}$
$\dfrac{1}{\sqrt{\pi t}}e^{-a^2/4t}, \quad a \geq 0$	$\dfrac{e^{-a\sqrt{s}}}{\sqrt{s}}$
$\dfrac{a}{\sqrt{4\pi t^3}}e^{-a^2/4t}, \quad a > 0$	$e^{-a\sqrt{s}}$
$\operatorname{erfc}\left(\dfrac{a}{\sqrt{4t}}\right), \quad a > 0$	$\dfrac{1}{s}\, e^{-a\sqrt{s}}$

(15.43)

Laplace transforms for some special functions

$f(t)$	$F(s) = L(f(t))$
$f(t) = \begin{cases} 0, & 0 \le t < T \\ 1, & t \ge T \end{cases}$	$\dfrac{e^{-Ts}}{s}$
$\theta(t-a) - \theta(t-b)$	$\dfrac{e^{-as} - e^{-bs}}{s}$
$f(t) = \begin{cases} 0, & 0 \le t \le T \\ c(t-T), & t \ge T \end{cases}$	$c\dfrac{e^{-Ts}}{s^2}$
$f(t) = n, \quad (n-1)T < t < nT, \quad n = 1, 2, \ldots$	$\dfrac{1}{s(1-e^{-Ts})}$
$f(t) = \begin{cases} ct, & 0 \le t < T \\ cT, & t \ge T \end{cases}$	$c\dfrac{(1-e^{-Ts})}{s^2}$
$\begin{cases} 1, & 0 \le t < 2T \\ -1, & 2T < t < 4T \end{cases} \quad f(t+4T) = f(t)$	$\dfrac{\tanh(Ts)}{s}$
$\begin{cases} 1, & 0 \le t < T \\ 0, & T < t < 2T \end{cases} \quad f(t+2T) = f(t)$	$\dfrac{1}{s(1+e^{-Ts})}$
$\dfrac{1}{2T}\begin{cases} t, & 0 \le t < 2T \\ 4T - t, & 2T < t < 4T \end{cases} \quad f(t+4T) = f(t)$	$\dfrac{\tanh(Ts)}{2Ts^2}$
$f(t) = \lvert \sin Tt \rvert$	$\dfrac{T}{s^2+T^2} \coth \dfrac{\pi s}{2T}$
$f(t) = \dfrac{t}{T}, \quad f(t+T) = f(t)$	$\dfrac{1}{Ts^2}\left(1 + \dfrac{s}{1-e^{Ts}}\right)$

(15.44)

The inverse Laplace transform $\mathcal{L}^{-1}$

$$\mathcal{L}^{-1}(F(s)) = \lim_{R\to\infty} \frac{1}{2\pi i}\int_{c-iR}^{c+iR} F(s)e^{st}dt = \int_{c-i\infty}^{c+i\infty} F(s)e^{st}ds, \tag{15.45}$$

where $c \in \mathbb{R}$ is chosen so that all singularities of $F(s)$ are to the left of the line $\text{Re}(z) = c$ in the complex plane.

Remark. In principle, the integral is a curve integral in the complex plane:

$$\mathcal{L}^{-1}(F(s)) = \lim_{R\to\infty} \frac{1}{2\pi i}\int_{\gamma} F(s)e^{st}ds, \tag{15.46}$$

where γ is the curve between $c - iy$ and $c + iy$ as in Figure 15.1 and Γ is the circular arc. The closed curve $\gamma + \Gamma$ is oriented counterclockwise. Computing of (15.45) can be performed as the following boundary-limit:

$$\lim_{R\to\infty}\left[\frac{1}{2\pi i}\oint_{\Gamma+\gamma} F(s)e^{st}ds - \frac{1}{2\pi i}\int_{\gamma} F(s)e^{st}ds\right]. \tag{15.47}$$

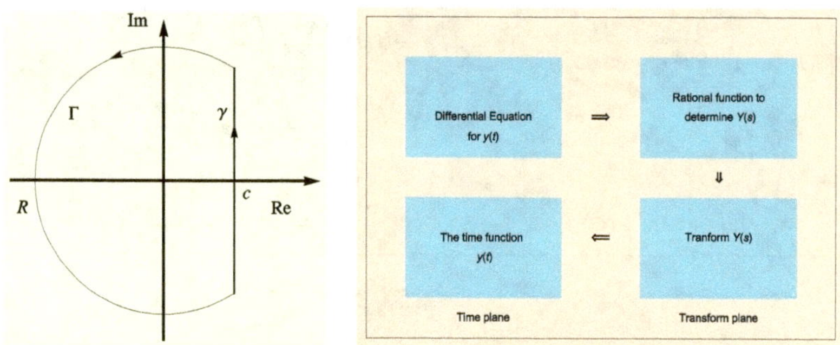

Figure 15.1: Left figure: Curves γ and Γ. Right figure: Steps to solve a DE.

Some inverse Laplace transforms

$F(s) = L(f(t))$	$f(t)$
$\frac{1}{s}e^{-k/s}$	$J_0(2\sqrt{kt})$
$\frac{1}{\sqrt{s}}e^{-k/s}$	$\frac{1}{\sqrt{\pi t}}\cos 2\sqrt{kt}$
$\frac{1}{\sqrt{s}}e^{k/s}$	$\frac{1}{\sqrt{\pi t}}\cosh 2\sqrt{kt}$
$\frac{1}{s^{3/2}}e^{-k/s}$	$\frac{1}{\sqrt{\pi t}}\sin 2\sqrt{kt}$
$\frac{1}{s^{3/2}}e^{k/s}$	$\frac{1}{\sqrt{\pi t}}\sinh 2\sqrt{kt}$
$\frac{1}{s^{\mu}}e^{-k/s}, \quad (\mu > 0)$	$\left(\frac{t}{k}\right)^{(\mu-1)/2} J_{\mu-1}(2\sqrt{kt})$
$\frac{1}{s^{\mu}}e^{k/s}, \quad (\mu > 0)$	$\left(\frac{t}{k}\right)^{(\mu-1)/2} I_{\mu-1}(2\sqrt{kt})$
$\operatorname{erf}\left(\frac{k}{\sqrt{s}}\right)$	$\frac{1}{\pi t}\sin(2k\sqrt{t})$
$\arctan\frac{k}{s}$	$\frac{1}{t}\sin kt$
$\frac{1}{s}\arctan\frac{k}{s}$	$\operatorname{Si}(kt) := \int_0^{kt}\frac{\sin x}{x}\,dx$
$\ln\frac{s-a}{s-b}$	$\frac{1}{t}(e^{bt} - e^{at})$
$\ln\frac{s^2+a^2}{s^2}, \ \log\frac{s^2-a^2}{s^2}$	$\frac{2}{t}(1-\cos at), \ \frac{2}{t}(1-\cosh at)$

(15.48)

15.5 Distributions

The classes $\mathcal{S}$ and $\mathcal{S}'$[1]

Definition 15.9. The class of test functions $\mathcal{S}$ is the class of complex valued functions $f : \mathbb{R} \longrightarrow \mathbb{C}$ satisfying

$$\sup_{x\in\mathbb{R}} \Big| |x|^\alpha D^\beta f(x) \Big| < \infty, \qquad \text{for any choice of } \alpha \geq 0 \text{ and } \beta \geq 0.$$

Properties of $\mathcal{S}$

$$f \in \mathcal{S}, \text{ and } g(x) = x^\alpha D^\beta f(x), \quad (\alpha, \beta \in \mathbb{Z}^+) \quad \Longrightarrow \quad g \in \mathcal{S}.$$
$$f \in \mathcal{S} \quad \Longrightarrow \quad \hat{f} \in \mathcal{S}.$$

$\mathcal{S}$ is a linear space:

$$\varphi_1, \varphi_2 \in \mathcal{S}, \ \alpha_1, \alpha_2 \in \mathbb{C} \quad \Longrightarrow \quad \alpha_1\varphi_1 + \alpha_2\varphi_2 \in \mathcal{S}, \quad (\varphi \equiv 0 \in \mathcal{S}).$$

The class $\mathcal{S}'$

Notation:

$$\langle \varphi_1 \,,\, \varphi_2 \rangle = \int_{\mathbb{R}} \varphi_1(x)\varphi_2(x)\,dx.$$

(This is a inner product if φ_1 and φ_2 are real-valued functions.)

Definition 15.10. A linear map $T : \mathcal{S} \longrightarrow \mathbb{C}$ that fulfills

$$T(\varphi_1 + \varphi_2) = T\varphi_1 + T\varphi_2, \quad \varphi_1, \varphi_2 \in \mathcal{S}$$

$$T(\alpha\varphi_1) = \alpha T\varphi_1$$

is a member of the Schwartz class $\mathcal{S}'$.

---

[1] Distributions (generalized functions) were introduced by **Laurent Schwartz** (~1940) and **Sergei Sobolev** (~1935) to give a rigorous theory of mathematical objects like **Dirac's** δ-function.

Definition 15.11. A tempered distribution $T : \mathcal{S} \longrightarrow \mathbb{C}$ is a continuous linear map: $\mathcal{S} \longrightarrow \mathbb{C}$, if for each sequence $\{\varphi_n\}_{n=1}^{\infty}$, $\varphi_n \in \mathcal{S}$. $\alpha, \beta \in \mathbb{Z}^+$,

$$\lim_{n\to\infty} \sup_{x\in\mathbb{R}} \left| |x|^\alpha \, D^\beta \varphi_n(x) \right| = 0.$$

This is denoted by

$$\lim_{n\to\infty} T(\varphi_n) = 0.$$

Example 15.2. Take f so that $f(x)/(1+x^2)^\alpha$ is integrable for some $\alpha \geq 0$, and let

$$T(\varphi) = \langle f, \varphi \rangle = \int_{-\infty}^{\infty} f(x)\, \varphi(x)\, dx.$$

Then T is a tempered distribution.

Observe that we can identify f with T (and write $f(\varphi)$ for $T(\varphi)$). This should not be confused with $f(x)$ where one considers the function $f : \mathbb{R} \longrightarrow \mathbb{C}$.

Definition 15.12.

Convergence in $\mathcal{S}$

$$\varphi_n \longrightarrow \varphi \text{ i } \mathcal{S} \quad \Longleftrightarrow \quad \lim_{n\to\infty} \sup_{x\in\mathbb{R}} \left| |x|^\alpha D^\beta \Big(\varphi_n(x) - \varphi(x) \Big) \right| = 0. \quad (\star)$$

Topology in $\mathcal{S}$: The family of limits in $(\star)$ (indexed by α and β) defines a topology in $\mathcal{S}$.
Let X and Y be two topological vector spaces and $f : X \longrightarrow Y$. Then, f is said to be continuous if

$$f(x_n) \longrightarrow f(x), \quad \text{as} \quad x_n \longrightarrow x \quad \text{in } X.$$

Theorem 15.4. *If $f \in C(\mathbb{R})$, $g \in C(\mathbb{R})$, and $f = g$ i $\mathcal{S}'$, then, as functions, $f(x) = g(x)$ for all $x \in \mathbb{R}$.*

$$\begin{aligned} T : \mathcal{S} &\longrightarrow \mathbb{C} \\ \varphi &\longmapsto \varphi(a) \ (a \in \mathbb{R}, \quad : \varphi \textit{ is evaluated at the point } a). \end{aligned}$$

Here is an example of the Dirac δ-function in a : $\delta_a(\varphi) = \varphi(a)$.

Example 15.3 (Evaluation).

(i) T is a linear map.
(ii) Let $\varphi_n \in \mathcal{S}$, $\varphi_n \longrightarrow 0$ in $\mathcal{S}$ and thus

$$\sup_{x\in\mathbb{R}} \left| x^\alpha D^\beta \varphi_n(x) \right| \longrightarrow 0, \qquad \text{as } n \longrightarrow \infty.$$

In particular, if $\varphi_n(a) \longrightarrow 0$ as $n \longrightarrow \infty$, then $T(\varphi_n) = \varphi_n(a) \longrightarrow 0$.

Hence, T is continuous.

(i) and (ii) imply that T is a tempered distribution.

Note that in this case it is only necessary to evaluate φ at 0, but none of its derivatives.

Example 15.4. Let $f_n(x) = \sqrt{n}e^{-n\pi x^2}$, $n = 1, 2, \ldots$. Then, $f_n \in \mathcal{S}'$ and for all $\varphi \in \mathcal{S}$, we have

$$f_n(\varphi) = \langle f_n, \varphi\rangle \longrightarrow \varphi(0) = \delta_0(\varphi), \quad \text{as } n \longrightarrow \infty.$$

Then, we say that $f_n \longrightarrow \delta_0$ i $\mathcal{S}'$.

Example 15.5. $f(x) = e^x$ is not a tempered distribution, since it grows too fast as $x \longrightarrow \infty$.

Take for instance $\varphi(x) = e^{-\sqrt{1+x^2}} \in \mathcal{S}$, then $\langle e^{(\cdot)}, \varphi\rangle = \int_{-\infty}^{\infty} e^x e^{-\sqrt{1+x^2}}\, dx$, which is divergent.

Example 15.6. Let $\delta_n : \varphi \longmapsto \varphi(n)$, and put $\text{III} = \sum_{n=-\infty}^{\infty} \delta_n$, so that $\langle \text{III}, \varphi\rangle = \sum_{-\infty}^{\infty} \varphi(n)$. Then III is a tempered distribution.

One can prove that tempered distributions share many properties with common functions: They can be differentiated, Fourier transformed, etc.

Derivatives of distributions: Let $f \in C^1(\mathbb{R})$, and suppose that f does not grow fast (e.g., it can be bounded). Then,

$$\langle f', \varphi\rangle = \int_{-\infty}^{\infty} f'(x)\varphi(x)\, dx = -\int_{-\infty}^{\infty} f(x)\varphi'(x)\, dx,$$

which is well-defined for all $\varphi \in \mathcal{S}$.

Definition 15.13. Let T be a tempered distribution, then

$$\langle DT, \varphi\rangle = -\langle T, D\varphi\rangle, \qquad \forall \varphi \in \mathcal{S}.$$

Multiplication by function: Let $f(x) \in C(\mathbb{R})$ and $g(x) \in C^\infty(\mathbb{R})$, and assume that there is a positive integer $\alpha \in \mathbb{Z}^+$ such that $|g(x)|/(1+x^2)^\alpha$ is bounded. Then,

$$\langle fg, \varphi\rangle = \int_{-\infty}^{\infty} f(x)g(x)\varphi(x)\,dx = \langle f, g\varphi\rangle, \qquad \forall\varphi \in \mathcal{S}.$$

Definition 15.14. Let $T \in \mathcal{S}'$, and g as above. Then, we define gT according to

$$\langle gT, \varphi\rangle = \langle T, g\varphi\rangle, \qquad \forall\varphi \in \mathcal{S}.$$

This is allowed because $g\varphi \in \mathcal{S}$, if $\varphi \in \mathcal{S}$.

Translation: Let $f(x) \in C(\mathbb{R})$ and put $f_\tau(x) = f(x-\tau)$. Then,

$$\begin{aligned}\int_{\mathbb{R}} f_\tau(x)\varphi(x)\,dx &= \int_{\mathbb{R}} f(x-\tau)\varphi(x)\,dx = \{y = x-\tau\}\\ &= \int_{\mathbb{R}} f(x)\varphi(x+\tau)\,dx = \langle f, \varphi_{-\tau}\rangle.\end{aligned}$$

Definition 15.15. For $T \in \mathcal{S}'$ we define T_τ according to

$$\langle T_\tau, \varphi\rangle = \langle T, \varphi_\tau\rangle.$$

But these definitions are useful only if DT, gT, T_τ satisfy some good properties.

Theorem 15.5. *If $T \in \mathcal{S}'$, then also DT, gT, $T_\tau \in \mathcal{S}'$.*

Theorem 15.6 (The Structure theorem). Let $T \in \mathcal{S}'$, then there exist functions $f_j \in C(\mathbb{R})$ such that

$$T = \sum_j D^{\beta_j} f_j.$$

Remark. Any temperate distribution can be written as a linear combination of (distribution) derivatives of continuous functions.

Example 15.7. Let

$$f(x) = \begin{cases} x, & \text{for } x > 0, \\ 0. & \text{for } x < 0. \end{cases}$$

Then,

$$\begin{aligned} \langle D^2 f, \varphi \rangle &= \int_{\mathbb{R}} f(x) D^2 \varphi(x)\, dx = \int_0^\infty x D^2 \varphi(x)\, dx = -\int_0^\infty D^2 \varphi(x)\, dx \\ &= \varphi(0), \end{aligned}$$

i.e., $D^2 f = \delta_0$.

Example 15.8. Let $f(x) = \frac{1}{2} x^{-3/2} H(x)$, where

$$H(x) = \begin{cases} 0, & \text{for } x \leq 0, \\ 1 & \text{for } x > 0. \end{cases}$$

Note: Normally, the value of $H(0)$ makes no sense. Define a distribution T according to

$$\langle T, \varphi \rangle = \lim_{\varepsilon \to 0+} \Big(-\frac{1}{2} \int_\varepsilon^\infty x^{-3/2} \varphi(x)\, dx + \frac{1}{\varepsilon^{1/2}} \varphi(0) \Big).$$

Then, T is a tempered distribution. In fact, $T = D^2 g$, where $g(x) = 2x^{1/2} H(x) \in \mathcal{S}'$ (but as a function $g(x)$ is not in $\mathcal{S}'$, because, generally, the integral $\int_{-\infty}^\infty g(x)\varphi(x)\, dx$ is divergent). T is called the finite part of f.

Theorem 15.7 (Plancherel's formula). Let f, $\varphi \in \mathcal{S}$. Then

$$\int_{\mathbb{R}} \hat{f}(x) \varphi(x)\, dx = \int_{\mathbb{R}} f(x) \hat{\varphi}(x)\, dx.$$

Fourier transform of distributions

Definition 15.16. Since $\langle \hat{f}, \varphi \rangle = \langle f, \hat{\varphi} \rangle$, we can define the Fourier transform of $T \in \mathcal{S}'$ as

$$\langle \hat{T}, \varphi \rangle = \langle T, \hat{\varphi} \rangle, \qquad \text{for all } \varphi \in \mathcal{S}.$$

Theorem 15.8. $\hat{T} \in \mathcal{S}$. *One needs to show*

(i) $\hat{T}$ *is linear: clear, since* $\varphi \longmapsto \hat{\varphi}$ *is linear.*
(ii) *Take* $\varphi_n \longrightarrow 0$ *i* $\mathcal{S}$. *Then* $\hat{\varphi}_n \longrightarrow 0$ *in* $\mathcal{S}$, *and therefore*

$$\langle \hat{T}, \varphi_n \rangle = \langle T, \hat{\varphi}_n \rangle \longrightarrow 0, \qquad \textit{as } n \longrightarrow 0.$$

Remark. That $\hat{\varphi}_n \longrightarrow 0$ in $\mathcal{S}$ follows from analysis needed to prove that $\hat{\varphi}_n \in \mathcal{S}$.

Example 15.9. Let $\beta \in \mathbb{Z}^+$. Compute $\mathcal{F}(D^\beta \delta_0)$.
Solution:

$$\begin{aligned} \langle \mathcal{F}(D^\beta \delta_0), \varphi \rangle &= (-1)^\beta \langle \delta_0, D^\beta \hat{\varphi} \rangle = \left\langle \delta_0, \mathcal{F}\Big((2\pi i \cdot)^\beta \varphi\Big) \right\rangle \\ &= \int_{\mathbb{R}} e^{-2i\pi\xi x} (2\pi i x)^\beta \varphi(x)\, dx \Big|_{\xi=0} = \int_{\mathbb{R}} (2\pi i x)^\beta \varphi(x)\, dx. \end{aligned}$$

Corollary: One obtains

$$\mathcal{F}(D^\beta \delta_0) = (2\pi i \cdot)^\beta, \quad \text{where} \quad \beta = 0 \quad \Longrightarrow \quad \mathcal{F}(\delta_0) = 1.$$

Theorem 15.9 ($\mathcal{F}$-inversion formula for tempered distributions). *Let* $\check{\varphi}(x) = \varphi(-x)$, *where* $\varphi \in \mathcal{S}$, *and let* $\check{T}$ *be defined as*

$$\langle \check{T}, \varphi \rangle = \langle T, \check{\varphi} \rangle, \qquad \textit{for} \quad T \in \mathcal{S}'.$$

Then for all $T \in \mathcal{S}'$

$$\mathcal{F}\mathcal{F}T = \check{T}.$$

Theorem 15.10 (Properties of the $\mathcal{F}$-transform of tempered distributions).

(1) *The Fourier transform is linear:*

$$\begin{aligned} \mathcal{F}(T_1 + T_2) &= \mathcal{F}(T_1) + \mathcal{F}(T_2) \\ \mathcal{F}(\alpha T_1) &= \alpha \mathcal{F}(T_1). \end{aligned}$$

(2) *Let* $T \in \mathcal{S}'$ *and* $f \in C^\infty$ *so that for all* $\beta > 0$ *there exists an* α *such that*

$$\sup_{x \in \mathbb{R}} \Big((1 + x^2)^{-\alpha} |D^\beta f(x)| \Big) < \infty,$$

Then,

$$\begin{array}{ll} \mathcal{F}(DT) = 2\pi i(\cdot)\mathcal{F}(T) & \mathcal{F}(-2\pi i(\cdot)T) = D\hat{T} \\ \mathcal{F}(fT) = \hat{f} * \hat{T} & \mathcal{F}(t * T) = \check{f}\mathcal{F}(T) \\ \mathcal{F}(\tau_s T) = e^{-2\pi i s(\cdot)}\hat{T} & \mathcal{F}\left(e^{2\pi i(\cdot)s}T\right) = \tau_s T. \end{array}$$

Furthermore,

$D(\hat{f} * T) = (D\hat{f}) * T + \hat{f} * DT$ *and* $\varphi_1, \varphi_2 \in \mathcal{S} \Longrightarrow \mathcal{F}(\hat{\varphi}_1 * \hat{\varphi}_2) \in \mathcal{S}$.

$\hat{T} \in \mathcal{S}'$, $f \in C^\infty$ *does not grow fast* $\Longrightarrow$ $fT \in \mathcal{S}$.

Definition 15.17.

$$\hat{f} * \hat{T} = \mathcal{F}(fT) \qquad \Longrightarrow$$

$$\langle \hat{f} * \hat{T}, \varphi \rangle = \langle \mathcal{F}(fT), \varphi \rangle = \langle fT, \hat{\varphi} \rangle.$$

$$\begin{aligned} \check{f}\check{T} &= \mathcal{F}\mathcal{F}(fT) = \mathcal{F}(\hat{f} * \hat{T}) \qquad \Longleftrightarrow \\ \check{f}\hat{T} &= \mathcal{F}(\hat{f} * T). \end{aligned} \tag{15.49}$$

$$\begin{aligned} \mathcal{F}\Big(D(\hat{f} * T)\Big) &= \Big(2\pi i(\cdot)\Big)\mathcal{F}(\hat{f} * T) = \Big(2\pi i(\cdot)\Big)\check{f}\hat{T} \\ &= \mathcal{F}(D\hat{f} * T) = \mathcal{F}(\hat{f} * DT). \end{aligned}$$

In real analysis, if $f \in C^1(\mathbb{R})$ and $f' = 0$, then f is constant.
What can we say about $T \in \mathcal{S}'(\mathbb{R})$ and $DT = 0$ i $\mathcal{S}'(\mathbb{R})$?
Let $T \in \mathcal{S}'$, and assume $(\cdot)T = 0$ in $\mathcal{S}'$. Then there is a constant $a \in \mathbb{C}$ such that $T = a\delta$.

Corollary. Let $T \in \mathcal{S}'$, assume $(\cdot)T = 0$ i $\mathcal{S}'$. Then $\hat{T} = a\delta$ for some $a \in \mathbb{C}$.

Example 15.10. Let H be *Heavisides'* step function $H(x) = \begin{cases} 1, & \text{for } x > 0 \\ 0, & \text{for } x < 0. \end{cases}$ Then

$$\mathcal{F}H = \frac{1}{2\pi i(\cdot)} + \frac{1}{2}\delta,$$

where

$$\left\langle \frac{1}{\cdot}, \varphi \right\rangle = \lim_{\varepsilon \to 0} \int_{|x|>\varepsilon} \frac{1}{2\pi i x} \varphi(x)\, dx, \quad \text{(Cauchy's principle value).}$$

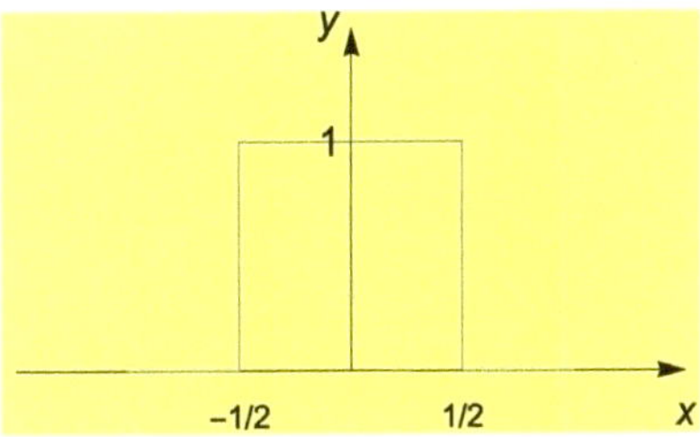

$\Pi(x)$

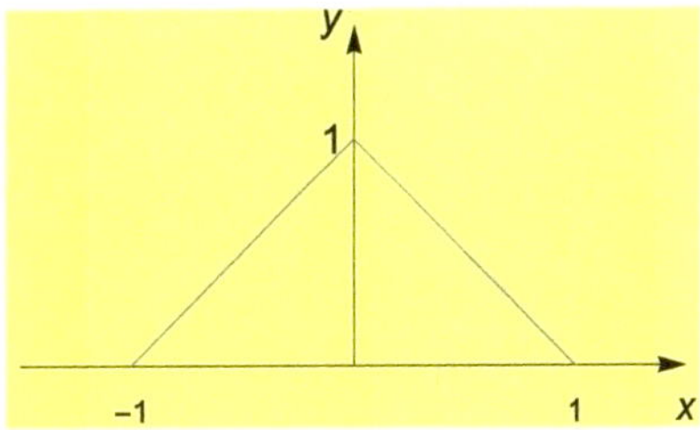

$\Lambda(x)$

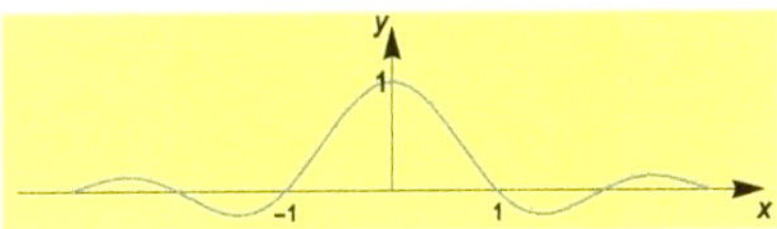

$Sinc(x) = \dfrac{\sin \pi x}{\pi x}$

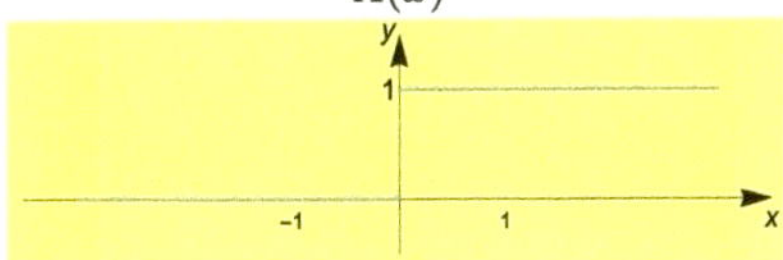

$H(x) = \theta(x)$

The Π-function:

$$\Pi(x) = \begin{cases} 1, & \text{for } \ |x| < 1/2, \\ 0, & \text{otherwise.} \end{cases}$$

The Λ function:

$$\Lambda(x) = \begin{cases} 1 + x, & \text{for } \ -1 < x < 1, \\ 1 - x, & \text{for } \ 0 < x < 1, \\ 0, & \text{otherwise.} \end{cases}$$

The Heaviside function:

$$H(x) = \begin{cases} 0, & \text{for } \ x < 0, \\ 1, & \text{for } \ x > 0. \end{cases}$$

The Signum function:

$$\operatorname{sgn}(x) = \begin{cases} -1, & \text{for } \ x < 0, \\ 0, & \qquad x = 0, \\ 1, & \text{for } \ x > 0. \end{cases}$$

The Sinc function:

$$\mathrm{Sinc}(x) = \frac{\sin \pi x}{\pi x}.$$

Special properties:

$$\mathcal{F}(\Pi) = \mathrm{Sinc}.$$

Since both Sinc and Π are even functions, so is

$$\mathcal{F}(\mathcal{F}(\Pi)) = \check{\Pi} = \Pi = \mathcal{F}(\mathrm{Sinc}) \quad \Longrightarrow \quad \Pi = \mathcal{F}(\mathrm{Sinc}(\cdot)).$$

Note that

$$1 = \Pi(0) = \int_{-\infty}^{\infty} \mathcal{F}(\Pi)\, d\xi = \int_{-\infty}^{\infty} \frac{\sin \pi\xi}{\pi\xi}\, d\xi.$$

This implies

$$\int_{-\infty}^{\infty} \mathrm{Sinc}(x)\, dx = 1.$$

Let $\varphi \in C(\mathbb{R})$ and $\tau > 0$. Then

$$\begin{aligned}\int_{-\infty}^{\infty} \frac{1}{\tau}\Pi\left(\frac{x}{\tau}\right)\varphi(x)\, dx &= \int_{-\infty}^{\infty} \Pi(x)\varphi(\tau x)\, dx \\ &= \int_{-1/2}^{1/2} \varphi(\tau x)\, dx \longrightarrow \varphi(0),\end{aligned}$$

as $\tau \to 0$.

Similarly, $\displaystyle\int_{-\infty}^{\infty} \frac{1}{\tau}\Lambda\left(\frac{x}{\tau}\right)\varphi(x)\, dx = \int_{-\infty}^{\infty} \Lambda(x)\varphi(\tau x)\, dx \longrightarrow \varphi(0),$ as $\tau \to 0$.

So one can consider the δ-function as the limit of $\frac{1}{\tau}\Pi\left(\frac{\cdot}{\tau}\right)$ or $\frac{1}{\tau}\Lambda\left(\frac{\cdot}{\tau}\right)$ as $\tau \to 0$.

Furthermore,

$$D\left(\frac{1}{\tau}\Lambda(\cdot/\tau)\right) \longrightarrow \delta' \quad \text{in} \quad \mathcal{S}'.$$

Chapter 16

Complex Analysis

16.1 Curves and Domains in the Complex Plane $\mathbb{C}$

Definition 16.1.

(i) Let $t \mapsto x(t)$ and $t \mapsto y(t)$ be continuous functions defined on an interval, $\{t : a \leq b\} = [a, b]$.

(ii) A curve

$$\gamma(t) := x(t) + i\,y(t), \quad a \leq t \leq b \text{ for } a \text{ and } b,\ a < b, \tag{16.1}$$

where $x(t)$ and $y(t)$ are real continuous functions.

A curve is closed if $\gamma(a) = \gamma(b)$.

A curve γ is a simple closed or a Jordan connected curve if $\gamma(a) = \gamma(b)$ and $\gamma(t_1) \neq \gamma(t_2)$ for all $a < t_1 \neq t_2 < b$.

A simple closed curve is positively oriented or counterclockwise oriented, as seen in the left figure on page 388.

A curve γ is *regular* if the mapping $\gamma : [a, b] \to \mathbb{C}$ is continuously differentiable and $\gamma'(t) \neq 0$ at all points $t \in [a, b]$.

A continuously differentiable curve γ with vanishing derivatives in at most a finite number of points $a \leq t_1 < t_2 < \cdots < t_n \leq b$ is called *piecewise regular*.

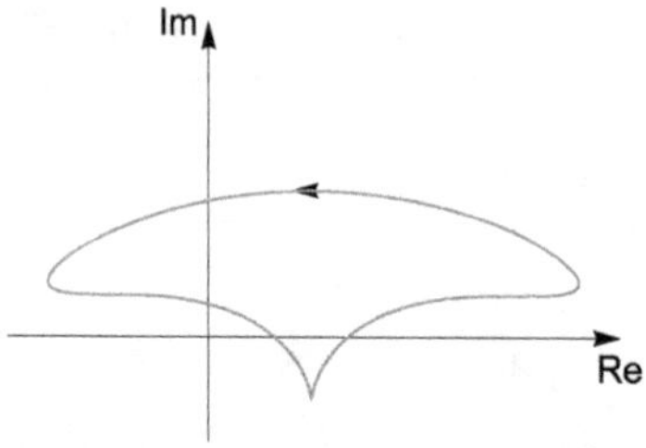

Simple closed curve, which also is counterclockwise or positively oriented.

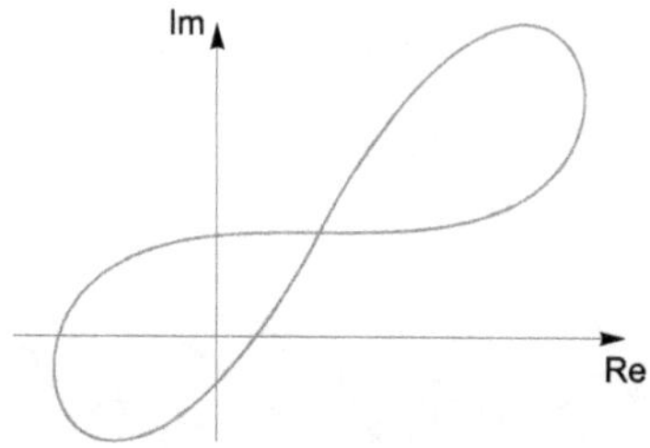

Closed but not simple closed curve.

Two curves γ_0 and γ_1 are homotopic if there exists a function $f(s,t)$, such that $f : [0,1] \times [a,b]$ is continuous in the variable s, $f(0,t) = \gamma_0(t)$ and $f(1,t) = \gamma_1(t)$.[1]

A curve γ *simply* surrounding a point z_0, if γ is homotopic with a circle defined as $\gamma_1(t) = z_0 + re^{it}$, $t \in [0, 2\pi]$ for some $r > 0$.

(iii) Domain in the complex plane $\mathbb{C}$

(a) $D(z_0; r) := \{z : |z - z_0| < r\} \subseteq \mathbb{C}$ is an open (circular) disc in $\mathbb{C}$.

(b) A set Ω is an open subset of $\mathbb{C}$ if for each $z_0 \in \Omega$ there is a radius $r = r(z_0) > 0$, such that

$$D(z_0; r) = \{z : |z - z_0| < r\} \subseteq \Omega.$$

(c) An open set $\Omega \subseteq \mathbb{C}$ is called *domain.*

(d) A domain Ω is called connected if for each pair of points z_1, $z_2 \in \Omega$ there exists a (continuous) curve such that $\gamma(a) = z_1$, $\gamma(b) = z_2$, and $\gamma(t) \in \Omega$ for all $t : a \leq t \leq b$.

(e) A domain is called simply connected if for each closed curve in the domain (i.e., $\gamma[a,b] \subseteq \Omega$) the curve is homotopic with a point in Ω.
In other words, there is no hole inside the domain.

[1]The interval $[0,1]$, due to bijectivity, may be replaced by any interval $[c,d]$, $c < d$.

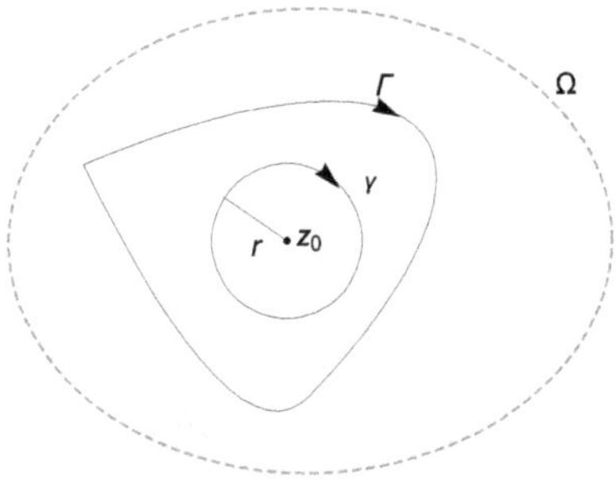

The open and simply connected set $\Omega \subset \mathbb{C}$ contains the curve Γ, the circle $\gamma : z_0 + r\,e^{it}$, $0 \le t \le 2\pi$ and the point z_0, where the three are homotopic with each other. The curves, Γ and γ are clockwise oriented.

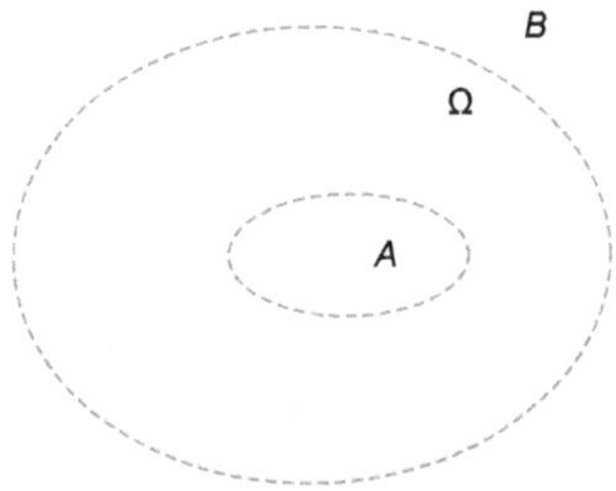

The open connected, but not simply connected, set $\Omega \subset \mathbb{C}$ (colored) has complement $\Omega^c = \mathbb{C} \setminus \Omega = A \cup B$.

16.2 Functions on the Complex Plane $\mathbb{C}$

Definition 16.2. Let Ω be a domain in $\mathbb{C}$ and $f(z)$ a function $f : \Omega \to \mathbb{C}$. Then the function f is called analytic (or holomorphic) if it is continuously differentiable in Ω, that is, if for each $z \in \Omega$, the limit

$$f'(z) := \lim_{\Delta z \to 0} \frac{f(z + \Delta z) - f(z)}{\Delta z} \tag{16.2}$$

is continuous. A function which is analytic in the whole complex plane $\mathbb{C}$ is called an *entire* function.

Remark. Examples of entire functions are polynomials, exponential functions: a^z, $a > 0$, and trigonometric functions: $\sin z$, $\cos z$, etc.

Theorem 16.1.

(i) *Let $f(z) = u(z) + iv(z) = u(x, y) + iv(x, y)$, where u and v are real. Further, x and y are real and imaginary parts of z. Then f is analytic $\iff$ u and v satisfy the Cauchy–Riemann equations:*

$$\frac{\partial u}{\partial x} = \frac{\partial v}{\partial y}, \qquad \frac{\partial u}{\partial y} = -\frac{\partial v}{\partial x}. \tag{16.3}$$

(ii) *An analytic function* f ($\frac{df}{dz}$ *exists in a domain* Ω) *is infinitely differentiable, which means that*

$$\frac{d^n f}{dz^n},\ n = 2, 3, \ldots, \qquad \textit{exist for all } z \in \Omega.$$

(iii) *An entire function* $w = f(z)$ *not assuming two complex values, say* w_1 *and* w_2*, is constant on whole* $\mathbb{C}$.

(iv) *Analytic functions obey the same rules of limits and differentiations as for real functions.*

16.2.1 Elementary functions

These are polynomials or, more generally, rational functions, defined as in the real case. For the transcendental functions, the following identities hold true:

$$e^z = e^{x+iy} = e^x(\cos y + i \sin y), \quad \log z = \ln|z| + i \arg z,$$

$$\sin z = \frac{e^{iz} - e^{-iz}}{2i}, \qquad \cos z = \frac{e^{iz} + e^{-iz}}{2}, \tag{16.4}$$

$$\tan z = \frac{\sin z}{\cos z}, \qquad \cot z = \frac{\cos z}{\sin z}.$$

Remark.

(i) The elementary functions e^z, $\sin z$, and $\cos z$ in (16.4) are analytic and also entire. The logarithm function is analytic, for example, at $\mathbb{C} \setminus \{z : \mathrm{Im}(z) = 0,\ \mathrm{Re}(z) \leq 0\}$, the so-called principal branch of the logarithm function.

(ii) In the case of non-integer exponents, the definition of power function depends on the definition of the logarithm function:

$$f(z) = z^\alpha = e^{\alpha(\ln|z| + i \arg z)}. \tag{16.5}$$

(iii) Note that, unlike the real case, $\sin z$, $z \in \mathbb{C}$ assumes also values outside the interval $[-1, 1]$ (likewise for $\cos z$).

(iv) The derivative of sum, product, ratio, and composition of two analytic functions $f(z)$ and $g(z)$ follow the same rules of calculus of real analysis, as presented in (9.12) page 194.

Definition 16.3.
The integral over a regular curve γ is defined as

$$\int_\gamma f(z)dz := \int_a^b f(\gamma(t))\gamma'(t)dt, \tag{16.6}$$

see the figure.

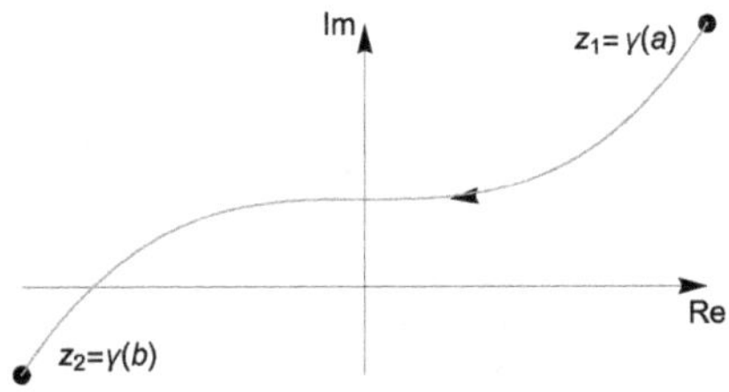

Theorem 16.2. *In the following, we assume that f is analytic in Ω.*

(i) *If γ_1 and γ_2 are two homotopic regular curves in Ω, then*

$$\int_{\gamma_1} f(z)dz = \int_{\gamma_2} f(z)dz. \tag{16.7}$$

(ii) *If Ω is simply connected and γ is a regular, closed curve in Ω, then*

$$\int_\gamma f(z)dz = 0. \tag{16.8}$$

In particular, for any two curves γ_1 and γ_2, connecting two points z_1, $z_2 \in \Omega$, the relation (16.7) holds true, i.e. the integral is independent of the trajectories connecting the two points. That is why the integral is written as $\int_\gamma f(z)dz = \int_{z_1}^{z_2} f(z)dz$, where z_1 is the starting point and z_2, the endpoint of the curve.

(iii) *If Ω is simply connected, then f has a primitive function F, i.e.,*

$$\int_\gamma f(z)dz = \int_{z_1}^{z_2} f(z)dz = F(z_2) - F(z_1). \tag{16.9}$$

(iv) (a) *Assume γ simply surrounds z_0 and is counterclockwise oriented. Then*

$$f(z_0) = \frac{1}{2\pi i}\int_\gamma \frac{f(z)}{z-z_0}dz, \quad \textit{more generally,}$$
$$\frac{\partial^n f}{\partial z^n}(z_0) \equiv f^{(n)}(z_0) = \frac{n!}{2\pi i}\int_\gamma \frac{f(z)}{(z-z_0)^{n+1}}dz. \tag{16.10}$$

The second formula is Cauchy's general integral formula.

(b) *f can be expanded in a power series about* z_0:

$$f(z) = \sum_{n=0}^{\infty} a_n(z - z_0)^n, \quad a_n = \frac{f^{(n)}(z_0)}{n!}, \quad n = 0, 1, 2, \ldots \tag{16.11}$$

with radius of convergence R being the radius of the largest circular disk $D(z_0; R) \subseteq \Omega$.

Theorem 16.3. *Assume that f is analytic in a connected domain* $D_f = \Omega$.

(i) ***Liouville's theorem:*** *If f is a bounded entire function* ($D_f = \mathbb{C}$), *then f is constant.*

(ii) *If* $(z_n)_{n=1}^{\infty} \subset \Omega$ *is a convergent sequence of* distinct points z_n, *with limit* z_0, *and* $f(z_n) = f(z_0) = A$ *for* $n = 0, 1, 2, \ldots$, *then* $f(z) \equiv A$ *for all* $z \in \Omega$.

(iii) ***The maximum principle:*** *If* $|f(z)|$ *assumes a maximum in* Ω. *Then f is constant in* Ω. *Consequently, if f is not constant, then* $|f(z)|$ *assumes its maximum at the boundary of* Ω.

Theorem 16.4.

(i) ***Schwarz lemma:*** *Assume that* $f(z)$ *is analytic in* $\Omega = \{z : |z| < 1\}$ *and that* $f(z)$ *satisfies*

$$f(0) = 0, \text{ and } |f(z)| \leq 1, \quad z \in \Omega.$$

Then

$$|f'(0)| \leq 1, \text{ and } |f(z)| \leq |z|, \quad z \in \Omega. \tag{16.12}$$

(ii) ***Rouche's theorem:*** *Assume that* $K \subset \Omega$ *is a compact set,* ∂K *is a piecewise regular closed curve, and that for two analytic functions f and g in* Ω, $|f(z)| > |g(z)|$ *for all* $z \in \partial K$, *then the functions f, g, and* $f + g$ *have the same number of zeros inside the curve* ∂K, *i.e., in the interior of K.*

16.3 Lines, Circles, and Möbius Transforms

16.3.1 *Preliminaries: The Riemann sphere*

Given the set $\mathbf{C}^* := \mathbb{C} \cup \{\infty\}$ and *the Riemann sphere*

$$\mathcal{R} := \left\{ (x_1, x_2, x_3) \in \mathbb{R}^3 : x_1^2 + x_2^2 + (x_3 - 1/2)^2 = \frac{1}{4} \right\},$$

i.e., the sphere with center at $(0, 0, 1/2)$ and radius $1/2$.

A bijection $\mathcal{F} : \mathbf{C}^* \to \mathcal{R}$ with $\mathcal{F}(z) = \mathcal{F}(x + iy) = (x_1, x_2, x_3)$ is given by

$$\begin{cases} \mathcal{F}(x+iy) = (x_1, x_2, x_3) \\ \qquad\qquad = \left(\frac{x}{1+x^2+y^2}, \frac{y}{1+x^2+y^2}, \frac{x^2+y^2}{1+x^2+y^2} \right) \\ \qquad\qquad\quad (x_1, x_2, x_3) \neq (0, 0, 1), \\ \mathcal{F}(\infty) \qquad = (0, 0, 1). \end{cases}$$

Geometrically, this corresponds to the line between the point $z = x + iy \in \mathbb{C}$ and the point $N = (0, 0, 1)$ which intersects the Riemann sphere at (x_1, x_2, x_3). The infinity point: $e^* = \infty$ is mapped to $(0, 0, 1)$.

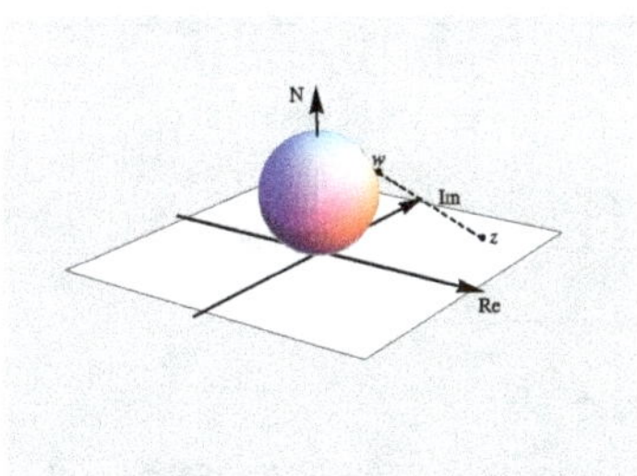

The Riemann sphere and the complex plane.

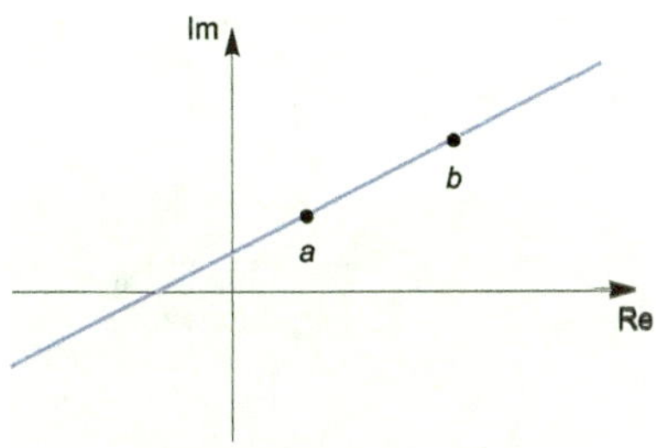

Line through two points a and b given by equation (16.14).

Theorem 16.5. *Let b be a real number. An equation for a line in $\mathbb{C}$ is*

$$z : \quad \overline{a}z + a\overline{z} + b = 0, \; a \neq 0. \tag{16.13}$$

Equation of a line through two different points $a = a_1 + i\,a_2$ and $b = b_2 + i\,b_2$, a_1, a_2, b_1, b_2 real, is given by

$$z = x + iy : \ \overline{a - b} \cdot z - (a - b) \cdot \overline{z} - \overline{a} \cdot b + a \cdot \overline{b} = 0. \qquad (16.14)$$

For $|a|^2 > b$, the following is the equation of a circle with radius $r = \sqrt{|a|^2 - b}$ and center $z_0 = -a$.

$$z : \quad z\overline{z} + \overline{a}\,z + a\,\overline{z} + b = 0.$$

Remark. The equation (16.14) can equivalently be written as $(a_2 - b_2)x - (a_1 - b_1)y + a_1 b_2 - a_2 b_1 = 0$.

16.4 Some Simple Mappings

(i) Translation: $w = z + \alpha, \quad \alpha \in \mathbb{C}$.
(ii) Rotation: $w = ze^{i\theta_0}$ rotates every point z_0 in z-plane by $|\theta_0|$.
(iii) Similarity: $w = az, \quad (a > 0)$.

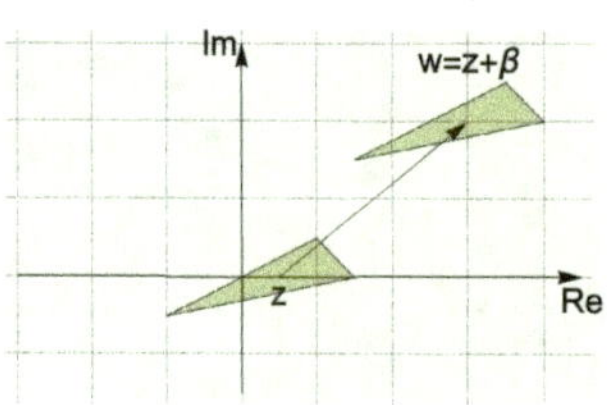

Translation: $w = z + \alpha, \quad \alpha \in \mathbb{C}$.

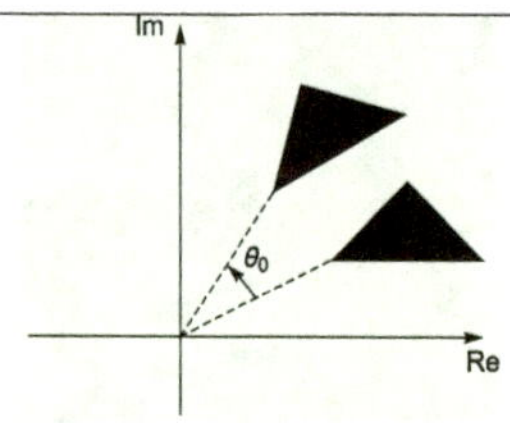

Rotation: $w = ze^{i\theta_0}$ rotates every point z_0 in z-plane by $|\theta_0|$.

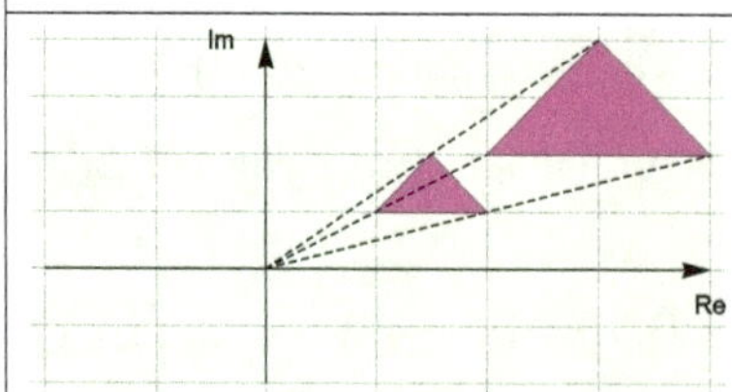

Similarity: $w = a\,z$ (stretch, $a > 1$).

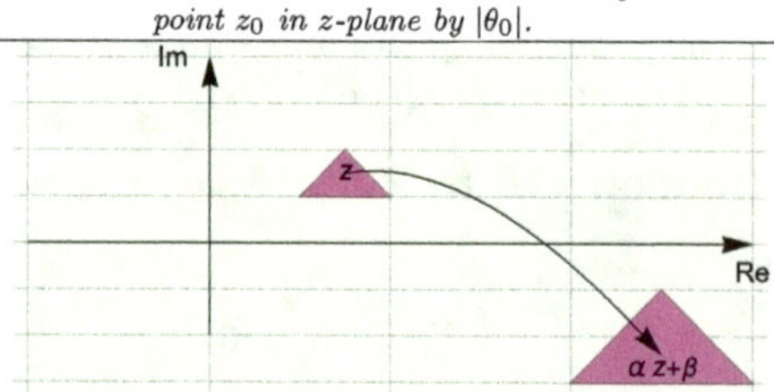

Linearity: $w = \alpha\,z + \beta$ here with $\alpha > 0$.

For similarity there are two cases

$$|w| = a|z|, \quad \begin{cases} a > 1 & \Longrightarrow |w| > |z| \quad \text{(stretch)}, \\ 0 < a < 1 & \Longrightarrow |w| < |z| \quad \text{(compress/contraction)}. \end{cases}$$

(iv) Linearity: $w = \alpha z + \beta, \quad (\alpha,\ \beta \in \mathbb{C},\ \text{constants})$.
Every linear mapping can be performed by a combination of similarity, rotation, and translation:

Set $\alpha = ae^{i\theta_0}$ where $a = |\alpha|$.
If

$$\begin{cases} w_1 = az & \text{(similarity)}, \\ w_2 = w_1 \cdot e^{i\theta_0} & \text{(rotation)}, \\ w = w_2 + \beta & \text{(translation)}, \end{cases} \qquad \text{then} \quad w = \alpha z + \beta.$$

(v) Inversion: $w = \dfrac{1}{z}$.

16.4.1 *Möbius mappings*

(i) A "circle" (or "circle line") means a circle or a line. A line is then considered a circle with infinite radius.

(ii) Assume that a, b, c, d are given real numbers such that $ad - bc \neq 0$.
A **Möbius transform or Möbius mapping** $T(z)$ is a function $\mathbb{C}^* \to \mathbb{C}^*$, given by

$$\begin{cases} T(z) = \dfrac{az+b}{cz+d}, \quad z \neq -d/c, \\ \\ T(-d/c) = e^* \quad (= \infty). \end{cases} \tag{16.15}$$

(iii) The Möbius transform T is bijective on $\mathbb{C}^*$. Its inverse function T^{-1} is also a Möbius transform, and is given by

$$z = T^{-1}(w) = \begin{cases} \dfrac{dw-b}{a-cw} & \text{for } w \neq a/c, \quad w \neq \infty, \\ \infty & \text{for } w = a/c, \\ -d/c & \text{for } w = \infty. \end{cases} \tag{16.16}$$

(iv) The class of Möbius transforms constitutes a non-commutative group under composition.
(v) If T has (at least) two fix points (A fix point satisfies $T(z) = z$), then $T(z) = z$ for all $z \in \mathbb{C}^*$.
(vi) A Möbius transform maps "circle lines" on "circle lines".
(vii) Composition of Möbius mappings is a Möbius mapping:

$$\text{If } w = T_1(z) = \frac{az+b}{cz+d} \quad \text{and} \quad u = T_2(w) = \frac{\alpha w+\beta}{\gamma w+\delta}$$

are two Möbius mappings, then the composition

$$u = T_2 \circ T_1(z) = T_2\Big(T_1(z)\Big) \quad \text{is also a Möbius mapping.} \tag{16.17}$$

Remark. The condition $ad \neq bc$ in (16.15) can be expressed with determinant. Let $\boldsymbol{A} = \begin{bmatrix} a & b \\ c & d \end{bmatrix}$. Then

$$ad - bc = \det \boldsymbol{A} \neq 0.$$

For the composition (16.17) and $\mathcal{A} = \begin{bmatrix} \alpha & \beta \\ \gamma & \delta \end{bmatrix}$, the numerator and denominator in the composition $T_2(T_1(z))$ are the first and second element in

$$\mathcal{A} \cdot \boldsymbol{A} \cdot \begin{bmatrix} z \\ 1 \end{bmatrix} = \begin{bmatrix} (a\alpha + \beta c)z + \alpha b + \beta d \\ (a\gamma + c\delta)z + b\gamma + d\delta \end{bmatrix}$$

i.e.,

$$T_2(T_1(z)) = \frac{(a\alpha + \beta c)z + \alpha b + \beta d}{(a\gamma + c\delta)z + b\gamma + d\delta}. \tag{16.18}$$

Some special cases are as follows:

$$c \neq 0 : T\left(-\tfrac{d}{c}\right) = \infty \quad (\text{or } |T(z)| \longrightarrow \infty \text{ as } z \to -d/c).$$

$$c = 0 : T(\infty) = \infty \quad (\text{or } |T(z)| \longrightarrow \infty \text{ as } |z| \longrightarrow \infty).$$

Theorem 16.6. *Any Möbius mapping can be obtained by successive combinations of translation, inversion, similarity, or rotation.*

Any Möbius mapping is uniquely determined by three points and their images:

For any pair of three distinct points $z_1,$, z_2, z_3 *and* w_1, w_2, w_3 *all in* $\mathbb{C}^*$, *there exists a unique Möbius mapping* $w = T(z)$ *such that* $w_k = T(z_k)$, $k = 1, 2, 3$. *The constants are obtained by*

$$\frac{(w - w_1) \cdot (w_2 - w_3)}{(w - w_3) \cdot (w_2 - w_1)} = \frac{(z - z_1) \cdot (z_2 - z_3)}{(z - z_3) \cdot (z_2 - z_1)} = \begin{cases} 0, & \text{if } z = z_1, \\ 1, & \text{if } z = z_2, \\ \infty, & \text{if } z = z_3. \end{cases} \tag{16.19}$$

If z_k *(and/or* w_k*) is* ∞, *then the above relation is modified by replacing the parentheses containing* z_k *(*w_k*, respectively) by ones.*

The mapping $w = \frac{1}{\bar{z}}$ *is not a Möbius mapping.*

16.4.2 *Angle preserving functions*

Definition 16.4. A conformal mapping, or angle-preserving transformation, is a complex function $w = f(z)$ that preserves local angles. An analytic function is conformal at any point where it has a non-zero derivative. Conversely, any conformal mapping of a complex variable which has continuous partial derivatives is analytic.

Given two regular curves with equations $z = \gamma_1(t)$ and $z = \gamma_2(t)$ in a domain Ω, intersecting at z_0, corresponding to parameters s and t, the angle between the curves at z_0 is

$$\alpha = \arg \gamma_2'(t) - \arg \gamma_1'(s).$$

Let $f(z)$ be an analytic function $f : \Omega \to \mathbb{C}$, and $\gamma_j(t)$, $j = 1, 2$ are curves, as above. Then $w = f(z)$ maps $\gamma_1(t)$ and $\gamma_2(t)$ on the curves $\Gamma_1(t)$ and $\Gamma_2(t)$, respectively, with equations

$$w = \Gamma_j(t) = f(\gamma_j(t)), \qquad j = 1, 2$$

and Γ_1 and Γ_2 intersect at $w_0 = f(z_0) = \Gamma_1(s) = \Gamma_2(t)$.

Assume that $f'(z_0) \neq 0$. Then the curves $\Gamma_j(t)$ are regular at $f(z_0)$, with angle

$$\beta = \arg \Gamma_2'(t) - \arg \Gamma_1'(s)$$

between the curves $\Gamma_2(t)$ and $\Gamma_1(s)$.

A function $f(z)$ is *angle preserving* if $\alpha = \beta$.

Theorem 16.7. *With the same conditions as above, the following equivalence holds true*

$$\alpha = \beta \Longleftrightarrow f(z) \text{ is analytic, where } f'(z) \neq 0.$$

A Möbius mapping (as in (16.15)*) is conformal except at* $z = -d/c$.

In the following, to the left: The curves $t \curvearrowright \gamma_1(t)$ *and* $t \curvearrowright \gamma_2(t)$ *with intersecting point* z_0 *and corresponding angle* α. *To the right: The curves* $t \curvearrowright f(\gamma_1(t)) = \Gamma_1(t)$ *and* $t \curvearrowright f(\gamma_2(t)) = \Gamma_2(t)$ *with intersecting point* $w_0 = f(z_0)$ *and corresponding angle* β.

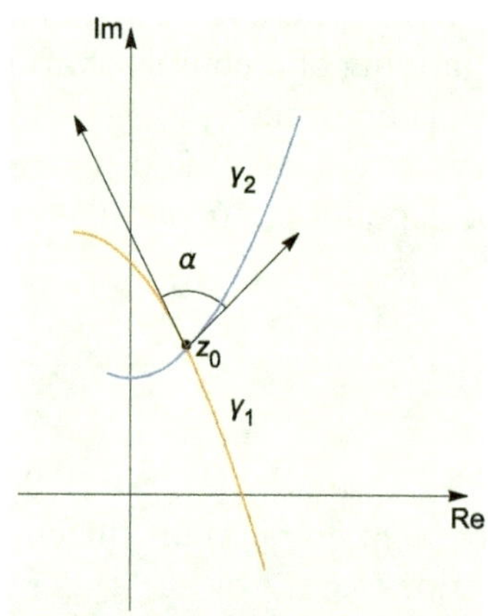

The angle $\alpha = \arg \gamma_2'(t) - \arg \gamma_1'(s)$.

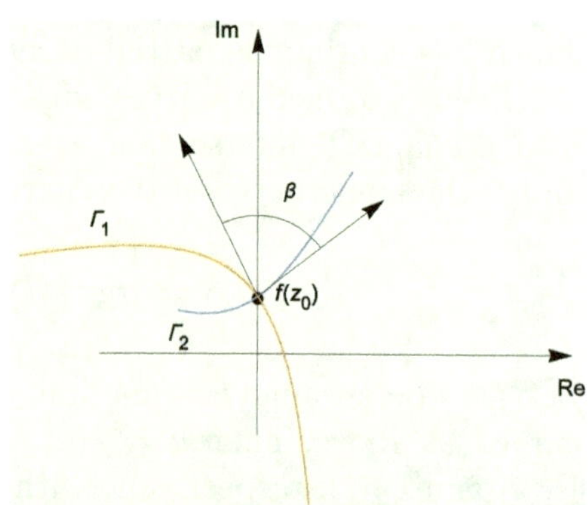

The angle $\beta = \arg \Gamma_2'(t) - \arg \Gamma_1'(s)$.

Theorem 16.8 (Riemann mapping theorem). *Let* R_1 *and* R_2 *be two arbitrary, simply connected domains;* $(R_1 \neq \mathbb{C},\ R_2 \neq \mathbb{C})$. *Then there is an analytic function that maps* R_1 *on* R_2.

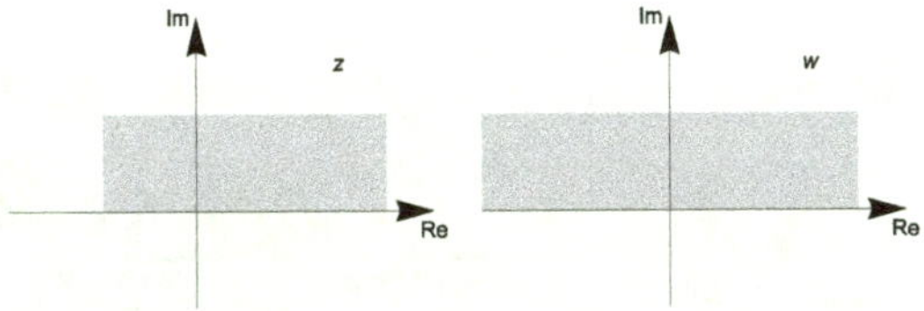

16.5 Some Special Mappings

16.5.1 *Applications in potential theory*

In the following figures, A in the z-plane corresponds to A' in the w-plane and so on.

(i) Upper half-plane to the unit disk $|w| \leq 1$ is made by the Möbius mapping $w = e^{i\theta_0} \dfrac{z - z_0}{z - \overline{z_0}}$.

z	z_0	$\overline{z_0}$	∞
w	w_0	∞	$e^{i\theta_0}$

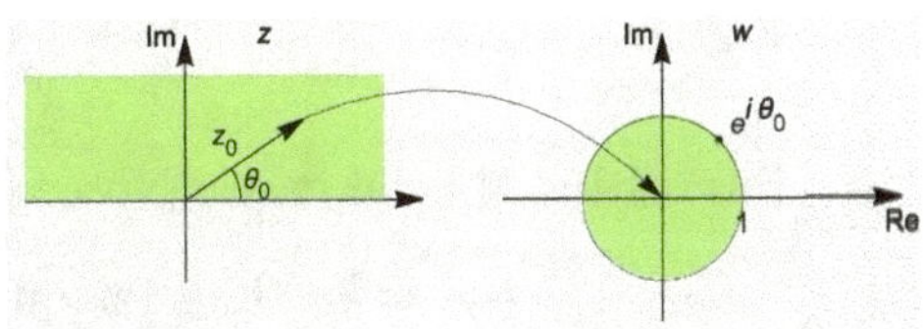

(ii) The region outside the unit disk to the unit disk: The Möbius mapping $w = \frac{1}{z}$.

z	∞	0	$re^{i\theta}$
w	0	∞	$\dfrac{1}{r}\, e^{-i\theta}$

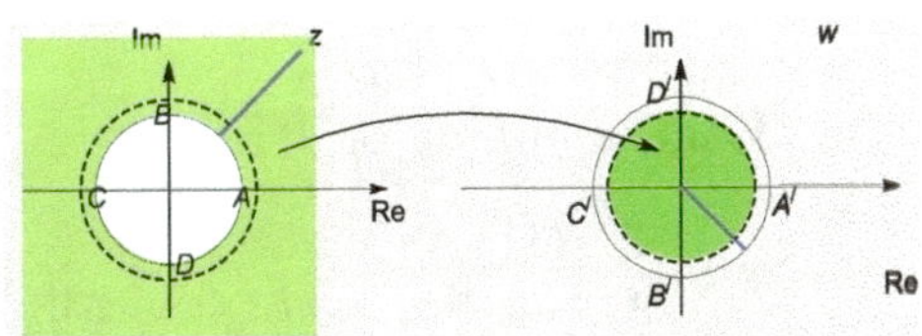

(iii) The region inside unit disk to the upper half-plane

$$w = \frac{z\,\overline{w_0} - w_0\, e^{i\theta_0}}{z - e^{i\theta_0}}.$$

z	0	∞	$e^{i\theta_0}$
w	w_0	$\overline{w_0}$	∞

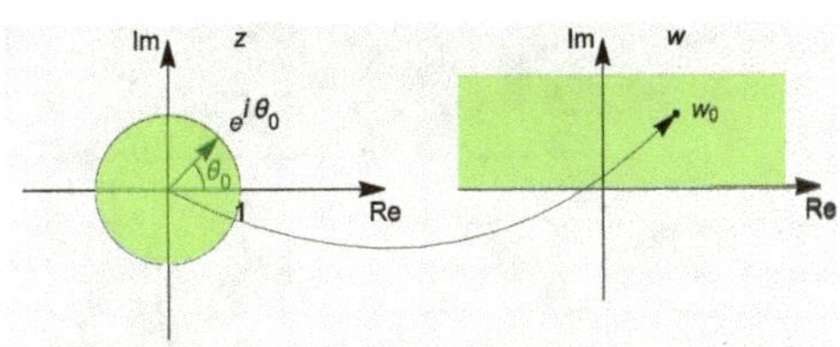

(iv) Angular domain to upper half-plane:

$$w = z^m, \quad m > \frac{1}{2}.$$

z	0	r	$re^{i\pi/m}$
w	0	r^m	$-r^m$

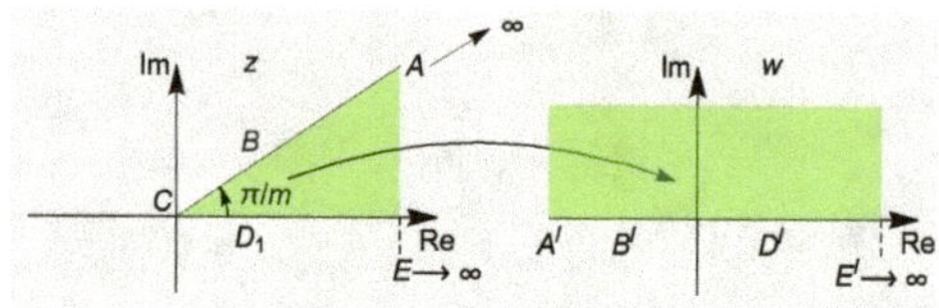

(v) Example of (iv), a composite mapping:

$$z \curvearrowright w_1 = z^2 \curvearrowright w = -i\, w_1 = -i\, z^2.$$

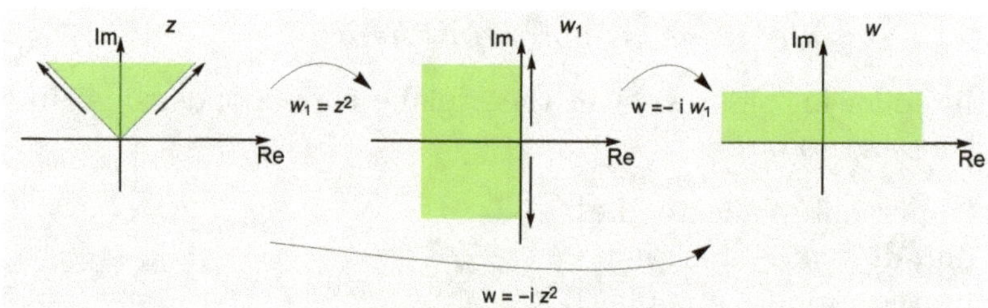

Interesting special cases: Note: Upper Half-Plane (UHP).

$$\begin{cases} m = 2 & \text{The 1st quadrant} \longrightarrow \text{UHP}, \\ m = 4 & \tfrac{1}{8}\text{th of plane} \longrightarrow \text{UHP}. \end{cases}$$

Note. For those m with multiple-defined z^m, choose the appropriate branch:
Example: $z = re^{\frac{3\pi}{4}i}$.

(vi) Band mapping on UHP:

$$w = e^{\frac{\pi}{a}z}, \qquad a = \text{ width of the band.}$$

z	0	$a\,i$	$\frac{a\,i}{2}$
w	1	-1	i

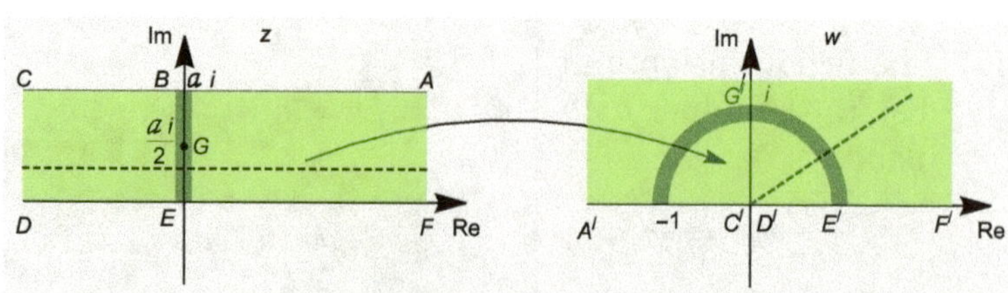

$$\begin{cases} z = x + ai \Longrightarrow \\ w = e^{\frac{\pi}{a}x} \cdot e^{\pi i} = -e^{\frac{\pi}{a}x} \longrightarrow \begin{cases} 0 & x \to -\infty \\ -\infty & x \to \infty \end{cases} \end{cases}$$

$$\begin{cases} z = x \Longrightarrow \\ w = e^{\frac{\pi}{a}x} \longrightarrow \begin{cases} 0 & x \to -\infty \\ \infty & x \to \infty \end{cases} \end{cases}$$

$$\begin{cases} z = x + bi, \quad 0 < b < a \ (b/a < 1) \Longrightarrow \\ w = e^{\frac{\pi}{a}x} \cdot e^{\frac{b}{a}i}, \ (\arg w = b/a), \longrightarrow \begin{cases} 0 & x \to -\infty \\ \infty & x \to \infty. \end{cases} \end{cases}$$

(vii) Mapping from upper half-band to upper half-plane:

$w = \sin \dfrac{\pi z}{a}$ $\qquad a =$ width of the band.

z	0	$a/2$	$b\,i$
w	0	1	$i \sinh \dfrac{b\pi}{a}$

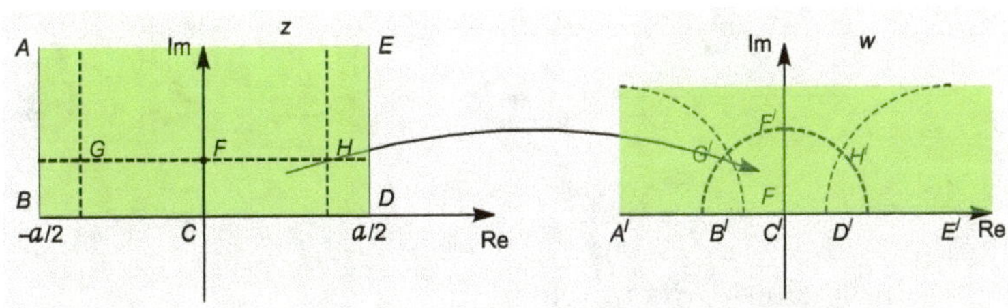

Remark. The mapping is conformal except for the points $z = \pm\dfrac{a}{2}$, i.e., B and D.

(viii) Example of (vii):

$$z \curvearrowright w_1 = i\,z \curvearrowright w_2 = w_1 + a/2 \curvearrowright w = \sin\left(\frac{\pi\, w_2}{a}\right)$$

or

$$\begin{aligned} w &= \sin \frac{\pi}{a} w_2 = \sin \frac{\pi}{a}\left(w_1 + \frac{a}{2}\right) \\ &= \sin\left(\frac{\pi}{a} w_1 + \frac{\pi}{2}\right) = \cos \frac{\pi}{a} w_1 = \cos(\frac{\pi}{a} i z) = \cosh\left(\frac{\pi}{a} z\right). \end{aligned}$$

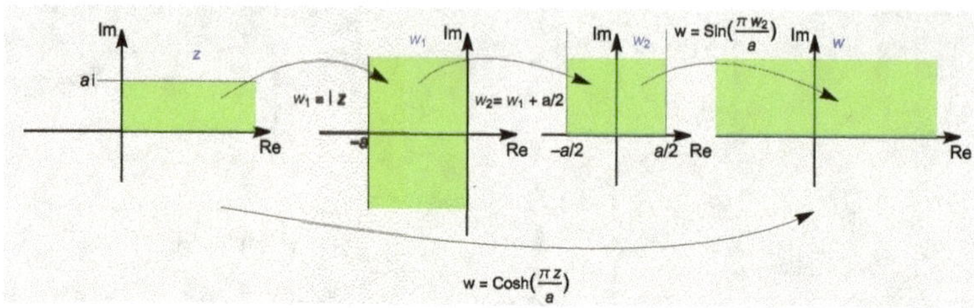

(ix) Special case: Mapping of half-circular disk to UHP:

$$w = \left(\frac{z+a}{z-a}\right)^2.$$

The mapping is not conformal at A.

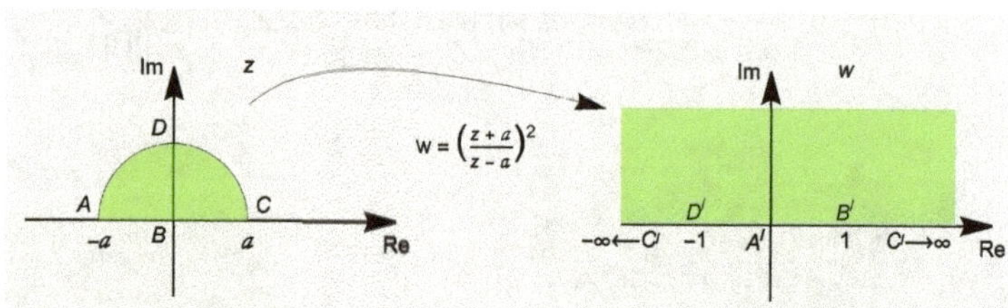

(x) Circle ring to rectangle:

$w = \log z$; (Suitable branch), $z = re^{i\theta}$, i.e.,
$w = \log z = \ln r + i\theta, \quad a \leq r \leq b, \quad w_0 \leq \theta \leq v_0 + 2\pi.$

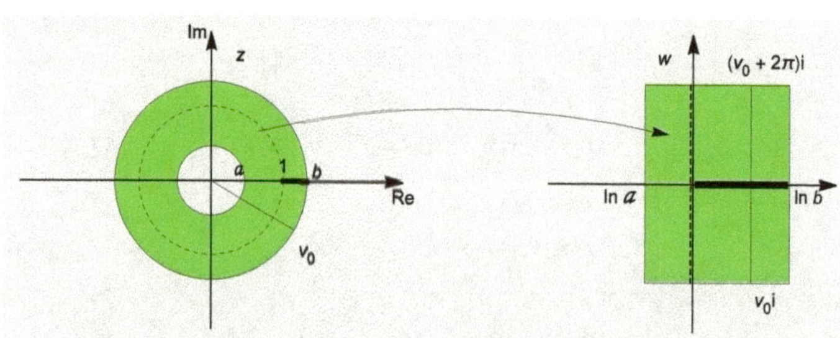

16.6 Harmonic Functions

Definition 16.5. A function $u = u(x,y)$ $(x,\, y \in \mathbb{R})$ is harmonic if it satifies the Laplace equation (Laplace PDE),

$$\Delta u = \frac{\partial^2 u}{\partial x^2} + \frac{\partial^2 u}{\partial y^2} = 0. \tag{16.20}$$

A real function v is a harmonic conjugate of u, if u is also real, harmonic, and $f = u + iv$ is analytic.

Theorem 16.9.

(i) *Real- and imaginary part of an analytic function are harmonic.*
(ii) *An analytic function f satisfies*

$$\Delta|f(z)|^2 = 4|f'(z)|^2. \tag{16.21}$$

(iii) ***Poisson's formula:*** *If $u(x,y)$ is harmonic in an (open) domain Ω, which contains $\{z : |z| \leq R\}$ and $z = re^{i\theta}$ is the polar representation of z, then*

$$u = u(r,\theta) = \frac{1}{2\pi}\int_0^{2\pi} \frac{R^2 - r^2}{R^2 - 2Rr\cos(\theta-\varphi) + r^2} u(R,\varphi)d\varphi, \tag{16.22}$$

for $0 \leq r < R$.

16.7 Laurent Series, Residue Calculus

Definition 16.6. A series

$$S(z; z_0) := \sum_{n=-\infty}^{\infty} c_n (z - z_0)^n \tag{16.23}$$

is a Laurent series about $z_0 \in \mathbb{C}$.

Theorem 16.10.

Given the series (16.23). With

$r := \liminf_{n\to-\infty} \sqrt[n]{|c_n|}$ *and*

$R := \limsup_{n\to\infty} \sqrt[n]{|c_n|}$,

the series (16.23) converges to an analytic function $f(z) := S(z; z_0)$ for $r < |z - z_0| < R$. In the figure, $z_0 = 0$.

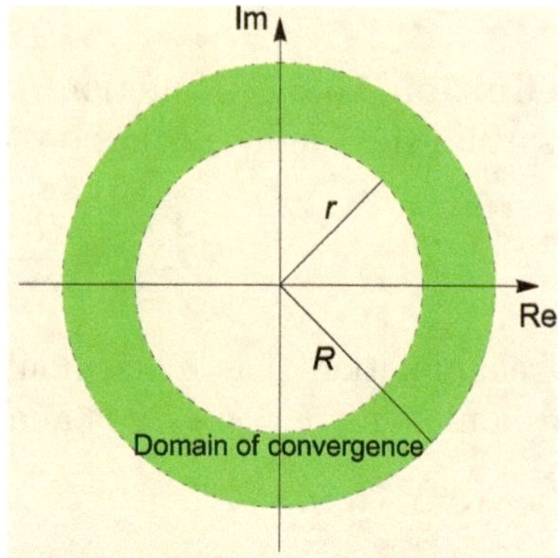

The coefficients are given by

$$c_n = \oint_\gamma \frac{f(z)}{(z-z_0)^{n+1}}\,dz, \quad n = 0, \pm1, \pm2, \ldots \tag{16.24}$$

where γ is a positively oriented curve that simply surrounds z_0 and is lying in the region $\{z : r < |z - z_0| < R\}$.

The series (16.23) converges uniformly to $f(z)$ in any compact subset of $\{z : r < |z - z_0| < R\}$, for instance in $K := \{z : r < r_1 \leq |z - z_0| \leq R_1 < R\}$.

- *If $c_n \neq 0$ for some $n < 0$, the function $f(z)$ has a singularity at z_0.*
- *If $c_n = 0$ for all $n \leq n_0$, for some $n_0 < 0$, then the singularity in z_0 is removable.*
- *If $c_n \neq 0$ for an infinite number of indices $n < 0$, then the singularity in z_0 is essential.*

(i) *Assume that γ is a counterclockwise-oriented (positively oriented) curve which simply surrounds z_0. Then the residue at z_0 is defined as*

$$c_{-1} = Res(f(z), z_0) = \frac{1}{2\pi i}\oint_\gamma f(z)\,dz,$$

where $f(z)$ is given by (16.23).

(ii) *For an analytic function $f(z)$, the residual at z_0 is*

$$Res\left(\frac{f(z)}{(z-z_0)^n}, z_0\right) = \frac{1}{(n-1)!}D^{n-1}f(z)_{|(z=z_0)}.$$

(iii) ***The Residue theorem:*** *Assume that $f(z)$ is analytic in an open and simply connected set $\Omega \subseteq \mathbb{C}$ except for $z_1, z_2, \ldots, z_k \in \Omega$. Further, assume that $\gamma \subset \Omega$ is a positively oriented curve which simply surrounds the points $\{z_1, z_2, \ldots, z_k\}$. Then*

$$\oint_\gamma f(z)dz = 2\pi i \sum_{r=1}^{k} Res\,(f(z), z_r). \tag{16.25}$$

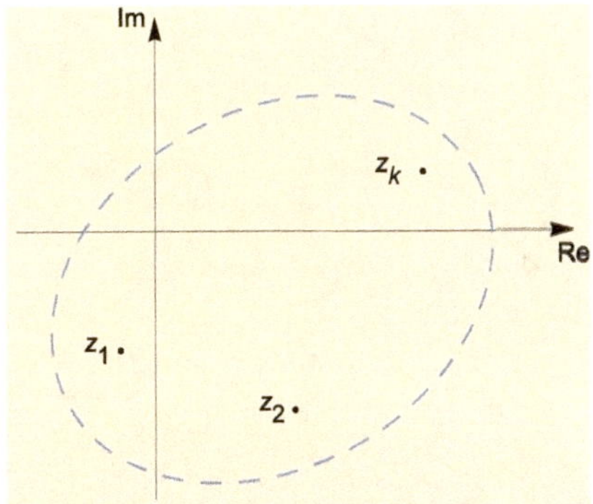

Chapter 17

Multidimensional Analysis

17.1 Topology in $\mathbb{R}^n$

17.1.1 *Subsets of* $\mathbb{R}^n$

An element in $\mathbb{R}^n$ is written as $\boldsymbol{x} = (x_1, x_2, \ldots, x_n)$.

(Also $(x_1, x_2, \ldots, x_n) = r$ is used).

$\boldsymbol{x} \in \mathbb{R}^n$ is considered as both point and location vector.

Definition 17.1.

(i) The length of $\boldsymbol{x} = (x_1, x_2, \ldots, x_n)$ is

$$|\boldsymbol{x}| := |(x_1, x_2, \ldots, x_n)| = \sqrt{x_1^2 + x_2^2 + \cdots + x_n^2}. \tag{17.1}$$

(ii) The distance between two points $\boldsymbol{x}$ and $\boldsymbol{y}$ is $|\boldsymbol{x} - \boldsymbol{y}|$,

$$|\boldsymbol{x} - \boldsymbol{y}| = \sqrt{(x_1 - y_1)^2 + (x_2 - y_2)^2 + \cdots + (x_n - y_n)^2}\,. \tag{17.2}$$

(iii) The diameter of a set $G \subseteq \mathbb{R}^n$ is given by

$$d(G) := \sup\{|\boldsymbol{x} - \boldsymbol{y}|,\ \boldsymbol{x}, \boldsymbol{y} \in G\}.$$

If $d(G) < \infty$, the set is bounded, otherwise, unbounded.

(iv) An open ball in $\mathbb{R}^n$ with center at $\boldsymbol{x}_0$ and radius r is the set

$$S_r(\boldsymbol{x}_0) := \{\boldsymbol{x} : \quad |\boldsymbol{x} - \boldsymbol{x}_0| < r\}.$$

(v) A subset G of $\mathbb{R}^n$ is open if for every $\boldsymbol{x}_0 \in G$, there is a radius $r > 0$, such that $S_r(\boldsymbol{x}_0) \subseteq G$.

(vi) A subset F is closed in $\mathbb{R}^n$ if its complement $F^c = \mathbb{R}^n \setminus F$ is open.

Theorem 17.1.

(i) (a) *Union of open sets is open: If G_i, $i \in I$ are open, so $\cup_{i \in I} G_i$ is open. In particular, the empty set $\emptyset$ is open.*
(b) *A finite intersection of open sets is open: If $G_1, G_2, \ldots, G_m$ are open, then $\cap_{i=1}^m G_i$ is open. In particular, $\mathbb{R}^n$ is open.*

(ii) (a) *Intersection of closed sets is closed: If F_i, $i \in I$ are closed, then $\cap_{i \in I} F_i$ is closed. In particular, $\mathbb{R}^n$ is closed.*
(b) *A finite union of closed sets is closed: If $F_1, F_2, \ldots, F_m$ are closed, so $\cup_{i=1}^m F_i$ is also closed. In particular, the empty set $\emptyset$ is closed.*

Definition 17.2.

(i) The interior of a subset $A \subseteq \mathbb{R}^n$ is the union of all open sets $G \subseteq A$. The interior of $\mathcal{A}$ is denoted by int $\mathcal{A}$. According to the previous theorem int $\mathcal{A}$ is open.

(ii) The closure of a set $A \subseteq \mathbb{R}^n$ is the intersections of all closed sets F such that $F \supseteq A$. The closure of A is denoted by $\overline{A}$. According to the previous theorem, $\overline{A}$ is closed.

(iii) The boundary of a A is the set $\partial A := \overline{A} \cap \overline{A^c}$.

(iv) A closed and bounded set is *compact*.

(v) (a) A set $D \subset \mathbb{R}^2$ given by $D = \{(x, y) : \phi(y) \leq x \leq \psi(y), \quad c \leq y \leq d\}$ is called *x-simple*.
(b) A set $D \subset \mathbb{R}^2$ given by $D = \{(x, y) : \phi(x) \leq y \leq \psi(x), \ a \leq x \leq b\}$ is called *y-simple*.

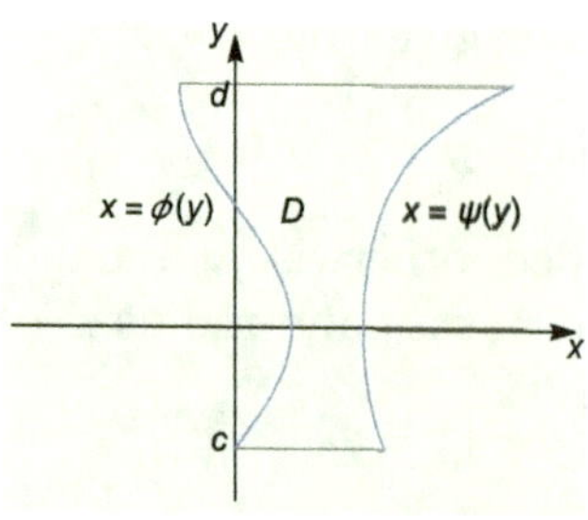

$D = \{(x, y) : \phi(y) \leq \psi(y), c \leq y \leq d\}$.

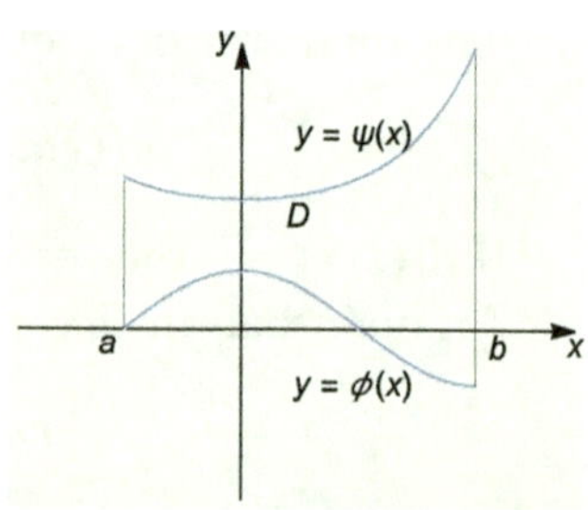

$D = \{(x, y) : \phi(a) \leq y \leq \psi(y), a \leq x \leq b\}$.

17.1.2 *Connected sets, etc.*

Definition 17.3.

(i) A set M is called connected if there are no two, non-empty, disjoint open sets G_1 and G_2 such that $M \subseteq G_1 \sqcup G_2$, i.e., M is not contained in a disjoint union of two non-empty open sets.

Alternatively, $\forall \boldsymbol{x}, \boldsymbol{y} \in M$, $\exists$ a curve $\gamma \subset M$, which connects $\boldsymbol{x}$ and $\boldsymbol{y}$.

(ii) A "domain" D means a connected open set in $\mathbb{R}^n$.

(iii) Let $\boldsymbol{x}$ and $\boldsymbol{y} \in \mathbb{R}^n$. The set $L(\boldsymbol{x}, \boldsymbol{y}) := \{\lambda \boldsymbol{x} + (1-\lambda)\boldsymbol{y}, \quad 0 \leq \lambda \leq 1\}$ is the line segment in $\mathbb{R}^n$ that connects $\boldsymbol{x}$ and $\boldsymbol{y}$.

(iv) A subset G of $\mathbb{R}^n$ is convex if for each pair $\boldsymbol{x}$ and $\boldsymbol{y}$ of points in G and every $0 \leq \lambda \leq 1$, $\lambda \boldsymbol{x} + (1-\lambda)\boldsymbol{y} \in G$. In other words, G is convex if all lines connecting any two points $\boldsymbol{x}, \boldsymbol{y} \in G$ lie in G: it contains the lines between all its points.

(v) Let $\boldsymbol{x}_k \in \mathbb{R}^n$, $\lambda_k \in \mathbb{R}^+ \cup \{0\}$; $k = 1, 2, \ldots, m$, and $\sum_{k=1}^{m} \lambda_k = 1$. Then $\boldsymbol{x} = \sum_{k=1}^{m} \lambda_k \boldsymbol{x}_k$ is a convex combination of the points $\boldsymbol{x}_k$, $k = 1, 2, \ldots, m$.

(vi) Let $A \subseteq \mathbb{R}^n$. The convex closure of A, is denoted $\mathrm{Conv}(A)$ and is the set of all convex combinations of the points $\boldsymbol{x}_k \in A, \quad k = 1, 2, \ldots, m$, where $m = 1, 2, \ldots$

(vii) A subset A of $\mathbb{R}^n$ is star shaped, if there is a point $\boldsymbol{x}_0 \in A$, such that the line $L(\boldsymbol{x}, \boldsymbol{x}_0) \subset A$ for every $\boldsymbol{x} \in A$.

(viii) Assume that $\boldsymbol{a}, \boldsymbol{x} \in \mathbb{R}^n$ (here considered as column vectors) and $c \in \mathbb{R}$. The set $\{\boldsymbol{x} : \boldsymbol{a}^T \cdot \boldsymbol{x} \leq c\}$ is a half space and the set $\{\boldsymbol{x} : \boldsymbol{a}^T \cdot \boldsymbol{x} = c\}$ is a hyperplane in $\mathbb{R}^n$.

Theorem 17.2.

(i) *Assume that $\boldsymbol{a}, \boldsymbol{x} \in \mathbb{R}^n$ (here considered as column vectors). Then, the set $\boldsymbol{a}^T \cdot \boldsymbol{x} \leq c$ is convex.*

(ii) *The intersection of convex sets is convex.*

(iii) *Let $\boldsymbol{A}$ be a real $m \times n-$matrix. The set*

$$M := \{\boldsymbol{x} \in \mathbb{R}^n : \boldsymbol{A} \cdot \boldsymbol{x} \leq \boldsymbol{c}\} \tag{17.3}$$

is the intersection between m half-spaces of $\mathbb{R}^n$ and thus is a convex set.

17.2 Functions $\mathbb{R}^m \longrightarrow \mathbb{R}^n$

Definition 17.4. In the following, notions of functions are described for the important cases for n and m.

$m = 1$: A function $\boldsymbol{f} = \boldsymbol{f}(t) : \mathbb{R} \to \mathbb{R}^n$ maps a real number t as a vector $\boldsymbol{f} \in \mathbb{R}^n$:

$$\boldsymbol{f}(t) = (y_1(t), y_2(t), \ldots, y_n(t)), \text{ where } y_j : \mathbb{R} \to \mathbb{R}. \quad (17.4)$$

Under some conditions of regularity on $y_j(t)$, $\boldsymbol{f}(t)$ is a *curve* in $\mathbb{R}^n$.

$m = 2$: A function $\boldsymbol{f} = \boldsymbol{f}(x_1, x_2) : \mathbb{R}^2 \to \mathbb{R}^n$ is a *surface* in $\mathbb{R}^n$. This also requires some regularity on $\boldsymbol{f}$.

$n = 1$: Such a function is called *real* or real valued.

A function f from $D_f \subseteq \mathbb{R}^m$ (where we assume that $D_f \neq \emptyset$) to $\mathbb{R}$ is continuous at $\boldsymbol{x}_0$ if for every $\varepsilon > 0$ there is a $\delta > 0$ such that

$$|\boldsymbol{x} - \boldsymbol{x}_0| < \delta \Longrightarrow |f(\boldsymbol{x}) - f(\boldsymbol{x}_0)| < \varepsilon. \quad (17.5)$$

A function is continuous in $M \subseteq D_f$ if it is continuous at every point $\boldsymbol{x}_0 \in M$.

$m = 2$,
$n = 1$: The function $f(x_1, x_2)$ describes, by the points $(x_1, x_2, f(x_1, x_2))$, *function surface*, in $\mathbb{R}^3$.

17.2.1 *Functions $\mathbb{R}^n \longrightarrow \mathbb{R}$*

Theorem 17.3. *Assume that f is continuous in a compact set $D \subset \mathbb{R}^n$.*

(i) *$f : D \to \mathbb{R}$ is uniformly continuous, if for every $\varepsilon > 0$ there is a $\delta > 0$ such that*

$$|\boldsymbol{x} - \boldsymbol{y}| < \delta \Rightarrow |f(\boldsymbol{x}) - f(\boldsymbol{y})| < \varepsilon, \qquad \forall\ \boldsymbol{x},\ \boldsymbol{y} \in D. \quad (17.6)$$

(ii) *f assumes a largest value ($f_{\max}$) and a smallest value ($f_{\min}$) on D. If, in addition, D is connected, then f assumes all values between $f_{\min}$ and $f_{\max}$.*

Definition 17.5. The function f has *partial derivative* in the coordinate x_i, $i = 1, 2, \ldots, n$,

$$\frac{\partial f}{\partial x_i} := \lim_{\Delta x \to 0} \frac{\begin{array}{c} f(x_1, x_2, \ldots, x_i + \Delta x, \ldots, x_n) \\ -f(x_1, x_2, \ldots, x_i, \ldots, x_n) \end{array}}{\Delta x}, \tag{17.7}$$

if the limit on the RHS exists. The limit is then the *partial derivative* in the coordinate x_i and is denoted as LHS. Higher-order derivatives with respect to x_i are defined inductively, *viz.*

$$\frac{\partial^m f}{\partial x_i^m} := \frac{\partial}{\partial x_i}\left(\frac{\partial^{m-1} f}{\partial x_i^{m-1}}\right), \quad m = 1, 2, \ldots \tag{17.8}$$

The *mixed* second-order derivative with respect to x_i and x_j (in this order), where $i \neq j$, is defined as

$$\frac{\partial^2 f}{\partial x_j \partial x_i} := \frac{\partial}{\partial x_j}\left(\frac{\partial f}{\partial x_i}\right), \tag{17.9}$$

as far as the right-hand side exists (is well-defined). For a multi-index $\alpha = (\alpha_1, \alpha_2, \ldots, \alpha_n)$, one defines $|\alpha| := \alpha_1 + \alpha_2 + \ldots + \alpha_n$ (Note! Not the length of a vector), where α_i are non-negative integers and the total derivative of order $|\alpha|$ is written as

$$\frac{\partial^{|\alpha|} f}{\partial x_1^{\alpha_1} \partial x_2^{\alpha_2} \ldots \partial x_n^{\alpha_n}} := \frac{\partial^{\alpha_1}}{\partial x_1^{\alpha_1}}\left(\frac{\partial^{\alpha_2}}{\partial x_2^{\alpha_2}}\left(\cdots\left(\frac{\partial^{\alpha_n}}{\partial x_n^{\alpha_n}} f\right)\cdots\right)\right) \tag{17.10}$$

as far as all derivatives exist and commute (see condition in Section 17.4).

If for a fixed $|\alpha|$, $\alpha = (\alpha_1, \alpha_2, \ldots, \alpha_n)$, all partial derivatives in (17.10) are continuous, then we write

$$f \in \mathcal{C}^{|\alpha|}(\mathbb{R}^n). \tag{17.11}$$

If all partial derivatives of all orders for f exist and are continuous, then f is said to be infinitely differentiable. This is written as

$$f \in \mathcal{C}^{\infty}(\mathbb{R}^n). \tag{17.12}$$

Remark. $\frac{\partial f}{\partial x_i}$ sometimes is written as f_i'.
$\frac{\partial}{\partial x}$ is also written, in short, as ∂_x and higher-order derivatives are written as $\frac{\partial^n}{\partial x^n} = \partial_x^n$.

Similarly, second derivatives are written as

$$\frac{\partial^2 f}{\partial x_i \partial x_j} = f_{ji}''.$$

If $n = 3$, then often variables $x_i, \quad i = 1, 2, 3$ are written as (x, y, z). One then writes

$$\frac{\partial^2 f}{\partial x} = f_x', \quad \frac{\partial f}{\partial x} = f_x', \text{ and } \frac{\partial f}{\partial z} = f_z'.$$

In two-dimensional case, with the variables (x, y) the mixed second derivatives may also be written in the following forms:

$$\frac{\partial}{x}\left(\frac{\partial f}{\partial y}\right) = \frac{\partial^2 f}{\partial x \partial y} = f_{21}'' = f_{yx}''.$$

In the following, we state a theorem as a sufficient condition for the two mixed derivatives to be equal, in the special case: $n = 2$.

The theorems are given in $\mathbb{R}^2$ but can be generalized to $\mathbb{R}^n$, $n = 2, 3, \ldots$.

Theorem 17.4. *If f_{xy}'' and f_{yx}'' exist in a neighborhood of (x, y) and are continuous in (x, y), then they are equal.*

Definition 17.6. A function is differentiable in (x, y) if

$$f(x+h, y+k) - f(x, y) = h\frac{\partial f}{\partial x}(x, y) + k\frac{\partial f}{\partial y}(x, y) + \sqrt{h^2 + k^2}\varepsilon(h, k), \tag{17.13}$$

where $\varepsilon(h, k) \to 0$, as $(h, k) \to (0, 0)$.

Theorem 17.5. *Assume that f is defined in a neighborhood of (x, y). Assume further that there are two numbers A and B such that*

$$f(x+h, y+k) - f(x, y) = hA + kB + \sqrt{h^2 + k^2}\,\varepsilon(h, k), \tag{17.14}$$

where $\varepsilon(h, k) \to 0$, as $(h, k) \to 0$. Then f has partial derivatives in (x, y): $A = \frac{\partial f}{\partial x}(x, y)$ and $B = \frac{\partial f}{\partial y}(x, y)$.

Theorem 17.6.

(i) *If f is differentiable in (x, y), then f is continuous in (x, y).*
(ii) *If f has continuous partial derivatives in a neighbourhood of (x, y) : that is, if $\frac{\partial f}{\partial x}(x, y)$ and $\frac{\partial f}{\partial y}(x, y)$ are continuous in (x, y), then f is differentiable in (x, y).*

Definition 17.7. Assume that f is defined in a neighborhood of (x, y) and that $\mathbf{v} = (\alpha, \beta)$ is a unit vector. Then, the directional derivative of f in the direction of $\mathbf{v}$ is given by

$$f'_{\mathbf{v}}(x, y) := \lim_{t \to 0} \frac{f(x + \alpha t, y + \beta t) - f(x, y)}{t}, \tag{17.15}$$

if the limit exists.

Theorem 17.7. *Assume that f is differentiable in (x, y). Then f:s directional derivatives exist in any direction $\mathbf{v} = (\alpha, \beta)$ and*

$$f'_{\mathbf{v}}(x, y) = \alpha \frac{\partial f}{\partial x}(x, y) + \beta \frac{\partial f}{\partial y}(x, y). \tag{17.16}$$

Observe that all directional vectors are normalized, i.e., here $|\mathbf{v}| = \sqrt{\alpha^2 + \beta^2} = 1$. Otherwise, one uses $\dfrac{1}{|\mathbf{v}|}\mathbf{v}$ as directional vector.

Definition 17.8. The gradient of f is defined by

$$\nabla f \equiv \text{grad } f = \left(\frac{\partial f}{\partial x}, \frac{\partial f}{\partial y} \right). \tag{17.17}$$

The tangent plane at (a, b), i.e., at the point $(a, b, f(a, b))$ is defined by

$$z - f(a, b) = \frac{\partial f}{\partial x}(a, b)(x - a) + \frac{\partial f}{\partial y}(a, b)(y - b). \tag{17.18}$$

Remark.

$\mathbf{n} := (f'_x(a, b), f'_y(a, b), -1)$ is a normal vector to the tangent plane at the point (a, b) (see Figure 17.1).
∇f can be expressed by the unit, coordinate, vectors: $\boldsymbol{e}_x$, $\boldsymbol{e}_y$, as $\nabla f = \mathbf{i}\frac{\partial f}{\partial x} + \mathbf{j}\frac{\partial f}{\partial y}$.

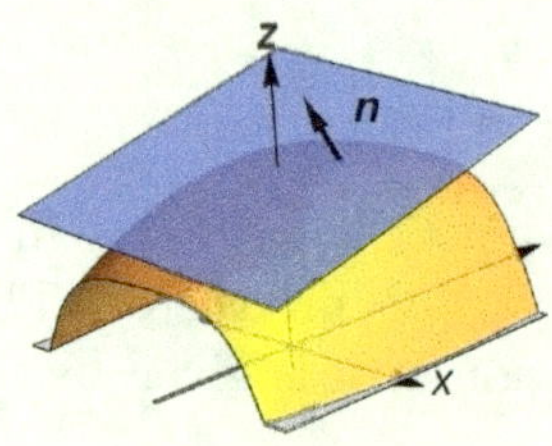

Figure 17.1: Tangent plane to the function surface with the normal vector $\boldsymbol{n}$.

A plane in $\mathbb{R}^n$ means a hyperplane of dimension $n-1$.

Theorem 17.8.

(i) $f'_{\mathbf{v}}(x,y) = \mathbf{v} \cdot \nabla f$.
(ii) *At every point* (a,b), *the function* $f(x,y)$ *grows fastest in the direction of its gradient vector at* (a,b): $\nabla f(a,b)$. *The maximal value is then* $|\nabla f(a,b)|$.
(iii) *Likewise at* (a,b), $f(x,y)$ *decreases fastest in the direction of* $-\nabla f(a,b)$. *The maximal decay is* $-|\nabla f(a,b)|$.
(iv) *The (momentary) change of* $f(x,y)$ *at the point* (a,b) *and in the direction of a tangent vector for the level curve through* (a,b) *is zero* $(=0)$.

17.2.2 *Some common surfaces*

Under reasonable conditions, e.g., continuity, on the function f, the mapping $(x,y) \curvearrowright f(x,y)$ presents a surface in $\mathbb{R}^3$ consisting of the points $(x,y,f(x,y))$. For instance, a paraboloid is such a surface. The function $f(x,y) = k(x^2+y^2)$ is a parabolic surface $(k \neq 0)$. Examples of one- and two-leaf hyperboloid are $z^2 = x^2+y^2+1$ and $z^2 = x^2+y^2-1$, equivalently, $z = \pm\sqrt{x^2+y^2+1}$ and $z = \pm\sqrt{x^2+y^2-1}, \quad x^2+y^2 \geq 1$, respectively.

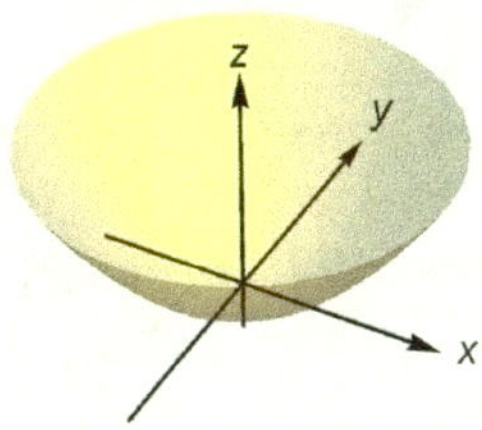

Paraboloid surface with equation $z = x^2 + y^2$.

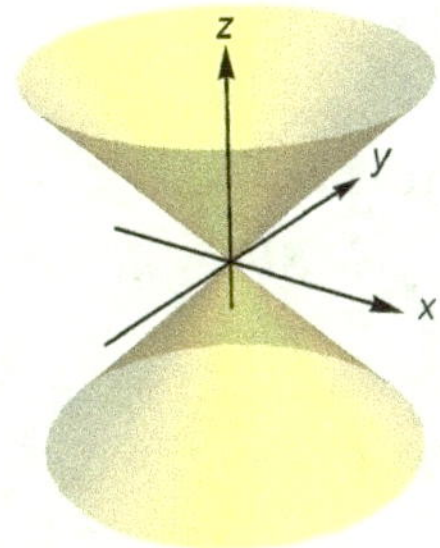

Double cone surface with equation $z = \pm\sqrt{x^2 + y^2}$.

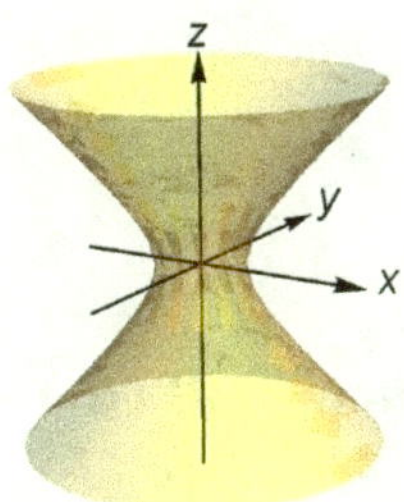

One-leaf hyperboloid surface with equation $z^2 + 1 = x^2 + y^2$.

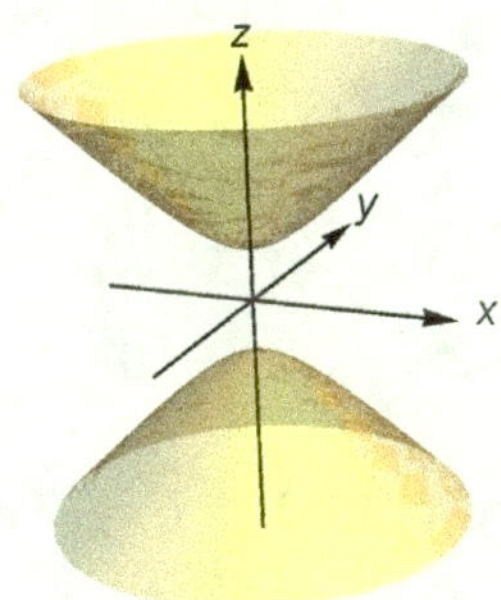

Two-leaf hyperboloid surface with equation $z^2 - 1 = x^2 + y^2$.

17.2.3 *Level curve and level surface*

Definition 17.9. Let C be a constant.

(i) For a function f of two variables $\{(x, y) : f(x, y) = C\}$ is a level curve.
(ii) For a function f of three variables $\{(x, y, z) : f(x, y, z) = C\}$ is a level surface.

17.2.4 *Composite function and its derivatives*

The expansions of derivatives of composite function can be interpreted as "chain rules", as in the one-dimensional case.

Theorem 17.9. *Let $t \mapsto (x(t), y(t))$ and consider the composite function $f(x, y) = f(x(t), y(t))$. If $x(t)$ and $y(t)$ have first-order derivatives and $f(x, y)$ is continuously differentiable (i.e., f'_x and f'_y are continuous), then the composite function $t \mapsto f(x(t), y(t))$ has derivatives in t and*

$$\frac{df}{dt} = \frac{\partial f}{\partial x}\frac{dx}{dt} + \frac{\partial f}{\partial y}\frac{dy}{dt}. \tag{17.19}$$

If $x = x(u, v)$ and $y = y(u, v)$, both are differentiable in u and v, then

$$\begin{aligned} \frac{\partial f}{\partial u} &= \frac{\partial f}{\partial x}\frac{\partial x}{\partial u} + \frac{\partial f}{\partial y}\frac{\partial y}{\partial u} \quad \text{and} \\ \frac{\partial f}{\partial v} &= \frac{\partial f}{\partial x}\frac{\partial x}{\partial v} + \frac{\partial f}{\partial y}\frac{\partial y}{\partial v}. \end{aligned} \tag{17.20}$$

Theorem 17.10. *Under continuity condition on all involved partial derivatives, it yields*

$$\begin{aligned} \frac{\partial^2 f}{\partial u^2} &= \left(\frac{\partial x}{\partial u}\right)^2 \frac{\partial^2 f}{\partial x^2} + \left(\frac{\partial y}{\partial u}\right)^2 \frac{\partial^2 f}{\partial y^2} + \frac{\partial^2 x}{\partial u^2}\frac{\partial f}{\partial x} + \frac{\partial^2 y}{\partial u^2}\frac{\partial f}{\partial y} \\ &\quad + 2\frac{\partial x}{\partial u}\frac{\partial y}{\partial u}\frac{\partial^2 f}{\partial x \partial y}. \\ \frac{\partial^2 f}{\partial u \partial v} &= \frac{\partial x}{\partial u}\frac{\partial x}{\partial v}\frac{\partial^2 f}{\partial x^2} + \frac{\partial y}{\partial u}\frac{\partial y}{\partial v}\frac{\partial^2 f}{\partial y^2} + \frac{\partial^2 f}{\partial x \partial y}\left(\frac{\partial x}{\partial u}\frac{\partial y}{\partial v} + \frac{\partial x}{\partial v}\frac{\partial y}{\partial u}\right) \\ &\quad + \frac{\partial f}{\partial x}\frac{\partial^2 x}{\partial u \partial v} + \frac{\partial f}{\partial y}\frac{\partial^2 y}{\partial u \partial v}. \end{aligned} \tag{17.21}$$

Coordinate transforms

Definition 17.10. A mapping $\mathbb{R}^m \to \mathbb{R}^n$ is also denoted by

$$\boldsymbol{u} \curvearrowright \boldsymbol{x}, \tag{17.22}$$

where $\mathbf{u} = (u_1, u_2, \ldots, u_m) \in \mathbb{R}^m$ and $\boldsymbol{x} = (x_1, x_2, \ldots, x_n) \in \mathbb{R}^n$, with $x_j = x_j(\mathbf{u})$, $j = 1, 2, \ldots, n$.

For a coordinate transform with $m = n$, the mapping (17.22) is bijective (one-to-one correspondence).

For a mapping, where all partial derivatives $\frac{\partial x_j}{\partial u_k}$ exist; $j = 1, 2, \ldots, n$ and $k = 1, 2, \ldots, m$, *the functional matrix* is defined as

$$\left(\frac{\partial x_j}{\partial u_k}\right)_{m\times n} = \begin{bmatrix} \dfrac{\partial x_1}{\partial u_1} & \dfrac{\partial x_2}{\partial u_1} & \cdots & \dfrac{\partial x_n}{\partial u_1} \\ \dfrac{\partial x_1}{\partial u_2} & \dfrac{\partial x_2}{\partial u_2} & \cdots & \dfrac{\partial x_n}{\partial u_2} \\ \vdots & \vdots & \ddots & \vdots \\ \dfrac{\partial x_1}{\partial u_m} & \dfrac{\partial x_2}{\partial u_m} & \cdots & \dfrac{\partial x_n}{\partial u_m} \end{bmatrix}. \tag{17.23}$$

For $m = n$, the functional determinant is defined as the determinant of the functional matrix.

Polar and cylindrical coordinates in $\mathbb{R}^2$ and in $\mathbb{R}^3$

Definition 17.11. Polar and spherical coordinates

$$\begin{aligned} &\mathbb{R}^2 : \begin{cases} x = r\cos\theta, \\ y = r\sin\theta, \end{cases} \qquad r = \sqrt{x^2 + y^2}, \\ &\mathbb{R}^3 : \begin{cases} x = r\sin\varphi\cos\theta, \\ y = r\sin\varphi\sin\theta, \\ z = r\cos\varphi, \end{cases} \qquad r = \sqrt{x^2 + y^2 + z^2}, \end{aligned} \tag{17.24}$$

where $0 \le \theta < 2\pi$, $0 \le \varphi < \pi$ and $r > 0$.

Cylindrical coordinates in $\mathbb{R}^3$:

$$\begin{cases} x = \rho\cos\theta, \\ y = \rho\sin\theta, \\ z = z, \end{cases} \tag{17.25}$$

where $0 \le \theta < 2\pi$ (same angle as for θ in polar coordinates) and

$$\rho = \sqrt{x^2 + y^2} > 0.$$

Below: The polar coordinates in $\mathbb{R}^2$, r and θ.
Right: The polar coordinates in $\mathbb{R}^3$, r, θ and φ.

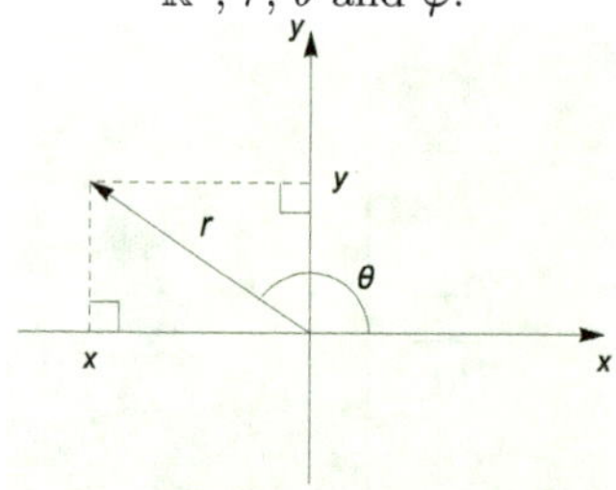

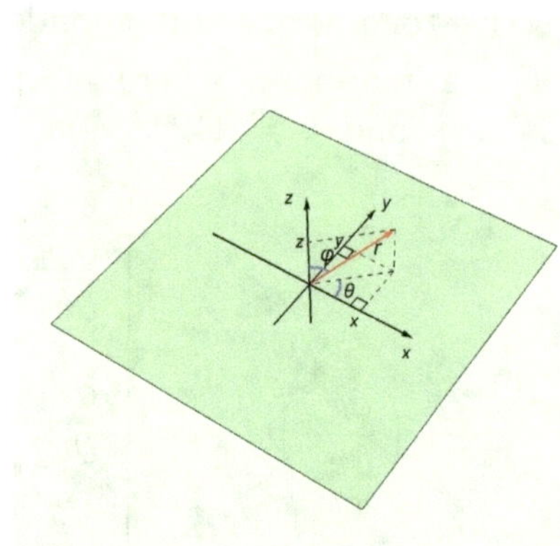

17.2.5 *Some special cases of chain rule*

Theorem 17.11. *In polar coordinates*

$$\begin{cases} x = r\cos\theta, \\ y = r\sin\theta. \end{cases}$$

$$\begin{aligned} \frac{\partial f}{\partial r} &= \frac{\partial f}{\partial x}\cos\theta + \frac{\partial f}{\partial y}\sin\theta. \\ \frac{\partial f}{\partial \theta} &= \frac{\partial f}{\partial x}(-r\sin\theta) + \frac{\partial f}{\partial y} r\cos\theta. \end{aligned} \tag{17.26}$$

Definition 17.12. The Laplace operator Δ is defined as

$$\begin{aligned} \Delta f &= \frac{\partial^2 f}{\partial x^2} + \frac{\partial^2 f}{\partial y^2} \qquad \text{in } \mathbb{R}^2. \\ \Delta f &= \frac{\partial^2 f}{\partial x_1^2} + \frac{\partial^2 f}{x_2^2} + \cdots + \frac{\partial^2 f}{x_n^2} \qquad \text{in } \mathbb{R}^n. \end{aligned} \tag{17.27}$$

The Laplace operator can be expressed using ∇, as $\nabla^2 = \Delta$.

Theorem 17.12. *The Laplace operator in polar coordinates in* $\mathbb{R}^2$*:*

$$\Delta f = \frac{\partial^2 f}{\partial r^2} + \frac{1}{r^2}\frac{\partial^2 f}{\partial \theta^2} + \frac{1}{r}\frac{\partial f}{\partial r}. \tag{17.28}$$

The Laplace operator in spherical coordinates in $\mathbb{R}^3$*:*

$$\Delta f = \frac{1}{r^2}\left[\frac{\partial}{\partial r}\left(r^2\frac{\partial f}{\partial r}\right) + \frac{1}{\sin\varphi}\frac{\partial}{\partial\varphi}\left(\sin\varphi\frac{\partial f}{\partial\varphi}\right) + \frac{1}{\sin^2\varphi}\frac{\partial^2 f}{\partial\theta^2}\right]. \tag{17.29}$$

The Laplace operator in cylindrical coordinates in $\mathbb{R}^3$*:*

$$\Delta f = \frac{1}{r}\frac{\partial}{\partial r}\left(r\frac{\partial f}{\partial r}\right) + \frac{1}{r^2}\frac{\partial^2 f}{\partial\theta^2} + \frac{\partial^2 f}{\partial z^2}. \tag{17.30}$$

The Laplace operator in $\mathbb{R}^n$ *:*

$$\Delta f = \frac{n-1}{r}\cdot\frac{df}{dr} + \frac{d^2 f}{dr^2}, \quad r = \sqrt{x_1^2 + x_2^2 + \cdots + x_n^2}. \tag{17.31}$$

17.3 Taylor's Formula

Theorem 17.13. *Assume that:*

(i) $f : D \to \mathbb{R}$, *where* D *is a non-empty, open, and connected subset of* $\mathbb{R}^2$, *containing the line segment between* (x_0, y_0) *and* $(x_0 + \Delta x, y_0 + \Delta y)$ *for some* Δx, $\Delta y \neq 0$, *and*
(ii) $f \in \mathcal{C}^{m+1}(D)$, *i.e., the partial derivatives of* f *: up to order* $m+1$ *are continuous.*

Then Taylor's formula is

$$\begin{aligned}
f(x_0+\Delta x, y_0+\Delta y) &= f(x_0,y_0) + \frac{\partial f}{\partial x}(x_0,y_0)\Delta x + \frac{\partial f}{\partial y}(x_0,y_0)\Delta y \\
&+\frac{1}{2}\left(\frac{\partial^2 f}{\partial x^2}(x_0,y_0)\Delta x^2 + 2\frac{\partial^2 f}{\partial x \partial y}(x_0,y_0)\Delta x\Delta y + \frac{\partial^2 f}{\partial y^2}(x_0,y_0)\Delta y^2\right) \\
&+\cdots+\frac{1}{m!}\sum_{k=0}^{m}\binom{m}{k}\frac{\partial^m f}{\partial x^k \partial y^{m-k}}(x_0,y_0)\Delta x^k\Delta y^{m-k}(x_0,y_0) \\
&+\underbrace{\frac{1}{(m+1)!}\sum_{k=0}^{m+1}\binom{m+1}{k}\frac{\partial^{m+1} f}{\partial x^k \partial y^{m+1-k}}(x_0+\theta\Delta x, y_0+\theta\Delta y)\Delta x^k\Delta y^{m+1-k}}_{\text{Lagrange's rest term}=R_m(\Delta x,\Delta y)}
\end{aligned} \tag{17.32}$$

for some $0 < \theta < 1$.

Theorem 17.14. *Under the same conditions as in the previous theorem and with* $D \subseteq \mathbb{R}^n$, $\boldsymbol{\Delta x} = (\Delta x_1, \Delta x_2, \ldots, \Delta x_n)$, *and* $\boldsymbol{x}_0 = (x_{01}, x_{02}, \ldots, x_{0n})$, *Taylor's formula is*

$$f(\boldsymbol{x}_0+\Delta \mathbf{x}) = f(\boldsymbol{x}_0) + \sum_{k=0}^{m}\frac{1}{k!}(\boldsymbol{\Delta x}\cdot\nabla)^k f(\boldsymbol{x}_0) + \frac{1}{(m+1)!}(\boldsymbol{\Delta x}\cdot\nabla)^{(m+1)} f(\boldsymbol{x}_0+\theta\boldsymbol{\Delta x}), \tag{17.33}$$

for some θ; $0 < \theta < 1$.

17.4 Maximum and Minimum Values of a Function

Definition 17.13. Consider a function $f : D_f \to \mathbb{R}$, where $D_f \subset \mathbb{R}^n$.

(i) The function has a local maximum at a point $\boldsymbol{x}_0 \in D_f$ if there is a neighborhood G of $\boldsymbol{x}_0$, $G := \{\boldsymbol{x} \in \mathbb{R}^n : |\boldsymbol{x} - \boldsymbol{x}_0| < \delta\}$, such that

$$f(\boldsymbol{x}) \le f(\boldsymbol{x}_0), \quad \text{for } \boldsymbol{x} \in D_f \cap G. \tag{17.34}$$

(ii) f has local minimum at $\boldsymbol{x}_0$ if $-f$ has a local maximum at $\boldsymbol{x}_0$.

(iii) If $f(\boldsymbol{x}_0) \geq (\leq) f(\boldsymbol{x})$ for all $\boldsymbol{x} \in D_f$, then $f(\boldsymbol{x}_0)$ is the largest (smallest) value of the function f.

(iv) If f has partial derivatives in an open neighborhood of $\boldsymbol{x}_0 \in D_f$ and $\frac{\partial f}{\partial x_i}(\boldsymbol{x}_0) = 0$ for $i = 1, 2, \ldots, n$, then f is said to have a stationary (or critical) point at $\boldsymbol{x}_0$.

(v) If $\boldsymbol{x}_0$ is a stationary point, and f does not have a local maximum or minimum at $\boldsymbol{x}_0$, then $\boldsymbol{x}_0$ is called a saddle point.

(vi) "Maximi"- and "minimi-points" are collectively called "extreme values". Corresponding function values are called "maximum" (shorter: max) or "minimum" (shorter: min) or collectively "extreme points".
A stationary point is of the form $(\boldsymbol{x}_0, f(\boldsymbol{x}_0))$, with $\frac{\partial f}{\partial x_i} = 0$, $i = 1, 2, \ldots n$.

(vii) The quadratic form of f at $\boldsymbol{x}_0$ is defined as

$$Q(\boldsymbol{x}_0; \boldsymbol{h}) = Q(\boldsymbol{h}) = \sum_{i,j} f''_{x_i,x_j}(\boldsymbol{x}_0) h_i h_j, \tag{17.35}$$

where $\boldsymbol{h} = (h_1, h_2, \ldots, h_n)$ and $f''_{x_i x_j} = \frac{\partial^2 f}{\partial x_j \partial x_i}$.

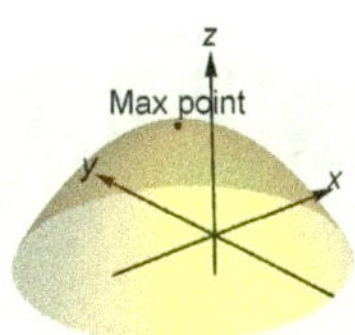

Function with max point

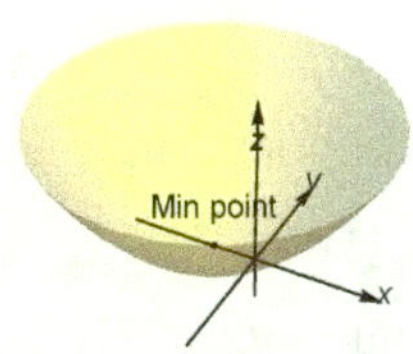

Function with min point

Function with saddle point

Theorem 17.15.

(i) *Assume that a differentiable function f has an extreme point at an interior point, $\boldsymbol{x}_0$, of D_f. Then $\frac{\partial f}{\partial x_i}(\boldsymbol{x}_0) = 0$ for $i = 1, 2, \ldots, n$.*

(ii) *Assume that* $\frac{\partial f}{\partial x_i}(\boldsymbol{x}_0) = 0$ *and that* f *has continuous second-order derivatives in a neighborhood of* $\boldsymbol{x}_0$. *Then the following hold true:*

(a) $Q(\boldsymbol{h}) > 0 \Longrightarrow$ f *has local minimum at* $\boldsymbol{x}_0$.
(b) $Q(\boldsymbol{h}) < 0 \Longrightarrow$ f *has local maximum at* $\boldsymbol{x}_0$.
(c) $Q(\boldsymbol{h})$ *assume both positive and negative values in every neighborhood of* $\boldsymbol{x}_0 \Longrightarrow$ f *has a saddle point at* $\boldsymbol{x}_0$.

If $Q(\boldsymbol{h}) \equiv 0$ *in a neighborhood of* $\boldsymbol{x}_0$, *one cannot decide the nature of the stationary point.*

The following are the criteria for stationary points of a function f, with $D_f \subseteq \mathbb{R}^2$.

Theorem 17.16.

(i) *Let* $f''_{xx}f''_{yy} - (f''_{xy})^2$ *be evaluated at the point* (x_0, y_0), $f'_x = f'_y = 0$ *at* (x_0, y_0), *and the second-order partial derivatives be continuous at* (x_0, y_0). *Then*

(i) *If* $f''_{xx}f''_{yy} - (f''_{xy})^2 > 0$ *and* $f_{xx} > 0$, *then* f *has a local minimum at* (x_0, y_0).
(ii) *If* $f''_{xx}f''_{yy} - (f''_{xy})^2 > 0$ *and* $f_{xx} < 0$, *then* f *has a local maximum at* (x_0, y_0).
(iii) *If* $f''_{xx}f''_{yy} - (f''_{xy})^2 < 0$, *then* f *has a saddle point at* (x_0, y_0).

Remark. For a function $f = f(x, y)$ defined over a sufficiently regular domain $M \subseteq \mathbb{R}^2$, the local max and min points of f at (x_0, y_0) are determined *viz.*

(i) An interior stationary point $f'_x(x_0, y_0) = f'_y(x_0, y_0) = 0$.
(ii) A boundary point, where the function is differentiable with derivative $= 0$, or
(iii) A point where f has no derivatives, e.g., "corners" of M.

17.4.1 *Max and min with constraints*

Definition 17.14. Let $f : D_f \to \mathbb{R}$, where $D_f \subseteq \mathbb{R}^n$.

(i) A constraint is a function $g(\boldsymbol{x}) = 0$, where $\boldsymbol{x} \in D_f$.

(ii) That f has a local minimum at $\boldsymbol{x}_0 \in D_f$, under the constraints $g_i(\boldsymbol{x}) = 0$, $i = 1, 2, \ldots, k$, means that:

(a) $g_i(\boldsymbol{x}_0) = 0$ for all $i = 1, 2, \ldots, k$, i.e., $\boldsymbol{x}_0$ satisfies the constraints and

(b) There is a neighborhood V of $\boldsymbol{x}_0$ such that

$$f(\boldsymbol{x}_0) \leq f(\boldsymbol{x}) \text{ for all } \boldsymbol{x} \in V \cap_{i=1}^{k} \{\boldsymbol{x} : g_i(\boldsymbol{x}) = 0\}.$$

(c) f has a local maximum, if $-f$ has a local minimum.

A necessary condition for extreme point

Definition 17.15. *The functional determinant* of a map $\boldsymbol{u} \to \boldsymbol{x}$, where
$\boldsymbol{u} = (u_1, u_2, \ldots, u_n)$ and $\boldsymbol{x} = (x_1, x_2, \ldots, x_n)$ is given by

$$\begin{vmatrix} \frac{\partial x_1}{\partial u_1} & \frac{\partial x_1}{\partial u_2} & \cdots & \frac{\partial x_1}{\partial u_n} \\ \frac{\partial x_2}{\partial u_1} & \frac{\partial x_2}{\partial u_2} & \cdots & \frac{\partial x_2}{\partial u_n} \\ & & \ddots & \\ \frac{\partial x_n}{\partial u_1} & \frac{\partial x_n}{\partial u_2} & \cdots & \frac{\partial x_n}{\partial u_n} \end{vmatrix} =: \frac{d(x_1, x_2, \ldots, x_n)}{d(u_1, u_2, \ldots, u_n)}. \tag{17.36}$$

Theorem 17.17. *Assume that f has a local extreme point under the constraints given in the definition, where $k < n$, and that f and g_i have continuous gradients in a neighborhood of $\boldsymbol{x}_0$. Then all functional determinants vanish as follows:*

$$\frac{d(f, g_1, g_2, \ldots, g_k)}{d(x_{i_1}, x_{i_2}, \ldots, x_{i_{k+1}})} = 0, \tag{17.37}$$

for every sub-index set $\{i_1, i_2, \ldots, i_{k+1}\} \subseteq \{1, 2, \ldots, n\}$ with $k+1$ elements.

Lagrange's multiplier method

The previous method can be formulated as follows:

Theorem 17.18 (Lagrange's multiplier method). *Assume that f has a local extreme point at $\boldsymbol{x}_0$. Then, either there are real numbers $\lambda_i, \quad i = 1, 2, \ldots, n$, such that*

$$\frac{\partial}{\partial x_i}[f + \lambda_1 g_1 + \lambda_2 g_2 + \cdots + \lambda_n g_n] = 0, \tag{17.38}$$

for $i = 1, 2, \ldots, n$ at the point $\boldsymbol{x}_0$,

Or all functional determinants satisfy

$$\frac{d(g_1, g_2, \ldots, g_k)}{d(x_{i_1}, dx_{i_2}, \ldots, x_{i_k})} = 0.$$

17.5 Optimization Under Constraints for Linear or Convex Function

Definition 17.16. (Optimization of linear function).

(i) With $\boldsymbol{x} = [x_1\, x_2\, \ldots\, x_n]^T$, $\boldsymbol{x}' = [x_1'\, x_2'\, \ldots\, x_n']^T \in \mathbb{R}^n$

$$\boldsymbol{x} \leq \boldsymbol{x}' \text{ means that } x_1 \leq x_1',\ x_2 \leq x_2',\ \ldots, x_n \leq x_n'.$$

(ii)

$$\begin{cases} \max(b_1 x_1 + b_2 x_2 + \cdots + b_n x_n), \text{ under constraints:} \\ x_1 \geq 0, x_2 \geq 0, \ldots, x_n \geq 0, \\ a_{11}x_1 + a_{12}x_2 + \cdots + a_{1n}x_n \leq c_1, \\ \ddots \\ a_{m1}x_1 + a_{m2}x_2 + \cdots + a_{mn}x_n \leq c_m, \end{cases} \tag{17.39}$$

is a linear program (LP) on standard maximi-form.

(iii) With notations

$$\boldsymbol{A} = \begin{bmatrix} a_{11} & a_{12} & \ldots & a_{1n} \\ a_{21} & a_{12} & \ldots & a_{2n} \\ & & \ddots & \\ a_{m1} & a_{m2} & \ldots & a_{mn} \end{bmatrix}, \boldsymbol{x} = \begin{bmatrix} x_1 \\ x_2 \\ \vdots \\ x_n \end{bmatrix}, \boldsymbol{y} = \begin{bmatrix} y_1 \\ y_2 \\ \vdots \\ y_m \end{bmatrix}, \boldsymbol{b} = \begin{bmatrix} b_1 \\ b_2 \\ \vdots \\ b_n \end{bmatrix}, \boldsymbol{c} = \begin{bmatrix} c_1 \\ c_2 \\ \vdots \\ c_m \end{bmatrix}$$

(17.39) is written, in compact form, as

$$\begin{cases} \max(\boldsymbol{b}^T \boldsymbol{x}), \\ \boldsymbol{x} \geq 0, \\ \boldsymbol{A}\boldsymbol{x} \leq \boldsymbol{c}, \end{cases} \quad \text{LP on standard maximi-form.} \tag{17.40}$$

(iv) The dual program corresponding to (17.40) is defined as

$$\begin{cases} \min(\boldsymbol{c}^T\boldsymbol{y}), \\ \boldsymbol{y} \geq 0, \\ \boldsymbol{A}\boldsymbol{y} \geq \boldsymbol{b}, \end{cases} \quad \text{LP on standard minimi-form.} \tag{17.41}$$

(v) $\boldsymbol{c}^T\boldsymbol{y}$ is called target function and the inequalities $\boldsymbol{A}\boldsymbol{y} \geq \boldsymbol{b}$, etc., are called constraints. A $\boldsymbol{y}$ that satisfies the constraints is called a permitted point/value.

Theorem 17.19 (The duality theorem). *For the dual programs* (17.40) *and* (17.41) *yield*

(i) (17.40) *lacks permitted points* $\Longrightarrow$ (17.41) *lacks optimal solution.*
(ii) (17.41) *lacks permitted points* $\Longrightarrow$ (17.40) *lacks optimal solution.*
(iii) *If both programs have permitted points, then both* (17.40) *and* (17.41) *have optimal solutions and the optimal values are equal.*

17.5.1 *Convex optimization*

Definition 17.17. The Hessian of f, denoted by $\boldsymbol{H}(f)$, is

$$\boldsymbol{H}(f)(\boldsymbol{x}) := \begin{bmatrix} \frac{\partial^2 f}{\partial x_1 \partial x_1} & \frac{\partial^2 f}{\partial x_1 \partial x_2} & \cdots & \frac{\partial^2 f}{\partial x_1 \partial x_n} \\ \frac{\partial^2 f}{\partial x_2 \partial x_1} & \frac{\partial^2 f}{\partial x_2 \partial x_2} & \cdots & \frac{\partial^2 f}{\partial x_2 \partial x_n} \\ \vdots & \vdots & \ddots & \vdots \\ \frac{\partial^2 f}{\partial x_n \partial x_1} & \frac{\partial^2 f}{\partial x_n \partial x_2} & \cdots & \frac{\partial^2 f}{\partial x_n \partial x_n} \end{bmatrix}. \tag{17.42}$$

A function f defined in a convex set M is convex if

$$f(\lambda \boldsymbol{x} + (1-\lambda)\boldsymbol{y}) \leq \lambda f(\boldsymbol{x}) + (1-\lambda) f(\boldsymbol{y}), \quad \forall \boldsymbol{x}, \boldsymbol{y} \in M, \ 0 < \lambda < 1. \tag{17.43}$$

If the inequality is strong, then the function is strongly convex.

Theorem 17.20.

(i) *Let M denote a convex set. Assume that $f : M \to \mathbb{R}$ is a convex function. Then, the set $\{\boldsymbol{x} \in M : f(\boldsymbol{x}) \leq a\}$ is a convex subset of M.*
(ii) *Let M be an open and convex set.*

(a) *Assume that f is differentiable on M. Then the following equivalence holds*

$$f \text{ convex } \Longleftrightarrow f(\boldsymbol{x}+\Delta\boldsymbol{x}) - f(\boldsymbol{x}) \geq \nabla f(\boldsymbol{x})\cdot\Delta\boldsymbol{x}.$$

(b) *Assume that f is two times continuously differentiable in M. Then the following equivalence holds true*

$$f \text{ convex } \Longleftrightarrow \boldsymbol{h}^T\boldsymbol{H}(f)(\boldsymbol{x})\boldsymbol{h} \geq 0 \text{ for all } \boldsymbol{h} = (h_1, h_2, \ldots, h_n)^T,$$

i.e., $\boldsymbol{h}^T\boldsymbol{H}(f)(\boldsymbol{x})\boldsymbol{h}$ is a positive semi-definite quadratic form.

Theorem 17.21. *Consider*

$$\begin{cases}\min(f(\boldsymbol{x})),\\ \boldsymbol{x}\in M,\end{cases} \tag{17.44}$$

where f is a convex function. Assume that (17.44) has an optimal solution.

(i) *If f is strongly convex, then the optimal solution is unique.*
(ii) *If $\boldsymbol{x}$ and $\boldsymbol{x}'$ both are optimal solutions, then all vectors in the set $\{\lambda\boldsymbol{x} + (1-\lambda)\boldsymbol{x}' : 0 \leq \lambda \leq 1\}$ are optimal solutions.*

The convex Kuhn–Tucker theorem:
Assume that $f, g_1, g_2, \ldots, g_m$ are convex functions in M. Consider

$$\begin{cases}\min(f(\boldsymbol{x})),\\ \\ g_k(\boldsymbol{x}) \leq c_k, \quad k = 1, 2, \ldots, m.\end{cases} \tag{17.45}$$

Suppose that there is a vector $\boldsymbol{y}\in\mathbb{R}^m$, $\boldsymbol{y}\geq\boldsymbol{0}$ so that the following conditions are satisfied:

$$\begin{aligned} &y_k(g_k(\boldsymbol{x}) - c_k) = 0, \quad k = 1, 2, \ldots, m \\ &\qquad\qquad\qquad\qquad\qquad\qquad \textit{and} \\ &\frac{\partial f}{\partial x_l} + y_1\frac{\partial g_1}{\partial x_k} + \cdots + y_m\frac{\partial g_m}{\partial x_m} = 0, \quad l = 1, 2, \ldots, n. \end{aligned} \tag{17.46}$$

Then $\boldsymbol{x}$ is an optimal solution of (17.44).

In particular, for (17.40) and (17.41) it yields that: $\boldsymbol{x}$ and $\boldsymbol{y}$ are permitted solutions to each program, moreover with

$$\boldsymbol{y}^T(\boldsymbol{A}\boldsymbol{x} - \mathbf{c}) = \boldsymbol{x}^T(\mathbf{b} - \boldsymbol{A}^T\boldsymbol{y}) = 0, \tag{17.47}$$

$\boldsymbol{x}$ and $\boldsymbol{y}$ are optimal for each program.

17.6 Integral Calculus

Definition 17.18. Let $f(x, y)$ be a bounded function in

$$M = [a, b] \times [c, d] = \{(x, y) : a \le x \le b, \quad c \le y \le d\}.$$

Split $[a, b]$ into m sub-intervals $[x_{i-1}, x_i) = \Delta x_i$ $i = 1, 2, \dots, m$, where $a = x_0 < x_1 < \cdots < x_m = b$ and the similar partition of $[c, d]$ as $[y_{j-1}, y_j) = \Delta y_j$ $j = 1, 2, \dots, n$, where $c = y_0 < y_1 < \cdots < y_n = d$. Furthermore, let

$$\begin{aligned} s_{ij} &= \inf\{f(x, y) : (x, y) \in \Delta x_i \times \Delta y_j\}, \text{ and} \\ S_{ij} &= \sup\{f(x, y) : (x, y) \in \Delta x_i \times \Delta y_j\}. \end{aligned}$$

Now set

$$\begin{aligned} s &:= \sum_{i=1, j=1}^{m,n} s_{ij} \Delta x_i \Delta y_j, \\ S &:= \sum_{i=1, j=1}^{m,n} S_{ij} \Delta x_i \Delta y_j. \end{aligned} \tag{17.48}$$

s and S are called lower and upper sums for f on D, respectively. f is called Riemann integrable if

$$\sup s = \inf S \quad \text{taken over all lower and upper sums.}$$

The common value is called the (double-)integral of f over D and is denoted by

$$\iint_D f(x, y) dx dy. \tag{17.49}$$

A multiple integral on $D \subseteq \mathbb{R}^n$, where D is compact, is defined similarly:

$$\iint \cdots \int_D f(x_1, x_2, \dots, x_n) dx_1 dx_2 \dots dx_n. \tag{17.50}$$

Theorem 17.22 (Fubini's theorem). *If f is continuous in $D = [a, b] \times [c, d]$, then f is integrable and the integration can be performed iteratively:*

$$\iint_D f(x, y)dxdy = \int_c^d \left(\int_a^b f(x, y)dx \right) dy$$
$$= \int_a^b \left(\int_c^d f(x, y)dy \right) dx. \qquad (17.51)$$

For a compact $x -$ simple set D given by $D = D = \{(x, y) : \phi(y) \leq x \leq \psi(y), \quad c \leq y \leq d\}$, where the functions ϕ and ψ are assumed to be continuous, the integral can be computed, see left figure on page 408,

$$\int_D f(x, y)dxdy = \int_c^d \left(\int_{\phi(y)}^{\psi(y)} f(x, y)dx \right) dy.$$

Remark. Fubini's theorem can be generalized to Riemannian integrable functions which are not necessarily continuous, and also to multiple integrals, i.e., integrals defined in $D \subset \mathbb{R}^n$.

The interval $[a, b]$ can be replaced by $[\phi(y), \psi(y)]$ if these functions are continuous in $y \in [c, d]$. In the same way $[c, d]$ can be replaced by $[\phi(x), \psi(x)]$, where $\phi \leq \psi$ are continuous functions.

The area A of a domain $D = \{(x, y) : \phi(y) \leq x \leq \psi(y), c \leq y \leq d\}$ is given by

$$A(D) = \iint_D dxdy.$$

Definitions and theorems can analogously be extended to $\mathbb{R}^n$.

The volume of

$$D = \{(x, y) : \phi_1(y, z) \leq x \leq \psi_1(y, z), \ \phi_2(z) \leq y \leq \psi_2(z), \ c \leq z \leq d\}$$

is given by

$$V(D) = \iiint_D dxdydz.$$

For two integrable functions f and g, the following holds true:

$$\left(\iint_D f(x,y)g(x,y)dxdy\right)^2 \leq \iint_D f(x,y)^2dxdy \iint_D f(x,y)^2dxdy. \tag{17.52}$$

This is called Schwarz's inequality (for integrals). It can easily be generalized to $\mathbb{R}^n$.

If $D = \prod_{k=1}^{n}[a_k, b_k]$, then the integral in (17.50) is computed iteratively:

$$\int_{a_1}^{b_1}\left(\int_{a_2}^{b_2}\left(\cdots\int_{a_n}^{b_n} f(x_1, x_2, \ldots, x_n)dx_n \ldots\right)dx_2\right)dx_1. \tag{17.53}$$

17.6.1 *Variable substitution in multiple integral*

Theorem 17.23 (Variable substitution *in* double integral). *Let x and y be two real-valued functions of (u, v), i.e.,* $\begin{cases} x = x(u,v) \\ y = y(u,v). \end{cases}$
If $(x, y) : D \to E$ (bijectively) and

$$\frac{d(x,y)}{d(u,v)} := \begin{vmatrix} \dfrac{\partial x}{\partial u} & \dfrac{\partial y}{\partial u} \\ \dfrac{\partial x}{\partial v} & \dfrac{\partial y}{\partial v} \end{vmatrix} \tag{17.54}$$

is continuous, then

$$\iint_D f(x,y)dxdy = \iint_E f(x(u,v), y(u,v))\left|\frac{d(x,y)}{d(u,v)}\right|dudv. \tag{17.55}$$

Remark. The expression (17.55) $\frac{d(x,y)}{d(u,v)}$ is called *functional determinant.*

The double integral is "non-oriented" in the sense that when divided in two single integrals $(\int_c^d(\int_a^b \ldots dx)dy)$, it is assumed that $a \leq b$ and $c \leq d$.

It suffices that the mapping $(u, v) \mapsto (x, y)$ exists, i.e., with $\boldsymbol{r} = (x, y)$, it suffices that $\boldsymbol{r}(D) = E$. One does not need to have a bijection between the domains D and E.

Generally, in a multiple-integral one can make a substitution of variables. Then, the ratio of substitution is determined by the functional determinant in $\mathbb{R}^n$:

$$\frac{d(x_1, x_2, \ldots, x_n)}{d(u_1, u_2, \ldots, u_k)}. \tag{17.56}$$

Theorem 17.24. *The polar/spherical and cylindrical coordinate transforms can be used for variable substitutions.*

The functional determinants are given as follows

Polar subst. in $\mathbb{R}^2$	*Spherical subst. in* $\mathbb{R}^3$	*Cylindrical subst. in* $\mathbb{R}^3$
r	$r^2 \sin\varphi$	ρ^2.

(17.57)

Improper double integral

Definition 17.19.

(i) Let $D \subseteq \mathbb{R}^n$ be an unbounded measurable set (in Riemann meaning). A nested sequence $(D_k)_{k=1}^{\infty}$ of subsets of D satisfies
  (a) $D_k \subseteq D_{k+1}$ for $k = 1, 2, \ldots$,
  (b) $\cup_{k=1}^{\infty} D_k = D$,
  (c) for every bounded set $D' \subseteq D$, there exists a D_k such that $D'_k \subseteq D_k$.

(ii) If for a Riemann integrable function $f \geq 0$, such that

$$\iint_D f(x, y)dxdy := \lim_{k\to\infty} \iint_{D_k} f(x, y)dxdy \tag{17.58}$$

exists for every nested sequence $(D_k)_{k=1}^{\infty}$. Then this limit is called the improper integral of f over D.

Theorem 17.25. *For the improper integral, above, it suffices that there exists* one *nested sequence* $(D_k)_{k=1}^{\infty}$, *such that the limit* (17.58) *exists.*

Remark. The concept of improper double integral can easily be generalized to multiple integrals in higher dimensions than 2.

Chapter 18

Vector Analysis

18.1 Differential Calculus in $\mathbb{R}^n$

Definition 18.1.

(i) A vector field is a function $\boldsymbol{f} : D \to \mathbb{R}^m$, where $D \subseteq \mathbb{R}^n$.
Let $\boldsymbol{x} = (x_1, x_2, \ldots, x_n) \in D$ and $\boldsymbol{x}_0 = (x_{1,0}, x_{2,0}, \ldots, x_{n,0}) \in D$.
$\boldsymbol{f}$ has the limit $\boldsymbol{A}$ at $\boldsymbol{x}_0$, if for each $\varepsilon > 0$ there is a $\delta > 0$, such that

$$|\boldsymbol{x} - \boldsymbol{x}_0| < \delta \Longrightarrow |\boldsymbol{f}(\boldsymbol{x}) - \boldsymbol{A}| < \varepsilon. \tag{18.1}$$

(ii) If $\boldsymbol{A} = \boldsymbol{f}(\boldsymbol{x}_0)$, so is $\boldsymbol{f}$ continuous in $\boldsymbol{x}_0$.

(iii) The Nabla operator is defined as

$$\begin{aligned} \nabla &= (\partial_{x_1}, \partial_{x_2}, \ldots, \partial_{x_n}) \\ &= e_1 \frac{\partial}{\partial x_1} + e_2 \frac{\partial}{\partial x_2} + \cdots + e_n \frac{\partial}{\partial x_n}. \end{aligned} \tag{18.2}$$

(iv) The Laplace operator is defined as

$$\nabla \cdot \nabla = \nabla^2 = \Delta = \frac{\partial^2}{\partial x_1^2} + \frac{\partial^2}{\partial x_2^2} + \cdots + \frac{\partial^2}{\partial x_n^2}. \tag{18.3}$$

The total derivative (or functional matrix) of $\boldsymbol{f} = \boldsymbol{f}(\boldsymbol{x}) = (f_1, f_2, \ldots, f_m)$, where $\boldsymbol{x} \in \mathbb{R}^n$ and $f_j = f_j(\,\boldsymbol{x}) \in \mathcal{C}^1(\mathbb{R}^n)$, is given by

$$\boldsymbol{f}'(\boldsymbol{x}) := \begin{bmatrix} \frac{\partial f_1}{\partial x_1} & \frac{\partial f_1}{\partial x_2} & \cdots & \frac{\partial f_1}{\partial x_n} \\ \frac{\partial f_2}{\partial x_1} & \frac{\partial f_2}{\partial x_2} & \cdots & \frac{\partial f_2}{\partial x_n} \\ \vdots & \vdots & \ddots & \vdots \\ \frac{\partial f_m}{\partial x_1} & \frac{\partial f_m}{\partial x_2} & \cdots & \frac{\partial f_m}{\partial x_n} \end{bmatrix}, \tag{18.4}$$

insofar as each individual derivative exists.

In the case $m = n$, the functional determinant corresponding to $\boldsymbol{f}$, is the determinant of the quadratic total derivative, $\boldsymbol{f}'(\boldsymbol{x})$ above.

$$\nabla f = \frac{\partial f}{\partial x_1}\mathbf{e}_1 + \cdots + \frac{\partial f}{\partial x_n}\mathbf{e}_n, \quad f : \mathbb{R}^n \to \mathbb{R}$$

(gradient of $\boldsymbol{F}$)

$$\operatorname{div} \boldsymbol{F} = \nabla \cdot \boldsymbol{F} = \frac{\partial F_1}{\partial x_1} + \cdots + \frac{\partial F_n}{\partial x_n}, \quad \boldsymbol{F} : \mathbb{R}^n \to \mathbb{R}^n$$

(divergence of $\boldsymbol{F}$)

$$\begin{aligned} \operatorname{curl} \boldsymbol{F} = \operatorname{rot} \boldsymbol{F} = \nabla \times \boldsymbol{F} &= \begin{vmatrix} \boldsymbol{i} & \boldsymbol{j} & \boldsymbol{k} \\ \frac{\partial}{\partial x} & \frac{\partial}{\partial y} & \frac{\partial}{\partial z} \\ F_1 & F_2 & F_3 \end{vmatrix} \\ &= \left(\frac{\partial F_3}{\partial y} - \frac{\partial F_2}{\partial z}, \frac{\partial F_1}{\partial z} - \frac{\partial F_3}{\partial x}, \frac{\partial F_2}{\partial x} - \frac{\partial F_1}{\partial y} \right), \\ &\text{(rotation of } \boldsymbol{F} = (F_1, F_2, F_2) : \mathbb{R}^3 \to \mathbb{R}^3). \end{aligned} \tag{18.5}$$

In three dimensions one writes the nabla operator as

$$\nabla = (\partial_x, \partial_y, \partial_z) = \mathbf{e}_x \frac{\partial}{\partial x} + \mathbf{e}_y \frac{\partial}{\partial y} + \mathbf{e}_z \frac{\partial}{\partial z}. \tag{18.6}$$

Theorem 18.1. *The following laws hold true:*

$$\begin{aligned} &(\boldsymbol{f} + \boldsymbol{g})'(\boldsymbol{x}) = \boldsymbol{f}'(\boldsymbol{x}) + \boldsymbol{g}'(\boldsymbol{x}), \\ &(k\boldsymbol{f})'(\boldsymbol{x}) = k\boldsymbol{f}'(\boldsymbol{x}), \quad (k \textit{ complex constant}). \\ &(\boldsymbol{f} \circ \boldsymbol{g})'(\boldsymbol{x}) = \boldsymbol{f}'(\boldsymbol{g}(\boldsymbol{x}))\boldsymbol{g}'(\boldsymbol{x}). \end{aligned} \tag{18.7}$$

$$\nabla(a\,f + b\,g(x)) = a\nabla f + b\nabla g.$$

$$\nabla\cdot(a\,\boldsymbol{F} + b\,\boldsymbol{G}) = a\nabla\cdot\boldsymbol{F} + b\nabla\cdot\boldsymbol{G}.$$

$$\nabla\times(a\,\boldsymbol{F} + b\,\boldsymbol{G}) = a\nabla\times\boldsymbol{F} + b\nabla\times\boldsymbol{G}.$$

$$\nabla(f\cdot g) = g\nabla f + f\nabla g. \quad \nabla(\boldsymbol{F}\cdot\boldsymbol{G}) = (\nabla\boldsymbol{F})\boldsymbol{G} + (\nabla\boldsymbol{G})\boldsymbol{F} + \boldsymbol{F}\times(\nabla\times\boldsymbol{G}) + \boldsymbol{G}\times(\nabla\times\boldsymbol{F}).$$

$$\nabla\cdot(f\,\boldsymbol{F}) = f\nabla\boldsymbol{F} + [\nabla f]\cdot\boldsymbol{F}. \quad \nabla\cdot(\boldsymbol{F}\times\boldsymbol{G}) = \boldsymbol{G}\cdot\nabla\times\boldsymbol{F} - \boldsymbol{F}\cdot\nabla\times\boldsymbol{G}.$$

$$\nabla\times(f\,\boldsymbol{F}) = f\,\nabla\times\boldsymbol{F} + (\nabla f)\times\boldsymbol{F}. \quad \nabla\times(\boldsymbol{F}\times\boldsymbol{G}) = (\boldsymbol{G}\cdot\nabla)\boldsymbol{F} - (\boldsymbol{F}\cdot\nabla)\boldsymbol{G} + \boldsymbol{F}(\nabla\cdot\boldsymbol{G}) - \boldsymbol{G}(\nabla\cdot\boldsymbol{F}).$$

$$\nabla\cdot(\nabla\times\boldsymbol{F}) = 0. \quad \nabla\times(\nabla f) = \boldsymbol{0}.$$

$$\nabla\times(\nabla\times\boldsymbol{F}) = \nabla(\nabla\cdot\boldsymbol{F}) - \nabla^2\boldsymbol{F}. \tag{18.8}$$

Theorem 18.2 (The implicit function theorem). *Let Ω be an open subset of $\mathbb{R}^n\times\mathbb{R}^k$. Assume that $\boldsymbol{f}:\Omega\to\mathbb{R}^k$ is continuously differentiable (or infinitely differentiable). Assume that $\boldsymbol{f}(\boldsymbol{x}_0,\boldsymbol{y}_0) = \boldsymbol{0}$. Assume further that the matrix $\frac{\partial\boldsymbol{f}}{\partial\boldsymbol{y}}$ is invertible at $(\boldsymbol{x}_0,\boldsymbol{y}_0)$. Then there is neighborhood U of $(\boldsymbol{x}_0,\boldsymbol{y}_0)$ and a continuously differentiable (or infinitely differentiable) function $\boldsymbol{g}$ such that $\boldsymbol{f}(\boldsymbol{x},\boldsymbol{y}) = \boldsymbol{0}$ on U if and only if $\boldsymbol{y} = \boldsymbol{g}(\boldsymbol{x})$. $\boldsymbol{g}$ is called the implicit function and is continuously differentiable as $\boldsymbol{f}$ with*

$$\boldsymbol{g}'(\boldsymbol{x}) = -[\boldsymbol{f}'_{\boldsymbol{y}}(\boldsymbol{x},\boldsymbol{y})]^{-1}\boldsymbol{f}'_{\boldsymbol{x}}(\boldsymbol{x},\boldsymbol{y}). \tag{18.9}$$

Definition 18.2.

(i) **Polar unit vectors** in $\mathbb{R}^3$ expressed in cartesian coordinates

$$(x,y,z),\ \text{with } r = \sqrt{x^2+y^2+z^2} \text{ and } \rho = \sqrt{x^2+y^2}:$$

$$e_r = \frac{(x,y,z)}{r}, \quad e_\theta = \frac{(-y,x)}{\rho}, \quad e_\varphi = \frac{(xz,yz,-(x^2+y^2))}{r\rho}. \tag{18.10}$$

(ii) **Cylindrical unit vectors** in $\mathbb{R}^3$ expressed in cartesian coordinates

$$(x, y, z), \text{ with } \rho = \sqrt{x^2 + y^2}:$$

$$e_\rho = \frac{(x, y, 0)}{\rho}, \quad e_\theta = \frac{(-y, x, 0)}{\rho}, \quad e_z = (0, 0, 1). \tag{18.11}$$

Theorem 18.3. *In polar coordinates in $\mathbb{R}^3$ the rotation becomes*

$$\nabla \times \boldsymbol{F} = \frac{1}{r^2 \sin\varphi} \begin{vmatrix} e_r & re_\varphi & (r\sin\varphi)e_\theta \\ \frac{\partial}{\partial r} & \frac{\partial}{\partial\varphi} & \frac{\partial}{\partial\theta} \\ F_r & rF_\varphi & (r\sin\varphi)F_\theta. \end{vmatrix}. \tag{18.12}$$

Theorem 18.4. *In cylindrical coordinates the operators in* (18.5) *are*

$$\nabla f = \frac{\partial f}{\partial\rho} e_\rho + \frac{1}{\rho}\frac{\partial f}{\partial\theta} e_\theta + \frac{\partial f}{\partial z}$$

$$\nabla \cdot \boldsymbol{F} = \frac{1}{\rho}\frac{\partial \rho F_\rho}{\partial\rho} + \frac{1}{\rho}\frac{\partial F_\theta}{\partial\theta} + \frac{\partial F_z}{\partial z} \tag{18.13}$$

$$\nabla \times \boldsymbol{F} = \frac{1}{\rho} \begin{vmatrix} e_\rho & \rho e_\theta & e_z \\ \frac{\partial}{\partial\rho} & \frac{\partial}{\partial\theta} & \frac{\partial}{\partial z} \\ F_\rho & \rho F_\theta & F_z \end{vmatrix}.$$

Definition 18.3.

(i) A curve γ in $\mathbb{R}^n$ is a map $\gamma(t) = (x_1(t), x_2(t), \ldots, x_{n(t)})$, where $t \in [a, b]$ and γ is continuously differentiable in all but a finite number of points $t_1, \ldots, t_m \in [a, b]$. It is assumed that right and left derivatives exist, i.e., $\gamma_L'(t_j)$ and $\gamma_H'(t_j)$ exist.

(a) A domain/subset of $\mathbb{R}^n$ is (path-wise) connected, if for each pair of points $\boldsymbol{x}$ and $\boldsymbol{y}$ in $\mathbb{R}^n$ there exists a curve $\gamma : [a, b]$ with $\gamma(a) = \boldsymbol{x}$ and $\gamma(b) = \boldsymbol{y}$.

(b) A curve is closed if $\gamma(a) = \gamma(b)$.

(c) Two curves γ_0 and γ_1 in $\mathbb{R}^n$ are homotopic if there is a map

$$f(s, t) : [0, 1] \times [a, b] \to \mathbb{R}^n, \tag{18.14}$$

such that $f(0, t) = \gamma_0(t)$, $f(1, t) = \gamma_1(t)$, and $f(s, t)$ is continuous in the variable s for each t.

(d) In a domain/subset of $\mathbb{R}^n$ each simply connected curve is homotopic with a point.

(ii) Let γ be a curve defined by the function $\boldsymbol{r}(t) = (x_1(t), x_2(t), \ldots, x_n(t))$, $t \in [a, b]$. Then the curve integral is defined as

$$\int_\gamma \boldsymbol{F} \cdot d\boldsymbol{r} = \int_a^b \boldsymbol{F} \cdot \frac{d\boldsymbol{r}}{dt} dt. \tag{18.15}$$

(iii) A curve γ in $\mathbb{R}^2$ simply encloses a domain D if $\gamma = \partial D$, and the mapping $\gamma : [a, b) \to \partial D$ is bijective if $\gamma(a) = \gamma(b)$.

(iv) $D \subseteq \mathbb{R}^2$ is simple in $x-$direction (or $x-$simple) if

$$D = \{(x, y) : \phi(y) \leq x \leq \psi(y),\ c \leq y \leq d\},$$

where $\phi(y)$ and $\psi(y)$ are continuous in $[c, d]$.

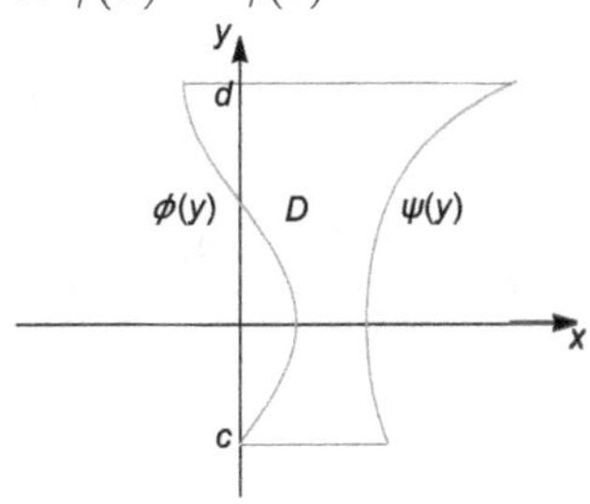

(v) Potential and gradient field

(a) A function f such that $\nabla f = \boldsymbol{F}$ is called potential and the corresponding $\boldsymbol{F}$ is called a gradient field.

(b) A function $\boldsymbol{A}$ such that $\text{rot}\boldsymbol{A} = \boldsymbol{F}$ is called a vector potential.

Remarks. For instance, in $\boldsymbol{R}^3$ one writes

$$\int_\gamma \boldsymbol{F} \cdot d\boldsymbol{r} = \int_a^b \boldsymbol{F} \cdot \frac{d\boldsymbol{r}}{dt} dt = \int_a^b \left(F_x \frac{dx}{dt} + F_y \frac{dy}{dt} + F_z \frac{dz}{dt} \right) dt.$$

Theorem 18.5.

(i) *If $\boldsymbol{F}$ is continuous in $\gamma([a, b])$ and γ is continuously differentiable in $[a, b]$, then the value of the curve integral is independent of the parametrization of $\gamma([a, b])$.*

(ii) *If f and $\boldsymbol{F}$ have continuous partial derivatives of second order, then*

$$\text{rot}\,(\nabla f) = \mathbf{0} \quad \textit{and} \quad \text{div}\,(\text{rot}\boldsymbol{F}) = 0. \tag{18.16}$$

(iii) *Let D be an open simply connected domain in $\mathbb{R}^3$ and assume $\boldsymbol{F}$ to have continuous partial derivatives. Then, the following equivalence holds:*

$$\exists f : \nabla f = \boldsymbol{F} \Longleftrightarrow \mathrm{rot}\boldsymbol{F} = \boldsymbol{0}. \tag{18.17}$$

Theorem 18.6 (Green's formula). *Consider the following three conditions:*

(i) *The curve γ simply encloses a simply connected domain $D \subseteq \mathbb{R}^2$, is counterclockwise-oriented and $D = \sqcup_{k=1}^{p} D_k$, $\gamma = \cup_{k=1}^{p} \gamma_k$ where $\gamma_k = \partial D_k$ and D_k are simple in x or y direction and $\overline{D}$ is compact,*
(ii) *the $\gamma : [a, b] \to \partial D$ and*
(iii) $F_x, F_y, \dfrac{\partial F_y}{\partial x}, \dfrac{\partial F_x}{\partial y}$ *are continuous on $\overline{D}$.*

Under these conditions

$$\oint_\gamma F_x dx + F_y dy = \iint_D \left(\frac{\partial F_y}{\partial x} - \frac{\partial F_x}{\partial y} \right) dxdy. \tag{18.18}$$

Definition 18.4. $F_x dx + F_y dy$ is an exact differential form if there exists a function G such that

$$\frac{\partial G}{\partial x} = F_x, \text{ and } \frac{\partial G}{\partial y} = F_y. \tag{18.19}$$

Theorem 18.7. *Assume that F_x and F_y have continuous partial derivatives in a simply connected domain $D \subseteq \mathbb{R}^2$ and the curve $\gamma \subset D$. Then, the following properties are equivalent.*

(i) *$F_x dx + F_y dy$ is an exact differential form.*
(ii) $\dfrac{\partial F_y}{\partial x} - \dfrac{\partial F_x}{\partial y} = 0$ *on D.*
(iii) *$\int_\gamma F_x dx + F_y dy$ is independent of the integration path, i.e., only depends on start and endpoint of curves.*

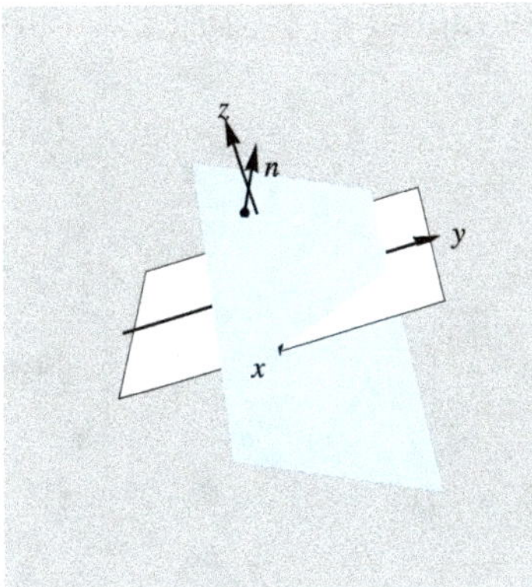

Positively oriented surface

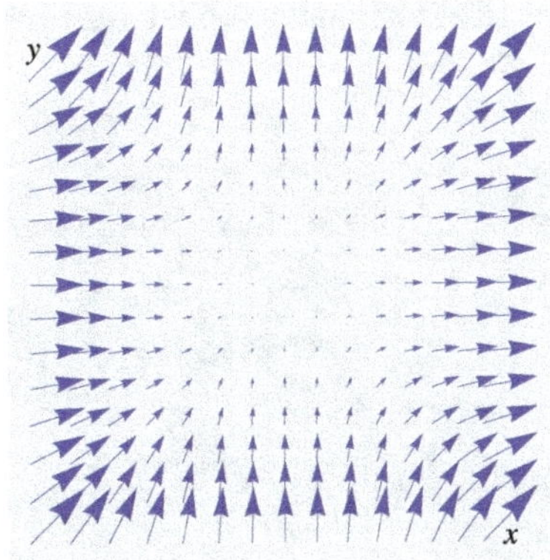

The vector field $(x, y) \curvearrowright \boldsymbol{F} = (x^2, y^2)$

Definition 18.5.

(i) A surface in $\mathbb{R}^3$ is a (piece-wise) continuously differentiable function S from a compact set $D \subseteq \mathbb{R}^2$ to $\mathbb{R}^3$. The mapping of S is denoted S.

(ii) A closed surface is a surface which is homotopic with $\{\boldsymbol{x} : |\boldsymbol{x} - \boldsymbol{x}_0| = r\}$ for some $\boldsymbol{x}_0$ and some $r > 0$. An outward unit normal for the latter surface is $\boldsymbol{n} = \dfrac{\boldsymbol{x} - \boldsymbol{x}_0}{|\boldsymbol{x} - \boldsymbol{x}_0|}, \quad |\boldsymbol{x} - \boldsymbol{x}_0| \neq 0.$

A parametrization is denoted by $(u, v) \curvearrowright (x(u, v), y(u, v), z(u, v)) = \boldsymbol{r}(u, v)$.

Generally, if $(x, y) \curvearrowright f(x, y)$, one has a function surface and map

$$(x, y) \curvearrowright (x, y, f(x, y)).$$

The normal vector of the surface S is given by

$$\boldsymbol{n} = \boldsymbol{r}_u \times \boldsymbol{r}_v. \tag{18.20}$$

Equation of the tangent plane to the surface S at the point $\boldsymbol{r}_0 = (x_0, y_0, z_0)$ is given by the "zero"-determinant

$$\begin{vmatrix} x - x_0 & y - y_0 & z - z_0 \\ \dfrac{\partial x}{\partial u} & \dfrac{\partial y}{\partial u} & \dfrac{\partial z}{\partial u} \\ \dfrac{\partial x}{\partial v} & \dfrac{\partial y}{\partial v} & \dfrac{\partial z}{\partial v} \end{vmatrix} = 0, \tag{18.21}$$

where the derivatives exist and are computed at (x_0, y_0, z_0). Further

$$\boldsymbol{r}_u \times \boldsymbol{r}_v \neq \mathbf{0}. \tag{18.22}$$

The area of the surface S is given by

$$A(S) = \int_M \left| \frac{\partial \boldsymbol{r}}{\partial u} \times \frac{\partial \boldsymbol{r}}{\partial v} \right| dudv. \tag{18.23}$$

The surface integral of $\boldsymbol{F}(x, y, z) = \boldsymbol{F}(\boldsymbol{r})$ is

$$\iint_S \boldsymbol{F}(x, y, z) dS = \iint_D \boldsymbol{F}(x(u, v), y(u, v), z(u, v)) \left| \frac{\partial \boldsymbol{r}}{\partial u} \times \frac{\partial \boldsymbol{r}}{\partial v} \right| dudv. \tag{18.24}$$

The normal-surface integral of a vector field $\boldsymbol{F}$ ($\boldsymbol{n}$ a unit normal) is defined as

$$\int_S \boldsymbol{F}(\boldsymbol{r}) \cdot \boldsymbol{n}\, dS = \int_S \boldsymbol{F}(\boldsymbol{r}) \cdot d\boldsymbol{S} = \int_D \boldsymbol{F}(\boldsymbol{r}) \cdot \left(\frac{\partial \boldsymbol{r}}{\partial u} \times \frac{\partial \boldsymbol{r}}{\partial v} \right) dudv. \tag{18.25}$$

Theorem 18.8 (Stokes' and Gauss' theorems). *Assume that γ is a curve that simply encloses the positively oriented, surface S with unit normal $\boldsymbol{n}$. Then*

$$\iint_S \mathbf{rotF} \times \mathbf{n}\, dS = \oint_\gamma \boldsymbol{F} \cdot dr \quad \textit{(Stokes' theorem)}. \tag{18.26}$$

Assume that S is a closed surface containing the domain V and $\boldsymbol{n}$ is an outward unit normal to S. Then

$$\iiint_V \mathrm{div}\boldsymbol{F} dxdydz = \iint_S \boldsymbol{F} \cdot \mathbf{n} dS, \quad \begin{pmatrix} \textit{Gauss' theorem, or} \\ \textit{the divergence theorem} \end{pmatrix}. \tag{18.27}$$

18.2 Types of Differential Equations

Definition 18.6.

$$\nabla^2 f = \Delta f = \sum_{k=1}^{n} \frac{\partial^2 f}{\partial x_k^2} = 0 \qquad \text{Laplace's equation}$$

$$\frac{\partial f}{\partial t} - \nabla^2 f = 0 \qquad \text{Heat conduction equation}$$

$$\frac{\partial^2 f}{\partial t^2} - \nabla^2 f = 0 \qquad \text{Wave equation}$$

$$\begin{cases} V(\boldsymbol{x})\psi - \dfrac{\hbar^2}{2m}\nabla^2\psi = E\psi & \text{The Schrödinger equation} \\ V(\boldsymbol{x})\psi - \dfrac{\hbar^2}{2m}\nabla^2\psi = i\hbar\dfrac{\partial \psi}{\partial t} & \text{Time-dependent Schrödinger equation.} \end{cases} \tag{18.28}$$

Remarks. In the Schrödinger equations

- $\Delta = \dfrac{\partial^2}{\partial x^2} + \dfrac{\partial^2}{\partial y^2} + \dfrac{\partial^2}{\partial z^2}$
- $\hbar = 1.0545727 \cdot 10^{-34}$ Js, is Planck's constant divided by 2π
- m = mass of the considered particle
- E = amount of energy
- V = potential energy.

Chapter 19

Topology

19.1 Definitions and Theorems

Definition 19.1. Let $\mathcal{T}$ be a class of subsets of a set X. $(X, \mathcal{T})$ is a topology (alt. $\mathcal{T}$ is a topology on X), if

(i) $G_i \in \mathcal{T} \implies \cup_i G_i \in \mathcal{T}$, where $\{G_i\}$ is an arbitrary class of elements in $\mathcal{T}$. G_i:s are referred to as open sets.
(ii) $G_i \in \mathcal{T} \implies \cap_i G_i \in \mathcal{T}$, where $\{G_i\}$ is a finite class of open sets.
(iii) The complement $F = X \setminus G = G^c$ of an open set G is said to be *closed*.
(iv) Let $A \subseteq X$. The relative topology on A consists of the class of all $\mathcal{T}_A := \{H\}$ where, for each H there exists a $G \in \mathcal{T}$ such that $H = G \cap A$.

Theorem 19.1.

(i) *$\emptyset$ and X, being complements of each other, are both open and closed.*
(ii) *The relative topology T_A on a subset (subspace) A is a topology on A.*

Definition 19.2.

(i) Assume that $A \subseteq X$.
   (a) The *interior* of A is denoted int $(A) = \cup G$, where the union is taken over all open $G \subseteq A$.
   (b) The *closure* of A is $\overline{A} = \cap F$, where the intersection is taken over all closed sets $F \supseteq A$.

(c) (i) A set A is everywhere dense in X if $\overline{A} = X$.
(ii) A set A is nowhere dense in X if $\text{int}\,(\overline{A}) = \emptyset$.
(iii) If there is a countable everywhere dense set, the space is called separable.

(ii) An open set G containing x is called a neighborhood of x (some literature define a neighborhood of x as a set H, such that $x \in G \subseteq H$, and G is open).

The separation axioms (Die Trennungsaxiomen).

(i) If for each pair of different elements x and y there are neighborhoods G_x and G_y of x and y, respectively, such that $x \notin G_y$ and $y \notin G_x$, then the space X is called a T_1-space.
(ii) If the above G_x and G_y can be chosen to be disjoint, then X is called a T_2-space, or most commonly a *Hausdorff space.*
(iii) If $x \notin F$ for a closed set F and there are disjoint open sets G_1 and G_2, such that $G_1 \supseteq F$ and $G_2 \ni x$, then X is called a T_3-space. Moreover, if X is both T_1- and T_3-space, then it is called a regular space.
(iv) If for each pair of disjoint closed sets F_1 and F_2 there exist a pair of two disjoint open sets G_1 and G_2 such that $F_1 \subset G_1$ and $F_2 \subset G_2$, the underlying space is called a T_4-space. Furthermore, if X is both T_1 and T_4, then it is called a normal space.

(a) A (local) base for an element $x \in X$ is a class of open neighborhoods $\mathcal{B}_x = \{B_{x,i}\}$ of x, such that for every open set G containing x, there exists a $B_{x,i}$ such that $x \in B_{x,i} \subseteq G$. If for each x there exists a *countable* local base, then the space X is said to be first countable.
(b) An (open) base of X is a class of open sets $\mathcal{B} = \{B_i\}$, such that each open set G can be expressed as a union of $B_i \in \mathcal{B}$. With a countable number of base sets, the space X is said to be second countable.
(c) An open cover of the space X is a union of open sets, such that $\cup_i G_i = X$.
If the union can be reduced to a countable cover, the space is called *Lindelöf space.*

(d) A sub-base is a class of (open) sets, such that the class of its finite intersections constitutes a base.

(e) An open cover of a set $E \subseteq X$ is a class of open sets G_i, such that $\cup_{i \in I} G_i \supseteq E$.

  (i) If each open cover of E can be reduced to a finite cover of E, then the set E is said to be compact. If $E = X$ has this property, then X is said to be compact.

  (ii) If for each $x \in X$ there is a neighborhood G of x, such that $\bar{G}$ is compact, then X is called locally compact.

(f) If there exists a distance function, a metric d in a set X such that $d : X \times X \to [0, \infty)$ with properties

$$d(x,y) = 0 \iff x = y, \quad d(x,y) = d(y,x),$$
$$d(x,z) \leq d(x,y) + d(y,z),$$

then the set $B = \{y : d(x,y) < \delta, \quad \delta > 0, \quad x \in X\}$, being open, generates a topology τ on X. The space is then called metric. If the topology $\mathcal{X}$ can be generated by a metric d, the topology is said to be metrizable.

(g) If $(X, \mathcal{T})$ and $(Y, \mathcal{U})$ are two topological spaces, then the product topology on $X \times Y$ is generated by the class $\{G \times H : \quad G \in \mathcal{T}, H \in \mathcal{U}\}$, which is an open base.

Theorem 19.2.

(i) *Every "singleton-"set $\{x\}$ is closed $\iff$ X is a T_1-space.*

(ii) *A second countable regular space is metric (metrizable).*

(iii) *A metric space is first countable and normal.*

(iv) *A locally compact Hausdorff space is regular.*

(v) *A compact Hausdorff space is normal.*

(vi) *Assume that X is a compact Hausdorff space. Then X is metric $\iff$ X is second countable.*

(vii) *If X is second countable, any open cover of X can be reduced to a countable cover (Lindelöf's theorem).*

(viii) *A regular Lindelöf space is normal.*

(ix) *X is a T_1-space $\Longrightarrow$ every compact set is closed.*

(x) *If X and Y are two compact spaces, then $X \times Y$ is compact (Tychonoff's theorem).*

Definition 19.3.

(i) A function $f : X \to Y$, where X and Y are topological spaces, is continuous if for each open set G in Y, $f^{-1}(G)$ is open in X.

(ii) A class $\mathcal{E} := \{E_i\}$, not necessarily open, is called locally finite if for each $x \in X$ there is a neighborhood G of x such that only a finite number of the E_i intersects G.

(iii) Assume that $\mathcal{G} = \{G_i\}$ is an open cover of X. $\mathcal{B} = \{B_j\}$ is called a locally finite refinement if

(a) $\mathcal{B}$ is locally finite, which means that for every $x \in X$, there exists a neighborhood G of x which intersects only a finite number of the $B_i \in \mathcal{B}$.

(b) $\mathcal{B}$ is a refinement of $\mathcal{G}$, which means that B_j is included in at least one of the sets G_i.

(iv) A Hausdorff space X is para-compact if for each open cover $\mathcal{G} = \{G_i\}$ of X there exists a locally finite refinement.

(v) A partition of unity is a class (set) of continuous functions $f_k : X \to [0,1]$, such that for every $x \in X$, there exists a neighborhood B of x such that, except for a finite number of f_k:s, $f_k \equiv 0$ on B, and

$$\sum_k f_k(x) \equiv 1 \text{ for every } x \in X. \qquad (19.1)$$

The partition is subordinate $\mathcal{B}$, if each set (each support) $\overline{\{x : f_k(x) \neq 0\}}$ is a subset of some $B \in \mathcal{B}$.

Theorem 19.3.

(i) *Assume that X is a normal space.*

(a) *(Urysohn's lemma) If F_0 and F_1 are two closed non-empty disjoint sets, then there exists a continuous function $f : X \curvearrowright [0,1]$, such that $f(F_0) = 0$ och $f(F_1) = 1$.*

(b) *(Tietze's Extension theorem) Assume that $f : F \curvearrowright [a,b]$ is a continuous function where F is closed in X. Then, f can be extended to a continuous function $f : X \curvearrowright [a,b]$.*

(ii) *If X is a locally compact Hausdorff space, K and F are two disjoint sets, which are compact and closed, respectively, then there exist disjoint open sets G and H such that $K \subseteq G$ and $F \subseteq H$.*

(iii) *For a locally compact Hausdorff space, the following version of Urysohn's lemma holds:*
For compact K and open G such that $K \subseteq G$, there exists a continuous function $f : X \curvearrowright [0, 1]$ with $f(K) = 1$ and $f(x) = 0$ outside G.

(iv) *A space is metric (metrizable) $\iff$ f is regular and has a locally finite base.*

(v) *Assume that X is a Hausdorff space.*
X is paracompact $\iff$ Every open cover $\mathcal{B}$ has a subordinate partition of unity.

(vi) *A metric space is paracompact (Stone). In particular, $\mathbb{R}^n$ is paracompact.*

(vii) *A paracompact Hausdorff space is normal.*

(viii) *Let X have two topologies, $\mathcal{T}_1$ and $\mathcal{T}_2$, both making X locally compact spaces with corresponding classes of compact sets $\mathcal{K}_1$ and $\mathcal{K}_2$.*
Furthermore, assume that $\mathcal{T}_1 \subseteq \mathcal{T}_2$, and $\mathcal{K}_1 = \mathcal{K}_2$. Then $\mathcal{T}_1 = \mathcal{T}_2$.

(ix) *Given a locally compact Hausdorff space X with $\mathcal{K}$, the corresponding class of compact sets, and an element $\infty \notin X$. Define the space*

$$X_\infty := X \cup \{\infty\} \text{ with topology } \mathcal{T}_\infty = \{G : X_\infty \setminus G \in \mathcal{K}\}.$$

Then, the space

$$(X_\infty, \mathcal{T}_\infty),$$

is a compact Hausdorff space where the restriction of X_∞ to X yields the restriction of $\mathcal{T}_\infty$ to $\mathcal{T}$.

(x) *Assume that $f : X \to Y$ is continuous. Then, for each compact set K in X, the set $f(K)$ is compact.*

(xi) **Urysohn's embedding theorem:** *If X is normal and second countable, then X is homeomorphic to a subset of*

$$\mathbb{R}^{\aleph_0} := \mathbb{R} \times \mathbb{R} \times \dots$$

and thus metrizable.

Definition 19.4.

(i) A sequence $(x_n)_{n=1}^{\infty}$ is convergent if there is an $x \in X$ such that for each neighborhood V of x there exists an index n_0 such that $n \geq n_0 \Longrightarrow x_n \in V$. In a metric space (X, d), this is expressed as for each $\varepsilon > 0$, there exists an n_0 such that $n \geq n_0 \Longrightarrow d(x, x_n) < \varepsilon$.

(ii) A sequence $(x_n)_{n=1}^{\infty}$ is called a Cauchy sequence if for each $\varepsilon > 0$ there is an n_0 such that $m, n \geq n_0 \Longrightarrow d(x_m, x_n) < \varepsilon$. The metric is complete if every Cauchy sequence is convergent.

Theorem 19.4.

(i) *Any metric space X can be extended to a complete metric space X' so that $\overline{X} = X'$.*

(ii) *Baire category theorem: Assume that there exists a complete metric on X and that $X = \cup_{k=1}^{\infty} A_k$. Then at least one of the sets in A_k is not nowhere dense, i.e., int $(\overline{A_k}) \neq \emptyset$ for at least one A_k.*

19.1.1 *Variants of compactness*

Definition 19.5. Let $(X, \mathcal{T})$ be a topological space.

(i) The space is compact if each open cover can be reduced to a finite subcover.

(ii) The space is sequentially compact if each sequence has a convergent subsequence.

(iii) The space is countably compact if each open and countable cover has a finite subcover.

(iv) The space has the Bolzano–Weierstrass property, if any set with an infinite number of elements has a limit-point.

Theorem 19.5. *For a topological space X, the following relations for compactness hold true.*

$$\begin{aligned} &\text{(i) } X \textit{ Compact} &&\Longrightarrow X \textit{ Countably compact,} \\ &\text{(ii) } X \textit{ Sequentially compact} &&\Longrightarrow X \textit{ Countably compact,} \\ &\text{(iii) } X \textit{ Countably compact} &&\Longrightarrow X \textit{ Bolzano Weierstrass.} \end{aligned} \tag{19.2}$$

The following converses hold:

$\Longleftarrow$ *holds in (i) if X has the Lindelöf property.*
$\Longleftarrow$ *holds in (ii) if X is first countable.*
$\Longleftarrow$ *holds in (iii) if X is T_1.*

19.2 The Usual Topology on $\mathbb{R}^n$

Definition 19.6. The usual topology, $\mathcal{T}$, on $\mathbb{R}$ is defined as the class of sets $G \subset \mathbb{R}$, such that

(i) $\emptyset \in \mathcal{T}$ and $\mathbb{R} \in \mathcal{T}$.
(ii) For every open set G, i.e. $G \in \mathcal{T}$, and $x \in G$, there is an open interval $I = (a, b) =: \{y : a < y < b\}$, such that

$$x \in (a, b) \subset G.$$

Remark. It can be shown that the usual topology meets the conditions of the general definition.

The usual topology in $\mathbb{R}^n$ $n = 2, 3, ...$ can either be defined as the product topology generated by open base-sets

$$B_j = \Pi_{j=1}^{n}(a_j, b_j), \quad a_j < b_j, \quad a_j, b_j \in \mathbb{R},$$

or by open spheres

$$B_j = \{x \in \mathbb{R}^n : |x - x_j| < r_j\}.$$

19.2.1 *A comparison between two topologies*

To compare the concepts presented above, we can consider usual topology $\mathcal{T}$ on $\mathbb{R}$ generated by the metric $d(x, y) = |x - y|$, or alternatively, the intervals (a, b), $a, b \in \mathbb{R}$ and the so-called right topology $\mathcal{T}_h$ generated by open sets $(a, b]$, $a, b \in \mathbb{R}$.

	$(\mathbb{R}, \mathcal{T})$	$(\mathbb{R}, \mathcal{T}_h)$	$(\mathbb{R}^2, \mathcal{T})$	$(\mathbb{R}^2, \mathcal{T}_h)$
Hausdorff	yes	yes	yes	yes
Compact	no	no	no	no
Locally compact	yes	no	yes	no
Regular	yes	yes	yes	no
Normal	yes	yes	yes	no
Metric	yes	no	yes	no
Lindelöf	yes	yes	yes	yes
Second countable	yes	no	yes	no
First countable	yes	yes	yes	yes
Paracompact	yes	no	yes	no
Every open union can be written as a disjoint union of intervals	yes	yes	no*	no*

Note: *Interval in $\mathbb{R}^2$ is interpreted as $(a, b) \times (c, d)$ and $(a, b] \times (c, d]$, respectively.

19.3 Axioms

An axiom is a statement that cannot be proved.

In mathematics there are a number of axioms, some of them are introduced in the following.

19.3.1 *The parallel axiom*

Given two different points P and Q, there exists exactly one line through them.

19.3.2 *The induction axiom*

Given the set $\mathcal{N} := \{n_0, n_1, ...\} \subset \mathbb{Z}$, let $P(n)$ be a statement for $n \in \mathcal{N}$.

$$\begin{cases} P(n_0) \text{ true and} \\ \\ P(n) \text{ true } \Longrightarrow P(n+1) \text{ true} \end{cases} \Longrightarrow P(n) \text{ true for } n = n_0, n_{0+1}, \dots \tag{19.3}$$

19.3.3 *Axiom of choice*

Given a class $\mathcal{X} = \{X_j : j \in I\}$ of non-empty sets X_j. Then there is a function

$$f : \mathcal{X} \to \cup_j X_j \text{ such that for every } X_j \in \mathcal{X},\ f(X_j) \in X_j.$$

19.4 The Supremum Axiom with Some Applications

With the supremum axiom follows a number of theorems (see the following section).

19.4.1 *The supremum axiom*

Definition 19.7. A non-empty subset $\mathcal{A}$ of $\mathbb{R}$ is said to be:

- Bounded above if there is a real number x, such that $x \geq a$ for all $a \in \mathcal{A}$, x is called a majorant to $\mathcal{A}$.
- Bounded below if there is a real number x, such that $x \leq a$ for all $a \in \mathcal{A}$, x is called a minorant to $\mathcal{A}$.
- Bounded, if it is both bounded above and below.

These notions coincide with the definition of bounded intervals in Chapter 1.

The supremum axiom

Every non-empty set $\mathcal{A} \subset \mathbb{R}$, which is bounded above, has a smallest majorant.

Definition 19.8. The smallest majorant is called the supremum of $\mathcal{A}$ and is denoted by $\sup \mathcal{A}$.

Theorem 19.6.

(i) *Every non-empty set, bounded below, has a largest minorant. This number is called infimum of* $\mathcal{A}$ *and is denoted by* $\inf \mathcal{A}$.

(ii) *Let the non-empty set* $\mathcal{A}$ *be bounded above, then* $x_0 = \sup \mathcal{A}$ *is equivalent to the following two conditions:*

$x_0 \geq x$ *for all* $x \in \mathcal{A}$ *and*
$x \leq y$ *for all* $x \in \mathcal{A}$ *imply* $x_0 \leq y$.

(iii) *The Supremum axiom is equivalent to the Dedekind property. The Dedekind property can thus be an alternative axiom, which one can start with. It says that:*
For two non-empty sets A *and* B *of* $\mathbb{R}$ *such that* $a \leq b$ *for all* $a \in A$ *and* $b \in B$ *there is a number* x *such that* $a \leq x \leq b$.

$$\textit{The Supremum axiom} \Longleftrightarrow \textit{The Dedekind property.} \tag{19.4}$$

Denote $s := \sup \mathcal{A}$. *For each* $\varepsilon > 0$, *there is an* $x \in \mathcal{A}$, *such that*

$$s - \varepsilon < x \leq s. \tag{19.5}$$

19.4.2 Compact set in $\mathbb{R}^n$

Definition 19.9. An open cover of a subset $\mathcal{A}$ of $\mathbb{R}$ is a union of open intervals $V_j = (c_j, d_j)$, such that

$$\mathcal{A} \subseteq \bigcup_{j \in J} V_j.$$

An interval (or more generally a subset of $\mathbb{R}$) is said to be compact, if each open cover can be reduced to a finite subcover.

Theorem 19.7. *Consider* $\mathbb{R}^n$ *with the usual topology.*
$\mathbb{R}^n$ *with the usual topology is second countable. More precisely,*

$$\bigcup_{j \in J} V_j = \bigcup_{j=1}^{\infty} V_j = V_1 \cup V_2 \cup \ldots$$

Every closed and bounded set $K \subset \mathbb{R}^n$ is compact (Heine–Borel theorem). Every multi interval $\Pi_{j=1}^n[a_j, b_j] \subset \mathbb{R}^n$ is compact (which follows from the above claim).

19.4.3 *Three theorems about continuity on compact, connected set $K \subseteq \mathbb{R}^n$*

This section contains theorems about $f(x)$, a continuous function defined on a compact set $K \subseteq D_f \subseteq \mathbb{R}^n$ (with the usual topology).

Uniform continuity

Definition 19.10. A function f is *uniformly continuous* on $\mathcal{A} \subseteq D_f \subseteq \mathbb{R}^n$, if for each $\varepsilon > 0$ there exists a $\delta > 0$ such that

$$|x - x'| < \delta \Longrightarrow |f(x) - f(x')| < \varepsilon, \quad x \text{ and } x' \in \mathcal{A}.$$

Uniform continuity differs from the "usual" continuity by the way of choosing δ, which here is independent of the point x.

In general, of course, uniform continuity $\Longrightarrow$ continuity, but not the other way around.

Instead, we have the following inversion of the claim.

Theorem 19.8. *Assume that f is continuous on $\mathcal{A}$ and $\mathcal{A}$ is compact. Then, f is uniformly continuous on $\mathcal{A}$.*

The theorem of largest and smallest value

Theorem 19.9. *A continuous function f on a compact $K \subset \mathbb{R}^n$ assumes a largest and a smallest value.*

The theorem of intermediate value

Theorem 19.10. *A continuous function f on a compact and connected set $K \subset \mathbb{R}^n$, assumes all values between its largest and smallest values.*

19.5 Map of Topological Spaces

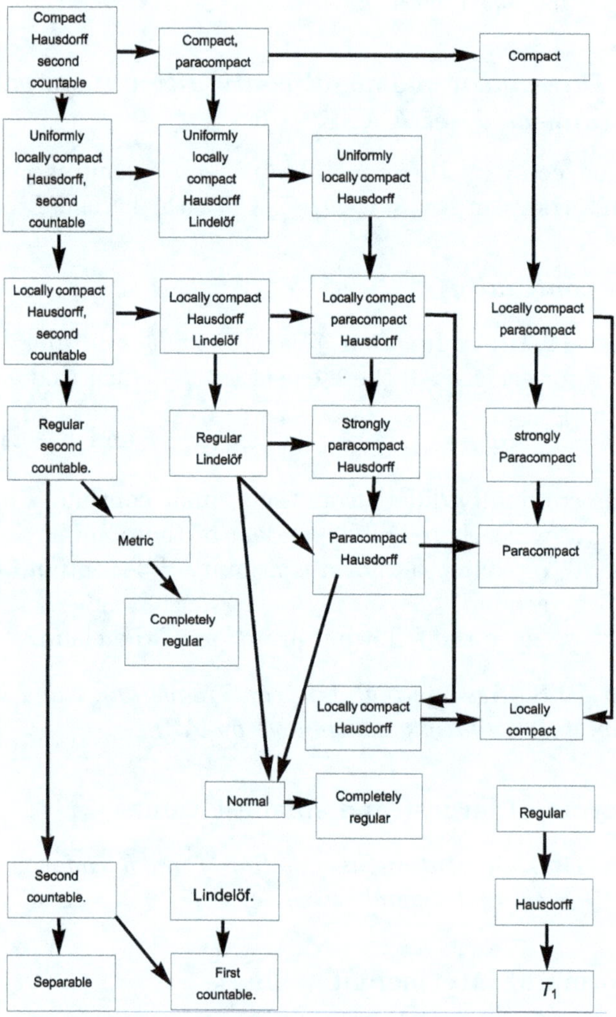

Schedule of the most common topological classes. "$A \to B$" means that $A \subset B$. For instance, a regular Lindelöf space is a normal space.

Chapter 20

Integration Theory

The first section treats essentially the same theory, as in Section 10.1. Then, the Riemann integral is defined for bounded functions $f : \mathbb{R}^n \curvearrowright \mathbb{R}$ defined on a compact set $I = \Pi_{j=1}^n [a_j, b_j] \subset \mathbb{R}^n$.

20.1 The Riemann Integral

Definition 20.1. Let $I = \Pi_{j=1}^n [a_j, b_j] = [a_1, b_1] \times [a_2, b_2] \times \cdots \times [a_n, b_n] \subset \mathbb{R}^n$, $a_j \leq b_j$, $j = 1, 2, \ldots, n$, denote a parallelepiped with sides parallel to the coordinate axes, which is a compact set in $\mathbb{R}^n$.

The measure of I is defined as the volume of the parallelepiped

$$m(I) := \prod_{j=1}^{n} (b_j - a_j).$$

Let I_k, $k = 1, 2, \ldots p$ be such parallelepipedes in $\mathbb{R}^n$ with pairwise disjoint interiors, i.e.,

$$\operatorname{Int} I_{k_1} \cap \operatorname{Int} I_{k_2} = \emptyset, \qquad \text{if } k_1 \neq k_2,$$

and define

$$J := \bigcup_{k=1}^{p} I_k.$$

Let f be a bounded real function defined on I and

$$\ell_k = \inf\{f(x) : x \in I_k\}, \quad u_k = \sup\{f(x) : x \in I_k\}.$$

A *lower sum* L and an *upper sum* U are defined as

$$L = \sum_{k=1}^{p} \ell_k \, m(I_k), \qquad U = \sum_{k=1}^{p} u_k \, m(I_k). \tag{20.1}$$

Using these concepts we define lower and upper integrals:

$$\underline{\int_J} f(x)dx := \sup\{L\} \quad \text{supremum taken over all lower sums } L. \tag{20.2}$$

$$\overline{\int_J} f(x)dx := \inf\{U\} \quad \text{infinimum taken over all upper sums } U. \tag{20.3}$$

Remarks. To be concise, the integral symbol $\int$ stands also for the multiple integral of $f : \mathbb{R}^n \to \mathbb{R}$, *viz.*

$$\underbrace{\int\int \cdots \int}_{n \text{ integral signs}} .$$

The integration variable $x := (x_1, x_2, \ldots x_n)$, where $x_k \in \mathbb{R}$ and $dx = dx_1 \, dx_2 \ldots dx_n$.

For more on Riemann sum, see page 207. Obviously, lower and upper integrals satisfy

$$\underline{\int_J} f(x)dx \leq \overline{\int_J} f(x)dx.$$

20.1.1 *Definition of the Riemann integral*

Definition 20.2. A bounded function f defined on I is integrable in the Riemann[1] sense if

$$\underline{\int_J} f(x)dx = \overline{\int_J} f(x)dx. \tag{20.4}$$

[1]Bernhard Riemann, (1826–1866).

Definition 20.3. The common value in (20.4) is called the integral of f (over the set J) and is denoted by

$$\int_J f(x)dx. \tag{20.5}$$

The definition is equivalent to the statement that for every $\varepsilon > 0$, there are lower and upper sums L and U, such that $U - L < \varepsilon$.

Theorem 20.1 (The linearity). *If f and g are Riemann integrable, then*

$$\int_J k\, f(x)dx = k \int_J f(x)dx, \quad k \in \mathbb{R}. \tag{20.6}$$

$$\int_J [f(x) + g(x)]dx = \int_J f(x)dx + \int_J g(x)dx. \tag{20.7}$$

20.1.2 *Integrability of continuous functions*

Theorem 20.2. *A continuous function f defined on a compact set J is Riemann integrable.*

Theorem 20.3 (Substitution of variables). *Let G_1 and G_2 be two open sets in $\mathbb{R}^n$ and φ be a continuously differentiable, bijective, function*

$$\varphi : G_1 \to G_2.$$

Then

$$\int_{G_1} f(x)dx = \int_{G_2} f(\varphi(x))|\det D|dy, \tag{20.8}$$

where

$$D = \left(\frac{\partial x_j}{\partial y_k}\right)_{n\times n} \quad \textit{is the functional matrix.}$$

20.1.3 *Comments about the Riemann integral*

- For instance, the integral $\int e^{-x^2} dx$ cannot be expressed by means of elementary functions (a non-elementary integral). The function is however continuous and thus integrable (in Riemann sense) over any compact interval $[a, b]$.

- For integrability, a function need not be continuous, i.e., continuity is a sufficient but not necessary criterion for integrability.
- An improper Riemann integral is not included in the very definition (20.4), but there is a "build-up" for it. Loosely speaking, an improper integral is referred to either as an integral over unbounded domain or an integration of an unbounded integrand/function.
- All real functions are not Riemann integrable, even if they are bounded. As an example we take

$$f(x) = \begin{cases} 0, & \text{if } x \text{ is rational,} \\ 1, & \text{if } x \text{ is irrational.} \end{cases}$$

An attempt to integrate f over $J := [0, 1]$ yields

$$\underline{\int_J} f(x)dx = 0, \quad \text{whereas} \quad \overline{\int_J} f(x)dx = 1.$$

20.2 The Lebesgue Integral

A more general integration concept, which relies on measure theory, was developed at the beginning of the twentieth century by Henri Lebesgue (1875–1941).

20.2.1 *General theory*

Definition 20.4.

(i) A σ−algebra in set X is a class $\mathcal{M}$ of measurable subsets of X, such that

(a) $X \in \mathcal{M}$.
(b) $E \in \mathcal{M} \Longrightarrow E^c = X \setminus E \in \mathcal{M}$.
(c) $E_n \in \mathcal{M}, \quad n = 1, 2, 3, \ldots \Longrightarrow \cup_{n=1}^{\infty} E_n \in \mathcal{M}$.

(ii) X (above) is called a *measure space*.
(iii) Let X be a measure space and Y, a topological space. A function $f : X \to Y$ is called measurable if for every open set $V \subseteq Y$, $f^{-1}(V)$ is a measurable set in X. Usually, $Y = \mathbb{R}$ or $\mathbb{C}$.

(iv) A positive measure μ is a function $\mathcal{M} \overset{\mu}{\to} [0, \infty]$ with the property

$$\mu\left(\bigcup_{k=1}^{\infty} E_k\right) = \sum_{k=1}^{\infty} \mu(E_k), \quad E_\ell \cap E_j = \emptyset, \quad \forall \ell \neq j. \tag{20.9}$$

Further, we assume $\mu(E) < \infty$ for at least one $E \in \mathcal{M}$.

(v) A measurable set E with measure $\mu(E) = 0$ is called a null set for μ.

(vi) Two measurable functions $f(x)$ and $g(x)$ which are equal in $X \setminus E$, where $m(E) = 0$, are said to be equal a.e. (almost everywhere).

(vii) That a sequence $(f_n(x))_{n=1}^{\infty}$ converges to $f(x)$ a.e. means that $f_n(x) \to f(x)$, pointwise, as $n \to \infty$, except for a set of measure zero.

The characteristic function $\mathcal{X}_E$ for a (measurable) set E is defined as

$$\mathcal{X}_E = \begin{cases} 1 & \text{if } x \in E, \\ 0 & \text{if } x \in E^c. \end{cases} \tag{20.10}$$

A non-negative simple function s is defined as

$$s(x) = \sum_{k=1}^{n} a_k \mathcal{X}_{E_k}(x), \quad \text{where} \quad a_k \geq 0. \tag{20.11}$$

The Lebesgue integral with respect to the measure μ of a simple function $s(x)$ is defined as

$$\int_X s(x) d\mu(x) = \sum_{k=1}^{n} a_k \mu(E_k), \tag{20.12}$$

where E_k are measurable.

Definition 20.5.

$$f_+(x) := \max(f(x), 0), \quad : f_-(x) := -\min(f(x), 0). \tag{20.13}$$

Then, $f = f_+ - f_-$, $|f| = f_+ + f_-$, $f_+ \geq 0$ and $f_- \geq 0$.
The Lebesgue integral of a non-negative measurable function f is defined as

$$\int_X f(x) d\mu := \sup \int_X s(x) d\mu, \tag{20.14}$$

where supremum is taken over all simple functions s such that $0 \leq s \leq f$. Supremum may assume all values in $[0, \infty]$.

A function f is integrable in the Lebesgue sense if not both $\int_X f_+(x)d\mu$ and $\int_X f_-(x)d\mu$ assume the value ∞. The Lebesgue integral is then defined as

$$\int_X f(x)d\mu := \int_X f_+(x)d\mu - \int_X f_-(x)d\mu.$$

If in addition $\int |f(x)|d\mu < \infty$, the function f is said to be an L^1 function, written as $f \in L^1(\mu)$.

For a measurable set E, the integral over E is defined as

$$\int_E f(x)d\mu := \int_X \mathcal{X}_E \cdot f(x)d\mu. \tag{20.15}$$

Theorem 20.4. *Let μ be a positive measure over the σ-algebra $\mathcal{M}$. Then*

(a) $\mu(\emptyset) = 0.$

(b) $\mu(E_1 \cup \cdots \cup E_n) = \mu(E_1) + \cdots + \mu(E_n), \quad E_i \cap E_j = \emptyset$ *for* $i \neq j$ *and* $E_j \in \mathcal{M},\ 1 \leq j \leq n.$

(c) $E,\ F \in \mathcal{M}$ *and* $E \subseteq F \Longrightarrow \mu(E) \leq \mu(F).$

(d) $E = \bigcup_{n=1}^{\infty} E_n,\ E_n \in \mathcal{M},\ E_1 \subset E_2 \subset \ldots,$
$\Longrightarrow \quad \mu(E_n) \longrightarrow \mu(E), \quad n \to \infty.$

(e) $E = \bigcap_{n=1}^{\infty} E_n,\ E_n \in \mathcal{M},\ E_1 \supset E_2 \supset \ldots,$ *and* $\mu(E_1) < \infty,$
$\Longrightarrow \quad \mu(E_n) \longrightarrow \mu(E), \quad n \to \infty.$

Theorem 20.5.

(i) *The equality $f(x) = g(x)$ a.e. is an equivalence relation.*

(ii) *For a measurable function $f \geq 0$, there is a sequence of simple functions $s_k(x) \nearrow f(x)$.*

(iii) *For each class S of subsets of a set X there exists a smallest $\sigma-$algebra $\mathcal{M}$. It is denoted $\sigma(S)$ and is called the Borel-algebra with respect to S.*

(iv) *The Lebesgue integration is linear:*

$$\int_X [af(x) + bg(x)]d\mu = a\int_X f(x)d\mu + b\int_X g(x)d\mu, \tag{20.16}$$

if f and g are L^1- functions and a and b are real or complex coefficients.

(v) *For functions $f(x)$ and $g(x)$, such that $f(x) = g(x)$ a.e.*

$$\int_X f(x)d\mu = \int_X g(x)d\mu,$$

if at least one of the integrals is well defined.

(vi) **Fubini's theorem:** *Let (X, μ) and (Y, ν) be two positive measure spaces. If $\varphi_y(x) = f(x, y)$ is μ−measurable for almost all $y \in Y$, $\psi_x(y) = f(x, y)$ is ν−measurable for almost all $x \in X$, then $f(x, y)$ is $\mu \times \nu$-measurable.*
If $\int_{X\times Y} |f(x, y)|d(\mu \times \nu) < \infty$, then

$$\int_{X\times Y} f(x, y)d(\mu \times \nu) = \int_X \left(\int_Y f(x, y)d\nu\right) d\mu. \tag{20.17}$$

Definition 20.6. A measurable function f such that

$$\int_X |f(x)|^p d\mu < \infty \tag{20.18}$$

is called an L^p−function, and is denoted by $f \in L^p(\mu)$.
For $1 \leq p < \infty$, the L^p-norm of f is defined as

$$\|f\|_p := \left(\int_X |f(x)|^p d\mu\right)^{1/p}. \tag{20.19}$$

The L^∞-norm: $||f||_\infty$ is defined by

$$||f||_\infty := \inf\{a : \mu\{x : |f(x)| \geq a\} = 0\}, \text{ if } ||f||_\infty < \infty. \tag{20.20}$$

Theorem 20.6.

(i) *An L^p−space is a complete metric space.*
(ii) $f(x) = g(x)$ *a.e.* $\Longrightarrow \|f\|_p = \|g\|_p$.
(iii) *Lebesgue's monotone convergence theorem: If f_n are an increasing sequence of measurable functions: $0 \leq f_n \leq f_{n+1}$, then $(f_n)_{n=1}^\infty$ has a limit f (f may assume the value ∞). Furthermore, f is measurable and*

$$\lim_{n\to\infty} \int_X f_n(x)d\mu = \int_X \lim_{n\to\infty} f_n(x)d\mu = \int_X f(x)d\mu. \tag{20.21}$$

(iv) *Fatou's lemma: If $f_n : X \to [0, \infty]$ are measurable, $n = 1, 2, \ldots$, then*

$$\int_X \liminf_{n\to\infty} f_n(x)d\mu \leq \liminf_{n\to\infty} \int_X f_n(x)d\mu. \tag{20.22}$$

(v) *Lebesgue dominated convergence theorem: If f_n are measurable and converge pointwise to f, a.e., and there is a function $g \in L^1(\mu)$, such that $|f_n| \leq g$, then the limit $f \in L^1(\mu)$ and*

$$\lim_{n\to\infty} \int_X f_n(x)d\mu = \int_X \lim_{n\to\infty} f_n(x)d\mu = \int_X f(x)d\mu$$

and (20.23)

$$\lim_{n\to\infty} \int_X |f_n(x) - f(x)|d\mu = 0.$$

Theorem 20.7.

(i) *The triangle inequality for the Lebesgue integral:*

$$\left|\int_X f(x)d\mu\right| \leq \int_X |f(x)|d\mu, \text{ if } f \in L^1(\mu). \tag{20.24}$$

(ii) $\|f\|_p$ *satisfies the properties of a metric $d(f, g)$ where $d(f, g) = ||f - g||_p$ for $1 \leq p \leq \infty$.*

(iii) *Jensen's inequality: If $\mu(X) = 1$ and φ is a convex function on $(a, b) \supseteq V_f$, where $f : X \to V_f$ is measurable, then*

$$\varphi\left(\int_X f(x)d\mu\right) \leq \int_X \varphi(f(x))d\mu. \tag{20.25}$$

(iv) *Assume $\frac{1}{p} + \frac{1}{q} = 1, \ldots, 1 < p,\ q < \infty$, f and g are measurable. Then,*

Hölder's inequality

$$\int_X |f(x)g(x)|\, d\mu \leq \left(\int_X |f(x)|^p d\mu\right)^{1/p} \cdot \left(\int_X |g(x)|^q\, d\mu\right)^{1/q},$$

which can be written as $\|fg\|_1 \leq \|f\|_p\|g\|_q$

Minkowski's inequality

$$\left(\int_X |f(x)+g(x)|^p d\mu\right)^{1/p} \le \left(\int_X |f(x)|^p d\mu\right)^{1/p} + \left(\int_X |g(x)|^p d\mu\right)^{1/p}, \tag{20.26}$$

which can be written as $\|f+g\|_p \le \|f\|_p + \|g\|_p$.

Young's inequality
$\|f * g\|_1 \le \|f\|_q \|g\|_p$

The generalized Young inequality
$\|f * g\|_r \le \|f\|_q \|g\|_p$, $1 < p, q, r < \infty$ and $1/p + 1/q = 1/r + 1$.

Remarks.

(i) For the Lebesgue measure on $\mathbb{R}$ the set of rational points is a null set.

(ii) (The author R.E. 1993) If $(X, \mathcal{T})$ is a second countable topological space and $\mathcal{M} = \sigma(\mathcal{T})$ is equipped with a positive measure μ, then the "essential" support of f is defined as

$$\text{essupp} f := \cap_i \text{supp} f_i, \tag{20.27}$$

where the intersection is taken over all *pointwise defined* f_i:s such that $f = f_i$ a.e.
Then the following hold true:

(a) There is a pointwise defined function f_0 with $f_0 = f$ a.e. such that $\text{essupp} f = \text{supp} f_0$.

(b) If $\mu(G) > 0$ for each non-empty open set G in X and g is continuous, then

$$\text{supp } g = \text{essupp } g. \tag{20.28}$$

(iii) A complex measure μ defined on a $\sigma-$algebra assumes values in $\mathbb{C}$. The total variation $|\mu|$ of a complex measure μ is defined as

$$|\mu|(E) := \sup \sum_{k=1}^{\infty} |\mu(E_k)|, \tag{20.29}$$

where supremum is taken over all disjoint unions of E.

(iv) $|\mu|$ is a positive finite measure.

(v) If μ is a finite, positive, measure, then

$$L_p \subset L_q, \quad \text{for } p > q.$$

(vi) If $\|f\|_p < \infty$ for some p, then $\|f\|_p \to \|f\|_\infty$, as $p \to \infty$.

(vii) If $1 \leq r < p < s$, then $L_r \cap L_s \subseteq L_p$.

20.2.2 *The Lebesgue integral on* $\mathbb{R}^n$

The general theory does not assume that $X = \mathbb{R}^n$ but the Lebesgue integral can be defined on this space/set in a natural way.

Definition 20.7.

(i) Put $B := \prod_{k=1}^{n} I_k$ where I_k is an interval in $\mathbb{R}$ with endpoints a_k and b_k, $a_k \leq b_k$, $k = 1, 2, \dots, n$. The measure μ is written m and is defined as

$$m(B) = \prod_{k=1}^{n} (b_k - a_k). \tag{20.30}$$

(ii) (a) $\mathcal{F}_\sigma$ is the class of sets which are countable unions of closed sets.

(b) $\mathcal{G}_\delta$ is the class of sets which are countable intersections of open sets.

(iii) The general $L^p(\mu)$ is now written as $L^p(\mathbb{R}^n)$.

(iv) A function is locally integrable if $\mathcal{X}_K f \in L^1(\mathbb{R}^n)$ for each compact set $K \in \mathbb{R}^n$. The class of locally integrable functions is denoted by $L^1_{\text{loc}}(\mathbb{R}^n)$.

Theorem 20.8.

(i) *The measure m given by* (20.30) *can be extended to a positive measure on a $\sigma-$algebra $\mathcal{M}$ on $\mathbb{R}^n$ including the usual topology τ.*

(ii) *$\mathcal{M}$ consists of just those sets E such that there exist $A \in \mathcal{F}_\sigma$ and $B \in \mathcal{G}_\delta$, where $A \subseteq E \subseteq B$ and $m(B \setminus A) = 0$.*

Theorem 20.9. *Assume that f is bounded in the interval $[a, b]$ and Riemann integrable. Then f is also Lebesgue integrable, Furthermore,*

$$\int_a^b f(x)dx = \int_{[a,b]} f(x)dm. \tag{20.31}$$

Remarks. Since the integrals coincide, one writes even the Lebesgue integral as LHS in (20.31).

A question is whether there are Riemann integrable functions which are not Lebesgue integrable in $\mathbb{R}^n$?

An improper conditionally convergent integral in the Riemann sense is not Lebesgue integrable, but measurable in the meaning of Lebesgue.

Chapter 21

Functional Analysis

21.1 Topological Vector Space

Definition 21.1. A vector space X over the field of real or complex numbers $\mathbb{K} = \mathbb{R}$ or $\mathbb{C}$ has the following properties:

(i) $\boldsymbol{x}, \boldsymbol{y} \in X \Rightarrow \boldsymbol{x} + \boldsymbol{y} \in X$, where $+$ is a commutative and associative binary operation.

(ii) Further there is a $\mathbf{0}$ element, such that $\boldsymbol{x} + \mathbf{0} = \mathbf{0} + \boldsymbol{x} = \boldsymbol{x}$.

(iii) For each $\boldsymbol{x}$ there is an element $-\boldsymbol{x}$, such that $\boldsymbol{x} + (-\boldsymbol{x}) = \mathbf{0}$.

(iv) $a \in \mathbb{K}$ och $\boldsymbol{x} \in X \Longrightarrow a\boldsymbol{x} \in X$, where a is called scalar.

(v) Given a set $G \subset X$ and $x \in X$.
$G + x$ is the set $G + x = \{g + x : g \in G\}$.

(vi) A topological vector space X is a vector space with a topology such that the maps $(x, y) \curvearrowright x + y$ and $(a, x) \curvearrowright ax$ are continuous.

(vii) Furthermore, X is a T_1−space if every element considered as set (singleton set) is closed.

(viii) X is a metrizable topological vector space if it is equipped with a topology given by a metric d.

(ix) A norm $\|\cdot\|$ in a vector space X is a map $\|\cdot\| : X :\rightarrow [0, \infty)$, with the property that for each x and y in X and for every scalar a:

$$\begin{aligned} &\text{(a) } \|x\| = 0 \Longleftrightarrow x = 0, \quad \text{(b) } \|ax\| = |a|\|x\|, \\ &\text{(c) } \|x + y\| \leq \|x\| + \|y\|. \end{aligned} \tag{21.1}$$

(x) X is a normable, topological, vector space if its topology is generated by a norm, e.g., the metric $d(x, y) = \|x - y\|$.

(xi) If X is a normable vector space and the norm is complete, i.e., every Cauchy sequence is convergent with respect to the norm $\|\cdot\|$, then X is called *Banach space.*

(xii) If $\|\cdot\|$ fulfills (ii) and (iii) in (21.1), then it is called semi-norm.

(xiii) A Frechet space X is a Hausdorff space (page 442) associated with the metric

$$d(x, y) := \sum_{n=1}^{\infty} \frac{1}{2^n} \frac{\|x - y\|_n}{1 + \|x - y\|_n}.$$

Here, $\{\|\cdot\|_n\}$ constitutes a countable class of semi-norms, such that for every pair of elements x and y, there is a semi-norm $\|x - y\|_n > 0$, where the corresponding metric d is complete.

(xiv) A linear map $\Lambda : X \to Y$ between two vector spaces is called *linear transformation.*

Λ is a bounded map between two normable spaces X and Y if there is a constant $k \geq 0$ such that $|\Lambda(x)| \leq k\|x\|$ for each $x \in X$. The norm of Λ is defined as

$$\|\Lambda\| := \sup_{x \in X} \frac{|\Lambda(x)|}{\|x\|}.$$

(xv) If $\Lambda : X \to \mathbb{R}$ (or $\mathbb{C}$) is linear, then it is called *linear functional.*

Theorem 21.1.

(i) *A topological vector space X is Hausdorff if for each subset $G \subseteq X$ and each $x \in X$, G open $\iff G + x$ is open.*

(ii) *With the above notation, the following statements are equivalent:*

 (a) *Λ is bounded.*

 (b) *Λ is continuous.*

 (c) *Λ is continuous at a point x.*

21.1.1 *Examples of topological vector space*

(i) Examples of Banach spaces

 (a) L^p–space ($p \in [1, \infty]$), i.e., the class of measurable functions $f : X \to \mathbb{C}$ with $||f||_p < \infty$ (page 459).

(b) l^p–space (page 315).
(c) $\mathcal{C}[a,b]$, The class of continuous functions

$$f : [a,b] \to \mathbb{R} \text{ with norm } \|f\| = \max\{|f(x)| : \quad a \le x \le b\}.$$

(ii) The Schwartz class
(a) $\mathcal{S}(\mathbb{R})$ or the class of test functions $\varphi : \mathbb{R} \to \mathbb{C}$ satisfying

$$\sup_{x\in\mathbb{R}} ||x|^{\alpha} D^{\beta}\varphi(x)| < \infty \text{ for all integers } \alpha,\ \beta = 0, 1, 2, \ldots$$

in other words $\varphi \in \mathcal{C}^{\infty}(\mathbb{R})$, i.e. an infinitely, differentiable function.
(b) $\mathcal{S}(\mathbb{R}^n)$ or the class of test functions $\varphi : \mathbb{R}^n \to \mathbb{C}$ satisfying

$$\sup_{\boldsymbol{x}\in\mathbb{R}^n} ||\boldsymbol{x}|^{\alpha} D^{\beta}\varphi(\boldsymbol{x})| < \infty \text{ for each } \alpha,\ \beta \in \mathbb{N}^n,$$

where $\alpha = (\alpha_1, \alpha_2, \ldots, \alpha_n)$, $\beta = (\beta_1, \beta_2, \ldots, \beta_n)$, are multi-indices.

$$\boldsymbol{x}^{\alpha} = x_1^{\alpha_1} \cdot x_2^{\alpha_2} \cdot \ldots \cdot x_n^{\alpha_n}$$

and

$$D^{\beta}\varphi(\boldsymbol{x}) = \frac{\partial^{\beta_1}\varphi}{\partial x_1^{\beta_1}} \cdot \frac{\partial^{\beta_2}\varphi}{\partial x_2^{\beta_2}} \cdot \ldots \cdot \frac{\partial^{\beta_n}\varphi}{\partial x_n^{\beta_n}}.$$

In other words, $\varphi \in \mathcal{C}^{\infty}(\mathbb{R}^n)$, i.e., an infinitely differentiable function.
The class $\mathcal{S}(\mathbb{R}^n)$ is an example of a Frechet space.

21.2 Some Common Function Spaces

Definition 21.2.

(i) The class of continuous functions defined on $\mathbb{R}^n$, denoted by $\mathcal{C}(\mathbb{R}^n)$.
(ii) The class of functions in $\mathbb{R}^n$ with continuous partial derivatives up to order k, denoted $\mathcal{C}^k(\mathbb{R}^n)$.
(iii) The class of measurable functions in $\mathbb{R}^n$.

(iv) The class of integrable functions (In Lebesgue sense) defined in X with measure μ: $L^1(\mu)$.

(v) If $X = \mathbb{R}^n$, the class is denoted $L^1(\mathbb{R}^n)$.

(vi) $L_{\text{loc}}(\mathbb{R}^n)$, denoting the class of locally integrable functions, i.e., the set of functions such that $\int_K |f(x)|dx < \infty$, for any compact set $K \subset \mathbb{R}^n$.

(vii) Given a measurable function $f : \mathbb{R}^n \to \mathbb{R}$. Consider the vector space

$$L^{1,\infty}(\mathbb{R}^n) = \{f : ||f||_{1,\infty} < \infty\},$$

where

$$||f||_{1,\infty} := \sup_{\alpha}(\alpha|\{x \in \mathbb{R}^n : |f(x)| > \alpha\}|),$$

defines a *quasi-norm* with

$$||f+g||_{1,\infty} \le 2(||f||_{1,\infty} + ||g||_{1,\infty}).$$

21.2.1 *Hilbert space*

Definition 21.3. A vector space X is an inner product space if for all x, y, and z in X and $a \in \mathbb{C}$, a scalar

$$\begin{aligned} &1.\ (x,y) = \overline{(y,x)} && 2.\ (x+y,z) = (x,z) + (y,z) \\ &3.\ a(x,y) = (ax,y) && 4.\ (x,x) \ge 0 \\ &5.\ (x,x) = 0 \Longleftrightarrow x = 0 && 6.\ \sqrt{(x,x)} =: \|x\|. \end{aligned} \tag{21.2}$$

Theorem 21.2. *From* $1-6$ *it follows that*

$$\begin{aligned} &\|x\| = 0 \Longleftrightarrow x = 0 \\ &\|ax\| = |a|\|x\| && \textit{for every } a \in \mathbb{C} \\ &|(x,y)| \le \|x\|\|y\| && (\textit{Schwarz inequality}) \\ &\|x+y\| \le \|x\| + \|y\| && (\textit{Triangle inequality}). \end{aligned} \tag{21.3}$$

$\|x-y\|$ *defines a metric* d *in* X, $d(x,y) = \|x-y\|$ *and (thus)* X *is a topological space.*

Definition 21.4.

(i) A metric d of the form $d(x,y) = \|x - y\|$ with $\|ax\| = |a|\|x\|$ is called a norm.
(ii) An inner product space X which is complete with respect to $\|\cdot\|$ is called a *Hilbert space.*

Definition 21.5.

(i) Two elements x and y in a Hilbert space are called orthogonal (or an orthogonal pair) if $(x,y) = 0$. We assume both x, $y \neq 0$.
(ii) A subset $\{x_\alpha\}$ of a Hilbert space H is an orthonormal set if

$$(x_\alpha, x_\beta) = \begin{cases} 0 & \text{if} \quad \alpha \neq \beta, \\ 1 & \text{if} \quad \alpha = \beta. \end{cases}$$

(iii) A separable Hilbert space has a countable dense subset.

Theorem 21.3.

(i) *A Hilbert space has an orthonormal base $\{e_n, n = 1, 2, 3 \ldots\}$ in the sense that each element $x \in H$ can be written as*

$$x = \lim_{n\to\infty} \sum_{k=1}^{n} (e_n, x) e_n = \sum_{k=1}^{\infty} (e_n, x) e_n. \tag{21.4}$$

The convergence is of course in the Hilbert norm $\|\cdot\|$ sense. Furthermore,

(a) $(x,x) = \sum_{k=1}^{\infty} |(e_n, x)|^2$ *(Parseval's formula).*
(b) (21.4) *is the Fourier series of x.*

(ii) *Assume that X is a Hilbert space with the induced norm $\|x\| = \sqrt{(x,x)}$. Then, the following equivalence holds:*
Λ is a bounded linear functional in $X \iff$ There exists a unique $y \in X$ such that $\Lambda(x) = (y, x)$.
This is also known as Lax–Milgram or Riesz representation theorem.

21.2.2 Hilbert space and Fourier series

$L^2([-T/2,T/2])$ is a Hilbert space, where $\Omega = \frac{2\pi}{T}$. The class

$$\left\{\frac{1}{\sqrt{T}}, \sqrt{\frac{2}{T}}\cos n\Omega t, \sqrt{\frac{2}{T}}\sin n\Omega t,\ n = 1,2,\dots\right\} \tag{21.5}$$

is an orthonormal base on $L^2([-T/2,T/2])$, where the scalar product (or inner product) is given by

$$(f,g) = \frac{2}{T}\int_{-T/2}^{T/2} f(x)\overline{g(x)}dx.$$

This means that its Fourier series converges to f in L^2–norm. We assume that f is a real function and define its Fourier coefficients as

$$\begin{aligned} a_n &:= \frac{2}{T}\int_{-T/2}^{T/2} f(t)\cos n\Omega t dt, \quad n = 0,1,2,\dots \\ b_n &:= \frac{2}{T}\int_{-T/2}^{T/2} f(t)\sin n\Omega t dt, \quad n = 1,2,\dots \end{aligned} \tag{21.6}$$

Then the Fourier series of f is defined as

$$\mathcal{F}(f) :\sim \frac{a_0}{2} + \sum_{n=1}^{\infty}(a_n \cos n\Omega t + b_n \sin n\Omega t). \tag{21.7}$$

Theorem 21.4.

(i) *The above $\sim$ is an equality $(=)$ at the points on continuity of f.*
(ii) *The Fourier series $\mathcal{F}(f) \longrightarrow f$ in L^2-norm.*
(iii) *If $\mathcal{F}(f) \longrightarrow f$, a.e. and its partial sums are bounded by an integrable function, then $f \in L^1([-T/2,T/2])$.*

21.2.3 A criterion for Banach space

A normed vector space $(X, \|\cdot\|)$ is a Banach space (with the same norm) if and only if for each sequence $(a_k, k = 1,2,\dots) \subseteq X$ the following holds true

$$\sum_{k=1}^{\infty} \|a_k\| < \infty \Longrightarrow \sum_{k=1}^{\infty} a_k \text{ is convergent with respect to the norm } \|\cdot\|. \tag{21.8}$$

21.2.4 *Fourier transform*

Let $t = (t_1, t_2, \dots, t_n)$ and $x = (x_1, x_2, \dots, x_n) \in \mathbb{R}^n$. The scalar- or inner product is written as $\langle t, x \rangle = t_1x_1 + t_2x_2 + \cdots + t_nx_n$.

The Fourier transform is defined as the map $\mathcal{F}$

$$\mathcal{F}(f)(x) := \int_{\mathbb{R}^n} f(t)e^{-i<t,x>}dt. \tag{21.9}$$

The Fourier transform is a continuous linear map

$$\mathcal{F} : L^p \longrightarrow L^q, \text{ where } \frac{1}{p} + \frac{1}{q} = 1, \quad 1 \leq p \leq 2. \tag{21.10}$$

If $f \in L^1$, then $\mathcal{F}(f)$ is continuous.

21.3 Distribution Theory

Definition 21.6. Assume that $f : \mathbb{R}^n \to \mathbb{R}$ is continuous.

The support of f is the closure:

$$\text{supp} f := \overline{\{x \in \mathbb{R}^n : f(x) \neq 0\}}.$$

Let $\alpha = (\alpha_1, \alpha_2, \dots, \alpha_n)$ with $\alpha_j = 0, 1, 2, \dots,$ $(1 \leq j \leq n)$ be a multi-index and put $|\alpha| := \alpha_1 + \alpha_2 + \dots + \alpha_n$, the partial derivative of f of order $|\alpha|$ is given by

$$\frac{\partial^{|\alpha|} f}{\partial x^\alpha} := \frac{\partial^{\alpha_1}}{\partial x_1^{\alpha_1}} \frac{\partial^{\alpha_2}}{\partial x_2^{\alpha_2}} \cdots \frac{\partial^{\alpha_n}}{\partial x_n^{\alpha_n}} f.$$

$\mathcal{C}_0^\infty(\mathbb{R}^n) \equiv \mathcal{D}(\mathbb{R}^n)$ denotes the class of real-valued functions $f : \mathbb{R}^n \to \mathbb{R}$, such that supp$f$ is compact and $\frac{\partial^{|\alpha|} f}{\partial x^\alpha}$ is continuous for all α.

An example of a function $f \in \mathcal{C}_0^\infty(\mathbb{R})$ is

$$f(x) = \begin{cases} e^{\frac{1}{(x-a)^2} - \frac{1}{(x-b)^2}}, & a < x < b, \\ 0, & \text{else.} \end{cases} \tag{21.11}$$

21.3.1 *Generalized function*

Generalized function or *distribution.*

Definition 21.7. The Schwartz class is defined as

$$\{f \in C_0^\infty(\mathbb{R}) : \sup_{\boldsymbol{x}\in\mathbb{R}^n} |\boldsymbol{x}^\alpha D^\beta \varphi(\boldsymbol{x})| < C_{\varphi,\alpha,\beta} < \infty\}. \tag{21.12}$$

Theorem 21.5. *The Schwartz class is a vector space, i.e., for each α och $\beta \in \mathbb{C}$, φ and $\psi \in \mathcal{S}(\mathbb{R}^n)$ it yields that*

$$\alpha\,\varphi(\boldsymbol{x}) + \beta\,\psi(\boldsymbol{x}) \in \mathcal{S}(\mathbb{R}^n).$$

Definition 21.8. The space of tempered distributions in $\mathcal{S}(\mathbb{R}^n)$, is denoted $\mathcal{S}'(\mathbb{R}^n)$ and is the class/set of linear maps $\Gamma : \mathcal{S}(\mathbb{R}^n) \mapsto \mathbb{C}$.

Definition 21.9. The Fourier transform in $S(\mathbb{R})$ is

$$\mathcal{F}(\varphi(s)) = \hat{\varphi}(s) := \int_{-\infty}^{\infty} e^{-2i\pi x\, s}\varphi(x)dx. \tag{21.13}$$

The notation $\vee$ defines a change of sign of the argument:

$$\overset{\vee}{f}(x) := f(-x).$$

Theorem 21.6. *Some important properties of the Fourier transform.*

The Fourier transform is a map $\mathcal{S}(\mathbb{R}) \to C(\mathbb{R})$. The inverse Fourier transform returns φ in the region (at the points) of its continuity, i.e., for the points x satisfying.

$$\mathcal{F}^{(-1)}(\hat{\varphi}(s)) = \int_{\mathbb{R}} e^{2i\pi x s}\hat{\varphi}(s)ds = \varphi(x). \tag{21.14}$$

Further,

$$\frac{d}{ds}\mathcal{F}(\varphi(s)) = -2\pi i x \mathcal{F}(\varphi(s)), \tag{21.15}$$

$$\varphi \in \mathcal{S} \Longrightarrow \frac{d}{ds}\varphi \in \mathcal{S}.$$

$$\mathcal{F}^2(f) \;=\; \overset{\vee}{f}. \tag{21.16}$$

And thus, $\mathcal{F}^4(f) = f$.

The Fourier transform of odd/even function is odd/even.

Definition 21.10. The even and odd parts of a function f are defined as

$$\mathcal{E}(f) := \frac{1}{2}(f(x) + f(-x)) \text{ and } \mathcal{O}(f) := \frac{1}{2}(f(x) - f(-x)), \quad (21.17)$$

respectively. The convolution of two functions f and g is defined by

$$f * g(x) := \int_{\mathbb{R}} f(y)g(x-y)dy. \quad (21.18)$$

The auto-correlation of a function f is defined as

$$\mathcal{C}(f;x) := f \star f(x) = \int_{\mathbb{R}} \overline{f}(u)f(u-x)du. \quad (21.19)$$

Theorem 21.7. *Assume that f and g have compact supports, $K_1 \subset [a,b]$ and $K_2 \subset [c,d]$, respectively. Define a convolution, viz.*

(i)
$$f * g(x) = \int_{\max(d,x-a)}^{\min(c,x-b)} f(x-y)g(y)dy$$

(ii) *Then $f * g$ has compact support and*

$$supp(f * g) \subset [a+c, b+d].$$

The Fourier transform of the convolution of two functions is the product of their Fourier transforms:

$$\mathcal{F}(f * g) = \hat{f} \cdot \hat{g}. \quad (21.20)$$

Theorem 21.8. *Fix points of the Fourier transform:*

$$\mathcal{F}\left(e^{-\pi x^2}\right) = e^{-\pi s^2}.$$

$$\mathcal{F}\left(\sum_{k\in\mathbb{Z}} \delta_k\right) = \sum_{k\in\mathbb{Z}} \delta_k.$$

The Fourier transform of a function $\varphi \in \mathcal{S}(\mathbb{R})$ also belongs to $\mathcal{S}(\mathbb{R})$:

$$\varphi \in \mathcal{S}(\mathbb{R}) \Longleftrightarrow \hat{\varphi} \in \mathcal{S}(\mathbb{R}). \quad (21.21)$$

Some results of auto-correlation

Theorem 21.9. *In the following, g^* denotes the complex conjugate of g. Then*

$$C^*(f;x) = \int_{\mathbb{R}} f(u-x)f^*(u)du = \int_{\mathbb{R}} f(t)f^*(t+x)dt = C(f :, -x). \tag{21.22}$$

If $f(x)$ is a real function, then $C^*(f;x) = C(f :, -x)$, i.e., an even function.

21.4 Distributions

Definition 21.11. Let $\Omega \subseteq \mathbb{R}^n$ be an open set.

A *distribution* u in Ω is a linear functional defined on $C_0^\infty(\Omega)$, such that for each compact set $K \subset \Omega$ there are constants C and k, such that

$$|u(\varphi)| \leq C \sum_{\alpha:\, |\alpha| \leq k} ||\partial^\alpha \varphi||_\infty, \tag{21.23}$$

for all $\varphi \in \mathcal{S}(\mathbb{R}^n)$ with support in K.

The class of distributions is denoted by $\mathcal{D}'(\Omega)$.

If the same k can be used for all $K \subset \Omega$, then k is called the order of u. The set of these distributions is denoted by $\mathcal{D}'_k(\Omega)$.

The smallest $k = 0, 1, 2, ...$ for which (21.23) makes sense, is called the *order of the distribution.*

The class of distributions of finite orders is written as

$$\mathcal{D}'_F = \cup_k \mathcal{D}'_k.$$

Theorem 21.10. $f \in L^1_{loc}(\Omega)$ *is a distribution of order* 0.

A complex measure is a distribution of order 0.
Let $x_0 \in \Omega$.

$$u(\varphi) = \partial^\alpha \varphi(x_0) \in \mathcal{D}'_{|\alpha|}(\Omega)$$

is a distribution of order $|\alpha|$.

21.4.1 *Tempered distribution*

Definition 21.12. A *tempered* distribution is a linear functional

$$\tau : \mathcal{S}(\mathbb{R}) \to \mathbb{C}, \tag{21.24}$$

which is continuous in the following sense: given a sequence $(\varphi_k(x))_{k=1}^{\infty} \subseteq \mathcal{S}(\mathbb{R})$, τ is continuous on $\mathcal{S}(\mathbb{R})$ if

$$\lim_{k\to\infty} \max ||x|^{\alpha} D^{\beta}\varphi_k(x)| \to 0 \Longrightarrow \tau(\varphi_k) \to 0. \tag{21.25}$$

for all pairs α, β.

The class of tempered distributions is denoted by $\mathcal{S}'(\mathbb{R}^n)$.

A function $f = f(x)$, $x \in \Omega \subset \mathbb{R}$ defines a map

$$f : \mathcal{S}(\mathbb{R}^n) \to \int_{\mathbb{R}} f(x)\varphi(x)dx.$$

The derivative of a function f is defined as the map

$$f' : \mathcal{S}(\mathbb{R}^n) \to -\int_{\mathbb{R}} f(x)\varphi'(x)dx. \tag{21.26}$$

Any polynomial p is a tempered distribution in the following sense:

$$\tau_p(\varphi) = \int_{\mathbb{R}} p(x)\varphi(x)dx. \tag{21.27}$$

Comments

The minus sign in (21.26) depends on integration by parts: If f' exists in the classical sense, one gets

$$\int_{\mathbb{R}} f'(x)\varphi(x)dx = [f(x)\varphi(x)]_{\pm\infty} - \int_{\mathbb{R}} f(x)\varphi'(x)dx.$$

with 0 contributions from the boundary.

"The impulse sequence" τ, defined by $\tau(\varphi) = \sum_{n\in\mathbb{Z}} \varphi(n) \in \mathcal{S}'(\mathbb{R})$ is a tempered distribution.

The Frechet-topology on $\mathcal{S}(\mathbb{R}^n)$ is defined as follows:

(i) Let m be a non-negative integer, $\alpha, \beta \leq m$ and

$$\|\varphi\|_m := \sup_{\alpha,\beta\leq m} |x^\alpha D^\beta \varphi|.$$

(ii) The metric on $\mathcal{S}(\mathbb{R}^n)$ is given by

$$d(\varphi, \psi) = \sum_m 2^{-m} \frac{\|\varphi - \psi\|_m}{1 + \|\varphi - \psi\|_m}. \tag{21.28}$$

d is a *complete* metric on $\mathcal{S}(\mathbb{R}^n)$.

Theorem 21.11. *For the Fourier transform of a tempered distribution, the following hold true (easily verified):*

$$\mathcal{F}(T(\varphi)) := T(\mathcal{F}(\varphi)),$$

$$\hat{T}(\varphi) : = T(\overset{\wedge}{\varphi}), \tag{21.29}$$

$$\text{where} \quad \overset{\wedge}{\varphi}(x) = \varphi(-x).$$

Theorem 21.12.

$$\mathcal{F}^2(T) = T(\mathcal{F}^2) = \overset{\vee}{T}. \tag{21.30}$$

Chapter 22

Mathematical Statistics

22.1 Elementary Probability Theory

Definition 22.1 (The elementary probability definition). Assume that Ω is a non-empty set (a *finite* sample space) containing a finite number of *members/outcomes* ω, i.e., $|\Omega| = m$ for some positive integer m.

m is the number of *possible outcomes.*

The probability p for each outcome $\omega \in \Omega$ is $p = \dfrac{1}{m}$.

An *event* A is a subset of Ω.

The probability of an event $A \subseteq \Omega$ is

$$p = \frac{|A|}{|\Omega|} = \frac{g}{m}, \tag{22.1}$$

where $g = |A|$ is the number of favorable outcomes.

Definition 22.2.

(i) Let Ω be a non-empty set. A $\sigma-$algebra $\mathcal{M}$ is a set/class of subsets of Ω (as its elements) with the properties

$A_j \in \mathcal{M}$, $j = 1, 2, \ldots \Longrightarrow \cup_{j=1}^{\infty} A_j \in \mathcal{M}$ (closed under countable union),

$A \in \mathcal{M} \Longrightarrow A^c \equiv \Omega \setminus A \in \mathcal{M}$ (closed under complement), $\emptyset \in \mathcal{M}$.

The *measurable* sets A are (as above) called *events.*

(ii) A *probability measure* P is a function (a positive measure) defined on $\mathcal{M}$ with the property $0 \leq P(A) \leq 1$ for all $A \in \mathcal{M}$. The event A is called *measurable*, the elements $\omega \in \Omega$ are called *outcomes*, and Ω is called a probability space. The empty event $\emptyset$ contains no elements and is also referred to as an *impossible event*.

(iii) For two events A and B with $\emptyset \subseteq A \subseteq B \subseteq \Omega$, the probability measure P satisfies

$$0 = P(\emptyset) \leq P(A) \leq P(B) \leq P(\Omega) = 1. \tag{22.2}$$

(iv) The triple $\{\Omega, \mathcal{M}, P\}$ is called *probability space* and P is called *probability measure*. The set/event Ω is referred to as a *sample space* (as mentioned above).

(v) Two events $A \subseteq \Omega$ and $B \subseteq \Omega$ are *non-coincident* or disjoint, if $A \cap B = \emptyset$.

(vi) An event A, with $P(A) = 0$ is called a null-event.

(vii) For a function $X : \Omega \to \mathbb{R}$, for which the set $\{\omega \in \Omega : X(\omega) \leq x\}$ is measurable for each $x \in \mathbb{R}$, the variable x is called a *random* or *stochastic variable.*

(viii) For simplicity, in the sequel, the event $\{\omega \in \Omega : X(\omega) \leq x\}$ is written as $\{X(\omega) \leq x\}$ or even $\{X \leq x\}$.

Remarks. Events follow the same rules as for sets, see page 4 and further.

A space X (or Ω) as on page 6 equipped with various sets (events) is called a *Venn-diagram.* The three events A, B, and Ω as well as $A \setminus B$, $A \cap B$, and $B \setminus A$ are present.
There are no conditions on the cardinality of Ω.

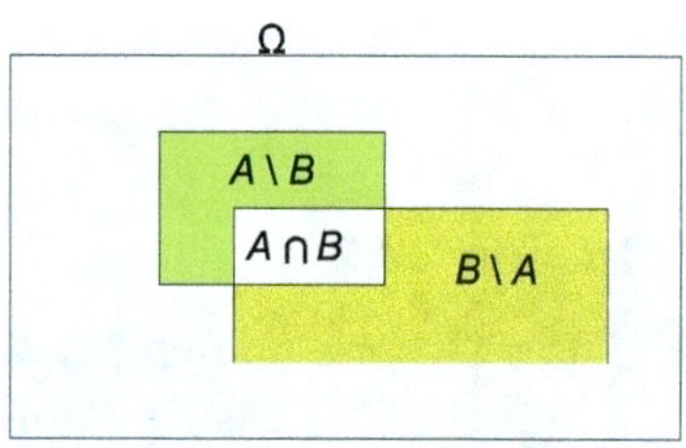

Theorem 22.1.

(i) *For events A, $B \subseteq \Omega$, also $A \setminus B = A \cap B^c$ is an event.*
(ii) *A $\sigma-$algebra $\mathcal{M}$ is closed under countable intersections.*

Theorem 22.2. *A probability measure P in a probability space Ω with events A, $B \subseteq \Omega$, satisfies the following properties:*

$$P(A) + P(A^c) = P(\Omega) = 1, \tag{22.3}$$

and

$$\begin{aligned} P(A \cup B) &= P(A) + P(B) - P(A \cap B), \\ P(A \cup B) &= P(A) + P(B), \text{ if } A \cap B = \emptyset, \end{aligned} \tag{22.4}$$

where $\Omega = A \cup A^c$ denotes the probability space and A, $B \subset \Omega$ are events in Ω.

Definition 22.3. The *conditional* probability that B occurs if A occurs is

$$P(B|A) = \begin{cases} \frac{P(A\cap B)}{P(A)}, & \text{if} \quad P(A) > 0, \\ 0, & \text{if} \quad P(A) = 0. \end{cases} \tag{22.5}$$

Theorem 22.3.

$$P(A|B)P(B) = P(B|A)P(A) = P(A \cap B)(\textit{Bayes' theorem})$$

$$P(A^c|B) + P(A|B) = 1$$

$$\begin{aligned} P(A) &= P(A \cap B) + P(A \cap B^c) \\ &= P(A|B)P(B) + P(A|B^c)P(B^c). \end{aligned} \tag{22.6}$$

If $\{B_i,\ i = 1, 2, \ldots, n\}$ is a partition of Ω, i.e. a class of pairwise disjoint events such that $\cup_{i=1}^{n} B_i = \Omega$ and $B_i \cap B_j = \emptyset$, $i \neq j$, then

$$P(A) = \sum_{i=1}^{n} P(A \cap B_i) = \sum_{i=1}^{n} P(A|B_i)P(B_i), \tag{22.7}$$

where n is a positive integer or $n = \infty$.

Theorem 22.4. *For events A, B, and C*

$$P(A \cap B|C) = P(A|B \cap C) \cdot P(B|C). \tag{22.8}$$

For events A_1, A_2, $\ldots$, A_n

$$P(A_1 \cap A_2 \cap \ldots \cap A_{n-1}|A_n) = \prod_{k=1}^{n-1} P(A_k|(\cap_{j=k+1}^{n} A_j). \tag{22.9}$$

Independence

Definition 22.4.

(1) Let $\{A_i,\ i \in I\}$ be a class of events. The class is said to be independent if

$$P(\cap_{i\in J} A_i) = \prod_{i\in J} P(A_i), \tag{22.10}$$

for each finite sub-class $J \subseteq I$. In particular, two events A and B are independent if

$$P(A \cap B) = P(A)P(B). \tag{22.11}$$

(ii) X_1 and X_2 are *independent* random variables if

$$P(X_1 \leq x_1 \cap X_2 \leq x_2) = P(X_1 \leq x_1) \cdot P(X_2 \leq x_2), \tag{22.12}$$

for all numbers x_1 and x_2.

Theorem 22.5. *That two events A and B are independent is equivalent with the following statements:*

(i) *A and B^c are independent, i.e., $P(A \cap B^c) = P(A) \cdot P(B^c)$.*
(ii) *$P(A|B) = P(A)$.*

Theorem 22.6 (Borel–Cantelli's lemma). *Let $\{A_n, n = 1, 2, \ldots\}$ be a class of events and $A = \limsup_{n\to\infty} A_n$(Definition of $\limsup$ is on page 7). Then*

$$\sum_{n=1}^{\infty} P(A_n) < \infty \Longrightarrow P(A) = 0$$

and

$$\sum_{n=1}^{\infty} P(A_n) = \infty \Longrightarrow P(A) = 1, \text{ if } A_k, k = 1, 2, \ldots \text{ are independent.} \tag{22.13}$$

22.2 Descriptive Statistics

Suppose you make n observations with values assigned in a finite set of observed values $Y := \{y_1, y_2, \ldots, y_k\}$. Then one may obtain a sample of size $n \leq k$.

The number of observations assuming a specific value y_i is called its frequency $f = f_i$.

The *relative* frequency number i is f_i/n.

The cumulative frequency is the sum of frequencies up to some index $m : 1 \leq m \leq k$.

The cumulative *relative* frequency is the cumulative frequency divided by n.

Relative frequency	Cumulative relative frequency	Mean value	Variance
$\frac{f_i}{n}$	$\sum_{i=1}^{m} \frac{f_i}{n}$	$\overline{x} = \frac{1}{n}\sum_{i=1}^{n}$ $x_i = \sum_{i=1}^{k} \frac{y_i f_i}{n}$	$\frac{1}{n-1}\sum_{i=1}^{n}(\overline{x} - x_i)^2$ $= \frac{1}{n-1}\sum_{i=1}^{k}(\overline{x} - y_i)^2 f_i$

22.2.1 *Class sample*

With a large number of observations, it is convenient to sort them into classes.

Example 22.1.

Class	$10 \leq x < 15$	$15 \leq x < 20$	$20 \leq x < 25$
Frequency	1	5	11

Class	$25 \leq x < 30$	$30 \leq x < 35$	$35 \leq x < 40$
Frequency	23	17	3

The class interval middles are 12.5, 17.5, …, 37.5. One can now present them in a histogram (see the following).

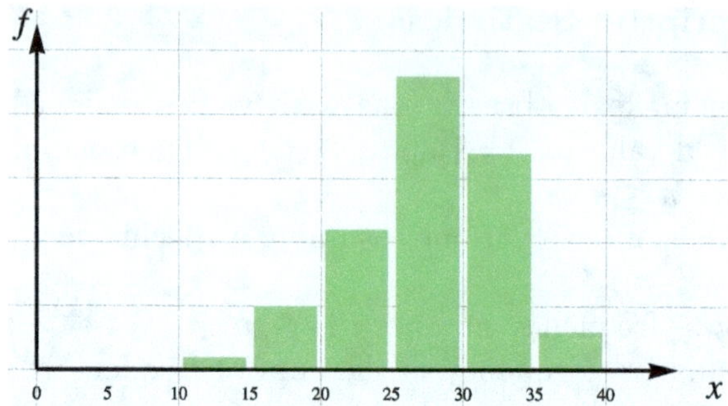

From a histogram one can calculate the pth percentile. This means that one has p percent of the observations to the left of that point on the horizontal axis.

For instance, to calculate the 80th percentile p_{80} for 60 observations, we have $0.80 \cdot 60 = 48$ observations to the left of p_{80}. We realize that p_{80} must fulfill $30 \leq p_{80} < 35$ since $1 + 5 + 11 + 23 = 40 < 48$ and $40 + 17 = 57 > 48$. To the right of $x = 30$ we take further eight observations from the staple with frequency 17 and get the supplement $\frac{8}{17} \cdot 5$ to the number 30. Hence,

$$p_{80} = 30 + \frac{8}{17} \cdot 5 \approx 32.4.$$

Histogram

(i) To make a histogram of a sample with size n we divide the sample into classes (intervals) $[k_i, k_{i+1}), \quad i = 0, 1, \ldots, m$. The class boundaries are $k_i, \quad i = 0, 1, \ldots, m$ with "midpoints" $\frac{k_{i+1}+k_i}{2}$. Frequency of class i is the number of observations that lie in the class i, i.e. the observations that lie in the interval $[k_i, k_{i+1})$.

(ii) Calculating a percentile p_α: Let i_0 be the index, such that

$$\sum_{i=0}^{i_0} f_i \leq n \cdot \frac{\alpha}{100} < \sum_{i=0}^{i_0+1} f_i.$$

Then

$$p_\alpha = k_{i_0} + \frac{n \cdot \dfrac{\alpha}{100} - \displaystyle\sum_{i=0}^{i_0} f_i}{f_{i_0+1}} \cdot (k_{i_0+1} - k_{i_0}). \tag{22.14}$$

22.2.2 *Simple regression analysis LS method*

$$S_{xx} = \sum (x_i - \bar{x})^2 = \sum x_i^2 - \tfrac{1}{n}(\sum x_i)^2$$

$$S_{yy} = \sum (y_i - \bar{y})^2 = \sum y_i^2 - \tfrac{1}{n}(\sum y_i)^2 \tag{22.15}$$

$$S_{xy} = \sum (x_i - \bar{x})(y_i - \bar{y}) = \sum x_i y_i - \tfrac{1}{n}(\sum x_i)(\sum y_i).$$

The line, which, by means of the Least-square method, is best adapted to the points (x_i, y_i), $i = 2, 3, ..., n$, is given by

$$y = a + bx \text{ where } a = \bar{y} - b\bar{x} \qquad b = \frac{S_{xy}}{S_{xx}}. \tag{22.16}$$

The correlation coefficient is given by

$$r_{xy} = \frac{S_{xy}}{\sqrt{S_{xx} \cdot S_{yy}}}. \tag{22.17}$$

22.3 Distributions

A (probability) distribution with countable (finite or infinite) number of outcomes is called *discrete*. A distribution where the random variable X can take all values in one or several intervals (a, b) is called *continuous*.

22.3.1 *Discrete distribution*

Definition 22.5. Let X be a discrete random variable, assuming the values $x_1 < x_2 < x_3 < \cdots < x_k < x_{k+1} < \cdots$.
A *probability density function* (PDF) $p = f(x_k) \geq 0$ is defined as

$$f(x_k) = P(X = x_k)(\geq 0). \tag{22.18}$$

With the corresponding *cumulative distribution function* (CDF) F to X means

$$F(x_k) = P(X \leq x_k) = \sum_{i=1}^{k} P(X = x_i). \tag{22.19}$$

Let $x_1, x_2, x_3, \ldots$ be *all possible outcomes* (finite or countably infinite).

Then,

$$\sum_k P(X = x_k) = \sum_k f(x_k) = 1. \tag{22.20}$$

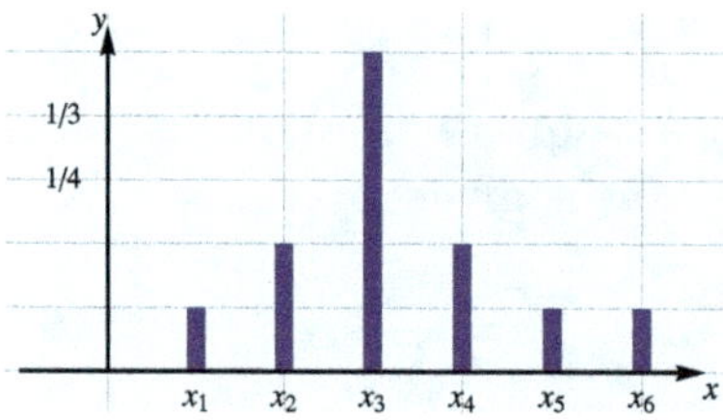

A discrete probability distribution function (PDF) with six outcomes.

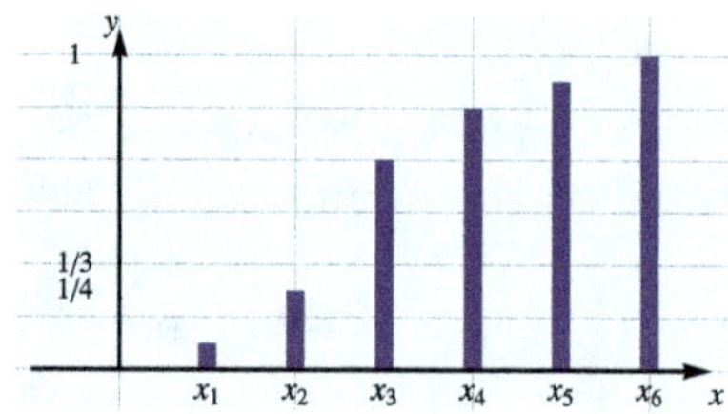

The corresponding cumulative distribution function (CDF) to the discrete distribution.

22.3.2 *Some common discrete distributions*

With a random variable means the class of all random variables having the same distribution, e.g., $X \in \mathrm{Po}(\lambda)$, etc.

Distribution with notation	Freq. function $f(x) = P(X = x)$	Expectation	Variance	Parameters
Bernoulli $\mathrm{I}_A(p)$	$f(x) = \begin{cases} p, & \text{if } x \in A \\ 1-p, & \text{if } x \in A^c \end{cases}$	p	$p(1-p)$	p
Uniform $\mathrm{U}(N)$	$\frac{1}{N}$	$\frac{N+1}{2}$	$\frac{N^2-1}{12}$	N
Binomial $\mathrm{Bin}(n, p)$	$\binom{n}{x} p^x (1-p)^{n-x}$	np	$np(1-p)$	n, p
Hyper-geometric $\mathrm{Hyp}(N, n, p)$	$\frac{\binom{Np}{x}\binom{N(1-p)}{n-x}}{\binom{N}{n}}$	np	$\frac{N-n}{N-1} np(1-p)$	N, n, p
Geometric $\mathrm{Ge}(p) = \mathrm{Neg}(1, p)$	$(1-p)^{x-1} p$	$\frac{1}{p}$	$\frac{1-p}{p^2}$	p
Negative binomial $\mathrm{Neg}(k, p)$	$\binom{x-1}{k-1} (1-p)^{x-k} \cdot p^k$	$\frac{k}{p}$	$\frac{k(1-p)}{p^2}$	k, p
Poisson $\mathrm{Po}(\lambda)$	$\frac{e^{-\lambda}\lambda^x}{x!}$	λ	λ	λ

(22.21)

Remarks. A Bernoulli- or *indicator distribution* is a binomial distribution with $n = 1$, that is X is Bernoulli distributed with parameter $p \Leftrightarrow X \in \mathrm{Bin}(1, p)$.

If $X_k \in \mathrm{I}_A(p)$, $k = 1, 2, \ldots, n$ are independent, then the sum $\sum_{k=1}^{n} X_k \in \mathrm{Bin}(n, p)$.

For the hypergeometric distribution, in some literature, Np is denoted by a parameter without the subscript p.

If $X_j, j = 1, 2, \ldots, k$ are independent geometric random variables with the same parameter p, then $X_1 + X_2 + \cdots + X_k \in \text{Neg}(k, p)$, i.e., a negative binomial distributed with parameters k and p.

Uniform distribution is described only when the outcomes are $1, 2, \ldots, N$, for some N.

Hypergeometric distribution means choosing n among N units (without return), where Np is of a certain kind.

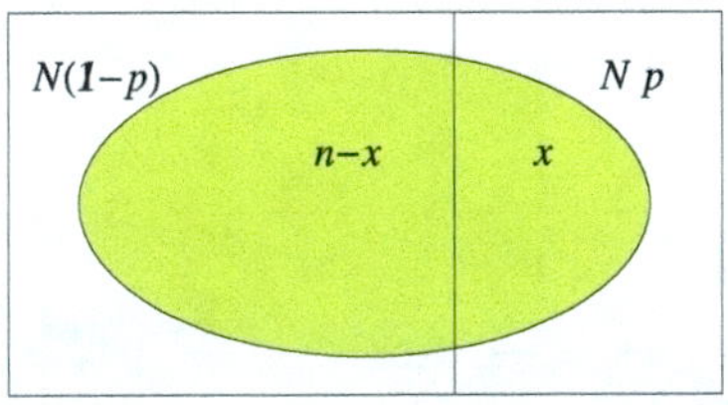

The parameters x, N, n, Np are integers ≥ 0 satisfying $0 \leq x \leq Np$, $0 \leq n - x \leq N - Np$.

The cumulative distribution functions (CDF) for the discrete distributions are not included in Table 22.21.

22.3.3 *Continuous distributions*

A simplified definition of continuous distribution:

Definition 22.6. A function f such that $f(x) \geq 0$ (Figure 22.1) for all x and

$$\int_{-\infty}^{\infty} f(x)dx = 1 \tag{22.22}$$

is a *probability density function* (PDF).

The corresponding *Cumulative distribution function* (CDF) (see Figure 22.2) is

$$F(x) := P(X \leq x) = \int_{-\infty}^{x} f(t)dt. \tag{22.23}$$

The Survival function is given by

$$R(x) := 1 - F(x) = P(X > x) = \int_{x}^{\infty} f(t)dt. \tag{22.24}$$

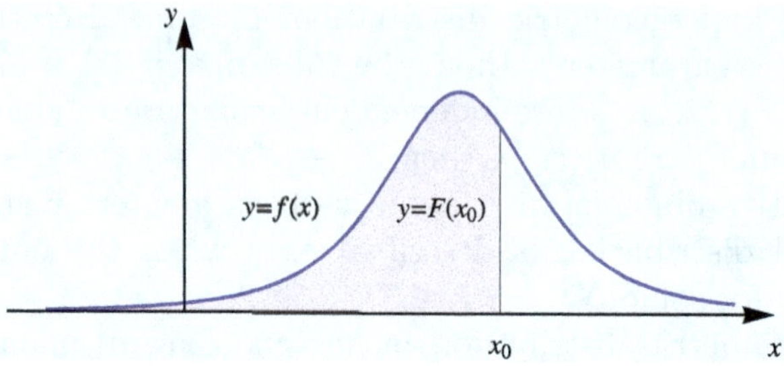

Figure 22.1: Curve given by a probability density function $y = f(x)$.

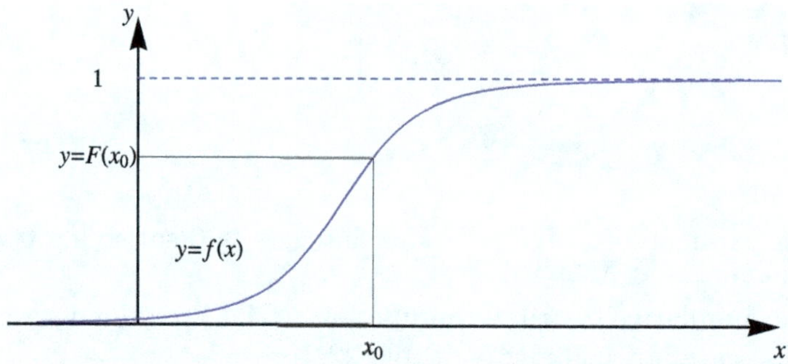

Figure 22.2: Curve given by a cumulative distribution function $y = F(x)$.

The *failure, or hazard rate*, is

$$\lambda(x) := \frac{f(x)}{1 - F(x)} = \frac{f(x)}{R(x)}, \tag{22.25}$$

where X is a *continuous* random variable.

Theorem 22.7. *Let X be a continuous random variable with PDF: f and CDF: F. Then, the following hold true:*

(a) $F'(x) = f(x)$ *(except for some isolated points)*

(b) $P(a < X \leq b) = \int_a^b f(x)dx = F(b) - F(a)$ *(see Figure 22.3)*

(c) $P(X > x) = \int_x^\infty f(t)dt = 1 - F(x)$ (22.26)

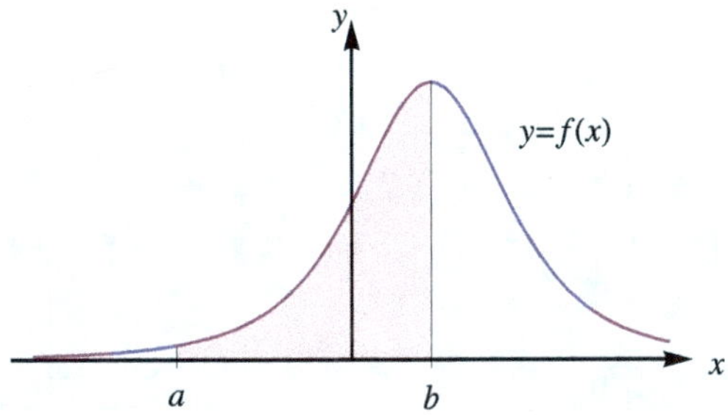

Figure 22.3: Illustration of Theorem (22.7b)

(d) $P(X = x) = 0$ *for all* x

(e) $\lim_{x\to-\infty} F(x) = 0, \quad \lim_{x\to\infty} F(x) = 1$

22.3.4 *Some common continuous distributions*

By designation for a random variable means the class of random variables that have same distribution, i.e., $X \in \exp(\lambda)$, etc.

Distribution notation	Probability density function : f	Parameters
Rectangle U(a, b)	$fx(x) = \begin{cases} 0 & \text{if } x < a \\ \dfrac{1}{b-a} & \text{if } a \le x \le b \\ 0 & \text{if } x > b \end{cases}$	$a < b$
Exponential $\exp(\lambda)$	$f(x) = \begin{cases} 0 & \text{if } x < 0 \\ \lambda e^{-\lambda x} & \text{if } x \ge 0 \end{cases}$	$\lambda > 0$
Beta B(α, β)	$f(x) = \begin{cases} 0 & \text{if } x < 0 \\ k_{\alpha,\beta} x^{\alpha-1}(1-x)^{\beta-1} & \text{if } 0 \le x \le 1 \\ 0 & \text{if } x > 1 \end{cases}$	$\alpha > 0, \beta > 0$

Distribution notation	Probability density function : f	Parameters
Chi-square $\chi^2(\nu)$	$f(x) = \begin{cases} 0 & \text{if } x < 0 \\ \dfrac{x^{\nu/2-1}e^{-x/2}}{\Gamma(\nu/2)} & \text{if } x \geq 0 \end{cases}$	$\nu = 1, 2, \ldots$
Gamma $\Gamma(\lambda, \gamma)$	$f(x) = \begin{cases} 0 & \text{if } x < 0 \\ \dfrac{x^{\gamma-1}\lambda^{\gamma}e^{-\lambda x}}{\Gamma(\gamma)} & \text{if } x \geq 0 \end{cases}$	$\lambda > 0, \gamma > 0$
Weibull $W(a, b)$	$f(x) = \begin{cases} 0 & \text{if } x < 0 \\ (a/b)\,(x/b)^{a-1}e^{-\left(\frac{x}{b}\right)^a} & \text{if } x \geq 0 \end{cases}$	$a > 0, \quad b \geq 1$
Normal $N(\mu, \sigma)$	$f(x) = \dfrac{1}{\sigma\sqrt{2\pi}}e^{-\frac{(x-\mu)^2}{2\sigma^2}}, \quad -\infty < x < \infty$	$-\infty < \mu < \infty, \sigma > 0$
$t-$ t_n	$f(x) = k_n\left(1 + \dfrac{x^2}{n}\right)^{-(n+1)/2}, \quad -\infty < x < \infty$	$n = 1, 2, \ldots$
F- $\mathcal{F}(m, n)$	$f_{m,n}(x) = \dfrac{m^{m/2}n^{n/2}x^{\frac{m}{2}-1}}{B\left(\frac{m}{2}, \frac{n}{2}\right)(mx+n)^{\frac{1}{2}(m+n)}}, \; x > 0$	$m, n = 1, 2, \ldots$
Rayleigh $R(\sigma)$	$f_\sigma(x) = \begin{cases} 0, & x < 0 \\ \dfrac{xe^{-\frac{x^2}{2\sigma^2}}}{\sigma^2}, & x \geq 0 \end{cases}$	σ
Gumbel	$f(x) = \dfrac{e^{\frac{a-x}{b} - e^{\frac{a-x}{b}}}}{b}, \quad -\infty < x < \infty$	$b > 0,\ a$
Laplace	$f(x) = \dfrac{1}{2b}e^{-\lvert x-a\rvert/b}, \quad -\infty < x < \infty$	$b > 0,\ a$
Sech	$f(x) = \dfrac{1}{\sigma}\dfrac{1}{e^{\frac{\pi(x-\mu)}{2\sigma}} + e^{\frac{\pi(\mu-x)}{2\sigma}}}, \quad -\infty < x < \infty$	$\mu,\ \sigma > 0$

(22.27)

Some of the PDF-graphs are reproduced on the following figure.

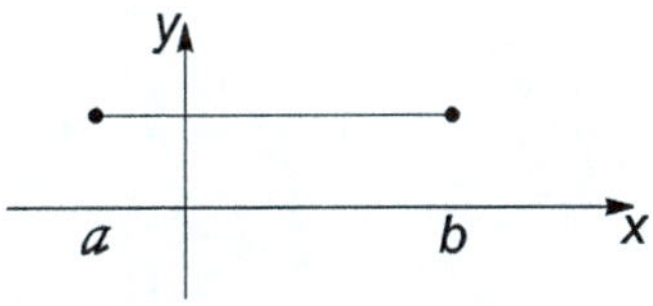

Rectangle distribution

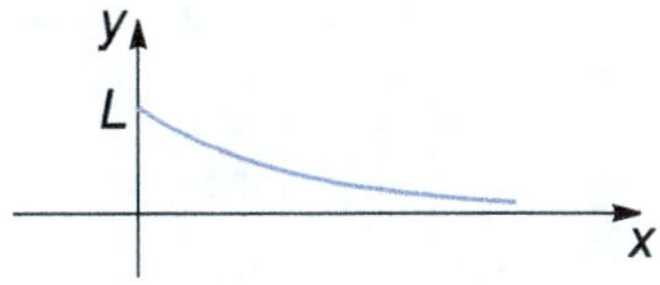

Exponential distribution

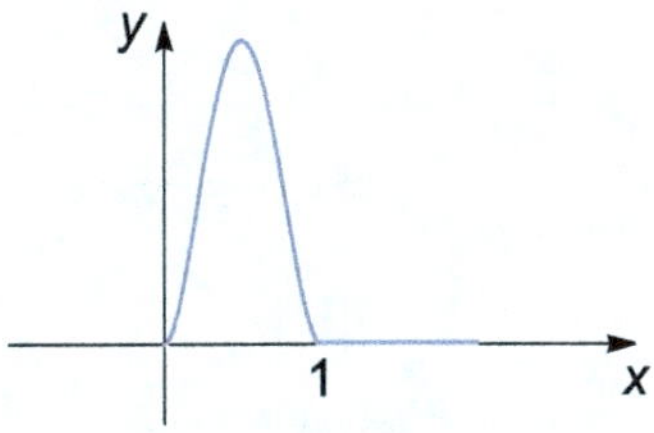

Beta distribution

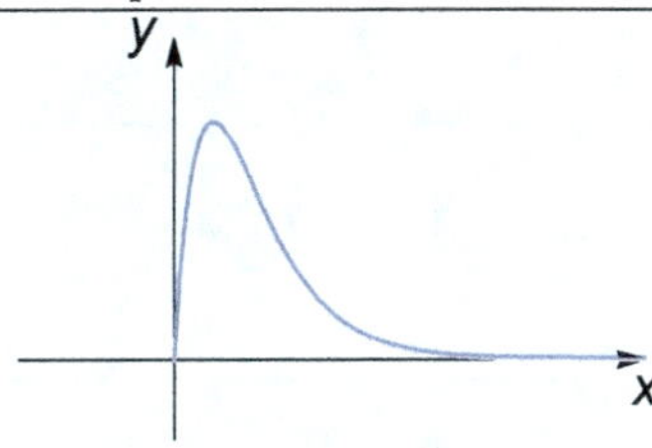

Gamma distribution

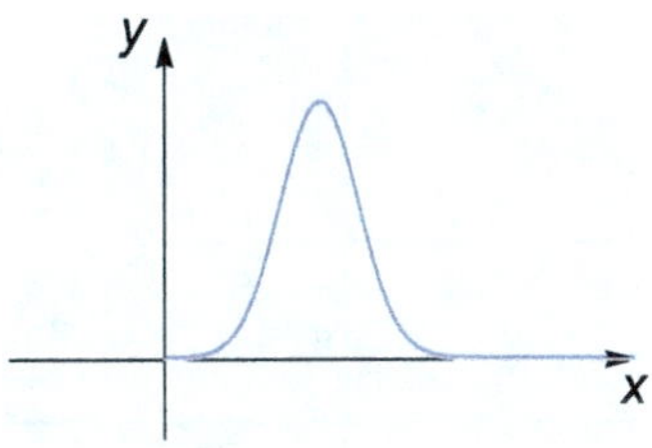

Normal distribution

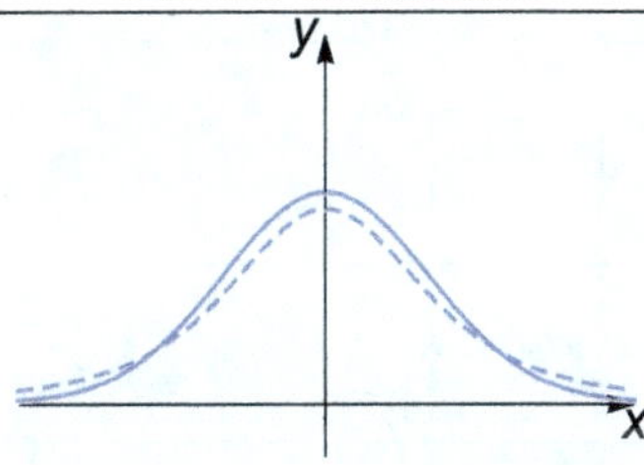

Standard normal distribution and t_3−distribution (dashed)

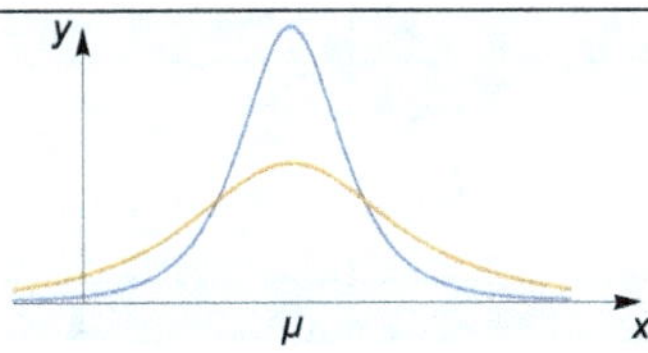

PDF for sech$(3, 1)$ *(i.e.* $\mu = 3$*)*

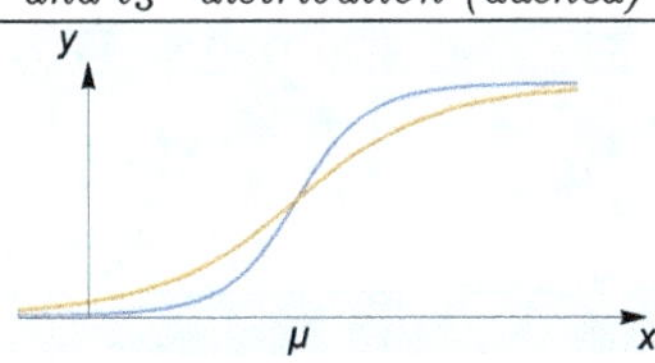

CDF for sech$(3, 1)$

Distribution	Cumulative distribution function : $F(x)$	Expectation μ and variance σ^2
Rectangle	$F(x) = \begin{cases} 0, & \text{if } x < a \\ \frac{x-a}{b-a}, & \text{if } a \le x \le b \\ 1, & \text{if } x > b \end{cases}$	$\mu = \frac{b+a}{2}$, $\sigma^2 = \frac{(b-a)^2}{12}$
Exponential	$F(x) = \begin{cases} 0, & \text{if } x < 0 \\ 1 - e^{-\lambda x}, & \text{if } x \ge 0 \end{cases}$	$\mu = \frac{1}{\lambda}$, $\sigma^2 = \frac{1}{\lambda^2}$
Beta	$F(x) = \begin{cases} 0, & \text{if } x < 0 \\ k_{\alpha,\beta} \int_0^x t^{\alpha-1}(1-t)^{\beta-1} dt, & \text{if } 0 \le x \le 1 \\ 1, & \text{if } x > 1 \end{cases}$	$\mu = \frac{\alpha}{\alpha+\beta}$, $\sigma^2 = \frac{\alpha\,\beta}{(\alpha+\beta)^2\ (1+\alpha+\beta)}$
Gamma	$F(x) = \begin{cases} 0, & \text{if } x < 0 \\ \frac{\lambda^\gamma}{\Gamma(\gamma)} \int_0^x t^{\gamma-1} e^{-\lambda t} dt, & \text{if } x \ge 0 \end{cases}$	$\mu = \frac{\gamma}{\lambda}$, $\sigma^2 = \frac{\gamma}{\lambda^2}$
Weibull	$F(x) = \begin{cases} 0, & \text{if } x < 0 \\ 1 - e^{-(x/b)^a}, & \text{if } x \ge 0 \end{cases}$	$\mu = b\Gamma\left(1 + \frac{1}{a}\right)$, $\sigma^2 = b^2\Gamma\left(1 + \frac{2}{a}\right) + -b^2\Gamma^2\left(1 + \frac{1}{a}\right)$
χ^2-	$F(x) = \begin{cases} 0, & \text{if } x < 0 \\ \frac{1}{2\Gamma(n/2)} \int_0^x (t/2)^{(n-2)/2} e^{-t/2} dt, & \text{if } x \ge 0 \end{cases}$	$\mu = n, \quad \sigma^2 = 2n$
Normal	$F(x) = \frac{1}{2\sqrt{\pi}} \left[1 + \mathrm{Erf}\left(\frac{x-\mu}{\sigma\sqrt{2}}\right)\right]$	$\mu, \quad \sigma^2$
t-	$F(x) = k_n \int_{-\infty}^x \left(1 + \frac{t^2}{n-1}\right)^{-n/2} dt$	$\mu = 0, \quad \sigma^2 = \frac{n-1}{n-3}$
F-	$F_{m,n}(x) = \begin{cases} 0, & x < 0 \\ \int_0^x \frac{m^{m/2} n^{n/2} t^{\frac{m}{2}-1}}{B\left(\frac{m}{2}, \frac{n}{2}\right)(mt+n)^{\frac{1}{2}(m+n)}} dt, & x > 0 \end{cases}$	$\mu = \frac{n}{n-2}$, $\sigma^2 = \frac{2n^2(m+n-2)}{m(n-4)(n-2)^2}$
Rayleigh	$F_\sigma(x) = \begin{cases} 0, & x < 0 \\ 1 - e^{-\frac{x^2}{2\zeta^2}}, & x \ge 0 \end{cases}$	$\mu = \sqrt{\frac{\pi}{2}}\zeta$, $\sigma^2 = \frac{1}{2}(4-\pi)\zeta^2$
Gumbel	$F(x) = e^{-e^{\frac{a-x}{b}}} \quad -\infty < x < \infty$	$\mu = a + \gamma\, b,\ \gamma \approx 0.577$, $\sigma^2 = b^2\,\pi^2/6$
Laplace	$F(x) = \begin{cases} \frac{1}{2} e^{(x-a)/b}, & -\infty < x < a \\ 1 - \frac{1}{2} e^{(a-x)/b}, & a \le x < \infty \end{cases}$	$\mu = a,\ \sigma^2 = 2b^2$
Sech	$F(x) = \frac{2}{\pi} \arctan\left(e^{\frac{x-\mu}{2\sigma}}\right), \quad -\infty < x < \infty$	$\mu,\ \sigma^2$

(22.28)

A random variable X is *log-normal distributed* if

$$Y := \ln X \in \mathrm{N}(\mu, \sigma), \text{ i.e., } X = e^Y. \tag{22.29}$$

The CDF is

$$F(x) := P(X \leq x) = \Phi\left(\frac{\ln x - \mu}{\sigma}\right).$$

Expectation and variance are $E(X) = e^{\mu+\sigma^2/2}$ and $e^{2\mu+\sigma^2}(e^{\sigma^2}-1)$, respectively.

A generalized gamma distribution has probability density function

$$f(x) = \begin{cases} 0, & \text{if } x < 0, \\ \dfrac{a\lambda^b\, x^{ab-1} e^{-\lambda\, x^a}}{\Gamma(b)}, & \text{if } x \geq 0. \end{cases} \tag{22.30}$$

A hyper exponential distribution has probability density function

$$\begin{aligned} &\text{PDF} \quad f(x) = F'(x) = e^{-x} \cdot e^{-e^{-x}} \text{ and CDF} \\ &F(x) = e^{-e^{-x}}, \quad x \in \mathbb{R}. \end{aligned} \tag{22.31}$$

Remarks. For Beta distribution, the normalization constant is

$$k_{\alpha,\beta} = \frac{\Gamma(\alpha+\beta)}{\Gamma(\alpha)\Gamma(\beta)}, \quad n = 2, 3, \ldots$$

In some literature γ corresponds to the Gamma distribution, of the parameter $\beta = 1/\gamma$ (page 490).

For the Weibull distribution, $a > 0$ and $b > 0$. For this distribution the parametrization varies. In some literature the parametrization is $F(x) = 1 - e^{-x^\beta/\alpha}$, $(x \geq 0)$. Sometimes, the same applies for the $\Gamma-$distribution: $\gamma = \alpha$ and $1/\lambda = \beta$.

For t-distribution, the normalization constant is

$$k_n = \frac{\Gamma(\frac{n+1}{2})}{\Gamma(\frac{n}{2})\sqrt{\pi}\sqrt{n}} = \frac{1}{\sqrt{n}B\left(\frac{n}{2}, \frac{1}{2}\right)}, \; n = 1, 2, \ldots$$

That X has the distribution $\mathrm{N}(\mu, \sigma)$ means that $P(X \leq x) = F(x)$.

The χ^2-distribution is a special case of gamma distribution with $\lambda = 1/2$ and $\gamma = n/2$.

The standard normal distribution is the normal distribution with $\mu = 0$ and $\sigma = 1$, i.e., $\mathrm{N}(0, 1)$.

With $n = 2$ in t-distribution, one obtains the *Cauchy distribution*, i.e., $f(x) = \dfrac{1}{\pi(1+x^2)}$. This distribution has no expectation.

Random variable for F-distribution is

$$\mathcal{F}_{m,n} = \frac{\chi_m^2/m}{\chi_n^2/n}.$$

CDF for F-distribution can be expressed as an elementary function, see the following figure on the right.

Rayleigh distribution is a special case of Weibull distribution, with $a = 2$ and $b^a = 2\sigma^2$.

In the Gumbel distribution, $\gamma \approx 0.577$ is Euler's constant.

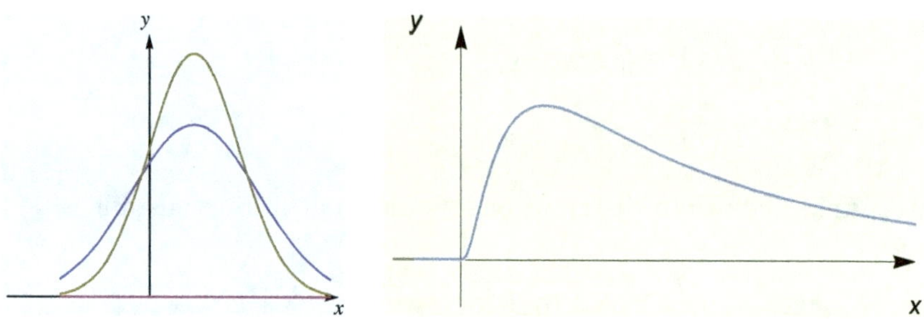

LHS: Two normally distributed PDF with same $\mu = 1$ and with $\sigma_1 = 1 < \sigma_2 = \sqrt{2}$.

RHS: The PDF lognormal $(0,1)$: $f(x) = \begin{cases} 0, & \text{if } x < 0, \\ \frac{e^{-\frac{1}{2}\ln^2 x}}{\sqrt{2\pi}x}, & \text{if } x \geq 0. \end{cases}$

22.3.5 *Connection between arbitrary normal distribution and the standard normal distribution*

Let $X \in \mathrm{N}(\mu, \sigma)$, F be its CDF, and Φ the CDF of $\mathrm{N}(0,1)$, i.e., of the standardized normal distribution (see the following). The connection is given by

$$F(x) = P(X \le x) = \Phi\left(\frac{x-\mu}{\sigma}\right). \tag{22.32}$$

The equality (22.32) implies that one can make use of the table on page 604 for N(0, 1) for *all* normal distributions. Corresponding PDF in here is denoted φ.

$$\Phi(b) = \int_{-\infty}^{b} \frac{1}{\sqrt{2\pi}} e^{-x^2/2} dx = \int_{-\infty} \varphi(x) dx. \tag{22.33}$$

22.3.6 *Approximations*

Approximations between distributions with rules of thumb are given in the following figure.

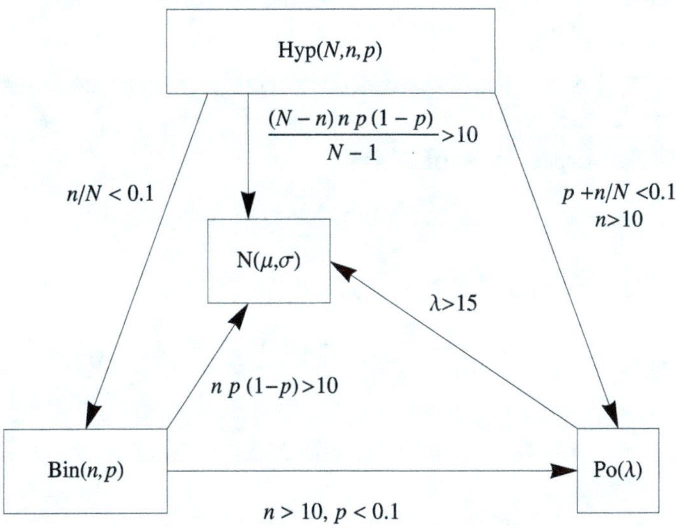

Approximations between distributions

Theorem 22.8. *Assume that* $\lambda = np$. *Then*

$$\lim_{n\to\infty} \binom{n}{x} p^x (1-p)^{n-x} \to \frac{e^{-\lambda}\lambda^x}{x!} \text{ as } n \to \infty. \tag{22.34}$$

$$\frac{\binom{Np}{x}\binom{N(1-p)}{n-x}}{\binom{N}{n}} = \frac{\binom{n}{x}\binom{N-n}{Np-x}}{\binom{N}{Np}} \to \binom{n}{x} p^x (1-p)^{n-x}, \quad \text{as } N \to \infty. \tag{22.35}$$

Remarks. The first limit value says that binomial distribution $\text{Bin}(n, p)$ can be approximated by Poisson distribution $\text{Po}(n{\cdot}p)$, when n large or p small.

The second limit value says that hypergeometric distribution can be approximated by binomial distribution $\text{Bin}(n, p)$, for large N.

22.4 Location and Spread Measures

Definition 22.7. Expectation and median are location measures. Variance and standard deviation are spread measures.

(i) Let X be a discrete random variable which assumes values $x_1, x_2, x_3, \ldots$.

(a) *The expectation* of X is

$$E(X) = \sum_i x_i P(X = x_i). \tag{22.36}$$

(b) *The variance* of X is defined as

$$V(X) = \sum_i (x_i - \mu)^2 P(X = x_i), \tag{22.37}$$

where $\mu = E(X)$.

(c) *The standard deviation* of X is defined as $\sigma = \sigma(X) = \sqrt{V(X)}$.

(ii) Let X be a continuous random variable and f, the corresponding PDF.

(a) The expectation of X or generally of $g(X)$ is defined as

$$E(X) = \int_{-\infty}^{\infty} x f(x) dx \text{ and } E(g(X)) = \int_{-\infty}^{\infty} g(x) f(x) dx, \tag{22.38}$$

insofar as the integral is convergent.

(b) The median is defined as the x-value, denoted md, such that

$$\int_{-\infty}^{\mathrm{md}} f(x)dx = 1/2. \tag{22.39}$$

For illustration of (a) and (b), see Figure 22.4.

(c) The variance equals

$$V(X) = E((X-\mu)^2) = \int_{-\infty}^{\infty} (x-\mu)^2 f(x)dx, \tag{22.40}$$

$\sigma = \sqrt{V(X)}$.

(d) A *quantile* k_α of a distribution is meant the "x-"value such that

$$P(X \leq k_\alpha) = 1-\alpha, \text{i.e.}, P(X > k_\alpha) = \alpha. \tag{22.41}$$

Quantiles for normal- (Student) t- χ^2-, and F-distrubitons can be found on page 604 and the following pages. For the sech distribution, see page 548.

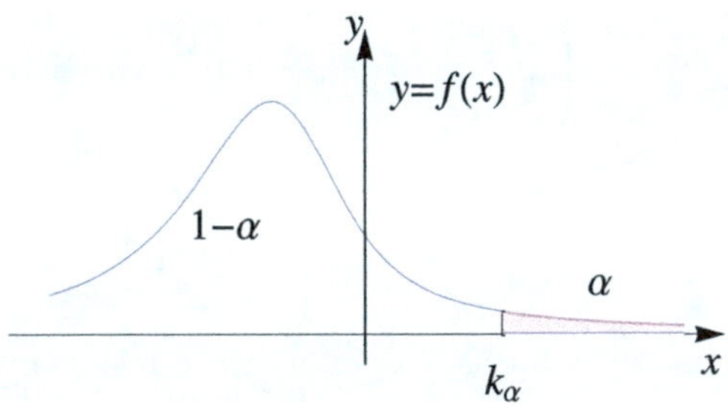

PDF $f(x)$ and CDF $F(x)$ with quantile k_α, i.e., the probability $1 - F(k_\alpha) = \alpha$.

Theorem 22.9.

(i) *If $X \geq 0$, then the expectation can be calculated by the corresponding CDF*:

$$E(X) = \int_0^{\infty} [1 - F(x)]\, dx = \int_0^{\infty} P(X > x)dx. \tag{22.42}$$

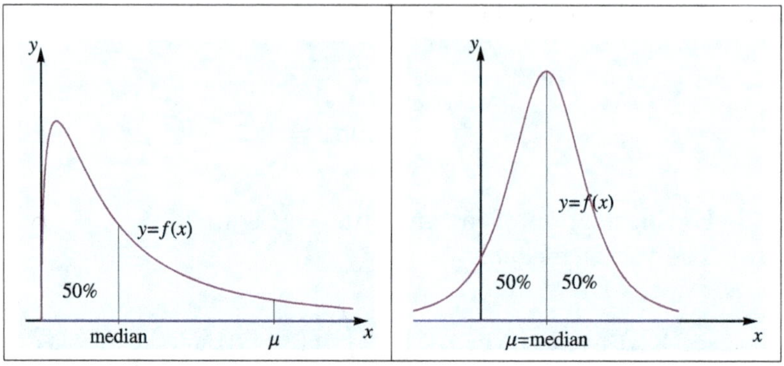

Figure 22.4: To the right: The median divides the surface into two parts, 50% of area on each side. The expectation μ does not in general coincide with the median. In contrast, if the PDF is symmetric and μ exists, we get the median $= \mu$.

(ii) *The variance can be calculated using the following alternative formula.*

$$V(X) = E(X^2) - \mu^2$$

$$\text{where } E(X^2) = \begin{cases} \displaystyle\sum_i x_i^2 P(X = x_i), & \text{if } X \text{ is discrete,} \\ \displaystyle\int_{-\infty}^{\infty} x^2 f(x)dx - \mu^2, & \text{if } X \text{ is continuous,} \end{cases}$$

$$\text{where } \mu = E(X). \tag{22.43}$$

Remarks.

(i) The expectation value is x-coordinate for the center of mass.
(ii) For a symmetric distribution, the median md and expectation coincide if the latter exists.
(iii) Two random variables X and Y with same distribution, i.e., $P(X \leq x) = P(Y \leq x)$ for all x, have the same expectation and variance.

Instead of the notation k_α, one uses

(i) λ_α for N(0, 1)-distribution (Figure 22.5).
(ii) $t_{n,\alpha}$ for t-distribution (Figure 22.6).
(iii) $\chi^2_{\alpha,n}$ for $\chi^2(n)$-distribution.

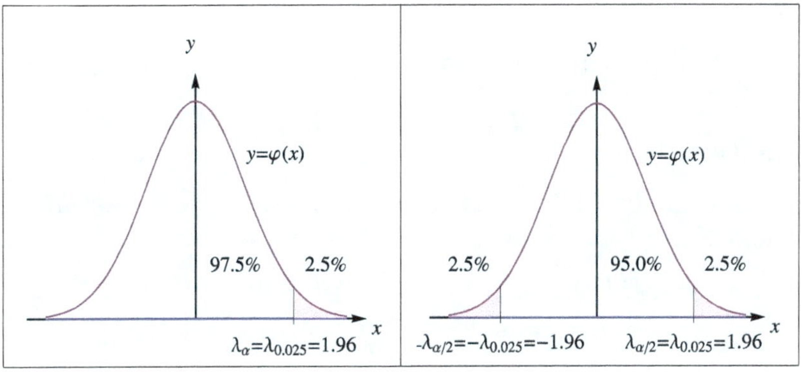

Figure 22.5: To the left: $x = \lambda_{0.025}$ a quantile of the standard normal distribution. To the right: $x = \lambda_{\alpha/2} = \lambda_{0.025} = 1.96$.

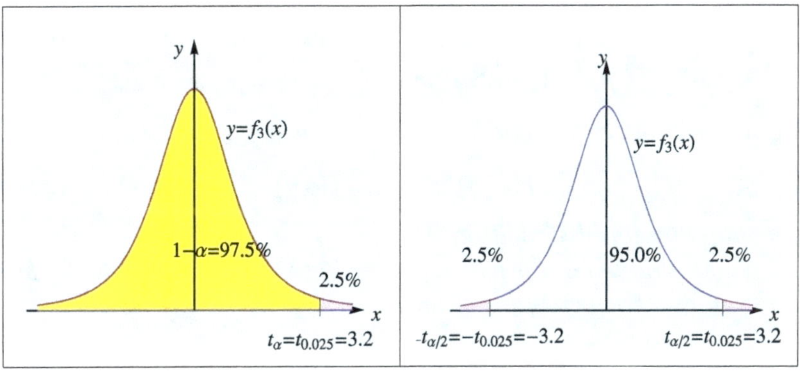

Figure 22.6: $x = t_\alpha = t_{3,0.025} = 3.2$, a quantile of the t-distribution (with $n = 3$).

22.5 Multivariate Distributions

Definition 22.8. Covariance and Correlation coefficient of two random variables X and Y are defined as

$$\operatorname{cov}(X, Y) := E[X - E(X)]E[Y - E(Y)]$$

and

$$\rho(X, Y) := \frac{\operatorname{cov}(Y)}{\sqrt{\operatorname{var}(X)\operatorname{var}(Y)}}, \tag{22.44}$$

respectively.

The covariance can be written $\operatorname{cov}(X, Y) = E(XY) - E(X)E(Y)$.

22.5.1 *Discrete distributions*

A discrete distribution depends on more than one discrete random variable.

Definition 22.9.

(i) The common PDF and CDF of two discrete random variables X and Y are

$$\begin{aligned} f(x,y) &:= P(X = x, Y = y) \text{ and} \\ F(x,y) &:= P(X \le x, Y \le y), \text{ respectively.} \end{aligned} \tag{22.45}$$

(ii) The PDF of the multinomial distribution is given by

$$P(X_1 = x_1, X_2 = x_2, \ldots, X_r = x_r) = \binom{n}{x_1 \quad x_2 \ldots x_r} p_1^{x_1} p_2^{x_2} \cdot \ldots \cdot p_r^{x_r}, \tag{22.46}$$

where $\sum_{j=1}^{r} p_j = 1$, $p_j > 0$ and $\sum_{j=1}^{r} x_j = n$, $x_j \ge 0$.

Theorem 22.10. *If X and Y are independent, their common PDF is*

$$f(x,y) = f_X(x) f_Y(y). \tag{22.47}$$

Bivariate Poisson distribution.

Assume that $X \in Po(\mu)$ and $Y \in Po(\lambda)$ are independent. Then their common frequency function is

$$f(x,y) = P(X = x, Y = y) = \frac{\mu^x \lambda^y}{x!y!} \cdot e^{-(\mu+\lambda)}. \tag{22.48}$$

22.5.2 *Bivariate continuous distribution*

Definition 22.10.

(i) Let X and Y be two continuous random variables. Their common cumulative distribution function is defined as

$$P(X \le x, Y \le y) := F(x,y). \tag{22.49}$$

If there is a function $f(x,y)$ such that

$$F(x,y) = \int_{-\infty}^{y} \int_{-\infty}^{x} f(u,v)\,du\,dv, \tag{22.50}$$

then this function is called the common PDF.

(ii) The probability of the event $\{a \leq X \leq b, c \leq Y \leq d\}$ is

$$P(a \leq X \leq b, c \leq Y \leq d) = \int_c^d \int_a^b f(u,v)dudv, \qquad (22.51)$$

(iii) The margin PDF with respect to X is

$$f_X(x) = \int_{-\infty}^{\infty} f(x,y)dy. \qquad (22.52)$$

22.6 Conditional Distribution

Definition 22.11.

(i) The PDF and CDF corresponding to the random variable X are here denoted $f_X(x)$ and F_X, respectively.
(ii) The *conditional* PDF and CDF for Y, with respect to $X = x$, are

$$f(y|x) = f_{Y|X}(y|x) = \frac{f(x,y)}{f_X(x)}$$

and

$$\begin{aligned} F_{Y|X}(y|x) &= \sum_y \frac{f(x,y)}{f_X(x)} \quad \text{(discrete case)} \\ F_{Y|X}(y|x) &= \int_{-\infty}^{\infty} \frac{f(x,v)dv}{f_X(x)} \quad \text{(continuous case)}, \end{aligned} \qquad (22.53)$$

respectively.
For those x such that $f_X(x) > 0$.
(ii) The conditional expectation value Y with respect to X is defined as $E(Y|X)$. Note that $E(Y|X)$ is a random variable.

Expectation and Variance

$$E(X) = E(E(X|Y)), \quad V(X) = E(V(X|Y)) + V(E(X|Y)).$$

Conditional Expectation

Discrete case	Continuous case
$E(X\|Y) = \sum_x x\, f(x,y)$	$E(X\|Y) = \int_{-\infty}^{\infty} x\, f(x,y)dx.$
$E(X) = \sum_y E(X\|y) f_Y(y)$	$E(X) = \int_{-\infty}^{\infty} E(X\|y) f_Y(y)dy.$

(22.54)

Theorem 22.11. *For X and Y independent, following hold true*

$$f(x|y) = f_X(x) \text{ and } \quad f(x,y) = f_X(x)\cdot f_Y(y). \tag{22.55}$$

22.7 Linear Combination of Random Variables

Definition 22.12. A linear combination of two random variables X_1 and X_2 is of the form $aX_1 + bX_2$, where a and b are constants.

Theorem 22.12. *Let X_1 and X_2 be two random variables with common PDF f. Let $X_1 + X_2 = Z$. Then the following hold true:*

$$X_1 \text{ and } X_2 \text{ discrete}: \begin{cases} P(Z=z) = \sum_x f(x, z-x). \\ P(Z=z) = \sum_x f_{X_1}(x) f_{X_2}(z-x) \\ \qquad \text{if } X_1 \text{ and } X_2 \text{ are independent.} \end{cases}$$

$$X_1 \text{ and } X_2 \text{ continuous}: \begin{cases} f_Z(z) = \int_{-\infty}^{\infty} f(x, z-x)dx. \\ f_Z(z) = \int_{-\infty}^{\infty} f_{X_1}(x) f_{X_2}(z-x)dx \\ \qquad \text{if they are independent.} \end{cases} \tag{22.56}$$

Theorem 22.13. *Let a and b be constants, X, X_1, and X_2 random variables. Then*

$$\begin{aligned}
&\text{(i) } E(aX+b) = aE(X)+b, \qquad \operatorname{cov}(X,X) = V(X).\\
&\text{(ii) } V(aX+b) = a^2 V(X), \quad \sigma(aX+b) = |a|\sigma(X).\\
&\text{(iii) } E(a\,X_1 + b\,X_2) = a\,E(X_1) + b\,E(X_2). \qquad (22.57)\\
&\text{(iv) } V(X_1+X_2) = V(X_1)+V(X_2)+2\operatorname{cov}(X_1,X_2).\\
&\text{(v) } V(a\,X_1+b\,X_2) = a^2\,V(X_1)+b^2\,V(X_2), \textit{if } X_1 \textit{ and } X_2 \textit{ indep.}
\end{aligned}$$

Theorem 22.14. *Let $c_1, c_2, \ldots, c_n$ be constants and $X_1, X_2, \ldots X_n$ be random variables. Then*

$$E(c_1X_1 + c_2X_2 + \cdots + c_nX_n) = c_1E(X_1) + c_2E(X_2) + \cdots + c_nE(X_n).$$

$$V(c_1X_1 + c_2X_2 + \cdots + c_nX_n) \tag{22.58}$$
$$= c_1^2V(X_1) + c_2^2V(X_2) + \cdots + c_n^2V(X_n), \text{ if } \xi_1, \xi_2, \ldots, \xi_n$$
independent.

Theorem 22.15. *Let $X_1, X_2, \ldots, X_n$ be independent random variables, with expectation $E(X_i)$ and variance $V(X_i) = \sigma^2$, $i = 1, 2, \ldots, n$. Put*

$$\overline{X} = \frac{1}{n}(X_1 + X_2 + \cdots + X_n) = \frac{1}{n}\sum_{i=1}^{n} X_i.$$

Then,

$$E(\overline{X}) = \mu \textit{ and } V(\overline{X}) = \frac{\sigma^2}{n}. \tag{22.59}$$

Remarks. From (22.57(ii)), one gets

$$\sigma(aX+b) = \sqrt{V(aX+b)} = \sqrt{a^2V(X)} = \sqrt{a^2}\sigma(X) = |a|\sigma(X).$$

The last equality follows since $\sqrt{a^2} = |a|$.

For the variances one has $V(X_1 \pm X_2) = V(X_1) + V(X_2)$. Put

$$V(X_1) = \sigma_1^2 \text{ and } V(X_2) = \sigma_2^2,$$

and the standard deviation for sum or difference to σ. Then

$$\sigma_1^2 + \sigma_2^2 = \sigma^2. \tag{22.60}$$

Theorem 22.16. *Let X_1 and X_2 be independent random variables.*

(i) *If $X_1 \in N(\mu_1, \sigma_1)$ and $X_2 \in N(\mu_2, \sigma_2)$ are normal distributed random variables, then also $a\,X_1 + b\,X_2$ is normal distributed, with*

$$\mu = E(a\,X_1 + b\,X_2) = a\,\mu_1 + b\,\mu_2 \text{ and } \sigma = \sqrt{a^2\sigma_1^2 + b^2\sigma_2^2}.$$

(ii) *If $X_1 \in Po(\lambda_1)$ and $X_2 \in Po(\lambda_1)$, i.e., Poisson distributed, then so is their sum $X := X_1 + X_2$. More precisely, $X \in Po(\lambda_1 + \lambda_2)$ with*

$$\mu := E(X_1 + X_2) = \lambda_1 + \lambda_2 \text{ and variance } \sigma^2 = V(X_1 + X_2) \\ = \lambda_1 + \lambda_2.$$

(iii) *If $X \in Po(\lambda)$ and $a > 0$, then*

$$aX \in Po(\lambda/a).$$

(iv) *If $X_1 \in Bin(n_1, p)$ and $X_2 \in Bin(n_2, p)$, i.e., binomial distributed with same p, then their sum*

$$X_1 + X_2 \in Bin(n_1 + n_2, p).$$

(v) *If $X_1 \in Exp(\lambda_1)$ and $X_2 \in Exp(\lambda_2)$, i.e., X_1 and X_2 are exponentially distributed, then*

$$\min(X_1, X_2) \in Exp(\lambda) \text{ with } \lambda = \lambda_1 + \lambda_2.$$

(vi) *If $X_1, X_2, \ldots, X_n$ are independent, $X_k \in \exp(\lambda_k)$, $k = 1, 2, \ldots, n$ and $p_k \geq 0$ with $\sum_{k=1}^{n} p_k = 1$, then*

$$\sum_{k=1}^{n} p_k X_k \text{ (by definition) is a hyper-exponential}$$

distributed variable.

$$\text{with PDF } f(x) = \begin{cases} \displaystyle\sum_{k=1}^{n} p_k\,\lambda_k\,e^{-\lambda\,x}, & x \geq 0, \\ 0, & x < 0. \end{cases} \tag{22.61}$$

Expectation and variance is

$$\mu = \sum_{k=1}^{n} \frac{p_k}{\lambda_k} \text{ and } \left[\sum_{k=1}^{n} \frac{p_k}{\lambda_k}\right]^2 + \sum_{j,k=1}^{n} p_j\, p_k \left[\frac{1}{\lambda_j} - \frac{1}{\lambda_k}\right]^2, \text{ respectively.}$$

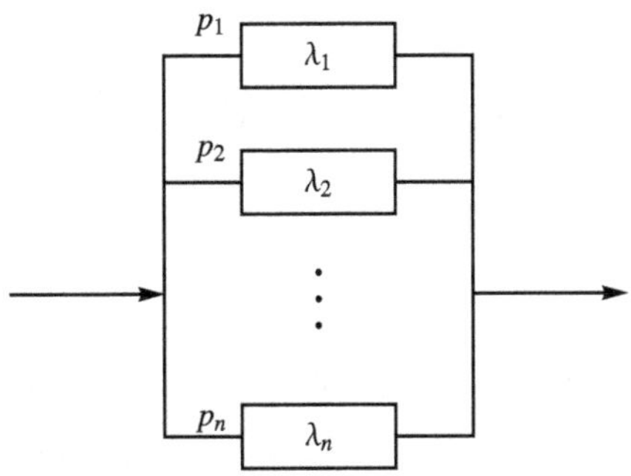

System with hyper exponential life span: Signal from the left enters component k with probability p_k and each component has life span $X_k \in \exp(\lambda_k)$, $k = 1, 2, \ldots, n$.

Theorem 22.17. *Assume that X_1, X_2, ..., X_n are independent and $N(\mu_j, \sigma_j)$ distributed, $j = 1, 2, \ldots, n$.*

(i) *Put $\overline{X} = \dfrac{X_1 + X_2 + \cdots + X_n}{n}$. Then, $\overline{X}$ and $\sum_{k=1}^{n}(X_k - \overline{X})^2$ are independent. Furthermore,*

$$\overline{X} \in N\left(\mu, \frac{\sigma}{\sqrt{n}}\right). \qquad (22.62)$$

(ii) *The random variable defined as*

$$\frac{1}{\sigma^2} \sum_{k=1}^{n} (X_k - \overline{X})^2 \in \chi^2_{n-1} \qquad (22.63)$$

is χ^2–distributed with $n-1$ degrees of freedom.

(iii) *The random variable defined as*

$$\frac{1}{\sigma^2} \sum_{k=1}^{n} (X_k - \mu)^2 \in \chi^2_{n} \qquad (22.64)$$

is χ^2–distributed with n degrees of freedom.

22.8 Generating Functions

Definition 22.13. Let X and Y be two random variables. Following functions are called generating:

Moment generating function:	$M_X(t) := E(e^{tX}) : \mathbb{R} \to [0, \infty)$
Characteristic function:	$\phi_X(t) := E(e^{itX}) : \mathbb{R} \to \mathbb{C}$
Two-dimensional common characteristic function:	$\phi_{X,Y}(s,t) := E(e^{isX}e^{itY}) : \mathbb{R}^2 \to \mathbb{C}$
Probability generating function:	$G_X(s) := E(s^X)$
Two-dimensional probability-generating function	$G_{X,Y}(s,t) := E(s^X t^Y).$

(22.65)

Table of some probability — and moment generating functions

Distribution	Probability generating function, $G(s)$	Moment generating function, $M(t)$
Uniform N	$\dfrac{s\left(s^N-1\right)}{N(s-1)}$	$\dfrac{e^t\left(e^{Nt}-1\right)}{N\left(e^t-1\right)}$
Binomial (n,p)	$(p(s-1)+1)^n$	$\left(p(e^t-1)+1\right)^n$
Geometric p	$\dfrac{p}{(p-1)s+1}$	$\dfrac{p}{(p-1)e^t+1}$
Negative binomial (k,p)	$p^k s^k((p-1)s+1)^{-k}$	$p^k e^{kt}\Big((p-1)e^t+1\Big)^{-k}$
Poisson λ	$e^{\lambda(s-1)}$	$e^{\lambda(e^t-1)}$
Rectangle $[a,b]$	$\dfrac{s^b-s^a}{(b-a)\ln s}$	$\dfrac{e^{bt}-e^{at}}{(b-a)t}$
Exponential λ	$\dfrac{\lambda}{\lambda-\ln s}$	$\dfrac{\lambda}{\lambda-t}$

(22.66)

Theorem 22.18. *Assume that $Y = aX + b$, X is a random variable and ϕ, a characteristic function. Then*

$$\phi_Y(t) = e^{itb}\phi_X(at). \tag{22.67}$$

If X and Y are independent, then

$$\begin{aligned}\phi_{X+Y}(t) &= \phi_X(t)\phi_Y(t)\\ \phi_{X,Y}(s,t) &= \phi_X(s)\phi_Y(t)\end{aligned} \tag{22.68}$$

$$\begin{aligned}\phi_X(t) &= (1+p(e^{it}-1))^n \iff X \in Bin(n,p)\\ \phi_X(t) &= \left(\frac{\lambda}{\lambda - it}\right)^{\gamma} \iff X \in \Gamma(\lambda,\gamma)\\ \phi_X(t) &= e^{i\mu t - \sigma^2 t^2/2} \iff X \in N(\mu,\sigma)\,.\end{aligned} \tag{22.69}$$

Theorem 22.19. *Let X be a random variable and $G_X(s) = E(s^X)$ its probability generating function. Then, $G(0) = P(X = 0)$, $G(1) = 1$ and*

$$E[X(X-1)\dots(X-k+1)] = \frac{d^kG(s)}{ds^k} \text{ for } s = 1, \quad k = 0,1,2,\dots \tag{22.70}$$

Assume that X and Y are independent random variables. Then

$$\begin{aligned}G_{X+Y}(s) &= G_X(s)G_Y(s)\\ G_{X,Y}(s,t) &= G_X(s)G_Y(t).\end{aligned} \tag{22.71}$$

If $X_1, X_2, \dots, X_N$ is a sequence of independent and equally distributed random variables and $N \in \{1,2,3,\dots\}$ is a random variable, then

$$G_{(X_1, X_2, \dots,)N}(s) = G_N(G_X(s)). \tag{22.72}$$

22.9 Some Inequalities

Theorem 22.20. *Assume that f is a non-negative measurable function and a a positive number. Then the following inequalities hold true:*

$$P(f(X) \geq a) \leq \frac{E(f(X))}{a}. \tag{22.73}$$

$$\begin{aligned}&\textit{Markov's inequality:} \quad P(|X| \geq a) \leq \frac{E(|X|)}{a}.\\ &\textit{Chebyshev's inequality:} \; P(|X| \geq a) \leq \frac{E(X^2)}{a^2}.\end{aligned} \tag{22.74}$$

Assume that f is a measurable function and $0 \leq f(x) \leq b$ for some positive real number. Then

$$P(f(X) \geq a) \leq \frac{E(f(X)) - a}{N - a} \quad \textit{if} \quad 0 \leq a < b. \tag{22.75}$$

$$\textit{Cauchy–Schwarz's inequality: } E(X\,Y)^2 \leq E(X^2)\,E(Y)^2, \tag{22.76}$$

with equality if and only if $P(a\,X = b\,Y) = 1$ for some real numbers a and b.

22.10 Convergence of Random Variables

Definition 22.14. Let X_1, X_2, $X_3, \ldots$ and X be random variables defined in some probability space Ω. Then the following four types of convergences take place.

(i) $X_n \to X$ almost surely (a.s.), written $X_n \overset{\text{a.s.}}{\to} X$ and means that

$$P(\{\omega \in \Omega : X_n(\omega) \to X(\omega)\}) = 1,$$

as $n \to \infty$.

(ii) $X_n \to X$ in $r-$mean, where $r \geq 1$, and is denoted $X_n \overset{r}{\to} X$ meaning

$$E(|X_n - X|^r) \to 0 \text{ as } n \to \infty.$$

(iii) $X_n \to X$ in probability, denoted $X_n \overset{P}{\to} X$ and means that

$$P(|X_n - X| > \varepsilon) \to 0,$$

for each $\varepsilon > 0$, as $n \to \infty$.

(iv) $X_n \to X$ in distribution sense, denoted $X_n \overset{\text{D}}{\to} X$ and means that

$$P(X_n \leq x) \equiv F_n(x) \to P(X \leq x) \equiv F(x),$$

for those x where $F(x)$ is continuous.

Remarks. Convergence in distribution is the same as $F_n(x) \to F(x)$ where F_n and F are CDF for X_n and X, respectively. If these random variables are continuous, this can be written as

$$\int_{-\infty}^{x} f_n(t)dt \to \int_{-\infty}^{x} f(t)dt,$$

if $F_n' = f_n$ and $F' = f$.

Theorem 22.21.

$$\left.\begin{array}{l} X_n \overset{a.s.}{\to} X \\ or \\ X_n \overset{r}{\to} X \end{array}\right\} \Longrightarrow X_n \overset{P}{\to} X \Longrightarrow X_n \overset{D}{\to} X. \tag{22.77}$$

Theorem 22.22.

(i) $X_n \overset{D}{\to} c \Longrightarrow X_n \overset{P}{\to} c$, *if c is a constant.*
(ii) *If $X_n \overset{P}{\to} X$ and there is a constant k such that $P(|X_n| \leq k) = 1$ for all n, so is $X_n \overset{r}{\to} X$ for all $r \geq 1$.*
(iii) *If for all $\varepsilon > 0$ the following holds: $\sum_n P(|X_n - X| > \varepsilon) < \infty$, then $X_n \overset{a.s.}{\to} X$.*

Theorem 22.23. *Assume that $X_1, X_2, X_3, \ldots$ are independent and equally distributed and $E(X_i) =: \mu < \infty$. Then the following rules hold:*

The law of large numbers:

$$\frac{X_1 + X_2 + \cdots + X_n}{n} \overset{D}{\longrightarrow} \mu, \ \ as\ n \to \infty.$$

The Central limit theorem:

$$\frac{X_1 + X_2 + \cdots + X_n - n\mu}{\sigma\sqrt{n}} \overset{D}{\longrightarrow} N(0,1)\,, \ \ as\ n \to \infty, \ \ if\ E(X^2) < \infty. \tag{22.78}$$

Theorem 22.24. *Assume that $X_1, X_2, X_3, \ldots$ are independent, equally distributed and $E(X^2) < \infty$. Then it applies that*

$$\frac{X_1 + X_2 + \cdots + X_n}{n} \xrightarrow{a.s.} \mu, \quad as\ n \to \infty.$$
$$\frac{X_1 + X_2 + \cdots + X_n}{n} \xrightarrow{r=2} \mu, \quad as\ n \to \infty. \tag{22.79}$$

The strong law of large numbers
Assume as above that X_1, X_2, $X_3, \ldots$ are independent and equally distributed. Then the following equivalence holds:

$$\lim_{n\to\infty} \frac{X_1 + X_2 + \cdots + X_n}{n} \overset{a.s.}{=} \mu \Longleftrightarrow E(|X_i|) < \infty. \tag{22.80}$$

If limit exists, then $\mu = E(X_i)$.

Theorem 22.25.

(i) *Assume that $X_n \overset{D}{\to} X$ or expressed with corresponding distributions: $F_n \to F$. Let $(\phi_n(t))_{n=1}^{\infty}$ and $\phi(t)$ be corresponding characteristic functions. Then*

$$\phi_n(t) \to \phi(t), \quad as\ n \longrightarrow \infty. \tag{22.81}$$

(ii) *Conversely, if the limit (22.81) of the characteristic functions exists for all t and $\phi(t)$ is continuous at $t = 0$, then $F_n \to F$, i.e., $X_n \overset{D}{\to} X$.*

Remarks. In practice, the Central limit theorem means that

$$P\left(\frac{\sum_{i=1}^n X_i - n\mu}{\sigma\sqrt{n}} \le x\right) \approx \Phi(x), \text{ if } n \text{ is large,}$$
i.e., when
$$\sum_{i=1}^n X_i \text{ is approximatively N}\left(n\mu, \sigma\sqrt{n}\right), \text{ for } n \text{ large.} \tag{22.82}$$

For $\overline{X}$ it holds that

$$P\left(\frac{\overline{X} - \mu}{\sigma/\sqrt{n}} \le x\right) \approx \Phi(x) \text{ if } n \text{ large.} \tag{22.83}$$

The convergence "$\xrightarrow{r=2}$" in (22.79) means convergence in "L^2-norm" $\|\cdot\|_2$, ($\|\cdot\|_2 = E(|\cdot|^2)$).

Theorem 22.26. *Assume that X_1, X_2, $X_3, \ldots$ and Y_1, Y_2, $Y_3, \ldots$ are two sequences of random variables, then*

$$\begin{aligned}
&X_n \overset{a.s.}{\to} X \,and\, Y_n \overset{a.s.}{\to} Y \Longrightarrow X_n + Y_n \overset{a.s.}{\to} X + Y \\
&X_n \overset{r}{\to} X \text{ and } Y_n \overset{r}{\to} Y \Longrightarrow X_n + Y_n \overset{r}{\to} X + Y \qquad (22.84)\\
&X_n \overset{P}{\to} X \text{ and } Y_n \overset{P}{\to} Y \Longrightarrow X_n + Y_n \overset{P}{\to} X + Y.
\end{aligned}$$

22.10.1 *Table of some probability and moment generating functions*

Distribution	Probability generating function, $G(s)$	Moment generating function, $M(t)$
Uniform N	$\dfrac{s\left(s^N-1\right)}{N(s-1)}$	$\dfrac{e^t\left(e^{Nt}-1\right)}{N\left(e^t-1\right)}$
Binomial (n,p)	$(p(s-1)+1)^n$	$\left(p(e^t-1)+1\right)^n$
Geometric p	$\dfrac{p}{(p-1)s+1}$	$\dfrac{p}{(p-1)e^t+1}$
Negative binomial (k,p)	$p^k s^k((p-1)s+1)^{-k}$	$p^k e^{kt}\left((p-1)e^t+1\right)^{-k}$
Poisson λ	$e^{\lambda(s-1)}$	$e^{\lambda(e^t-1)}$
Rectangle $[a,b]$	$\dfrac{s^b-s^a}{(b-a)\ln s}$	$\dfrac{e^{bt}-e^{at}}{(b-a)t}$
Exponential λ	$\dfrac{\lambda}{\lambda-\ln s}$	$\dfrac{\lambda}{\lambda-t}$

(22.85)

Theorem 22.27. *Assume that $Y = aX + b$, X is a random variable and ϕ, a characteristic function. Then*

$$\phi_Y(t) = e^{itb}\phi_X(at). \qquad (22.86)$$

If X and Y are independent, then

$$\begin{aligned}
\phi_{X+Y}(t) &= \phi_X(t)\phi_Y(t) \\
\phi_{XY}(s,t) &= \phi_X(s)\phi_Y(t).
\end{aligned} \qquad (22.87)$$

$$\phi_X(t) = (1 + p(e^{it} - 1))^n \Longleftrightarrow X \in Bin(n, p)$$

$$\phi_X(t) = \left(\frac{\lambda}{\lambda - it}\right)^\gamma \Longleftrightarrow X \in \Gamma(\lambda, \gamma) \tag{22.88}$$

$$\phi_X(t) = e^{i\mu t - \sigma^2 t^2/2} \Longleftrightarrow X \in N(\mu, \sigma) .$$

Theorem 22.28. *Let X be a random variable and $G_X(s) = E(s^X)$ its probability generating function. Then $G(0) = P(X = 0), G(1) = 1$ and*

$$E[X(X-1)\dots(X-k+1)] = \frac{d^k G(s)}{ds^k} \text{ for } s = 1, \quad k = 0, 1, 2, \dots \tag{22.89}$$

Assume that X and Y are independent random variables. Then

$$\begin{aligned} G_{X+Y}(s) &= G_X(s)G_Y(s) \\ G_{X,Y}(s,t) &= G_X(s)G_Y(t). \end{aligned} \tag{22.90}$$

If $X_1, X_2, \dots, X_N$ is a sequence of independent and equally distributed random variables and $N \in \{1, 2, 3, \dots\}$ is a random variable, then

$$G_{(X_1, X_2, \dots, X_N)}(s) = G_N(G_X(s)). \tag{22.91}$$

22.11 Point Estimation of Parameters

Definition 22.15.

(i) A sample of size n is a sequence $X_1, X_2, \dots, X_n$ of n independent equally distributed random variables. An *observed* sample is the set of corresponding observed values $x_1, x_2, \dots, x_n$.

(ii) Let θ be a parameter for X_i and let Ω be the sample space of X_i.

(iii) An estimation function $\mathcal{E}$ is given by $\mathcal{E} : \Omega^n \curvearrowright \mathbb{R}$.

  (a) $\mathcal{E}(X_1, X_2, \dots, X_n) = \theta^*$ is called a point estimation of θ.

  (b) $\mathcal{E}(X_1, X_2, \dots, X_n) = \theta^*_{\text{obs}}$ is called an *observed* point estimation of θ.

Estimation functions of expectation
For a sample of size n one can estimate the expectation of the distribution by

$$\mu^* = \frac{X_1 + X_2 + \cdots + X_n}{n} = \frac{1}{n}\sum_{i=1}^{n} X_i. \tag{22.92}$$

Corresponding observed point estimation is

$$\overline{x} = \mu^*_{\text{obs}} = \frac{x_1 + x_2 + \cdots + x_n}{n} = \frac{1}{n}\sum_{i=1}^{n} x_i. \tag{22.93}$$

22.11.1 *Expectancy accuracy and efficiency*

Definition 22.16. Let $(X_1, X_2, \ldots,)$ be equally distributed and independent random variables and $\mathcal{E}((X_1, X_2, \ldots,)) = \theta^*$, a point estimation of a parameter θ for X_i. θ^* is unbiased if

$$E(\mathcal{E}(X_1, X_2, \ldots, X_n)) = E(\theta^*) = \theta. \tag{22.94}$$

If $E(\theta^*) \neq \theta$, one says that there is a systematic error.

Definition 22.17. Let θ_1^* och θ_2^* be two unbiased point estimations of a parameter θ. If $V(\theta_1^*) < V(\theta_2^*)$, one says that θ_1^* is *more effective* than θ_2^*.

Let $X_1, X_2, \ldots, X_n$ be a sample of size n of a random variable X with $E(X_j) = E(X) = \mu$, $j = 1, 2, \ldots, n.$ and $V(X) = \sigma^2$. "obs" means "observed". Useful point estimations and corresponding observed point estimations, regardless of distribution are

$$\mu^* = \overline{X}, \qquad \mu^*_{\text{obs}} = \overline{x}$$

$$\sigma^{2*} = \frac{1}{n}\sum_{i=1}^{n}(X_i - \mu)^2\,,\ {\sigma^2_{\text{obs}}}^* = s^2 = \frac{1}{n}\sum_{i=1}^{n}(x_i - \mu)^2, \tag{22.95}$$

if μ is known.

$$\sigma^{2*} = \frac{1}{n-1}\sum_{i=1}^{n}(X_i - \overline{X})^2, {\sigma_{\text{obs}}^2}^* = s^2 = \frac{1}{n-1}\sum_{i=1}^{n}(x_i - \overline{x})^2,$$

if μ is unknown. (22.96)

Corresponding estimate of standard deviation is
$\sigma^* = \sqrt{\sigma^{2*}}, \quad \sigma_{\text{obs}}^* = \sqrt{s^2} = s.$

Remarks.

(i) One can easily show that $E(\overline{X}) = \mu$. Even σ^{2*} is unbiased. However, $\sqrt{\sigma^{2*}}$ is not unbiased, i.e., biased.
(ii) When applying point estimation of σ^2, in practice μ is unknown, i.e., it is only (22.96) which is used.

22.12 Interval Estimation

22.12.1 *Confidence interval for μ in normal distribution: $X \in N(\mu, \sigma)$*

Two-sided (symmetric) interval of confidence of (confidence) degree $1 - \alpha$:

$$\sigma \text{ known:} \quad \left[\bar{x} - \lambda_{\alpha/2}\frac{\sigma}{\sqrt{n}}, \bar{x} + \lambda_{\alpha/2}\frac{\sigma}{\sqrt{n}}\right]. \tag{22.97}$$

$$\sigma \text{ unknown:} \quad \left[\bar{x} - t_{n-1,\alpha/2}\frac{s}{\sqrt{n}}, \bar{x} + t_{n-1,\alpha/2}\frac{s}{\sqrt{n}}\right]. \tag{22.98}$$

The estimated standard deviation s in (22.98) is given by

$$s = s_{n-1} = \sqrt{\sigma_{\text{obs}}^{2*}} = \sqrt{\frac{1}{n-1}\sum(x_i - \bar{x})^2}.$$

One-sided upper bounded and lower bounded confidence intervals of degree $1 - \alpha$:

$$\sigma \text{ known:} \quad \left(-\infty, \bar{x} + \lambda_\alpha\frac{\sigma}{\sqrt{n}}\right], \quad \left[\bar{x} - \lambda_\alpha\frac{\sigma}{\sqrt{n}}, \infty\right). \tag{22.99}$$

$$\sigma \text{ unknown:} \quad \left(-\infty, \bar{x} + t_{n-1,\alpha}\frac{s}{\sqrt{n}}\right], \quad \left[\bar{x} - t_{n-1,\alpha}\frac{s}{\sqrt{n}}, \infty\right). \tag{22.100}$$

22.12.2 *Confidence interval for σ^2 in normal distribution*

Let X be a χ^2 distributed random variable with $n-1$ degrees of freedom. The quantile $\chi_{1-\alpha}(n-1)$ fulfills

$$P(X \leq \chi^2_{1-\alpha}(n-1)) = \alpha.$$

Similarly the quantile $\chi^2_\alpha(n-1)$ satisfies

$$P(X \leq \chi^2_\alpha(n-1)) = 1-\alpha.$$

$\left[0, \frac{(n-1)s^2}{\chi^2_{1-\alpha}(n-1)}\right]$	One-sided upper bounded conf. interval of degree $1-\alpha$ for σ^2.
$\left[\frac{(n-1)s^2}{\chi^2_\alpha(n-1)}, \infty\right)$	One-sided lower bounded conf. interval of degree $1-\alpha$ for σ^2.
$\left[\frac{(n-1)s^2}{\chi^2_{\alpha/2}(n-1)}, \frac{(n-1)s^2}{\chi^2_{1-\alpha/2}(n-1)}\right]$	Two-sided bounded conf. interval $1-\alpha$ for σ^2.

(22.101)

Remarks. Corresponding confidence interval for σ in (22.101) is obtained by taking the root of respective boundary value.

Also one can treat the case when σ^2 (and σ) interval is estimated with μ known. In practice, however, this is not the case.

22.12.3 *Sample in pair and two samples*

By sample in pair it is supposed we have pairwise observations (X_i, Y_i) , $i = 1, 2, \ldots, n$, where

$$X_i \in \mathrm{N}(\mu_i, \sigma_1) \text{ and } Y_i \in \mathrm{N}(\mu_i + \Delta\mu, \sigma_2) \tag{22.102}$$

and that the pairs (X_1, Y_1), $(X_2, Y_2), \ldots, (X_n, Y_n)$ are independent.

By *two samples* it is assumed that

$$\begin{array}{ll} X_1, X_2, \ldots, X_{n_1} & \text{is a sample of N}(\mu_1, \sigma) \\ Y_1, Y_2, \ldots, Y_{n_2} & \text{is a sample of N}(\mu_2, \sigma), \end{array} \tag{22.103}$$

and that the samples are independent.

Samples in pair

A confidence interval for $\Delta\mu$ is formed to detect significant difference between ξ_i and η_i. The interval estimation is then made for $\Delta\mu$ and

$$\frac{1}{n}\sum_{k=1}^{n}(\xi_k - \eta_k) = \overline{\xi} - \overline{\eta} \in \mathrm{N}\left(\Delta\mu, \sqrt{\sigma_1^2 + \sigma_2^2}\right). \tag{22.104}$$

Confidence intervals are created as in (22.98).

Two samples

Theorem 22.29. *If one has two observed samples, $x_1, x_2, \ldots, x_{n_1}$ of size n_1 from $N(\mu_1, \sigma)$ and $y_1, y_2, \ldots, y_{n_2}$ of size n_2 from $N(\mu_2, \sigma)$, i.e., from normal distributions with equal σ, then the best (most effective) observed point estimation of σ^2 is*

$$\sigma^{2*}_{\text{obs}} = \frac{(n_1 - 1)s_1^2 + (n_2 - 1)s_2^2}{n_1 - 1 + n_2 - 1}, \tag{22.105}$$

where

$$s_1^2 = \frac{1}{n_1 - 1}\sum_{i=1}^{n_1}(x_i - \overline{x})^2 \textit{ and } s_2^2 = \frac{1}{n_2 - 1}\sum_{i=1}^{n_2}(y_i - \overline{y})^2.$$

Then it is true that $$\frac{\overline{\xi} - \overline{\eta} - (\mu_1 - \mu_2)}{\sigma^*\sqrt{\frac{1}{n_1} + \frac{1}{n_2}}} \in t(n_1 - 1 + n_2 - 1). \tag{22.106}$$

An interval estimation for $\mu_1 - \mu_2$ of degree $1 - \alpha$ is

$$\left[\overline{\xi} - \overline{\eta} - t_{\alpha/2}(n_1 + n_2 - 2)\sigma^* r_{12}, \overline{\xi} - \overline{\eta} + t_{\alpha/2}(n_1 + n_2 - 2)\sigma^* r_{12}\right]. \tag{22.107}$$

A confidence interval for $\mu_1 - \mu_2$ *with degree* $1 - \alpha$ *is thus*

$$\left[\overline{x} - \overline{y} - t_{\alpha/2}(n_1+n_2-2)\sigma^*_{\text{obs}}r_{12}, \overline{x} - \overline{y} + t_{\alpha/2}(n_1+n_2-2)\sigma^*_{\text{obs}}r_{12}\right], \tag{22.108}$$

where $r_{12} = \sqrt{\frac{1}{n_1} + \frac{1}{n_2}}$.

22.13 Hypothesis Testing of μ in Normal Distribution σ Known

Definition 22.18.

(i) H_0 stands for the null hypothesis and in the same way H_1 stands for an alternative hypothesis. Let $X_1, X_2, \ldots, X_n$ be a sample of $N(\mu, \sigma)$ and $\overline{x}$, the corresponding observed mean value.

(ii) Strength function is defined as

$$P(\text{reject}\mathrm{H}_0|\mathrm{H}_1\text{true}). \tag{22.109}$$

The value of the strength function for a given value of H_1 is called its power.

(iii) One-sided hypothesis tests:

$$\begin{cases} \mathrm{H}_0 : \mu = \mu_0, \quad \mathrm{H}_1 : \mu > \mu_0 \\ \mathrm{H}_0 \text{ rejected at the significance level } \alpha \Longleftrightarrow \overline{x} > \mu_0 + \lambda_\alpha \dfrac{\sigma}{\sqrt{n}} \\ \hline \text{Strength function: } S(\mu) := \Phi\left(\sqrt{n} \cdot \frac{\mu - \mu_0}{\sigma} - \lambda_\alpha\right). \end{cases} \tag{22.110}$$

$$\begin{cases} \mathrm{H}_0 : \mu = \mu_0, \quad \mathrm{H}_1 : \mu < \mu_0 \\ \mathrm{H}_0 \text{ rejected at the significance level } \alpha \Longleftrightarrow \overline{x} < \mu_0 - \lambda_\alpha \dfrac{\sigma}{\sqrt{n}} \\ \hline \text{Strength function: } S(\mu) := \Phi\left(\sqrt{n} \cdot \frac{\mu_0 - \mu}{\sigma} - \lambda_\alpha\right). \end{cases} \tag{22.111}$$

(iv) Two-sided hypothesis test

$$\begin{cases} \mathrm{H}_0 : \mu = \mu_0, \quad \mathrm{H}_1 : \mu \neq \mu_0 \\ \mathrm{H}_0 \text{ rejected at the significance level } \alpha \\ \Longleftrightarrow \\ \overline{x} < \mu_0 - \lambda_{\alpha/2}\dfrac{\sigma}{\sqrt{n}} \text{ or } \overline{x} > \mu_0 + \lambda_{\alpha/2}\dfrac{\sigma}{\sqrt{n}} \\ \hline \text{Strength function:} \\ S(\mu) = \Phi\left(\sqrt{n} \cdot \frac{\mu-\mu_0}{\sigma} - \lambda_{\alpha/2}\right) + \Phi\left(\sqrt{n} \cdot \frac{\mu_0-\mu}{\sigma} - \lambda_{\alpha/2}\right). \end{cases} \tag{22.112}$$

22.13.1 σ *unknown*

For σ unknown, σ is changed to $s = \frac{1}{n-1}\sqrt{\sum_{k=1}^{n}(x_i - \overline{x})^2}$ in (22.110–22.112) and λ_α to $t_{n-1,\alpha}$. Corresponding strength functions are difficult to express.

Definition 22.19.

(i) One-sided hypothesis test

$$\begin{cases} \mathrm{H}_0 : \mu = \mu_0, \quad \mathrm{H}_1 : \mu < \mu_0 \\ \mathrm{H}_0 \text{ rejected at the significance level } \alpha \Longleftrightarrow \overline{x} < \mu_0 - t_{n-1,\alpha}\dfrac{s}{\sqrt{n}}. \end{cases} \tag{22.113}$$

$$\begin{cases} \mathrm{H}_0 : \mu = \mu_0, \quad \mathrm{H}_1 : \mu > \mu_0 \\ \mathrm{H}_0 \text{ rejected at the significance level } \alpha \Longleftrightarrow \overline{x} > \mu_0 + t_{n-1,\alpha}\dfrac{s}{\sqrt{n}}. \end{cases} \tag{22.114}$$

(ii) Two-sided hypothesis test

$$\begin{cases} \mathrm{H}_0 : \mu = \mu_0, \quad \mathrm{H}_1 : \mu \neq \mu_0 \\ \mathrm{H}_0 \text{ rejected at the significance level } \alpha \\ \Longleftrightarrow \\ \overline{x} < \mu_0 - t_{n-1,\alpha/2}\dfrac{s}{\sqrt{n}} \text{ or } \overline{x} > \mu_0 + t_{n-1,\alpha/2}\dfrac{s}{\sqrt{n}}. \end{cases} \tag{22.115}$$

22.14 F-Distribution and F-Test

(i) The PDF for the F-ratio distribution equals

$$f(n_1, n_2; x) = \begin{cases} \dfrac{n_1^{n_1/2} n_2^{n_2/2} x^{\frac{n_1}{2}-1} (n_1 x + n_2)^{-\frac{1}{2}(n_1+n_2)}}{B\left(\frac{n_1}{2}, \frac{n_2}{2}\right)}, & \text{if } x \geq 0, \\ 0, & \text{if} x < 0, \end{cases} \tag{22.116}$$

where $B\left(\dfrac{n_1}{2}, \dfrac{n_2}{2}\right)$ is the beta-function, given on page 175.

(ii) The CDF with n_1 and n_2 degrees of freedom, for the F-ratio distribution, fulfills

$$F(n_1, n_2; x) + F(n_2, n_1; 1/x) = 1. \tag{22.117}$$

Corresponding equality for quantiles:

$$\frac{1}{F_\alpha(n_1, n_2)} = F_{1-\alpha}(n_2, n_1).$$

(iii) Expectation and variance is

$$E = \frac{n_2}{n_2 - 2}, \quad n_2 > 2, \quad \text{and}$$

$$V = \frac{2n_2^2(n_1 + n_2 - 2)}{n_1(n_2 - 4)(n_2 - 2)^2}, \quad n_2 > 4, \text{ respectively.}$$

Hypothesis test for standard deviation
Given two samples of size n_1 and n_2 for a distribution, of roughly the shape of normal distribution, hypothesis test comparing their standard deviations, σ_1 and σ_2, is given in the following table.

(iv)

	H_1	Test statistic	Rejects H_0, if	Accept H_0, or reserve judgement, if
1.	$\sigma_1 < \sigma_2$	$F = \frac{s_2^2}{s_1^2}$	$F \geq F_\alpha$	$F < F_\alpha(n_1 - 1, n_2 - 1)$
2.	$\sigma_1 > \sigma_2$	$F = \frac{s_1^2}{s_2^2}$	$F \geq F_\alpha$	$F < F_\alpha(n_2 - 1, n_1 - 1)$
3.	$\sigma_1 \neq \sigma_2$	$F = \max\left(\frac{s_2^2}{s_1^2}, \frac{s_1^2}{s_2^2}\right)$	$F \geq F_{\alpha/2}$	$F < F_{\alpha/2}$

where

$$F_{\alpha/2} = \max(F_\alpha(n_1 - 1, n_2 - 1), F_\alpha(n_2 - 1, n_1 - 1)).$$

Here the quantile is $F_\alpha = F_\alpha(n_1 - 1, n_2 - 1)$, due to the numbers of freedom.

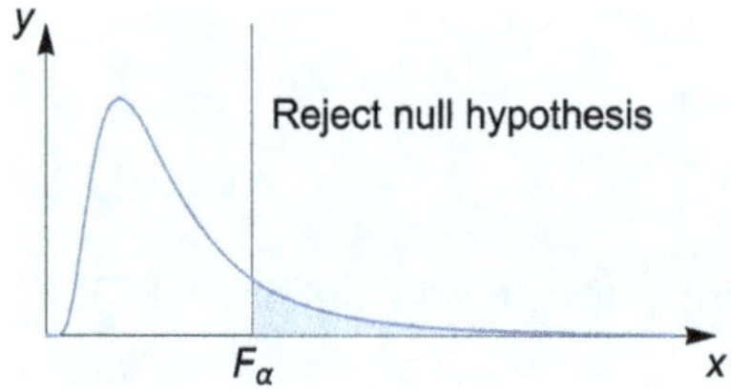

One-sided test

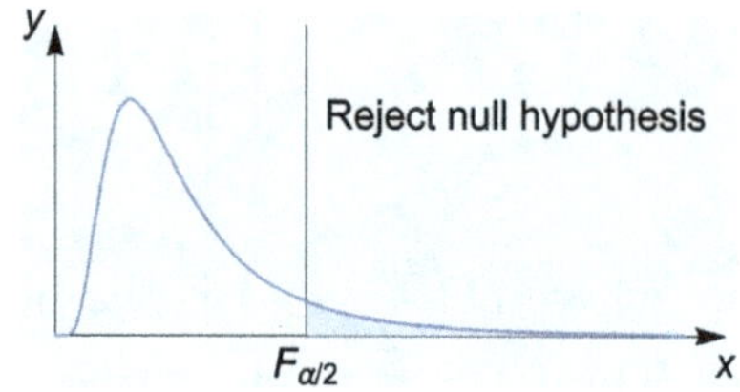

Two-sided test

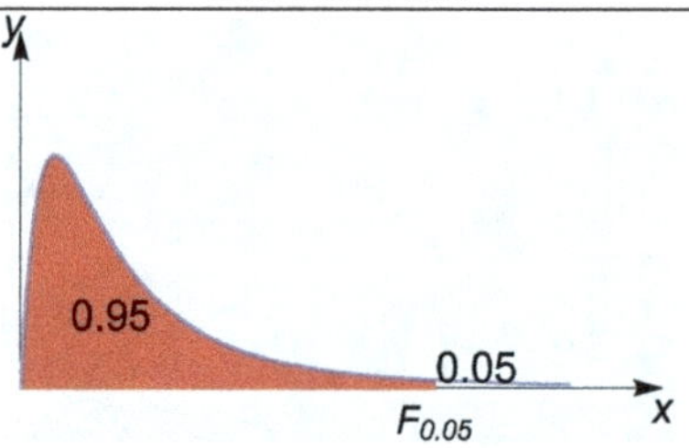

A PDF $f(x) = f(n_1, n_2, x)$ and its quantile $x_1 = F_{0.05}(n_1, n_2)$

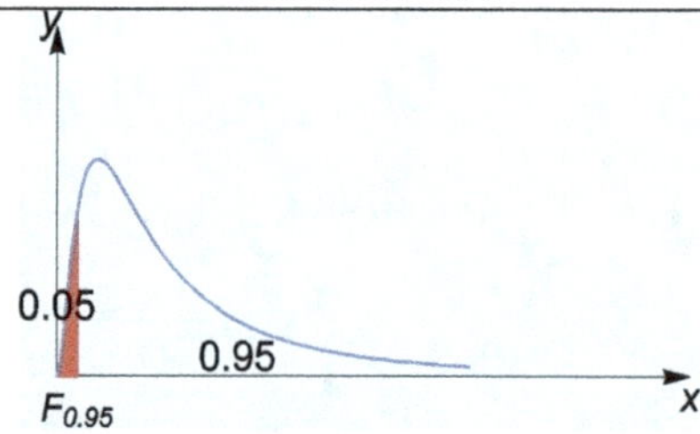

The PDF $f(x) = f(n_2, n_1, x)$ and its quantile $x_2 = F_{0.95}(n_2, n_1)$

In the two last figures, the quantiles fulfill $x_1 \cdot x_2 = 1$.

Convenient code in Mma (Wolfram Mathematica)

To get a desirable quantile $x = F_{0.05}(15, 20)$, here of significance 95%, one first define a CDF — Fratiodistribution:

```
F[n1_,n2_,x_]:=CDF[FRatioDistribution[n1,n2],x]

NSolve[F[15, 20, x] == 0.95, x] // Flatten
(*or*)
FindRoot[F[15, 20, x] == 0.95, {x, 1}]
```

Giving the outputs

```
{x -> 2.20327}            {x -> 2.20327}
```

This means that one gets the quantile $x = \mathcal{F}_{0.05}(15, 20) = 2.20327$.

22.15 Markov Chains

Definition 22.20.

(i) A class $(X_t;\, t \in T)$ of random variables is a *random process.* If $T = \{0, 1, 2, \ldots\}$, it is called *time-discrete* process. If $T = \mathbb{R}$, $T = [0, \infty)$ (or any other infinite interval in $\mathbb{R}$), it is called *time-continuous process.*

(ii) **Time-discrete process:** Let X_1, X_2, $X_3, \ldots$ be sequence of random variables, assuming values in a finite or infinitely countable sample, outcome, or state space Ω.
If

$$P(X_n = s|X_0, X_1, \ldots, X_{n-1}) = P(X_n = s|X_{n-1}) \qquad (22.118)$$

for all $s \in \Omega$ and all $n = 1, 2, \ldots$, the sequence X_1, X_2, $X_3, \ldots$ is a *discrete Markov chain.*

(iii) The chain is homogeneous if

$$P(X_{m+n} = j|X_n = i) = P(X_m = j|X_0 = i), \quad m, n, = 1, 2, \ldots \qquad (22.119)$$

Further notations are as follows:

$$p_{ij} = P(X_{m+1} = j|X_m = i), \qquad \text{transition probabilities.}$$

$$\boldsymbol{P} = (p_{ij})_{|\Omega|\times|\Omega|}, \qquad \text{transition matrix.}$$

$$p_{ij}(n) = P(X_{m+n} = j|X_m = i),$$

$$\boldsymbol{P}_n = (p_{ij}(n))_{|\Omega|\times|\Omega|}. \qquad (22.120)$$

(iv) A state i is recurrent if

$$P(X_n = i \,\text{for some } n \geq 1|X_0 = i) = 1. \qquad (22.121)$$

Otherwise, the state is transient.

Theorem 22.30.

(i) $\boldsymbol{P}$ *is a stochastic matrix due to the properties*

$$p_{ij} \geq 0 \text{ and } \sum_j p_{ij} = 1. \qquad (22.122)$$

(ii) ***Chapman–Kolmogorov equations***

$$p_{ij}(m+n) = \sum_k p_{ik}(m)\,p_{kj}(n), \ \textit{implying} \qquad (22.123)$$
$$\boldsymbol{P}_{m+n} = \boldsymbol{P}_n \cdot \boldsymbol{P}_m\,, \ \textit{and} \quad \boldsymbol{P}_n = \boldsymbol{P}^n.$$

Example 22.2. A discrete Markov chain/process described by its transition-matrix $\boldsymbol{P} = \begin{bmatrix} \frac{1}{2} & \frac{1}{4} & 0 & \frac{1}{4} \\ \frac{1}{2} & \frac{1}{3} & \frac{1}{12} & \frac{1}{12} \\ \frac{1}{2} & \frac{1}{3} & 0 & \frac{1}{6} \\ 0 & 0 & 0 & 1 \end{bmatrix}$ for four states/ outcomes and the corresponding graph, to the right. One observes that $p_{i1} + p_{i2} + p_{i3} + p_{i4} = 1$ for $i = 1, 2, 3, 4$, in accordance with (22.122).

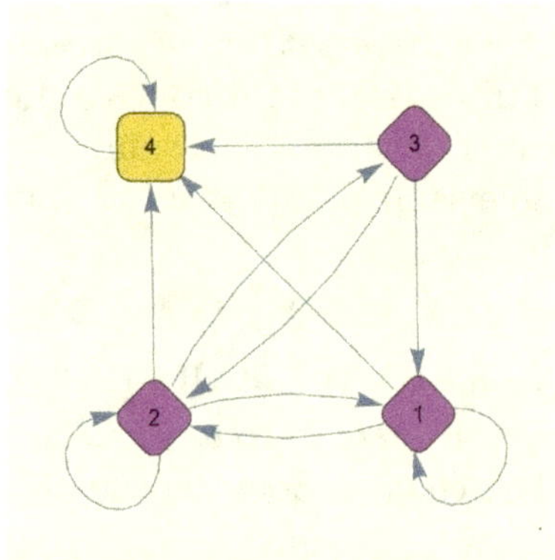

The element in position (3,4) in $\boldsymbol{P}$ is $p_{34} = 1/6 \neq 0$, and it corresponds to the vertical arrow from state 3 to state 4.

Definition 22.21. A *continuous* Markov chain $\{X(t) : t \geq 0\}$, where the random variable is denoted $X(t)$ rather than X_t, obeys the property

$$P(X(t_n) = j | X(t_1), X(t_2), \ldots, X(t_{n-1})) = P(X(t_n) = j | X(t_{n-1})), \qquad (22.124)$$

for each sequence $t_1 < t_2 < \cdots < t_n$ of times and for any $j \in \mathbb{Z}_+$.

Remarks. The condition (22.118) says that X_n only depends on the outcome of the immediately preceding random variable in the sequence X_1, X_2, $X_3, \ldots$

Equation (22.123) holds even for a continuous Markov chain. The elements in the transition matrix are

$$p_{ij}(s,t) = P(X(t) = j | X(s) = i), \quad s \leq t. \qquad (22.125)$$

For a homogeneous chain (per definition) yields $p_{ij}(s,t) = p_{ij}(0, t-s)$.

Here only homogeneous Markov chains are treated. The matrix containing elements $p_{ij}(t)$ is defined as $\boldsymbol{P}_t$, for which the following holds true:

$$\boldsymbol{P}_{s+t} = \boldsymbol{P}_s \boldsymbol{P}_t. \tag{22.126}$$

In particular, $\boldsymbol{P}_0 = \boldsymbol{I}$.

In applications the parameters s and t represent time.

The class $\{\boldsymbol{P}_t : t \geq 0\}$ is *standard* if $\lim\limits_{t\to 0_+} \boldsymbol{P}_t = \boldsymbol{I}$, e.g., right-continuity at $t = 0$.

The limits $\lim_{t\to 0_+} \frac{p_{ij}(t)}{t} =: g_{ij}$ of a standard chain exist and constitute the elements in the generator(-matrix) $\boldsymbol{G} := (g_{ij})_{|\Omega|\times|\Omega|}$.

Theorem 22.31.

(i) ***Kolmogorov equations***

$$\begin{aligned} &\tfrac{d}{dt}(\boldsymbol{P}_t) = \boldsymbol{P}_t\boldsymbol{G} \text{ (the forward equation)} \\ &\tfrac{d}{dt}(\boldsymbol{P}_t) = \boldsymbol{G}\boldsymbol{P}_t \text{ (the backward equation).} \end{aligned} \tag{22.127}$$

(ii) *With initial condition* $\boldsymbol{P}_0 = \boldsymbol{I}$, (22.127) *have the solution*

$$\boldsymbol{P}_t = e^{t\boldsymbol{G}} = \boldsymbol{I} + \frac{t\boldsymbol{G}}{1!} + \frac{(t\boldsymbol{G})^2}{2!} + \cdots \tag{22.128}$$

Part II

Appendices

Appendix A

Mechanics

A.1 Definitions, Formulas, etc.

Given time t, mass m, and $\boldsymbol{r}$ the (vectorial) location of a body.

Its velocity and acceleration are

$$\boldsymbol{v}(t) = \boldsymbol{r}'(t) \text{ and } \boldsymbol{a}(t) = \boldsymbol{v}'(t) = \boldsymbol{r}''(t), \text{ respectively.}$$

An angle (in radians) is denoted φ.

Let t denote time. The corresponding vectorial angular velocity $\boldsymbol{\omega}$ of φ is

$$\boldsymbol{\omega} = \frac{d\varphi}{dt}\boldsymbol{e}_{\varphi}, \tag{A.1}$$

where $\boldsymbol{e}_{\varphi}$ is the unit vector in the same direction as φ. att

$$\begin{aligned} \dot{\boldsymbol{r}} &= \boldsymbol{\omega} \times \boldsymbol{r} \\ \ddot{\boldsymbol{r}} &= \boldsymbol{\omega} \times (\boldsymbol{\omega} \times \boldsymbol{r}). \end{aligned} \tag{A.2}$$

$$\begin{aligned} &\text{Impulse} && \boldsymbol{P} = m\boldsymbol{v} \\ &\text{Force} && \boldsymbol{F} = \dot{\boldsymbol{P}} = \frac{d\boldsymbol{P}}{dt} = m\dot{\boldsymbol{v}} = m\boldsymbol{a} \\ &\text{Impulse momentum} && \boldsymbol{L} = \boldsymbol{r} \times \boldsymbol{P} = m\boldsymbol{r} \times \boldsymbol{v} \\ &\text{Torque} && \boldsymbol{M} = \frac{d\boldsymbol{L}}{dt} = \boldsymbol{r} \times \boldsymbol{F} \\ &\text{Kinetic energy} && W_k = \frac{1}{2} m\, v^2. \end{aligned} \tag{A.3}$$

A force $\boldsymbol{F}$ affects a body with mass m. The force is then drawn with starting point in the center of mass $\boldsymbol{r}_c$ of the body.

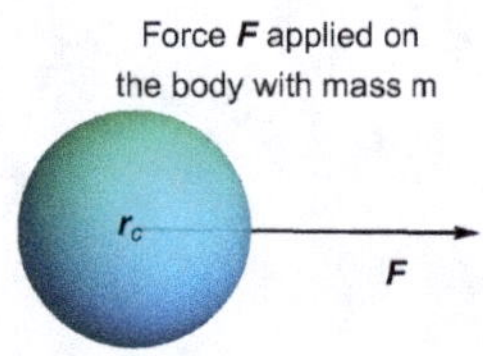

Definition A.1.

(i) Center of mass for a body is defined as

$$\boldsymbol{r}_c = \frac{1}{m}\int_D \boldsymbol{r}dm. \tag{A.4}$$

(ii) *Moment of inertia*, with respect to a line l (or axis), for a body D, is

$$I_l = \int_D r^2 dm, \tag{A.5}$$

where r is the perpendicular distance between the line and the location of dm.

Remarks. If the center of mass for body with mass m_j is $\boldsymbol{r}_j$, $j = 1, 2, \ldots, n$, their common center of mass is

$$\boldsymbol{r}_c := \frac{m_1\boldsymbol{r}_1 + m_2\boldsymbol{r}_2 + \cdots + m_n\boldsymbol{r}_n}{m_1 + m_2 + \ldots m_n}, \tag{A.6}$$

Rotation gives rotation energy (kinetic energy due to rotation)

$$E_k = \frac{I\omega^2}{2}. \tag{A.7}$$

(i) The vectorial sum of moments in an isolated system is constant. This implies that its time derivative $= \mathbf{0}$, more precisely

$$\dot{\boldsymbol{P}} = \boldsymbol{F} = \mathbf{0}, \quad \dot{\boldsymbol{L}} = \boldsymbol{M} = \mathbf{0}. \tag{A.8}$$

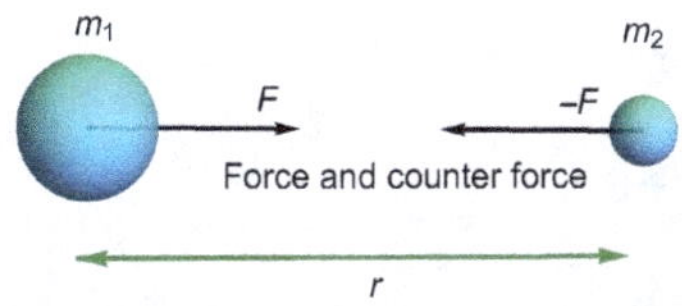

Figure A.1: Newton's third law.

(ii) The vectorial sum of the impulse moments is constant.
(iii) The sum of the total energy is constant.

A.1.1 *Newton's motion laws*

(1) **Newton's first law (Law of inertia).** A body not subject to external forces remains with constant vectorial velocity.
(2) **Newton's second law (Law of acceleration).** $\boldsymbol{F} = m \cdot \boldsymbol{a}$, where m is the mass of the body, $\boldsymbol{a}$, its acceleration, and $\boldsymbol{F}$, the net force applied on the body.
(3) **Newton's third law (The law of action and reaction, see Figure A.1).** The body with mass m_1 affects the body with mass m_2 with force $-\boldsymbol{F}$ while the body with mass m_2 affects the body with mass m_1 with force $\boldsymbol{F}$, that is with a force of same magnitude but opposite directed.
(4) **Newton's fourth law (The Superposition principle).** Assume that a body is affected by the (vectorial) forces

$$\boldsymbol{F}_1, \boldsymbol{F}_2, \ldots, \boldsymbol{F}_n.$$

This is the same as if the body is affected by the vectorial sum (the net force)

$$\boldsymbol{F} = \boldsymbol{F}_1 + \boldsymbol{F}_2 + \cdots + \boldsymbol{F}_n$$

of the forces.

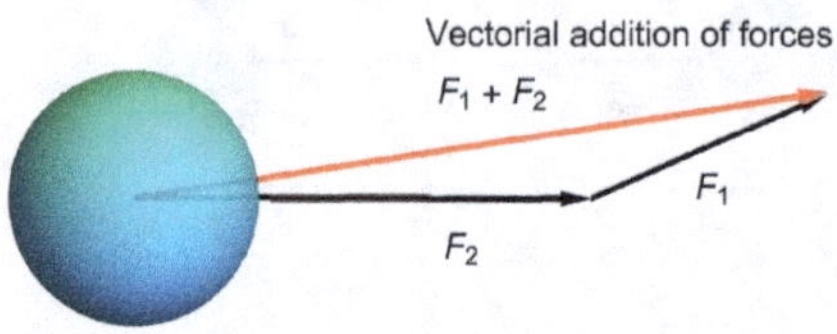

Newton's fourth law for $n = 2$ forces.

Newton's law of gravitation
For two bodies, as in the figure above, with mass m_1 and m_2, respectively, at distance r, affect each other with the force

$$\boldsymbol{F} = -\gamma \frac{m_1 m_2}{r^2} \boldsymbol{e}, \tag{A.9}$$

$\boldsymbol{e}$ is a unit vector parallel to $-\boldsymbol{F}$ and γ is the constant of gravity, see page 534.

Theorem A.1. *For a sphere with density, depending only on the distance to its center, the force $\boldsymbol{F}$ affected on a body with mass m_1 at distance r to its center of mass is given by* (A.9), *where m_2 is the mass inside the concentric sphere with radius r.*

A.1.2 *Linear momentum*

Collision between two bodies with mass m_1, to the left, and m_2, to the right (Figure A.2). Denote their velocities before collision by $\boldsymbol{v}_1$ and $\boldsymbol{v}_2$. Likewise, let $\boldsymbol{u}_1$ and $\boldsymbol{u}_2$ be the velocities after collision.

The law of conservation of linear momentum states that

$$m_1 \boldsymbol{u}_1 + m_2 \boldsymbol{u}_2 = m_1 \boldsymbol{v}_1 + m_2 \boldsymbol{v}_2. \tag{A.10}$$

If in addition the total kinetic energy is conserved, then

$$m_1 u_1^2 + m_2 u_2^2 = m_1 v_1^2 + m_2 v_2^2, \tag{A.11}$$

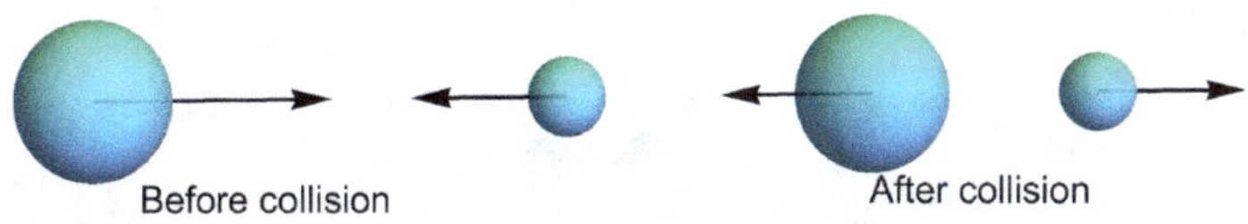

Figure A.2: Collision between two bodies.

and the collision is called *elastic*. Note that $v^2 = \boldsymbol{v} \cdot \boldsymbol{v} = |\boldsymbol{v}|^2$.

$$m_1 u_1 + m_2 u_2 = (m_1 + m_2) v, \tag{A.12}$$

$$E_{\text{before}} - E_{\text{after}} \equiv \Delta E = \frac{m_1 m_2 (u_1 - u_2)^2}{2(m_1 + m_2)} = \frac{\mu (u_1 - u_2)^2}{2}, \tag{A.13}$$

where

$$\mu = \frac{m_1 m_2}{m_1 + m_2} \text{ is } \textit{the reduced mass}. \text{ Equivalently: } \frac{1}{\mu} = \frac{1}{m_1} + \frac{1}{m_2}. \tag{A.14}$$

Imperically, there is a constant $e : 0 \leq e \leq 1$, a *shock coefficient*, such that

$$u_2 - u_1 = -e(v_2 - v_1). \tag{A.15}$$

The collision is elastic (inelastic) if $e = 1$ ($e < 1$).

A.1.3 *Impulse momentum and moment of inertia*

Definition A.2. Impulse momentum $\boldsymbol{L}$:

$$\boldsymbol{L} = \boldsymbol{r} \times \boldsymbol{p} \text{ for one particle.}$$

$$\boldsymbol{L} = \sum_{k=1}^{n} \boldsymbol{r}_k \times \boldsymbol{p}_k \text{ for } n \text{ particles.}$$

$$\begin{aligned} \boldsymbol{L} &= \int_D \boldsymbol{r} \times \boldsymbol{dp} = \int_D \boldsymbol{r} \times \boldsymbol{v} dm \\ &= \int_D \boldsymbol{r} \times (\boldsymbol{\omega} \times \boldsymbol{r}) dm = \int_D (\boldsymbol{\omega} r^2 - \boldsymbol{r}(\boldsymbol{\omega} \cdot \boldsymbol{r})) dm. \end{aligned} \tag{A.16}$$

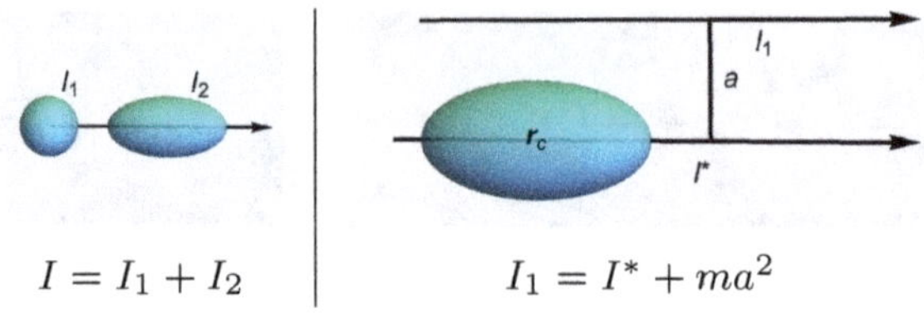

Figure A.3: Addition for moment of inertia and Steiner's theorem.

With $\boldsymbol{L} = (L_x, L_y, L_z)$, and

$$I_{xx} = \int (y^2 + z^2)dm, \qquad I_{yy} = \int (z^2 + x^2)dm, \qquad I_{zz} = \int (x^2 + y^2)dm,$$

$$I_{xy} = I_{yx} = -\int xydm, \quad I_{zx} = I_{xz} = -\int xzdm, \quad I_{zy} = I_{yz} = -\int zydm,$$

the following matrix equation holds true:

$$\begin{bmatrix} L_x \\ L_y \\ L_z \end{bmatrix} = \begin{bmatrix} I_{xx} & I_{xy} & I_{xz} \\ I_{yx} & I_{yy} & I_{yz} \\ I_{zx} & I_{zy} & I_{zz} \end{bmatrix} \begin{bmatrix} \omega_x \\ \omega_y \\ \omega_z \end{bmatrix}, \text{ or } \boldsymbol{L} = \boldsymbol{I}\,\boldsymbol{\omega}. \tag{A.17}$$

The rotational kinetic energy is then

$$W_k = \frac{1}{2}(\boldsymbol{\omega} \cdot \boldsymbol{L}). \tag{A.18}$$

$\boldsymbol{I}$ is the *inertia tensor*.

Steiner's theorem (Figure A.3)

$$I = I_1 + I_2 \text{ (Additivity)}, \tag{A.19}$$

$$I = I^* + ma^2, \tag{A.19'}$$

where I^* refers to moment of inertia of an axis through the center of mass.

A.1.4 Table of center of mass, and moment of inertia of some homogenous bodies

The moment of inertia, I_x, refers to rotation around the $x-$axis and (hence) around an axis through the center of mass m, $\boldsymbol{r}_c = (x_c, y_c)$ or $\boldsymbol{r}_c = (x_c, y_c, z_c)$. In the same way, I_z refers to rotation around the $z-$axis (The torus).

Name, Length/ Area/Volume	Center of mass	Moment of inertia	Form
Thin rod with length L	$\boldsymbol{r}_c = x = 0$	$I = \frac{mL^2}{12}$	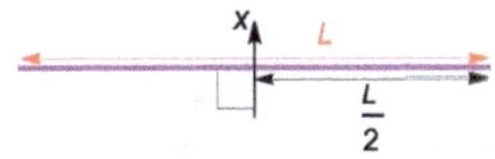
Circle with radius r	$\boldsymbol{r}_c = (0, 0)$	$I = mr^2$	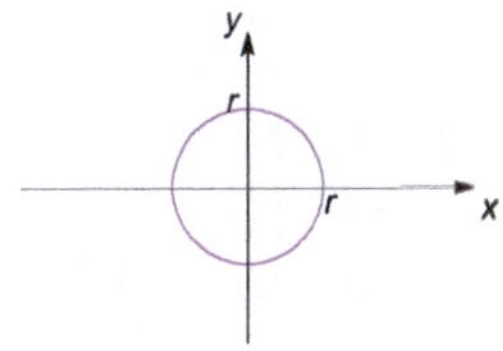
Rectangle with sides a, b. $A = a \cdot b$.	$\boldsymbol{r}_c = (0, 0)$	$I = \frac{m\,b^2}{12}$	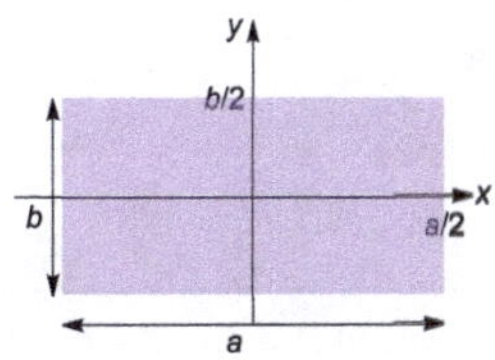
Circular disc with radius r $A = \pi r^2$	$\boldsymbol{r}_c = (0, 0, 0)$	$I_x = I_y = \frac{mr^2}{4}$	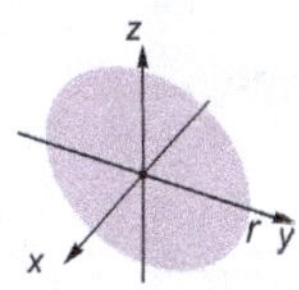
Circle sector $A = \frac{\theta r^2}{2}$	$\boldsymbol{r}_c = \left(\frac{4r\sin(\theta/2)}{3\theta}, 0\right)$	$I = \frac{mr^2}{4}\left(1 - \frac{\sin\theta}{\theta}\right)$	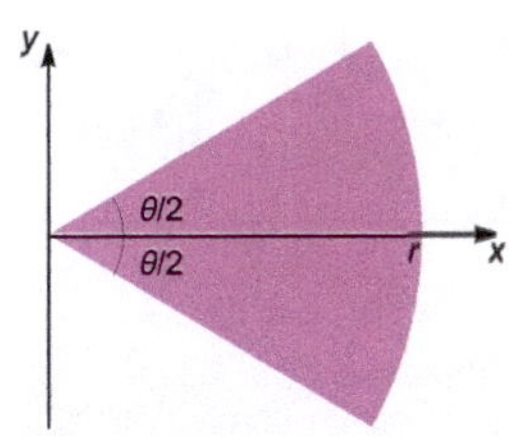

Name, Length/ Area/Volume	Center of mass	Moment of inertia	Form
Triangle $A = \frac{a \cdot y_1}{2}$	$\boldsymbol{r}_c = (x_c, y_c) = \left(\frac{1}{3}(a + x_1), \frac{y_1}{3}\right)$	$I_x = \frac{m y_1^2}{6}$ $I_t = I^* = \frac{m y_1^2}{18}$	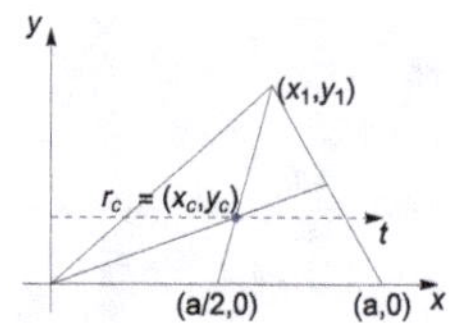
Circular band with radius r. $A = \pi r^2$	$\boldsymbol{r}_c = 0$	$I_x = m r^2$	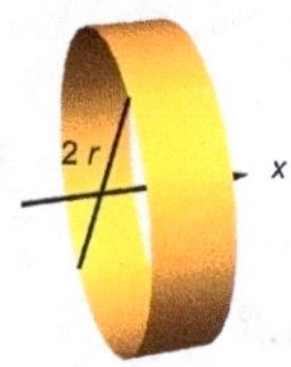
Circular cylinder $V = \pi r^2 h$	$\boldsymbol{r}_c = (h/2, 0, 0)$	$I = \frac{m r^2}{2}$	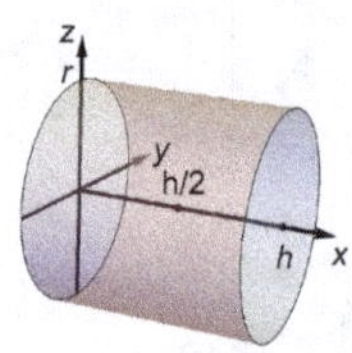
Rectangular cuboid ($\mathbb{R}^3$) With side lengths a, b, c and $V = a\,b\,c$	$\boldsymbol{r}_c = (0, 0, 0)$	$I = \frac{mbc}{2}$	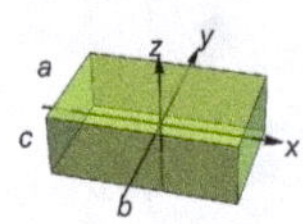
Circular cone with base radius r and height h $V = \frac{\pi r^2 h}{3}$	$\boldsymbol{r}_c = (x, y, z) = (3h/4, 0, 0)$	$I = \frac{3 m r^2}{10}$	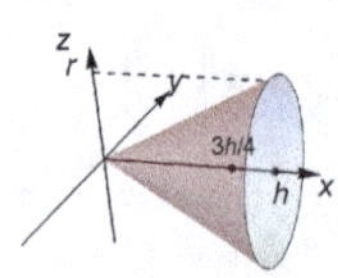

Name, Length/ Area/Volume	Center of mass	Moment of inertia	Form
Truncated circular cone $V = \frac{\pi h}{3}(r^2 + rR + R^2)$.	$x_c = \frac{h\,(r^2+2rR+3R^2)}{4\,(r^2+rR+R^2)}$, $y_c = z_c = 0$.	$\frac{3m(R^2 - r^2)}{10}$	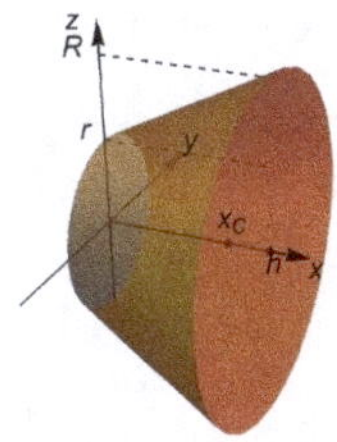
Sphere $V = \frac{4\pi r^3}{3}$	In its geometrical center	$I = \frac{2mr^2}{5}$	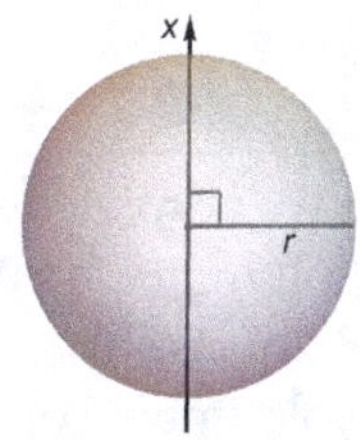
Ellipsoid $V = \frac{4\pi a\,b\,c}{3}$	$\boldsymbol{r}_c = (0, 0, 0)$	$I_x = \frac{m(b^2 + c^2)}{5}$	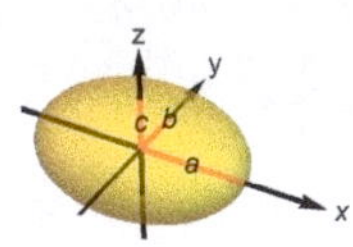
Torus $V = 2\pi^2 r^2 R$	$\boldsymbol{r}_c = (0, 0, 0)$	$I_z = \frac{m}{4}(3r^2 + 4R^2)$	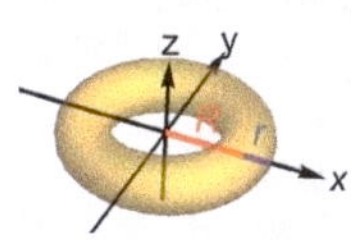

Remarks. With ρ as the constant density in all moments I, the mass $m = \rho V$, with V substituted with L for the rod and A for the rectangle, respectively.

For the triangle, by Steiner's theorem,

$$I_x = I^* + m(y_1/3)^2 = \frac{my_1^2}{18} + \frac{my_1^2}{9} = \frac{my_1^2}{6}.$$

The rectangular cuboid has corners at the points $(x, y, z) = (\pm a/2, \pm b/2, \pm c/2)$.

Setting $r = 0$ for the truncated cone, one gets the corresponding values for a cone.

Here, the torus is circular, and is also given on page 51.

A.1.5 *Physical constants*

Avogadro's constant	$6.02214 \cdot 10^{23}\,\text{mol}$	N_A	
Boltzmann's constant	$1.38065 \cdot 10^{-23}\,\text{J/K}$	k_B	
Bohr radius	$5.29177 \cdot 10^{-11}\,\text{m}$	a_0	
Coloumb's constant	$1.60218\,10^{-19}\,\text{C}$	e	
Faraday's constant	$96485.3\,\text{C/mol}$	F	
Constant of gravity	$6.6720 \cdot 10^{-11}\,\text{Nm}^2/\text{kg}^2$	γ	
Speed of light in vacuum	$2.99792 \cdot 10^{8}\,\text{m/s}$	c_0	(A.20)
The molar volume	$0.022414\,\text{m}^3/\text{mol}$	V_0	
Planck's constant	$6.62607 \cdot 10^{-34}\,\text{Js}$	h	
Planck mass	$2.1767 \cdot 10^{-8}\,\text{kg}$		
Rydberg's constant	$1.09737 \cdot 10^{7}\,\text{m}$	R_∞	
Solar constant	$1373.0\,\text{W/m}^2$		
Stefan–Boltzmann's constant	$5.6704 \cdot 10^{-8}\,\text{W}/(\text{m}^2 \cdot \text{T}^4)$	σ	

Appendix B

Varia

B.1 Greek Alphabet

B.1.1 *Uppercase*

A	Alpha	B	Beta	Γ	Gamma	Δ	Delta
E	Epsilon	Z	Zeta	H	Eta	Θ	Theta
I	Iota	K	Kappa	Λ	Lambda	M	My
N	Ny	Ξ	Xi	O	Omicron	Π	Pi
R	Ro	Σ	Sigma	T	Tau	Υ	Ypsilon
Φ	Fi	X	Chi	Ψ	Psi	Ω	Omega

B.1.2 *Lowercase*

α	alpha	β	beta	γ	gamma	δ	delta
ε	epsilon	ζ	zeta	η	eta	θ	theta
ι	iota	κ	kappa	λ	lambda	μ	my
ν	ny	ξ	xi	o	omicron	π	pi
ρ	ro	σ	sigma	τ	tau	υ	ypsilon
φ	fi	χ	chi	ψ	psi	ω	omega

B.1.3 *The numbers π and e*

The numbers π and e are transcendent and normal. By the latter is meant that on average the digits $0, 1, 2, 3, 4, 5, 6, 7, 8, 9$ are equally common in their decimal expansions. In addition, all finite decimal sequences are found in the numbers.

$\pi =$	$3.14159265358979323846264338327950288419716939937510582097494459230781646406286208998628034825342117068\ldots$
$e =$	$2.718281828459045235360287471352662497757247093699959574966967627724076630353547594571382178525166427\ldots$

B.1.4 *Euler constant*

Define

$$H_n = \sum_{k=1}^{n} \frac{1}{k}.$$

Euler constant is defined as the limit value

$$\gamma = \lim_{n \to \infty} (H_n - \ln n). \tag{B.1}$$

Numerically,

$$\gamma = 0.57721566\ldots$$

Appendix C

Programming Mathematica (Mma)

The aim with this text is to make the reader familiar with syntax in the program *Mathematica*. It is a program dealing with almost all known mathematics to date. Beside the syntax, Mma has a rich array of palettes, treating notions and equations in an easy way.

C.1 Elementary Syntax

The Mathematica syntax in this text are written in verbatim or in upright bold.

A particular command such as **Simplify** acts on a certain expression, say $\mathbf{3x^3 - x^3}$, by means of **square brackets**, "[" and "]". More precisely

$$\mathbf{Simplify}[\mathbf{3x^3 - x^3}]\,.$$

The command is then executed/activated by pressing the buttons

$$\boxed{\text{uppercase}}\quad \boxed{\text{enter}}, \tag{C.1}$$

in that order making Mma executes the command **Simplify** on the expression, giving the output $\mathbf{2x^3}$.

All the following commands are executed by (C.1).

The first letter of each meaningful part of a command is capitalized.

C.1.1 *Parentheses*

Square brackets are used to affect an expression with a command, as above.

Curly brackets, "{" and "}", are used to create a list, containing a finite number of elements as an ordered set, e.g.,

$$\{1, 2, 3, a, b, c\}.$$

Ordinary round brackets "(" and ")" are used for the laws of mathematics (elementary algebra), e.g., $a(b+c)$, and can be expanded by means of

$$\mathbf{Expand}[\mathbf{a}(\mathbf{b}+\mathbf{c})] \quad \text{activated by (C.1), with output} \quad \mathbf{a\,b}+\mathbf{a\,c}.$$

C.1.2 *Operations*

Operations between real (complex) numbers:
Multiplication between real or complex numbers a and b is made by space $a\ \ b$ or $a * b$.
Division is done by a/b or by (C.3), page 540.

Inner product between vectors, e.g., for $\mathbf{u} := \{\mathbf{a}, \mathbf{b}, \mathbf{c}\}$ and $\mathbf{v} := \{\mathbf{x}, \mathbf{y}, \mathbf{z}\}$, and is made by a dot:

$$\mathbf{a}\,.\,\mathbf{b} \quad \text{giving} \quad \mathbf{a\,x}+\mathbf{b\,y}+\mathbf{c\,z}.$$

The dot "." is, in more general form, an operation between list or lists, such that matrix/tensor multiplication.

The cross-product (existing only in $\mathbb{R}^3$) is made by the command

$$\mathbf{Cross}[\mathbf{u}, \mathbf{v}],$$

giving

$$\mathbf{Output:} \quad \{\mathbf{b\,z}-\mathbf{c\,y}, \mathbf{c\,x}-\mathbf{a\,z}, \mathbf{a\,y}-\mathbf{b\,x}\}.$$

C.1.3 *Equalities and defining concepts*

Equality

(i) Writing $\mathbf{a} = 25$ means that from now on $\mathbf{a}$ means just 25. Similarly, $\mathbf{a} := \mathbf{25}$ means almost the same thing (" : " stands for delayed equality). In the case above, and $\mathbf{a} = \mathbf{25}$; i.e., by a semicolon, the printing/output of "25" is suppressed.

(ii) Equality for a equation is written by *two* equalities in a row, for example, for equation

$$x^2 - 3x + 2 = 0 \text{ one writes } \mathbf{Solve[x^2 - 3x + 2 == 0, x]},$$

giving $x = 1$, $x = 2$. By factoring the polynomial $x^2 - 3x + 2$, one gets

$$\mathbf{Factor[x^2 - 3x + 2]} \text{ giving } \mathbf{(x - 1)(x - 2)}.$$

(iii) To define a new concept, for example, $\ln x$ from Mma:s natural logarithm, **Log** one may write

$$\mathbf{ln := Log}.$$

(iv) To define a concept, e.g., a function of a single variable, say x, one writes $\mathbf{f[x\_] := x^3}$ or just $\mathbf{f[x\_] = x^3}$. Underscore "_" of the independent variable used only initially in the very definition.

(v) Defining a function φ of *two* variables, one may write

$$\varphi[\mathbf{x\_}, \mathbf{y\_}] := \frac{\mathbf{x^2 + y^2}}{\mathbf{4}}.$$

(vi) With

$$\mathbf{a === b},$$

testing identity between a and b and gets answer **True** or **False**.

Making an own short command of an Mma-included one, e.g.,

tog := Together or even **tog = Together**;

The semi-colon suppresses text printing, as mentioned above. The effect of the command, see page 540.

It is worth noting, each part of a significant command begins with an uppercase letter, making it possible for the user to introduce own commands beginning with lower case letters.

C.1.4 *Elementary algebra*

Example C.1.

(i) Constructing a power a^b, one writes a∧b or using the symbol

$$\blacksquare^{\square} \tag{C.2}$$

found under the palette **Writing Assistant**.

Likewise, for a quotient, one uses

$$\frac{\blacksquare}{\square}. \tag{C.3}$$

(ii) Taking the square root of 18 one uses

$$\sqrt{\blacksquare}$$

giving

$$\mathbf{\sqrt{18}} \text{ and after activating the command, } \mathbf{3\sqrt{2}}.$$

To get a numeric value, one writes

$$\mathbf{N[\sqrt{18}]} \quad \text{or} \quad \mathbf{N[\sqrt{18}, 10]}$$

the last demanding an answer with ten digits.

(iii) To define the expression $\frac{x^2-4}{x+2}$, one uses (C.3) and (C.2) to get

$$\frac{\mathbf{x^2 - 4}}{\mathbf{x + 2}}.$$

The following commands are to be find under the palette **Other**.

(iv) **Simplify**$[\frac{\mathbf{x^2-4}}{\mathbf{x+2}}]$ gives the output $\mathbf{x - 2}$.

(v) To put two terms of expressions together, one uses the command **Together**. As an example:

$$\textbf{Together}\left[\frac{\mathbf{1}}{\mathbf{x-3}} + \frac{\mathbf{3x}}{\mathbf{x+3}}\right] \text{ giving}$$

$$\frac{\mathbf{3x^2 - 8x + 3}}{\mathbf{(x-3)(x+3)}}. \tag{C.4}$$

(vi) To expand the denominator $(x-3)(x+3)$, one writes

$$\mathbf{Expand}[(\mathbf{x}-\mathbf{3})(\mathbf{x}+\mathbf{3})], \quad \text{giving } \mathbf{x^2-9}.$$

(vii) To expand the expression (C.4), one writes

$$\mathbf{Apart}\left[\frac{\mathbf{3x^2-8x+3}}{(\mathbf{x}-\mathbf{3})(\mathbf{x}+\mathbf{3})}\right],$$

giving back $\mathbf{3}+\dfrac{\mathbf{1}}{\mathbf{x-3}}-\dfrac{\mathbf{9}}{\mathbf{x+3}}$.

(viii) Factorizing the expression x^2-9 is done by

$$\mathbf{Factor}[\mathbf{x^2-9}] \text{ giving } (\mathbf{x}-\mathbf{3})(\mathbf{x}+\mathbf{3}).$$

To factorize x^2-3, one writes

$$\mathbf{Factor}[\mathbf{x^2-3}\,,\ \mathbf{Extension}->\sqrt{\mathbf{3}}] \text{ getting } (\mathbf{x}-\sqrt{\mathbf{3}})(\mathbf{x}+\sqrt{\mathbf{3}}).$$

Remarks. Every introduced object gets an "input" number, for example,
In[25] $\mathbf{Factor}[\mathbf{x^2-9}]$.
The treated object gets a corresponding number
Out[25] $(\mathbf{x}-\mathbf{3})(\mathbf{x}+\mathbf{3})$.

From now on, **Out**[**25**] refers to the object $(x-3)(x+3)$ and so writing **Out**[**25**], one calls in the object.

To figure out how to use a certain command, e.g., "Factor", one writes

??**Factor**

getting

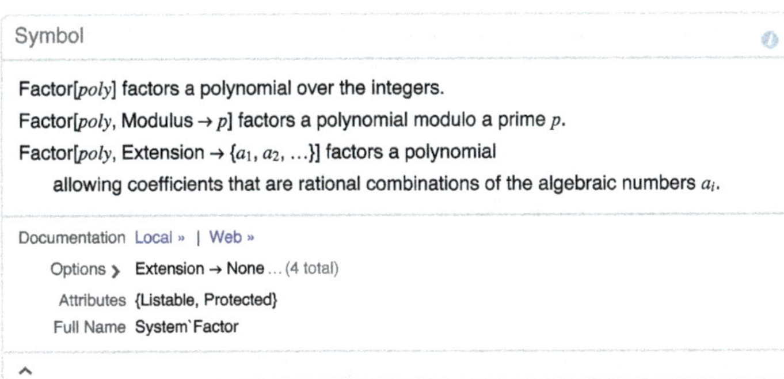

Symbol

Factor[*poly*] factors a polynomial over the integers.
Factor[*poly*, Modulus → *p*] factors a polynomial modulo a prime *p*.
Factor[*poly*, Extension → {a_1, a_2, ...}] factors a polynomial allowing coefficients that are rational combinations of the algebraic numbers a_i.

Documentation Local » | Web »
Options › Extension → None ... (4 total)
Attributes {Listable, Protected}
Full Name System`Factor

where "Local" and "Web" are clickable for further information.

C.2 Linear Algebra

Example C.2.

(i) Solving a system of equations

$$\begin{cases} x^2 - x = 0, \\ \\ y^2 - 2x = 1, \end{cases}$$

$$\mathbf{Solve}[\{\mathbf{x^2} == \mathbf{x}, \mathbf{y^2} - \mathbf{2x} == \mathbf{1}\}, \{\mathbf{x}, \mathbf{y}\}],$$

giving

```
{{x -> 0, y -> -1}, {x -> 0, y -> 1}, {x -> 1,
  y -> -Sqrt[3]}, {x -> 1, y -> Sqrt[3]}}
```

and meaning

$$(x, y) = (0, \pm 1), \quad (x, y) = (1, \pm\sqrt{3}).$$

The system is not linear, so let us look at linear systems.

(ii) Solving a linear equation system (see, for instance, page 86).

$$\begin{cases} 2x - y = 3, \\ 3x + 2y = 1. \end{cases} \tag{C.5}$$

This ES can be solved by all means given in Chapter 5.

(a) Direct:

```
Solve[{2x-y==3,3x+2y==1},{x,y}]
```

(b) Using matrix algebra:

$$\boldsymbol{A} := \begin{bmatrix} 2 & -1 \\ 3 & 2 \end{bmatrix}, \ \boldsymbol{B} = \begin{bmatrix} 3 \\ 1 \end{bmatrix} \text{ and } \boldsymbol{X} := \begin{bmatrix} x \\ y \end{bmatrix}.$$

The definitions of the matrices as well as the solution are as follows

```
A:={{2,-1},{3,2}}
B:={3,1}
X:={x,y}

Inverse[A].B
```

This is possible iff det $\boldsymbol{A} \neq 0$, which is checked by

$$\mathbf{Det}[\mathbf{A}], \text{ giving } \det \boldsymbol{A} = 7.$$

(c) Finally, making use of augmented matrix: Here one uses the command **Transpose**.

```
RowReduce[Transpose[Transpose[A],{B}]
```

giving $\{\{1, 0, 1\}, \{0, 1, -1\}\}$.
This means, using matrix notation,

$$\left[\begin{array}{cc|c} 1 & 0 & 1 \\ 0 & 1 & -1 \end{array}\right].$$

The rows are interpreted as follows:

$$\begin{cases} \text{Firs row:} & 1 \cdot x + 0 \cdot y = x = 1, \\ \\ \text{Second row:} & 0 \cdot x + 1 \cdot y = y = -1. \end{cases}$$

C.3 Calculus

Example C.3. To *define* a function, as on page 539, for example, $f(x) := x^2 - 3x + 2$, one writes

```
f[x_]:=x^2-3x+2
```

(i) To differentiate this function, one can choose between the following three methods:

$$\mathbf{f}'[\mathbf{x}], \quad \mathbf{D}[\mathbf{f}[\mathbf{x}], \mathbf{x}], \qquad \text{or} \qquad \mathbf{D}[\mathbf{x^2} - \mathbf{3x} + \mathbf{2}, \mathbf{x}]$$

all giving the desired result $2x - 3$.

(ii) Plotting the graph in the interval $\{x : -2 \leq x \leq 3\}$ goes as follows:

```
Plot[f[x], {x, -2, 3}, PlotStyle -> Thick,
 AxesStyle -> Thickness[0.003], Background
-> LightBlue]
```

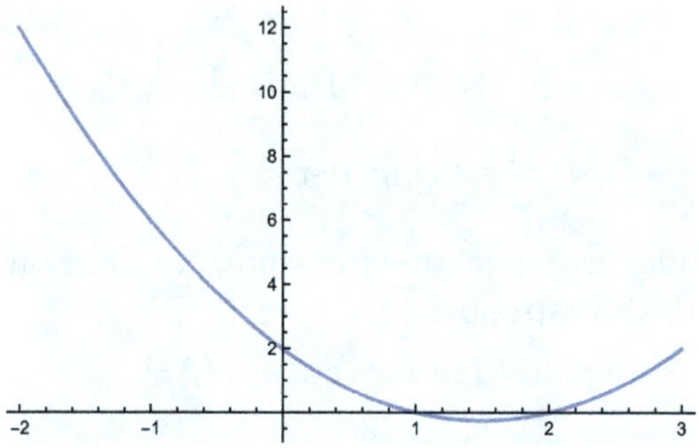

(iii) Integrating the function by pure commands over the same interval yields

```
Integrate[f[x], {x, -2, 3}]
```

giving the output $\frac{85}{6}$.
Numeric integration is obtained by

```
NIntegrate[f[x], {x, -2, 3}]
```

giving the output 14.1667. By adding, e.g., the number 5:

```
NIntegrate[f[x], {x, -2, 3},5]
```

one controls the digits in the decimal expansion.

(iv) On the palettes there are integral symbols, e.g.,

$$\int \blacksquare d\square$$

C.3.1 *Calculus in several variables*

As an example, we take the function $f(x, y) = \frac{1}{4}(x^2 + y^2)$.

(i) Defining and differentiating with respect to, e.g., the second variable:

```
f[x_,y_]:=(x^2+y^2)/4
D[f[x,y],y]
```

giving the output

$$\frac{\mathbf{y}}{\mathbf{2}}$$

(ii) Plotting (the graph of) the function over the rectangle $[-1,1]\times[-2,2]$:

```
Show[Plot3D[f[x, y], {x, -1, 1}, {y, -2, 2},
  PlotStyle -> {Opacity[0.8]}],
 Axes -> False, Boxed -> False, PlotRange -> All,
 Background -> LightBlue]
```

gives the following graph:

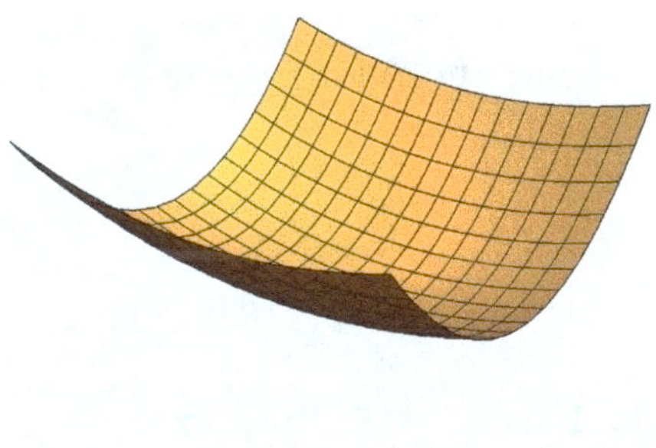

(iii) Plotting over the domain $D := \{(x,y) : f(x,y) \leq 4\}$, that is over the disk $\{(x,y) : x^2 + y^2 \leq 16\}$, one changes to polar coordinates, in this case

$$\begin{cases} x &= r\cos t, \\ y &= r\sin t. \end{cases}$$

The change requires the command **ParametricPlot3D**.

```
th := 0.007
Show[Graphics3D[{Thickness[th], Arrow[{{-4, 0, 0},
{4, 0, 0}}]}],
 (*Graphics3D[Arrow[{{0,0,0},{0,0,3}}]]*),
 Graphics3D[Text["x", {4.3, 0, 0}]],
   ParametricPlot3D[{r Cos[t], r Sin[t], r^2/4},
{r, 0, 4}, {t,
```

```
    0, 2 Pi}], Axes -> False, Boxed -> False,
PlotRange -> All,
 Background -> LightBlue]
```

yielding the graph

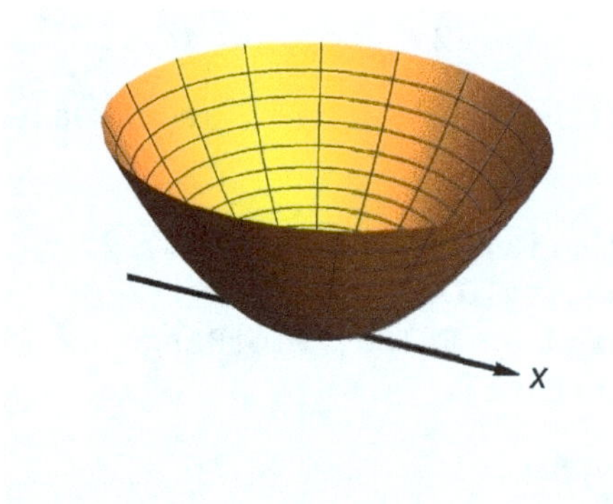

(iv) For the plot of a Möbius band, the following syntax is sufficient:

```
r := 4
ParametricPlot3D[
 r {Cos[t], Sin[t], 0} + s {Cos[t] Sin[t], Sin[t]^2,
 Cos[t]}, {t, 0,
  2 Pi}, {s, -1, 1}, Boxed -> False, Axes -> False,
   PlotRange -> All,
 Mesh -> False, Background -> LightGreen]
```

This amount of text creates the graph to the right.

(v) **Suppressing a command is done by (∗ and ∗):** We observe the suppressed command

(*Graphics3D[Arrow[{{0,0,0},{0,0,3}}]]*).

(vi) Integration over the same rectangle

```
Integrate[Integrate[f[x,y],{x,-1,1}],{y,-2,2}]
```

or, using the integral symbol in the palette **Writing assistant** :

$$\int_{-2}^{2}\left(\int_{-1}^{1} \mathbf{f[x,y]dx}\right)\mathbf{dy},$$

giving the output $\frac{10}{3}$.

(vii) Integrating over the disk $D := \{x, y : f(x,y) \leq 16\}$, one begins with classical mathematics, changing to variable polar coordinates.

$$D_1 := \{(r,\theta) : 0 \leq r \leq 4,\ 0 \leq \theta \leq 2\pi\}.$$

The variable substitution is

$$\begin{cases} x &= r\cos\theta, \\ y &= r\sin\theta, \end{cases}$$

with $0 \leq r \leq 4, \quad -\pi < \theta \leq \pi$, and functional determinant r.

$$\iint_D f(x,y)dxdy = \iint_{D_1} \frac{r^2}{4} r\,dr\,d\theta = \int_0^4 \frac{r^3}{4}dr \cdot \int_{-\pi}^{\pi} d\theta = 32\pi.$$

The corresponding solution with Mma is as follows, here only by means of commands.

```
Integrate[r^3/4,{r,0,2}] Integrate[1,{t,-Pi,Pi}]
```

C.4 Ordinary Differential Equations

The command is **DSolve**.

Example C.4.

(i) To solve the ordinary differential equation

$$2y' = 3xy, \quad y(0) = 1,$$

one writes

```
DSolve[{y'[x]==3 x y[x],y[0]==1},y[x],x]
```

The solution is $y(x) = e^{\frac{3x^2}{2}}$ or typed in Mma:

```
Out[116]={{y[x] -> E^((3 x^2)/2)}}
```

where $E = 2.71828\ldots$ is the Napier number.

(ii) The following DE with solution, written by Mma-commands, describes the mirror of a reflector (telescope).

```
DSolve[{(y'[x]^2 - 1)/(2 y'[x]) == (y[x] - F)/x,
 y[0] == 0},
y[x], x]

{{y[x] -> x^2/(4 F)}}
```

interpreting the solution as

$$y = \frac{x^2}{4F}.$$

C.5 Mathematical Statistics

Most of notions in this subject are to be found in Mathematica. Here follows a survey of some common expressions.

$$\begin{array}{rl} \textbf{PDF}[\textbf{NormalDistribution}[\mu,\sigma],\textbf{x}] & \text{PDF for the normal distr.} \\ \textbf{CDF}[\textbf{ExponenitalDistribution}[\lambda],\textbf{x}] & \text{CDF for the exponential distr.} \\ \textbf{Mean}[\textbf{PoissonDistribution}[\lambda]] & \text{Mean for Poisson distr.} \\ \textbf{Mean}[\{\mathbf{1,3,5,7,9}\}] & \text{Gives the mean value 5.} \end{array} \tag{C.6}$$

To determine the quantile q_α for the sech-distribution, see figure to the right, one can directly compute it by writing

$$\frac{2\sigma\ln\left(\tan\left(\frac{1}{2}\pi(1-\alpha)\right)\right)}{\pi}+\mu$$

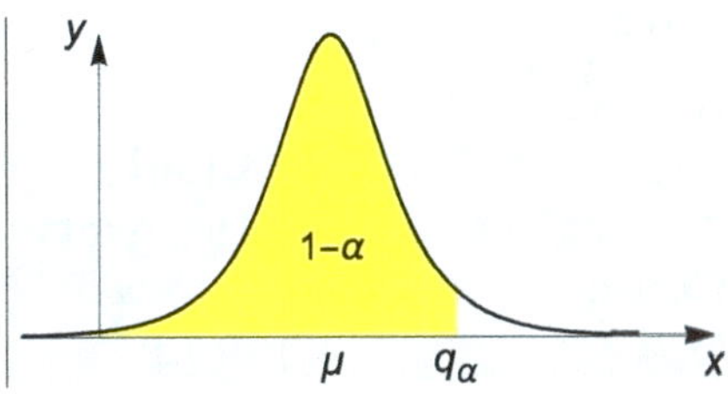

or in Mathematica code:

$$\mu+\frac{\mathbf{2}\sigma\,\mathbf{Log}[\mathbf{Tan}[\frac{\pi}{\mathbf{2}}(\mathbf{1}-\alpha)]]}{\pi}.$$

For example, with $\mu = 3.0$, $\sigma = 1.0$, and $\alpha = 0.05$, the quantile becomes $q_{0.05} = 4.61835$.

C.6 Difference or Recurrence Equations (RE)

An RE is solved by the command **RSolve**.

Example C.5. To solve the RE

$$a_{n+1}+a_n-6a_{n-1}=0,\quad \begin{cases} a_0 &= 1,\\ a_1 &= 1,\end{cases}$$

one writes

```
RSolve[{a[n + 1] + a[n] - 6 a[n - 1] == 0, a[0] == -5,
  a[1] == 5}, a[n], n]
```

obtaining

```
{{a[n] -> (-3)^(1 + n) - 2^(1 + n)}}
```

which means

$$a_n = (-3)^{n+1} - 2^{n+1}.$$

C.6.1 *List of common commands*

Apart
Arg
AspectRatio
Count
D
DeleteDuplicates
Det
DSolve
E
Evaluate
Expand
ExpandAll
Factor
FactorInteger
FullSimplify
Graphics
RSolve
Show
Simplify
Graphics3D
IdentityMatrix
Integrate
Inverse
Join
LeastSquares
Length
Limit
LinearSolve
MatrixForm
Merge
NSolve
ParametrictPlot
ParametrictPlot3D
Pi
Plot
Solve
Sum
Together
Plot3D
Polygon
Polyhedron
QPrime
Random is a prefix for a lot of commands, for instance: *RandomChoice, RandomColor, RandomComplex, RandomGraph, RandomInteger, Random, RandomReal, RandomWalkProcess.*
RealPart
Rectangle
RowReduce
Transpose

Random polygon

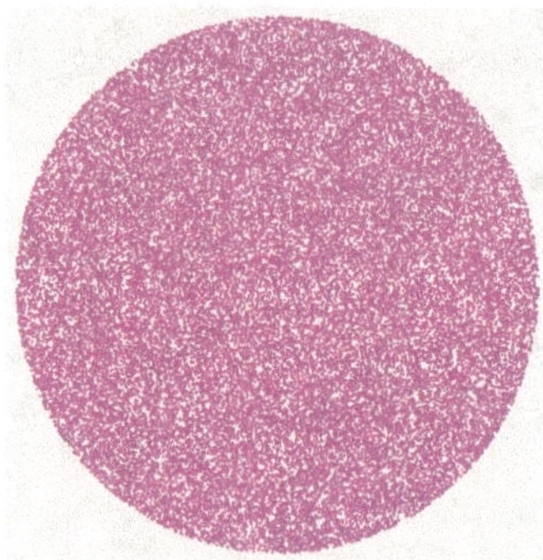

Random points within the unit disk

Appendix D

The Program Matlab

D.1 Introduction

MATLAB (MATrix LABoratory) is an interactive, matrix-based system for scientific and engineering computations. This note is based on version 5 of MATLAB, in order to lean on basic concepts compatible with most of the successive versions.

Conventional styles are as follows:

Boldface for commands in operational system (in the examples we choose UNIX as operation system).

Italic text for MATLAB comments and answers to these comments.

D.1.1 *Accessing MATLAB*

Starting MATLAB depends on the used soft- and hardware environment. Then, in our version one gets

>>

meaning that the program is ready to get instructions.

All logical chains of operations are performed using the *enter* key: RETURN.

The commands to leave the program are *quit* or *exit*.

Stopping an ongoing computation is done through *Control* — *c*, i.e., by pushing the *Control* key and the letter *c* simultaneously.

D.1.2 *Arithmetic operations*

The usual arithmetic operations are

+	addition	−	subtraction	
*	multiplication	/	right division	(D.1)
\	left division	′	transpose	
^	exponential			

The numbers π and e are written *pi* and *exp(1)*, respectively.
For example:
Writing
>>10* *pi*
yields
ans =
31.41 593 (*ans* is the latest answer)

In MATLAB, the usual arithmetical order of operational priority is applied:
>> 8 ^ 2/3
yields the answer 21.3333, i.e. 64/3, whereas
>> 8 ^ (2/3)
gives the answer 4.0000.

D.1.3 *Elementary functions*

MATLAB has all usual basic elementary functions, such as exponential functions, logarithm functions, trigonometric functions, square root functions, and many more. The following is a concise list of notation for some known functions in MATLAB.

$$\begin{array}{llll} exp, & log(=ln), & log10(=lg) & \\ sin, & cos, & tan, & sqrt \end{array}$$

There are many redefined functions as well.

Example D.1.
We compute $\ln(\sqrt{e})$:
>> $log(sqrt(exp(1)))$
ans= 0.5000.

D.1.4 *Variables*

Variables are named by combining letters and digits, where the first character *must be* a letter. One should avoid using the names of MATLAB functions and commands as variable names. The program is also case sensitive (distinguishes between lowercase and capital forms of the letters).

The equality sign "=" is used to assign values for a variable.

$>> x = pi/3$

gives the value $\pi/3$ for the variable x.

One may give instructions on the same line separating them by a comma sign or a semi-colon.

$>> x = pi/3, X = \sin(x)$

gives

$x = 0.7854$ and $X = 0.7071$.

Then the instruction

$>> x = x + 1$

gives

$>> x = 1.7854$.

D.1.5 *Editing and formatting*

One can stop the program to write the answer by ending the instruction by a semi-colon (before RETURN).

$>> z = exp(1);$

$>>$

One can control that whether the variable z has got the correct value e. To do so, just write

$>> z$

One may reach the previous command lines pressing the "sign-up" tangent. Then, the next command line is achieved by pressing the "downward-arrow" tangent.

Likewise, correcting a given command is easily done by moving the arrow-tangents to left or right in order to come to the correction site. Then take away the incorrect sign on the left of the marker either by a Back Space or using Del-tangent. New text may be inserted in the marker's position.

One may decide the number of written decimals using the *format* command. The most usual ones are

format short
which gives five significant digits, and
format long
which gives fifteen significant digits.

Example D.2.

>> *format long*; $10 \times pi$
ans= 31.4159263589793
Then
>> *format short; ans*
gives
ans= 31.4159.

D.2 Help in MATLAB

D.2.1 *Description of help command*

MATLAB is equipped with a direct on-screen help. The help function is a tool that can aid to extend your MATLAB skills rather than as an emergency rescuer. The help function is organized in levels (a description can be seen by typing the command *help*.)

The Command
>> *helpwin*
opens a separate help window.

The command *help* gives a list of subject titles, with each line possessing a subtitle.

Here is a list of few first lines appearing after the use of help command:
>> *help*
HELP topics:

matlab/general	– General purpose commands.
matlab/ops	– Operators and special characters.
matlab/lang	– Language constructs and debugging.
matlab/elmat	– Elementary matrices and matrix manipulation.
matlab/specmat	– Specialized matrices.
matlab/elfun	– Elementary math functions.

A double-click on a title in help window, or the command *help* followed by a library name, gives information on the content of the library and how to know more about it. For instance, clicking *help ops* you get (e.g., in version 5) a long list of MATLAB operations. Then, a double-click on "+" yields:

+ plus

$X + Y$ adds matrices X and Y. X and Y must have the same dimensions unless

one is a scalar (a 1-by-i matrix). A scalar can be added to anything.

To go back to the first list, click on box HOME in the help window.

D.2.2 *Example for how to use help*

Go to help function giving the command.

With a double-click on >> $helpelfun$ (or giving this command) you get a long list of functions. Most of them you will recognize. Perhaps not the command *fix*. To see its action, we start giving the command

>> $helpfix$

FIX Round toward zero.
FIX(X) rounds the elements of X
to the nearest integer toward zero.

For example, we get by

>> $fix(1.5)$

ans=1

and by

>> $fix(-1.5)$

ans=-1

The command

>> *help abs*

gives several lines of information. One of them is

ABS Absolute value and string to numeric conversion.
ABS(X) is the absolute value of the elements of X.
When X is complex, ABS(X) is the complex modulus
(magnitude) of the elements of X.

As an example

>> $abs(-pi)$

gives the expected output

ans= 3.1416.

D.2.3 *The error message*

Non-existing (in MATLAB) or undefined commands will result in error message. Same occurs when a variable is used before assigning value to it. Typos and missprints will return error messages, too.

Example D.3.

>> *plutone*

??? undefined function or variable "plutone".

MATLAB does not recognize *plotone*, and one needs to assign *plotone* a value, for example

>> $plutone = pi/2$;

Now it works to write >> *plutone* without getting an error message.

Here is another error generating example:

>> $x = \sin$

??? Error using == > sin

incorrect number of inputs.

One should of course give an argument so that the sine function can operate

>> $x = sin(plotone)$

ans=1.

Further examples of errors:

>> $x = cos(1, 4)$

??? Error using == > cos

incorrect number of inputs.

Here is an example of wrong use of capital letters

>> $X = SIN(3)$

??? undefined variable ...;

Caps Lock may be on

The description of the last line is as follows. Most common reason for getting capital letters is that one has hit the *Caps Lock* key. Hit it again!

D.2.4 *The command look for*

To know about the name of a function in MATLAB, for example, one may write

>> *look for logarithm*

to get info about which logarithms exist. The command *look for afg* searches all MATLAB routines that contain the text *afg* on the first line of the *help* text. The command *look for — all afg* searches through the whole *help* text.

The command

>> *look for -all arctan*

gives (among others) the answer

ATAN Inverse tangent

While only >> *look for arctan* would not return any information at all.

You stop MATLAB from searching by pressing *Control c.*

D.2.5 *Demos and documentation*

The command *intro* gives a short presentation of MATLAB.

The command *demo* gives the access to some demonstration programs. These are helpful after one is familiar with the basics.

There is a huge amount of documentations and instructions on the internet. To access them are via, e.g., the command *helpdesk* or your hard/soft-ware at hand.

D.3 Row Vectors and Curve Plotting

D.3.1 *Operations with row vectors*

MATLAB can interpret some data/input for multitask performs, e.g., $(x_1, x_2, \ldots, x_n)$. as a matrix or a row vector. The number n is the length of the vector and $(1, n)$, its matrix size. The following command defines two row vectors x and y of length 4:

>> $x = [0\ \ 1\ -1\ \ 2]$; $y = [7\ \ 5\ \ -2\ \ 9]$;

One may put commas between the numbers.

>> $x = [0\,,1\,,-1\,,2]$; $y = [7\,,5\,,-2\,,9]$.

If x and y are two row vectors of the same length, then one may perform coordinatewise addition and subtraction writing $x + y$ and $x - y$, respectively. Coordinatewise multiplication and division are performed by the vectors using commands $x. * y$ and $x./y$, respectively. One may even perform elementary functions, operations on row vectors. General guidance in this regard can be found in *help elmat* and *help ops.*

Example D.4.

(answers are in *format short*)

>> $x = [1\ \ 2\ \ 3]$; $y = [4\ \ 5\ \ 6]$;

>> $x + y$ gives 5 7 9

>> $x. \times y$ gives 4 10 18

>> $x./y$ gives 0.2500 0.4000 0.5000

>> $x.\hat{}\ y$ gives 1, 32 729

>> $exp(x)$ gives 7.3891 20.0855

Whereas

>> $x * y$ gives

Error message

??? Error using ==> ×

Inner matrix dimensions must agree.

The symbol $*$ means matrix multiplication, which is not the same as elementwise multiplication.

However, the instruction

>> x/y does not give an error message, but a rather bizarre answer.

$ans = 0.4156$.

This phenomenon is described in Section D.6.

Important special commands are *ones* and *zeros*. They are used to generate matrices with all elements being ones and zeros, respectively.

For instance, the command

>> $ones(1, 4)$ or $ones(size(x))$ for $x = [1\ 2\ 3\ 4]$ (or a vector x of length 4) gives the answer

$ans = 1\ 1\ 1\ 1$.

For this x:

>> $x = [1\ 2\ 3\ 4]$;

>> $ones(size(x))./x$

gives the answer 1.0000 0.5000 0.3333 0.2500

>> $x + ones(size(x))$ gives the answer 2 3 4 5

This answer is also obtained using the command $x + 1$.

>> $2 * ones(size(x))$ gives the answer 2 2 2 2

The command $ones(1, n)$ gives a row vector with n ones.

D.3.2 *Generating arithmetic sequences*

For the given numbers a, h, and b, one may build up the row vector $x = (a, a+h, a+2h, \dots, b)$, using the command

$>> x = a : h : b$

This is one of the most used commands in MATLAB.

$>> x = -5 : 2 : 5$

$x = -5 \quad -3 \quad -1 \quad 1 \quad 3 \quad 5$

Analogously, the command

$>> x = -pi : 01 : pi$

yields the row vector $x = (-\pi, -\pi+0.1, \dots, \pi)$, which is a vector of length 63. This you can check after you input x as above and then give the command

$>>$ *length*(x)

You can also let the computer write down the vector x, only with the command

$>> x$ (without semi-colon)

For the step size $h = 1$ one can only write

$>> x = a : b$

For instance,

$>> x = 0 : 10$

$x = 0 \quad 1 \quad 2 \quad 3 \quad 4 \quad 5 \quad 6 \quad 7 \quad 8 \quad 9 \quad 10$

If $b < a$, then one can take $h < 0$.

Further information is available through *help colon.*

D.3.3 *Plotting curves*

If $x = (x_1, \dots, x_n)$ and $y = (y_1, \dots, y_n)$ are two row vectors of the same length, then the command *plot(x,y)* will draw a plane curve connecting the points $(x_1, y_1), \dots, (x_n, y_n)$. If there are no particular commands, then MATLAB will choose coordinate axis so that all points are visible (this is called auto-scaling). One may direct how the curve should be plotted giving commands for special line types. It is also possible plot the points without connecting them with curves/lines. The following are some examples.

Example D.5.

$>> x = -3 : 0.5 : 3; \quad y = sin(x);$

$>> ploy(x, y)$

$>> ploy(x, y, ' :'))$ (plots pricked curve)

$>> ploy(x, y, 'o')$ (plots points as small circles.)

General help for two-dimensional graphic can be found in *help plotxy*. You may also find a list of line- and point-types in *help plot*.

D.3.4 *Plotting graphs of functions*

The above examples show how to plot the graphs of functions. Here is a general procedure, to plot the graph of a function $f(x)$, on the interval $a \leq x \leq b$. One starts choosing a suitable step size h, and builds the vector $x = a : h : b$. Then one writes the function in the form of "functions name=the expression of the function". After that the graph of the function will be plotted with the command: *plot*(x, functions name). For instance,

$>> x = -2 * pi : 0.1 : 2 * pi$

$>> f = sin(3 * x) + cos(5 * x); plot(x, f)$

Here the graph of the function $f(x) = \sin(3x) + \cos(5x)$ will be plotted on the interval $[-2\pi, 2\pi]$. Similarly, one can plot the graph of the function $g(x) = x \cos x^2$ on the same interval by the command

$>> g = x. * cos(x. * x); plot(x, g).$

Another way to plot a curve is through using the command *fplot* (this is described in the next section). See also the command *ezplot*.

D.3.5 *Several graphs/curves in the same figure*

Suppose that we want to plot the function f and g above in the same figure. To do so we can give the command

$plot(x, f, x, g)$

If we want both functions, curves to be plotted, we can write

$plot(x, f, `-', x, g`-')$

There is also another possibility: Suppose we want to plot first the graph of f using the command

$>> plot(x, f)$

Then, first one can give the command

$>> hold\ on$

Then the old curve/graph, i.e. (x, f) will be kept while the new one is plotted. For instance

>> $plot(x, g)$

When one is no longer interested in keeping the old figure, one gives the command

>> $hold\ off$

The command *hold* itself yields a shift from a former "hold on position" to "hold off position" and from a previous "hold off position" to "hold on position". You may check these actions by reading through *help hold.*

D.3.6 *Dimensioning of the coordinate axes*

Normally, MATLAB chooses a coordinate system where all points that should be plotted are visible on the screen. One may design the coordinate axis using the command *axis.* To see how *axis* works, you may lookup for

>> $helpaxis$

The following are some examples that you can work out yourself:

Example D.6.

>> $x = 0 : 0.1 : pi; y = sin(x); plot(x, y)$

>> $axis([-14\ \ -12])$

>> $axis('off')$

>> $axis([0\ pi\ 0\ 1])$

>> $axis('on')$

>> $axis(axis); hold\ on$

>> $x = -1 : 0.1 : 2; y = cos(x); plot(x, y)$

>> $hold\ off$

>> $x = -1 : 0.1 : 2; y = cos(x); plot(x, y)$

D.4 Good to Know

D.4.1 *Strings and the command eval*

Element of a matrix can also be "sign strings", i.e., sequences of signs placed within apostrophes:

>> $A =$ "Bicycle, Reimond!"

$A =$

Bicycle, Reimond!

This is applied, e.g., for inserting text in a graph (see the following). Also for building function expressions like $f =' x.\hat{}a.*exp(-x)'$;

The command *eval* is used to "open" a text string to an arithmetic expression

$>> f =' x.\hat{}a.*exp(-x)'$;

$>> x = 0 : 0.1 : 10$;

$>> a = 0; plot(x, eval(f))$

While f is a string, $eval(f)$ becomes a vector, namely $x.\hat{}a.*exp(-x)$.

A new value for a yields a new vector for $eval(f)$.

$>> a = 0.5; plot(x, eval(f))$;

$>> a = 1.0; plot(x, eval(f))$

>> *hold off*

D.4.2 *The command fplot*

This is a simple way to get the graph of a function defined by strings on an interval $[a, b]$. You can try

$>> fplot('sin(3 * x) + cos(5 * x)', [-2 * pi, 2 * pi])$

Note that the variable must be named x.

D.4.3 *Complex numbers*

Complex numbers are represented in the form $a + bi$ (or $a + bj$). The calculus is similar to that of the real numbers.

Example D.7.

$>> (1 + 2i) * (3 + 4i)$

$ans = -5.0000 + 10.0000i$

$>> (1 + 2i)/(3 + 4i)$

$ans = 0.4400 + 0.0800i$

One can also have complex entries in row vectors, matrices, and the elementary functions.

Complex conjugate, the absolute value, real and imaginary parts are obtained using the commands: *conj, abs, real,* and *image*, respectively.

A word of warning: The symbol i is reserved for the imaginary unit. Therefore, it is irrelevant to use i as a variable name. In case one uses i as a variable name, then its original value as imaginary unit is returned through the command $i = sqrt(-1)$.

D.4.4 *Polynomials*

One can enter a polynomial giving its coefficients as a row vector in reduced degree order (highest order coefficient first, ...). Then one can evaluate the value of the polynomial using the command *polyval*, and its zeros by the command *roots*.

Example D.8.

$>> myPol = [1\,2\,3]; roots(myPol)$

Here the polynomial $myPol = x^2 + 2x + 3$ is inserted and its roots are $-1.0000 \pm 1.4142i$.

The values for different x are computed using the command $polyval(myPol, x)$

$>> polyval = (myPol, [2\,3\,4])$

$ans = 11$ 18, 27

To plot the graph of a polynomial, evidently one needs to evaluate it at a number of points before using the command *plot*.

$>> x = -5 : 0.1 : 5; y = polyval(myPol, x); plot(x, y)$

Further, special commands for the polynomials can be found in *help polyfun*.

D.4.5 *To save, delete, and recover data*

The command *who* gives a list of the typical variables. When leaving MATLAB (*quit* or *exit*) these variables will disappear. One may save them (names and data) using the command *save*. They will be saved in a file named "matlab.mat". These data can be recovered using the command *load*.

In case one needs/wants to save the data in another name, it suffices to give this name after the save command. For example, using

$>>$ *save temp*

would save the data in a file named "temp.mat", in the current library.

Then, the same data is recovered entering the command *load temp*.

To save a particular data, e.g., P, Q, R, one uses

>> *save P Q R temp*

One can clear the memory from the current variables by giving the command *clear*. The same command followed by a list of certain variables will remove those variables, e.g., *clear P*. Such cleanings are adequate for avoiding mixing of the new and old variables when starting a new problem.

To clear the graphic window, one uses the command *cl f*, (clear figure).

D.4.6 *Text in figures*

One can insert text in the figure window with the command text:

>> *text(xpos, ypos,* "the text itself inside apostrophes")

Here, *(xpos, ypos)* gives the starting position in the coordinate system for the current figure. One may name the axes using the commands *xlabel, ylabel*. See also the command *title* and *gtext*.

D.4.7 *Three-dimensional graphics*

This is a huge subject. Here we only give a very short introduction, but in *help plotxyz* you get some more 3d plotting information.

One may plot the surface of a given function of two variables using the command *meshgrid* and *mesh*. The following is an example that plots the surface graph for the function.

Example D.9.

$$f(x, y) = xy^{-x^2-y^2}$$

over the interval

$$(x, y) : -2 \leq x \leq 2, \ -3 \ \leq y \leq 3.$$

>> $f =' x. * y. * exp(-x. * x - y. * y)';$

(Note the apostrophes ‘, ’)

>> $[x, y] = meshgrid(-2 : .2 : 2, -3 : .2 : 3);$

(defines the domain.)

>> $mesh(x, y, eval(f))$

(Plots the surface of the function).

One can get very nice figures using the command *print*. Using just *print* would give the actual graphic window set by the system administrator. Typing *print -Printer name* would print using the given printer.

One may save the figures to import them into other documents. To get a high quality picture/figure, it is recommended to save its graphic window in the so-called PostScript-file. This is done by using the command

\>\> *print -deps -epsi, file name*

(Here file name without apostrophes). The figure can then be imported to other document types, e.g., FrameMaker, LATEX, and so on. One can visualize the figure using the menu-driven program **ghostview** (the out-printed figure will have better quality than the one that appears on the screen.)

D.5 To Create Own Commands

One may enrich the MATLAB commands library defining own commands. This is done in two ways: either in the form of the so-called, script files or function files. Both file types are known as M-files, due to the names ending on ".m". Script files consist of a combination of usual MATLAB commands, while function files are your own-defined commands. Some general information is available through *help script* and *help function.*

To be able to write own M-files, you need to login to your own editor, open an editing window on the screen. You can write in your files therein and test them directly from the MATLAB-window, meanwhile you can give UNIX-commands from the terminal window.

Before testing a ".m"-file you must save it in the format "file name.m".

D.5.1 *Textfiles*

Text files (or script files) consist of a gathering of usual MATLAB commands. The following file is to draw the graph of $y = \sin(x),\hat{}N$. On interval $[a, b]$ for different values of a and b and different integer N. To begin with let $a = -4\pi$, $b = 4\pi$, and $N = 2$. The m.file looks as follows:

"mySinus.m"

$a = -4 * pi$; $b = 4 * pi$;
$N = 2$;
Step=(b-a)/1000;
x=a: Step: b;
y=sin(x).^N;
plot(x,y);
Comments:

The first two lines give values to the end points a and b of the interval and the integer N.

The next line sets the step size *Step* (as one promille of the length of the interval).

Then the row vectors x and $y = \sin(x)\hat{.}N$ are defined and plotted in the last line.

Note that elementwise exponenting is performed due to the fact that now both x and $\sin(x)$ can be row vectors.

Now write the above file in your editing window (excluding the comments) and save it under the name "mySinus.m".

Check the file sinusN.m in your matlab library through giving the command *what* in the matlab-window. Then test the following command. from the matlab-window:

>> *mySinus*;

Now you can return to your editing window and change, e.g., the value of N. Change the second line in the file "mySinus.m" to

N=3;

Save the file (do not forget this) and run it again:

>> *mySinus*;

You may of course change the end-points of the interval, likewise the step size, as well as the function *sin*.

If you like to have both curves on the same figure, use the command *hold on*. Then, remember to end with *hold off*.

An important observation is that all variables appearing in a script-file are global, in the sense that they are available outside of the file. You may see this by writing the command

>> *Step*

OBS! capital S

D.5.2 *Function files*

A function file should be named as “functions name.m”. The first word in a function file is *function.* Then, it follows by the description of the function’s output (if it exists), the name of the function and the input.

We start to write a simple function, called “myfunc”, having an input variable and an output variable. Thus, the file name will be “myFunc.m”. Then, here is the first line:

function out=myFunc(in)

Then follows the computational part. All computational instructions end by; to avoid getting the results interemediate calculus on the screen.

At the end the value of output variable will be given. And after that there shouldn’t be anything written on the file (except possibly comments).

The following is a first example of a function.

Example D.10.

“myFunc.m”

function y=myFunc(x)
numerator = x. * x -2*x. -1;
denominator =1+x.*x;
y=nominator./denominator;
Comments:
The function is $myfunc(x) = (x^2 - 2x - 1)/(1 + x^2)$.
In the first line, the nominator $x^2 - 2x + -1$ is computed.
One uses .* since the input can be a row vector.
The second line is for computing $1 + x^2$.
In the last line, the output y is computed.
(Note the division is denoted by ./)

Note that none of the in- or out-variables need to have the same name as in the definition in the function file. For instance, one may write

>> $t = 0 : 0.1 : 5; z = myFunc(t); plot(t, z);$

In contrast to the script-files, all variables that are used inside a function are local, i.e., they cannot be reached from outside. Check this through giving the command

>> *nominator*

The only variables that may be used from the outside are the ones that appear in the functions, output variables.

Observe that, given a value for a variable in the matlab-window, this value will not be effected even if the name of variable is used inside a function.

More on function files are given in Section D.7.

D.5.3 *How to write own help command*

One can write own comments anywhere in an M-file, only let them start with the % sign. MATLAB ignores the rest of the line. Writing such comments in the top of a script-file or just after declaring a function in a function-file, then such commands will be printed out the command *help* and then the file name. The following are two examples where the files "mySinus.m" and "myFunc.m" are described with additional comments.

"mySinus.m"

```
% mySinus
% This script file is to plot the graph of the function
% y=sin(x).^ N on the interval (a,b).
a = -4 * pi; b = 4 * pi;   % Choose the endpoints of the interval
N = 2;                     % Choose the exponent N
Step=(b-a)/1000;           % Choose the step size
x=a: Step: b;
y=sin(x).^N;               % OBS: elementwise power.
plot(x,y);
```

You may run the file to see whether everything is correct. Then give the command

>> help mySinus

The following is a similar function file "myFunc.m" written with comments and help text.

"myFunc.m"

```
function y=myFunc(x)
% myFunc
%      myFunc(x)=(x. * x -2*x. -1)./(1+x.*x)
nominator = x. * x -2*x. -1;
```

% use .* since the input x can be a row vector.
denominator =1+x.*x;
y=nominator./denominator; % ./

Write in the comments and run the program. Test also

>> *help myFunc*

It is recommended to supply the files with adequate comments. and help information. They help to remember the process of the program. Often it is totally cumbersome to see what an uncommented program does, even if one has written it oneself.

D.5.4 *Some simple but important recommendations*

It is recommended to put MATLAB-files in a special sublibrary called "matlab". If you have not already done this, then write **mkdir matlab** (from your main library).

When you start MATLAB running, first open a new terminal window. Then, go into the matlab library with the command **cd matlab**. Then, start MATLAB with the command

matlab.

If you have already started MATLAB (from your main library), then you can instead give the command

>> *cd matlab*

Every time you start MATLAB, it is better to do it from your matlab library.

D.6 Matrix Algebras

In this section, it is assumed that the reader is familiar with basic concepts such as matrix multiplication, linear system of equations, inverse of a matrix, and many more. Help for this section can be found in the libraries *ops, elmat, specmat, and polyfun.*

D.6.1 *Basic matrix operations*

One loads matrices row-wise with semi-colon between the rows and an empty space or comma sign between the elements in the rows. For instance, an input as

>> $A = [1,\ 2,\ 3;0,\ 7,\ 9;4,\ 6,\ 5]$

results

$$A =$$

$$\begin{matrix} 1 & 2 & 3 \\ 4 & 5 & 6 \\ 7 & 8 & 0 \end{matrix}$$

The symbol for transposing is ′. (Hence, the apostrophe has a double function.) Transposing a row vector yields a column vector as follows:

$>> x = [-1 \quad 6 \quad 2]'$

which results

$$x = \begin{matrix} -1 \\ 0 \\ 2 \end{matrix}$$

Addition, subtraction, and multiplication of matrices are denoted by $+$, $-$, and $*$, respectively.

Example D.11 (Example with A and x as above). $>> \; b = A * x$

$$b = \begin{matrix} 5 \\ 8 \\ -7 \end{matrix}$$

D.6.2 *System of equations (matrix division)*

There are two symbols for matrix division in MATLAB; namely \ and /. If A is a non-singular (i.e., invertible) square matrix, then one can get a unique solution for $A * X = b$ for a matrix b that has as many rows as A. The solution X is obtained using the command X=A \ b. For example, with above A and b, we get

>> X=A \ b

$$X = \begin{matrix} -1 \\ 0 \\ 2 \end{matrix}$$

Analogously, the command $Y = c/A$ gives the solution for the system of equations $Y * A = c$ for every matrix c having the same number of columns as A.

Example D.12.

```
>> c=[1  2  3];    Y=c/A
Y=100
```

We summarize this as follows:

X=A\ b gives a solution for $A * X = b$

Y=c/A gives a solution for $Y * A = c$

The division commands \ and / also can be used for non-square matrices. For instance, you may try to solve the following system of equations for some given right-hand side.

$$\begin{aligned} x_1 + 2x_2 &= b_1 \\ 3x_1 + 4x_2 &= b_2 \\ 5x_1 + 6x_2 &= b_3 \end{aligned}$$

Denoting the matrix of the system W, we have in matrix form $W * x = b$. Then

```
>> W=[1  2; 3  4; 5  6]; b=[1  3 5]'; W\b
ans  =
      1
      0
```

You may check whether we have the correct answer using the command

```
>> W * ans
ans  =
      1
      3
      5
```

In case a given system of equations does not have a unique solution, MATLAB computes an approximate solution using the *method of mints square.* We have described this technique in the chapter on linear algebra. Here, we give an example using the same system above:

If we let $b(1, 0, 0)$, then the system $W * x = b$ is not solvable. Nevertheless, MATLAB returns an answer:

```
>>  b = [1 0,  0]'; W\b
ans =
        -1.3333
         1.0833
```

which is not an exact solution. This can be seen as follows:

```
>> W * ans
```

$ans =$

$$\begin{matrix} 0.8333 \\ 0.3333 \\ -0.1667 \end{matrix}$$

Observe that if a system of equations has infinitely many solutions, MATLAB returns only one of them. Without any warning about the existence of the other solutions. However, one may use the command *rref* to get the extended matrix for the system in row-reduced trap step form and then decide all solutions. For instance, if we want to solve $A * X = b$, where $A = [1\ 1,\ 1\ ; 1\ 2\ 3]$ and $b = [2, 3]'$, then

>> A\b

returns only the solution (1.5, 0.0, 0.5). But, if we let $U = [A\ b]$ be the extended matrix, then we get

$>> rref(U)$

$ans\quad =$

$$\begin{matrix} 1 & 0 & -1 & 1 \\ 0 & 1 & 2 & 1 \end{matrix}$$

Through this one can conclude that the general solution for this equation system is: $(x, y, z) = (1, 1, 0) + t(1, -2, 1)$.

Now you may try

$>> rrefmovie(U)$

D.6.3 *Rows, columns, and individual matrix elements*

For a given matrix A, $A(r, :)$ denotes the row r of A, $A(:, k)$ its column k, and $A(r, k)$ is the element at position (r, k). If u and v are two vectors with integer components, then $A(u, v)$ is the submatrix of A having those rows in A whose indices are given by u and those columns of A whose indices are given of the vector v.

Example D.13.

$>> A = [11 : 15; 21 : 25; 31 : 35]$

$A =$

$$\begin{matrix} 11 & 12 & 13 & 14 & 15 \\ 21 & 22 & 23 & 24 & 25 \\ 31 & 32 & 33 & 34 & 35 \end{matrix}$$

$>> A(:, 1)$

$ans =$
11
21
31
$>> A(3,4)$
$ans = 34$
$>> B = A([2,3],[1,3,5])$
$B =$

21	23	25
31	33	35

One may change certain elements or whole rows and columns through giving them new values. For example,
$>> A(3,4) = 134$
$A =$

11	12	13	14	15
21	22	23	24	25
31	32	33	134	35

$>> A(:,5) = [1:3]'$
$A =$

11	12	13	14	1
21	22	23	24	2
31	32	33	134	3

One can change the size of a matrix with directly assigned commands. The matrix will probably be extended with additional zeros.

Example D.14. $>> B(4,:) = ones(1,3)$
$B =$

21	23	25
31	33	35
0	0	0
1	1	1

One may even build matrices using other matrices as "blocks". For example,
$>> C = [[A; (-1) .\hat{}(1:5)], B]$

$$C =$$

$$\begin{matrix} 11 & 12 & 13 & 14 & 1 & 21 & 23 & 25 \\ 21 & 22 & 23 & 24 & 2 & 31 & 33 & 35 \\ 31 & 32 & 33 & 34 & 3 & 1 & 1 & 1 \\ -1 & 1 & -1 & 1 & -1 & 1 & 1 & 1 \end{matrix}$$

Some other useful commands to build new matrices are

$tril(A)$	give the lower triangular matrix of the given matrix A
$triu(A)$	give the upper triangular matrix of the given matrix A
$ones(n)$	$n \times n$ matrix with only ones as elements
$ones(n, m)$	$n \times m$ matrix with only ones as elements
$zeros(n)$	$n \times n$ matrix with only zeros as elements
$zeros(n, m)$	$n \times m$ matrix with only zeros as elements.

For instance, to add the number 3 for all elements of the matric C above, one can give the command

$>> C + 3. * ones(size(C))$

The matrix $ones(size(C))$ is the matrix of size C which has only ones as its elements.

D.6.4 *A guide for a better way to work with matrices*

It is practical to write matrices in a special text file. To do so one can start an editing window and write in the matrix. For instance,

$A = [1 \quad 1; \quad 1 \quad -1; \quad 0 \quad 0];$

Or it is convenient to recognize:

$$\begin{matrix} A = [& 1 & 1 & \\ & 1 & -1 & \\ & 0 & 0 &]; \end{matrix}$$

Then you may save the file with a suitable name, for instance, "mymatrix.m". Then you can run the file using the command $>>$ *mymatrix*. You may control whether you have inserted the matrix correctly by using the command $>> A$. You may of course write several matrices and vectors in the file "mymatrix.m".

D.6.5 *Inverse and identity matrix*

The inverse of a square matrix is computed using the command *inv*. For the matrix

$A = [1, 2, 3; 4, 5, 6; 7, 8, 0]$, this will be the result.

$>> C = inv(A)$

$C =$

−1.7778	0.8889	−0.1111
1.5556	−0.7778	0.2222
−0.1111	0.2222	−0.1111

The identity (unit) matrix is denoted by *eye* as follows:

$eye(n)$	$n \times n$ unit matrix: ones in the diagonal and zeros else
$eye(n, m)$	$n \times m$ matrix with ones in the diagonal and zeros else

You may check the command $>> C * A - eye(3)$

and observe that the expected, correct answer, i.e., 0-matrix, is not obtained. The reason is that computing the inverse of a matrix in MATLAB is associated with certain numerical errors.

D.6.6 *Determinants, eigenvalues, and eigenvectors*

If A is a square matrix, its determinant is computed using the command $det(A)$. We check with an example

Example D.15.

$>> A = [7,\ 2,\ 0\ ; 2,\ 6,\ -2\ ;\ 0,\ -2,\ 5];$

$>> det(A)$

ans= 162

Eigenvalues of A (both real and complex) are computed using the command $eig(A)$.

$>> eig(A)$

$$ans = \begin{matrix} 9.0000 \\ 6.0000 \\ 3.0000 \end{matrix}$$

Hence, the eigenvalues are 9, 6, and 3. One can also compute the eigenvectors using the command *eig*. This is written in the form

$[V, D] = eig(A)$. The matrix D is a diagonal matrix containing eigenvalues in its diagonal whereas the columns of the matrix V are the corresponding eigenvectors.

Example D.16.
$>> [V, D] = eig(A)$
$V =$

0.6667	−0.6667	0.3333
0.6667	0.3333	−0.6667
−0.3333	−0.6667	−0.6667

$D =$

9.0000	0	0
0	6.0000	0
0	0	3.0000

Here, the vector $(.6667, = .6667, -0.3333)$ is an eigenvector corresponding to the eigenvalue 9. You may control that the matrix V is indeed diagonalizing A giving the command $inv(V) * A * V$. The result should be the matrix D. MATLAB utilizes general numerical routines which only give closer values. Therefore, for many matrices one does not get the exact eigenvalues and eigenvectors.

Example D.17. $>> A = [2,\ 0,\ -2,\ ;\ 1\,1\ -2,\ 2,\ ;\ -2,\ 2,\ 1]$
$>> format\ long; Eig(A)$

$$ans\ =$$
$$\begin{array}{l} 2.00000000000000 \\ 1.00000001924483 \\ 0.99999998075517 \end{array}$$

Thus, obviously the exact eigenvalues are 2, 1, 1. Computing the eigenvectors using the command $[V, D] = eig(A)$ yields the following answer (*in format short*):

$V =$

0.7071	−0.6667	−0.6667
0.7071	0.6667	−0.6667
0.0000	−0.3333	−0.3333

Observe that in this example the matrix A is not diagonalizable. This can be easily seen be computing manually (by hand). Here the exact matrix V is not invertible, but since MATLAB does use the closer values, it interprets V as being invertible. Therefore, the command $inv(V) * A * V$ will return the answer for D (with 7 correct decimals). One can find out that something is not correct through letting MATLAB compute the determinant of V. If this determinant has a very small value (here $-1.512e - 09$), one must become suspicious that the given matrix is not diagonalizable.

There are several other numerical routines for matrix factorizations.

See, e.g., *lu, svd, qr, rref.*

D.6.7 *Functions of matrices*

One can compute the power A^n of a given square matrix A, using the command $A\hat{}n$. Here, n is an arbitrary integer (even negative in case A has inverse). If

$$p(z) = a_n z^n + a_{n-1} z^{n-1} + \cdots + a_1 z + a_0.$$

Then one can easily compute

$$p(A) = a_n A^n + a_{n-1} A^{n-1} + \cdots + a_1 A + a_0 I.$$

We take an example with the polynomial $p(z) = z^2 + 2z + 3$ and a 2×2 matrix. First we insert A and the coefficients of the polynomial:

Example D.18. $>> A = [1,\ 2\ ;\ 3,\ 4];\ \ p = [1,\ 2,\ 3];$

Then, we compute $p(A)$ using the command *polyvalm* as follows

$>> polyvalm(p, A)$

$ans =$

12 14

21 33

If A is a square matrix, then the command $poly(A)$ will give a vector whose coordinates are the coefficients in the characteristic polynomial for A starting with higher degree coefficients. As a consequence, *roots(poly(A))* gives the eigenvalues of the matrix A.

It is also possible to compute with complicated functions of a quadratic matrix A. Let us here name only the exponential function

$$expA = I + A + \frac{A^2}{2!} + \frac{A^3}{3!} + \cdots + \frac{A^n}{n!} + \cdots$$

This is computed using the command $expm(A)$ (not $exp(A)$ which gives the elementwise exponential function). Observe that if c is a given column vector, then $x = expm(t * A) * c$ is the solution at time t (a given number) to the system of equations $x' = Ax, \quad x(0) = c$.

D.7 Programming in MATLAB

D.7.1 *General function files*

A function file can have none, one, or many in-variables. An in-variable can be a number, a row or, a column vector, or a matrix. In the same way, the functions can have none, one, or many out-variables. If the function will have many out-variables, then this must be given in the declaration through putting them within parentheses [ ]. The following is an example of a rather general form of a function named *example* with four in- and three out-variables.

Example D.19. function [ut1, ut2, ut3] =example(in1,in2,in3,in4)
% $[u, v, w]$ = example(a, b, c, d)computes
% function values $u, v,$ and w
% of the function example in points (a, b, c, d).
computations;
$ut1$=computation result1;
$ut2$=computation result2;
$ut3$=computation result3;

D.7.2 *Choice and condition*

The simplest condition looks like this:
if condition
instructions (statements) separated of;
end

The condition is that the expressions/orders are performed only if they are valid. The conditions are expressed using different comparison relations, for example, the relation ¡. Information about the relations are obtained using *help relop*. In the following, we list a few of them:

if a==b means "whether a is equal to b"
if a ~=b means "whether a is not equal to b"
if $a >= b$ means "whether a is greater than or equal to b"
if $a > b$ means "whether a is greater than b"
if $(a > 0)$ & $(b > 0)$ means "whether $a > 0$ and $b > 0$"
if $(a > 0)|(b > 0)$ means "whether $a > 0$ or $b > 0$"

One may use matrices in relation operations. Read about this in *help relop*.

The condition terms can have a more complex form, e.g.,

if condition
first statements1;
else
statemdents2;
end

If the conditions are satisfied, then the first statements1 will be performed. Otherwise, the second statements2 are performed.

Check further *help if*. See also *any, all*, and so on under *help ops*.

D.7.3 *Loops*

A program loop concerns repeating a computation several times. The simplest loops are performed with *for*. Program loops, in principal, look as follows:

for variable = vector
statements;
end

The most often used loop looks like:

for $k = 1 : n$
statements;
end

Here, n is a positive integer which is assumed given from the start. The result of the loop is that the statements will be performed for $k = 1, 2, \ldots, n$. (Examples will follow later.)

Another type of loop has the form

while condition

statements;
end
then the statements are performed as long as the condition is fulfilled.

D.7.4 *Input and output*

A MATLAB program (i.e., a function or a script file) can print out text, data, or error message. The command *disp* is used for printing out a matrix or a text.

```
disp('The matrix has eigenvalues');
disp([2  3]);
```

gives the out-print

```
The matrix has eigenvalues
2  3
```

With the command *input* one makes the program stop and wait until the user of the program enters a number. The program row:

$a = input$('Write in a positive number!');

Leads to that the program writes out the text and waits for users to respond. The entered number then becomes the value of the variable a.

One may also, temporarily, stop the program with *pause*. The execution of the program continues as soon as the user hits any arbitrary key.

A program row having the command *error* yields that MATLAB write out an error message and leaves the program.

Example D.20.

error('You shouldn't enter a negative number')

The command *nargin* (number of arguments in) and *nargout* (number of arguments out) controls how many in- and out-variables are given explicitly. For an example of using them, see the following.

There are, of course, a number of other commands that steer how the program should work. For information on them see *help lang*.

D.7.5 *Functions as in-variables*

In-variable in a MATLAB-function can be the name of a file for other functions. To compute, inside a function, the value of another function whose name is given, one uses the command *feval*. Let us give an example. We want to write down a function that computes sum of two other functions: $f(x) + g(x)$ where f and g are two functions named "$fname$" and "$gname$".

"sumfg.m"

```
function s = sumfg('fname' , 'gname', x)
% sumfg
%      s=sumfg('fname' ,'gname', x)
%      if f and g are two functions with the names
%      'fname' and 'gname', then  s=f(x)+g(x)
s=feval('fname', x)+feval('gname', x);
```

This function can now be used to compute sum of arbitrary functions, for instance:

$>> x = 0,\ 0.1\ \ : 2 * pi;$

$>> y = sumfg('myFunk','cos'x); plot(x, y)$

D.7.6 *Efficient programming*

We summarize this note with some useful hints to write effective MATLAB programs. The program loops are often very slow in MATLAB. Therefore, it is advised to take all possible opportunities to transform the loops to vector or matrix operations. For instance to compute $sin(n)$ for $n = 1, \ldots, 1{,}000$, one should not write

```
n=0;
for k=0:999
n=n+1;
y(n)=sin(k);
end;
```

Instead it is better to write the loop in vectorized form:

```
n=1:1000;
y=sin(n);
```

If one has to use program loops, it is advised to create memory space for the loop variables through, e.g., given the zero-values.

Example D.21.

```
y=zeros(1,100);
for n=1:100
y(n)=sum (x^n);
end;
```

If one does not create memory space in advance MATLAB must expand the vector y every time the loop performs.

D.7.7 *Search command and related topics*

When MATLAB reads a command which is not among those that are built-in, then MATLAB will search through the files with the command's name ending with ".m". MATLAB searches for M-files in the following order:

MATLAB's program library
the current library
your own matlab-library (if you have one)

Observe that if there are two M-files with the same name, MATLAB will use the one which will be found first.

One gets a list of all M-files in the actual library through giving the command *what* in the matlab-window. To change libraries, one can write *cd* followed by the name of the library (including the search root).

There are two types of MATLAB-functions, those that are, the so-called, built-in, and those defined as M-files. Example for a built-in function is *exp*, while *sinh* is given as M-file. Writing *type* and then the name of the function (e.g., *type sinh*), one gets either the written/printed M-file or an information telling that the function is built-in.

The command *path* gives a list (including search roots) of libraries where MATLAB searches for files. One can copy program library files over to ones own matlab-library for inspecting for possible modifications. Such copying is best done in a terminal window. Example (if your user name is "plutten" and the search root to MATLAB-functions has the name "stig"):

cd stig/matlab/elfun/cosh.m plutton/matlab/

D.7.8 *Examples of some programs*

Example D.22. **("myFunc2.m")**

```
function y=myfunc2(x)
% myfunc2
%       y=mydunc2(x) computes y=exp(-x)-log(1+x)
%       If all x-values are ¿ -1, then write an error message.
M=min(x);
```

```
if M ¡ = -1
error ('no variable should have a value less than 1');
% The function is not defined if min(x) ¡ = -1
else
y=exp(-x)-log(1+x);
end
```

Example D.23. ("mySum.m")

```
function = s = mySum(z, maxN)
% mysum
%       s = mySum(z, maxN)
%       computes the sum of geometric series
%       s = 1 + z + z^2 + ... + z^N
%       for n=1,2,..., maxN and writes out the sums
%       Stop for each new value for n.
%       To continue press an arbitrary key.
%       If you want to cancel press Ctrl-C
s=1;
for n=1:maxN
s=1+z*z; n=n+1;
disp('number of terms and the corresponding sum');
disp([n,s]); pause;
end
```

Example D.24. ("myPowers.m")

```
% myPowers
%       Script-file for plotting the graph of the function y = x^a
on the interval (0,1)
%       where a is given from the keyboard.
%       The power a must be positive
%
%
continue=1;
% continue =1 as long as the user wants to continue
a = 1
while continue == 1
     a=input('insert a power')
     if a >= 0
x = 0 : 0.05 : 1; y = x.^a; plot(x,y);
else
```

```
disp('The power must be positive');
end
disp('Press 1 if you want to continue, otherwise press 0');
% To end the user should press 0.
continue =input('Press 1 or 0');
if continue== 12
hold on;
end
hold off;
```

Example D.25. ("plotpol.m")

This example requires some (limited) knowledge from section D.7sec:matralg about the matrices.

```
function plotpoly (thePoly, linetype)
% PLOTPOLY
% plotpoly(P) plots the polygon defined by 2 × M-matrix P, where M > 2 using the linetype '-'
% The matrix P holds the coordinates of the vertices of the polygon in its columns
% plotpoly(P, 1type) plots the same polygon using the linetype 1type
[N M]=size(thePoly);
if N == 2 & M > 2
x=[thePoly(1,:) thePoly(1, 1)];
y=[thePoly(2,:) thePoly(2, 1)];
if nargin < 2
        plot(x,y, '-');
else
        plot(x, y, linertype);
end;
else
error('Input must be 2× M-matrix with M> 2');
end
```

D.7.9 *List of most important command categories*

These main categories are available via *helpwin* or directly by giving the command *help* and then the name of the command category, for example, >> *help ops*.

Command category	Content	Example
general	General commands	help, clear, load, save
ops	Elementary mathematical operations	$+, *, .*, /, .$
lang	Programming commands	if, else, end, feval
elmat	Basic matrix commands	zeros, ones, size
elfun	Elementary mathematical functions	sin, exp, abs
matfun	Matrix functions	det, inv, rref, eig
datafun	Functions for data analyzing	min, max, sum
polyfun	Polynom and interpolation	roots, polyval
funfun	Zeros, minimization, integration	fmin, fzero
graph2d	Two-dimensional graphic	plots, axis, title
graph3d	Three-dimensional graphic	mesh, surf
graphics	General graphic commands	figure, clf, subplot
strfun	Manipulation of loops	eval, num2str
demos	Demonstration files	demo, intro

D.8 Algorithms and MATLAB Codes

For the computational aspects, we have gathered suggestions for some algorithms and Matlab codes that can be used in implementations. These are specific codes on the concepts such as

- Finding a zero of a continuous function: Bisection, Secant and Midpoint rules.
- L_2-projection.
- Numerical integration rules: Midpoint, Trapezoidal, Simpson.
- Finite difference Methods: Forward Euler, Backward Euler, Crank-Nicolson.
- Matrices/vectors: Stiffness, Mass-, and Convection Matrices. Load vector.

The Matlab codes are not optimized for speed, but rather intended to be easy to read.

D.8.1 *The bisection method*

The following is a MATLAB routine that uses the bisection method to a zero of a given function f (defined as an inline function in the script) in the interval $[a, b]$. Note that in the bisection method $f(a)$ and $f(b)$ must have opposite signs. This routine localizes the root in subinterval of length as $1/2$ of the current length of the interval. The process stops if either:

1. The magnitude of the function at the current stage is less than a given tolerance *tol* or
2. The maximum number of iterations *kmax* has been reached.

D.8.2 *An algorithm for the bisection method*

```
f= inline (' x.^3-3*x.^2+1*)  % Bisection method for $f(x)= x^3-3x^2+1$.
a=0;    b=1;       kmax=7;        tol=0.00001;
ya=f(a);  yb=f(b);
if sign(ya)==sign(yb),    error('function has same sign at the end points'),
end
disp(' step     a     b      m     ym     bound')

for k=1:kmax
      m=(a+b)/2;      ym=f(m);     iter=k;        bound=(b-a)/2;
      out = [iter, a, b, m, ym, bound];    disp( out )
      if abs(ym) < tol,    disp('bisection has converged');    break;
       end
      if sign(ym)~=sign(ya)
               b=m;         yb=ym;
      else
               a=m;         ya=ym;
      end
      if (iter >= kmax),    disp('zero not found to desired
      tolerance'),.
       end  end
```

The following MATLAB function utilizes the secant method to find the zero of the function f (given as an inline function), using the starting values $x(1) = a$ and $x(2) = b$.

In contrast to the bisection method, $f(a)$ and $f(b)$ need not have opposite signs, and there is no guarantee that there is a zero in the interval between two successive approximations.

D.8.3 *An algorithm for the secant method*

```
function [xx, yy]  = Secant(f,  a,  b,  tol,  kmax)
% $f $ is an inline function
y(1)  =  f(a);
y(2)  =  f(b);
x(1)  = a;
x(2) =  b;
Dx(1) = 0;
Dx(2) = 0;
disp('   step        x(k-1)        x(k)        x(k+1)     y(k+1)       Dx(k+1)')
for k  =. 2:kmax
        x(k+1)    =  x(k)-y(k)*(x(k)-x(k-1))/(y(k)-y(k-1));
        y(k+1)    =  f(x(k+1));
        Dx(k+1) = x(k+1)-x(k);
        iter = k-1;
        out = [ iter,   x(k-1),       x(k),        x(k+1),   y(k+1),
Dx(k+1)'];
        disp( out  )
        xx =  x(k+1);
        yy =  y(k+1);
        if abs(y(k+1))< tol
             disp('secant method has converged');  break;
        end
        if (iter >= kmax)
              disp('zero not found to desired tolerance')
        end
 end
```

The following MATLAB function finds a zero of a function near the initial estimate x_1 using Newton's method. The procedure stops either

1. The change in successive iterates (which is also the estimate of the error) is less than a given tolerance /tol or
2. the maximum number of iterations, *kmax*, has been reached.

D.8.4 *An algorithm for the Newton's method*

```
function [x,   y]  = Newton(fun,    fundr,     x1,     tol,     kmax)
%  Input:
%            fun                   function (inline function or m-file function)
%            fundr                derivative function ( inline or m-file)
%            x1                    starting estimate
%            tol                    allowable tolerance in computed zero
%            kmax               maximum number of iterations
% Output:
%            x                        (row) vector of approximations to zero
```

```
%              y                       (row) vector  fun (x)
x(1) = x1;
y(1) = feval(fun,  x(1));
ydr(1)  = feval(fundr, x(1));
for   k = 2 : kmax
      x(k)  =  x(k-1) -y(k-1)/ydr(k-1);
      y(k)  =  feval( fun, x(k));
      if abs(x(k)-x(k-1)) <. tol
                disp(*Newton method has converged');   break;
       end
       ydr(k)= feval(fundr, x(k));
       iter = k;
end
if  (iter >= kmax)
                 disp('zero not found to desired tolerance');
end
n=length(x);
k = 1: n;
out = [k',   x',   y'];
disp('            step                    x       y ' )
disp (out)
```

D.8.5 *An algorithm for L_2-projection*

(i) $\mathcal{T}_h$ is a partition of the interval I into N subintervals, and $N+1$ nodes. Define the corresponding space of piecewise linear functions V_h.

(ii) Compute the $(N+1)\times(N+1)$ mass matrix M and the $(N+1)\times 1$ load vector $\mathbf{b}$:

$$m_{ij} = \int_I \varphi_j\varphi_i\,dx, \qquad b_i = \int_I f\varphi_i\,dx, \qquad i,j = 0,1,\ldots,N.$$

(iii) Solve the linear system of equations

$$M\xi = \mathbf{b}.$$

(iv) Set

$$P_h f = \sum_{j=0}^{N} \xi_j\varphi_j.$$

Here are two versions of Matlab codes for computing the mass matrix M:

```
function M = MassMatrix(p, phi0, phiN)

%------------------------------------------------------------------
% Syntax:   M = MassMatrix(p, phi0, phiN)
% Purpose:  To compute mass matrix M of partition p of an interval
% Data:     p -    vector containing nodes in the partition
%           phi0 - if 1: include basis function at the left endpoint
%                  if 0: do not include a basis function
%           phiN - if 1: include basis function at the right endpoint
%                  if 0: do not include a basis function
%------------------------------------------------------------------

N = length(p);     % number of rows and columns in M
M = zeros(N, N);   % initiate the matrix M

% Assemble the full matrix (including basis functions at endpoints)
for i = 1:length(p)-1
    h = p(i + 1) - p(i); % length of the current interval
    M(i, i)         = M(i, i)         + h/3;
    M(i, i + 1)     = M(i, i + 1)     + h/6;
    M(i + 1, i)     = M(i + 1, i)     + h/6;
    M(i + 1, i + 1) = M(i + 1, i + 1) + h/3;
end

% Remove unnecessary elements for basis functions not included
if ~phi0
    M = M(2:end, 2:end);
end
if ~phiN
    M = M(1:end-1, 1:end-1);
end
```

D.8.6 *A Matlab code to compute the mass matrix M for a non-uniform mesh*

Since now the mesh is not uniform (the subintervals have different lengths), we compute the mass matrix assembling the local mass matrix computation for each subinterval. To do so, we can easily

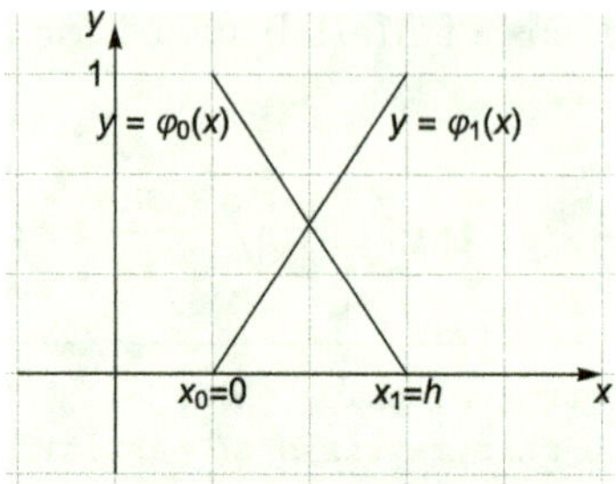

Figure D.1: Standard basis functions $\varphi_0 = (h - x)/h$ and $\varphi_1 = x/h$.

compute the mass matrix for the *standard interval* $I_1 = [0, h]$ with the basis functions $\varphi_0 = (h - x)/h$ and $\varphi_1 = x/h$ (Figure D.1):

Then, the *standard mass matrix* is given by

$$M^{I_1} = \begin{bmatrix} \int_{I_1} \varphi_0\varphi_0 & \int_{I_1} \varphi_0\varphi_1 \\ \int_{I_1} \varphi_1\varphi_0 & \int_{I_1} \varphi_1\varphi_1 \end{bmatrix}.$$

Inserting for $\varphi_0 = (h - x)/h$ and $\varphi_1 = x/h$, we compute M^{I_1} as

$$M^{I_1} = \begin{bmatrix} \int_0^h (h-x)^2/h^2\,dx & \int_0^h (h-x)x/h^2\,dx \\ \int_0^h x(h-x)/h^2\,dx & \int_0^h x^2/h^2\,dx \end{bmatrix} = \frac{h}{6}\begin{bmatrix} 2 & 1 \\ 1 & 2 \end{bmatrix}. \tag{D.2}$$

Thus, for an arbitrary subinterval $I_k := [x_{k-1}, x_k]$ of length h_k, and basis functions φ_k and φ_{k-1} (see Fig. 3.4), the *local mass matrix* is

$$M^{I_k} = \begin{bmatrix} \int_{I_k} \varphi_{k-1}\varphi_{k-1} & \int_{I_k} \varphi_{k-1}\varphi_k \\ \int_{I_k} \varphi_k\varphi_{k-1} & \int_{I_k} \varphi_{k1}\varphi_k \end{bmatrix} = \frac{h_k}{6}\begin{bmatrix} 2 & 1 \\ 1 & 2 \end{bmatrix}. \tag{D.3}$$

Note that, assembling, the diagonal elements in the *Global mass matrix* will be multiplied by 2 (see Example 4.1). These elements correspond to the interior nodes and are the result of adding their contribution for the intervals in their left and right.

D.8.7 *A Matlab routine to compute the load vector* b

To solve the problem of the L_2-projection, it remains to compute/assemble the load vector $\mathbf{b}$. Note that $\mathbf{b}$ depends on the unknown function f, and therefore will be computed by some of numerical integration rules (midpoint, trapezoidal, Simpson, or general quadrature). In the following, we shall introduce Matlab routines for these numerical integration methods.

```
function b = LoadVector(f, p, phi0, phiN)

%------------------------------------------------------------------------
% Syntax:   b = LoadVector(f, p, phi0, phiN)
% Purpose:  To compute load vector b of load f over partition p
%           of an interval
% Data:     f -    right hand side function of one variable
%           p -    vector containing nodes in the partition
%           phi0 - if 1: include basis function at the left endpoint
%                  if 0: do not include a basis function
%           phiN - if 1: include basis function at the right endpoint
%                  if 0: do not include a basis function
%------------------------------------------------------------------------

N = length(p);     % number of rows in b
b = zeros(N, 1);   % initiate the matrix S

% Assemble the load vector (including basis functions at both
   endpoints)
for i = 1:length(p)-1
    h = p(i + 1) - p(i); % length of the current interval
    b(i)     = b(i)     + .5*h*f(p(i));
    b(i + 1) = b(i + 1) + .5*h*f(p(i + 1));
end

% Remove unnecessary elements for basis functions not included
if ~phi0
    b = b(2:end);
end
if ~phiN
    b = b(1:end-1);
end
```

The data function f can be either inserted as `f=@(x)` followed by some expression in the variable `x`, or more systematically through a separate routine, here called "Myfunction" as in the following example:

Example D.26 (Calling a data function $f(x) = x^2$ of the load vector).

```
function y= Myfunction (p)

y=x.^2
```

Then, we assemble the corresponding load vector:

```
b = LoadVector (@Myfunction, p, 1, 1)
```

Or alternatively we may write

```
f=@(x)x.^2
b = LoadVector(f, p, 1, 1)
```

Now we are prepared to write a Matlab routine "My1DL2Projection" for computing the L_2-projection.

D.8.8 *Matlab routine to compute the L_2-projection*

```
function  pf = L2Projection(p, f)

M = MassMatrix(p, 1, 1);        % assemble mass matrix
b = LoadVector(f, p, 1, 1);     % assemble load vector
pf = M\b;                       % solve linear system
plot(p, pf)                     % plot the L2-projection
```

The above routine for assembling the load vector uses the *Composite trapezoidal rule* of numerical integration. In the following, we gather examples of the numerical integration routines:

D.8.9 *A Matlab routine for the composite midpoint rule*

```
function M = midpoint(f,a,b,N)

h=(b-a)/N
x=a+h/2:h:b-h/2;
M=0;
for i=1:N
```

```
  M = M + f(x(i));
end
M=h*M;
```

D.8.10 *A Matlab routine for the composite trapezoidal rule*

```
function T=trapezoid(f,a,b,N)

h=(b-a)/N;
x=a:h:b;

T = f(a);
for k=2:N
    T = T + 2*f(x(k));
end
T = T + f(b);
T = T * h/2;
```

D.8.11 *A Matlab routine for the composite Simpson's rule*

```
function S = simpson(a,b,N,f)

h=(b-a)/(2*N);
x = a:h:b;
p = 0;
q = 0;

for i = 2:2:2*N      % Define the terms to be multiplied
                       by 4
    p = p + f(x(i));
end

for i = 3:2:2*N-1    % Define the terms to be multiplied
                      by 2
    q = q + f(x(i));
end

S = (h/3)*(f(a) + 2*q + 4*p + f(b));  % Calculate final
 output
```

The precomputations for standard and local stiffness and convection matrices:

$$S^{I_1} = \begin{bmatrix} \int_{I_1} \varphi_0'\varphi_0' & \int_{I_1} \varphi_0'\varphi_1' \\ \int_{I_1} \varphi_1'\varphi_0' & \int_{I_1} \varphi_1{}'\varphi_1' \end{bmatrix} = \begin{bmatrix} \int_{I_1} \frac{-1}{h}\frac{-1}{h} & \int_{I_1} \frac{-1}{h}\frac{1}{h} \\ \int_{I_1} \frac{1}{h}\frac{-1}{h} & \int_{I_1} \frac{1}{h}\frac{1}{h} \end{bmatrix} = \frac{1}{h}\begin{bmatrix} 1 & -1 \\ -1 & 1 \end{bmatrix}.$$

As in the assembling of the mass-matrix, even here, for the global stiffness matrix, each interior node has contributions from both intervals that the node belongs to. Consequently, assembling we have $2/h$ as the interior diagonal elements in the stiffness matrix (rather than $1/h$ in the single interval computed above). For the convection matrix C, however, because of the skew-symmetry the contributions from the *two adjacent interior intervals* will cancel out. Hence,

$$C^{I_1} = \begin{bmatrix} \int_{I_1} \varphi_0'\varphi_0 & \int_{I_1} \varphi_0\varphi_1' \\ \int_{I_1} \varphi_1\varphi_0' & \int_{I_1} \varphi_1{}'\varphi_1 \end{bmatrix} = \begin{bmatrix} \int_{I_1} \frac{-1}{h}\frac{h-x}{h} & \int_{I_1} \frac{h-x}{h}\frac{1}{h} \\ \int_{I_1} \frac{x}{h}\frac{-1}{h} & \int_{I_1} \frac{x}{h}\frac{1}{h} \end{bmatrix}$$

$$= \frac{1}{2}\begin{bmatrix} -1 & 1 \\ -1 & 1 \end{bmatrix}.$$

A thorough computation of all matrix elements, for both interior and boundary nodes, in the case of continuous piecewise linear approximation, for Mass-, stiffness- and convection matrices, are demonstrated in the text.

D.8.12 *A Matlab routine assembling the stiffness matrix*

```
function S = StiffnessMatrix(p, phi0, phiN)

%--------------------------------------------------------------------
% Syntax:   S = StiffnessMatrix(p, phi0, phiN)
% Purpose:  To compute the stiffness matrix S of a partition p of an
%           interval
% Data:     p -    vector containing nodes in the partition
%           phi0 - if 1: include basis function at the left endpoint
%                  if 0: do not include a basis function
%           phiN - if 1: include basis function at the right endpoint
%                  if 0: do not include a basis function
%--------------------------------------------------------------------

N = length(p);     % number of rows and columns in S
```

```
S = zeros(N, N);  % initiate the matrix S

% Assemble the full matrix (including basis functions at endpoints)
for i = 1:length(p)-1
    h = p(i + 1) - p(i); % length of the current interval
    S(i, i)         = S(i, i)         + 1/h;
    S(i, i + 1)     = S(i, i + 1)     - 1/h;
    S(i + 1, i)     = S(i + 1, i)     - 1/h;
    S(i + 1, i + 1) = S(i + 1, i + 1) + 1/h;
end

% Remove unnecessary elements for basis functions not included
if ~phi0
    S = S(2:end, 2:end);
end
if ~phiN
    S = S(1:end-1, 1:end-1);
end
```

D.8.13 *A Matlab routine to assemble the convection matrix*

```
function C = ConvectionMatrix(p, phi0, phiN)

%--------------------------------------------------------------------------
% Syntax:   C = ConvectionMatrix(p, phi0, phiN)
% Purpose:  To compute the convection matrix C of a partition p of an
%           interval
% Data:     p -    vector containing nodes in the partition
%           phi0 - if 1: include a basis function at the left endpoint
%                  if 0: do not include a basis function
%           phiN - if 1: include a basis function at the right endpoint
%                  if 0: do not include a basis function
%--------------------------------------------------------------------------

N = length(p);    % number of rows and columns in C
C = zeros(N, N);  % initiate the matrix C

% Assemble the full matrix (including basis functions at both endpoints)
for i = 1:length(p)-1
    C(i, i)         = C(i, i)         - 1/2;
    C(i, i + 1)     = C(i, i + 1)     + 1/2;
    C(i + 1, i)     = C(i + 1, i)     - 1/2;
    C(i + 1, i + 1) = C(i + 1, i + 1) + 1/2;
end

% Remove unnecessary elementC for basis functions not included
if ~phi0
    C = C(2:end, 2:end);
```

```
end
if ~phiN
    C = C(1:end-1, 1:end-1);
end
```

Finally, in the following we gather the Matlab routines for finite difference approximations (also cG(1) and dG(0)) for the time discretizations.

D.8.14 *Matlab routines for Forward-, Backward-Euler and Crank–Nicolson*

```
function [] = three_methods(u0, T, dt, a, f,  exactexists, u)

% Solves the equation du/dt + a(t)*u = f(t)
% u0: initial value;  T: final time;  dt: time step size
% exactexists = 1 <=> exact solution is known
% exactexists = 0 <=> exact solution is unknown

timevector = [0];      % we build up a vector of
                       % the discrete time levels

U_explicit_E = [u0];   % vector which will contain the
                       % solution obtained using ''Forward Euler''

U_implicit_E = [u0];   % vector which will contain the
                       % solution with ''Backward Euler''

U_CN = [u0];           % vector which will contain the
                       % solution using ''Crank-Nicolson''

n = 1;                 % current time interval

t_l = 0;               % left end point of the current
                       % time interval, i.e. t_{n-1}

while t_l < T

  t_r = n*dt;          % right end point of the current
                       % time interval, i.e. t_{n}

  % Forward Euler:
  U_v = U_explicit_E(n);                 % U_v = U_{n-1}
  U_h = (1-dt*a(t_l))*U_v+dt*f(t_l);     % U_h = U_{n};
```

```
  U_explicit_E(n+1) = U_h;%
  % Backward Euler:
  U_v = U_implicit_E(n);                         % U_v = U_{n-1}
  U_h = (U_v + dt*f(t_r))/(1 + dt*a(t_r));    % U_h = U_{n}
  U_implicit_E(n+1) = U_h;

  % Crank-Nicolson:
  U_v = U_CN(n);     % U_v = U_{n-1}
  U_h = ((1 - dt/2*a(t_l))*U_v + dt/2*(f(t_l)+f(t_r))) ...
         / (1 + dt/2*a(t_r));    % U_h = U_{n}
  U_CN(n+1) = U_h;


  timevector(n+1) = t_r;
  t_l = t_r;  % right end-point in the current time interval
              % becomes the left end-point in the next time interval.

  n = n + 1;

end

% plot (real part (in case the solutions become complex))

figure(1)

plot(timevector, real(U_explicit_E), ':')
hold on
plot(timevector, real(U_implicit_E), '--')
plot(timevector, real(U_CN), '-.')

if (exactexists)
   % if known,  plot also the exact solution
  u_exact = u(timevector);
  plot(timevector, real(u_exact), 'g')
end

xlabel('t')
legend('Explicit Euler', 'Implicit Euler', 'Crank-Nicolson', 0)
hold off


if (exactexists)

  % if the exact solution is known, then plot the error:
  figure(2)%
```

```
  plot(timevector, real(u_exact - U_explicit_E), ':')
  hold on
  plot(timevector, real(u_exact - U_implicit_E), '--')
  plot(timevector, real(u_exact - U_CN), '-.')
  legend('Explicit Euler', 'Implicit Euler', 'Crank-Nicolson', 0)
  title('Error')
  xlabel('t')
  hold off

end


return
```

Example D.27. Solving $u'(t) + u(t) = 0$ with `three methods`

```
a= @(t) 1;
f= @(t) 0;
u= @(t) exp(-t)
u_0=1;
T= 1;
dt=0.01;
three_methods (u_0, T, dt, a, f, 1, u)
```

D.8.15 *A Matlab routine for mass-matrix in 2D*

```
function M=MassMatrix2D(p,t,h);

n = size(p,2);      % Number of nodes. (=number of columns in p)
ntri = size(t,2);   % Number of triangles. (= number of columns in t)
M = zeros(n,n);     % Initiate the mass matrix.

for el 0 1:ntri

  nodes = t(1:3,el);
  coords = p(:,nodes);
  Me = ElementmMassMatrix2D(h);
  M(nodes,nodes) = M(nodes,nodes) + Me;

end

% subroutines -----------------------------------------------------
```

```
function Me = ElementmMassMatrix2D(h)
Me = zeros(3,3);
% Complete Me, the element mass-matrix.

Me = 0.5*h^2*(ones(3,3) + eye(3,3))/12;
```

D.8.16 *A Matlab routine for a Poisson assembler in 2D*

```
function [S, M, R, v, r] = PoissonAssembler2D(p,e,t,h);

n = size(p,2);      % Number of nodes. (=number of columns in p)
ntri = size(t,2);   % Number of triangles. (= number of columns in t)

S = zeros(n,n);     % Initiate the Stiffness-matrix.
M = zeros(n,n);     % Initiate the Mass-matrix.
R = zeros(n,n);     % Initiate the Boundary-matrix.
v = zeros(n,1);     % Initiate the Load-vector.
r = zeros(n,1);     % Initiate the Boundary-vector.

for el 0 1:ntri

  nodes = t(1:3,el);
  coords = p(:,nodes);

  Me = ElementMassmatrix(h);
  Se = ElementStiffnessmatrix(h);
  ve = ElementLoadvector(coords,h);

  M(nodes,nodes) = M(nodes,nodes) + Me;
  S(nodes,nodes) = S(nodes,nodes) + Se;
  v(nodes) = v(nodes) + ve;

end


% The contribution from the boundary, OBS : DO NOT CHANGE!
%

for be1 = 1:size(e,2)

    nodes = e(1:2,be1);
    coords = p(:,nodes);
    g_N = 0.0;
    g_D = 0.0;
    gamma = 1e5;
    sidelength = norm(coords(:,1)-coords(:,2));
```

```
    phi = [.5 .5];
    R(nodes,nodes) = R(nodes,nodes) + gamma*phi'*phi*sidelength;
    r(nodes) = r(nodes) + (gamma*g_D - g_N)*phi'*sidelength;

end

% subroutines -----------------------------------------------


function Me = ElementMassmatrix(h)
%
% Complete with the correct values of the element mass-matrix Me.
%
Me = [ 0.0 , 0.0 , 0.0;
       0.0 , 0.0 , 0.0;
       0.0 , 0.0 , 0.0 ];
Me = 0.5*h^2*(ones(3,3)+eye(3,3))*1/12;

function Se = ElementStiffnessmatrix(h)
%
% Complete with the correct values of the element stiffness-matrix Se.
%
Se = [ 0.0 , 0.0 , 0.0;
       0.0 , 0.0 , 0.0;
       0.0 , 0.0 , 0.0 ];
Se = 1/2*[1.0  -1.0  0.0;
          -1.0  2.0 -1.0;
          0.0  -1.0  1.0];
%
function ve = ElementLoadvector(coords,h)
%
% Use quadrature to compute the element load-vector ve.
%
triangelarea = h^2/2;
%
x = coords(1,:);
y = coords(2,:);

ve = [ f(x(1),y(1)) ; f(x(2),y(2)) ; f(x(3),y(3)) ] * trianglearea/3;

% Load f. (An example of load function)

function load = f(x,y)
load = y^2*sin(7*x);
```

Part III
Tables

Appendix E

Tables

E.1 Some Mathematical Constants

Constant	Notation	Numerical value	Exact value
e	e	2.7182818284590452354	$\lim_{n\to\infty}(1+1/n)^n$
Euler	γ	0.57721566490153286061	$\lim_{n\to\infty}\left(\sum_{k=1}^{n}\frac{1}{k}-\ln n\right)$
Glaisher		1.28242712910062263 69	
Golden ratio		1.61803398874989484 82	$\frac{1+\sqrt{5}}{2}$
Catalan		0.91596559417721901505	$\sum_{k=0}^{\infty}(-1)^k\frac{1}{(2k+1)^2}$
Khinchin		2.68545200106530644 53	$\prod_{k=1}^{\infty}\left(1+\frac{1}{k(k+2)}\right)^{\log_2 k}$
pi	π	3.14159265358979323 85	$4\sum_{k=0}^{\infty}\frac{(-1)^k}{2k+1}$

(E.1)

E.2 Table of the CDF of N(0, 1)

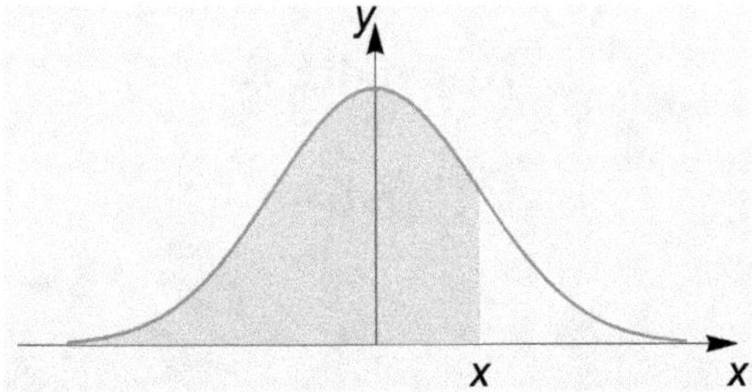

$$\Phi(-x) = 1 - \Phi(x) \text{ where } \Phi(x) = \frac{1}{\sqrt{2\pi}} \int_{-\infty}^{x} e^{-t^2/2} dt$$

x	0.00	0.01	0.02	0.03	0.04	0.05	0.06	0.07	0.08	0.09
0.0	0.5	0.504	0.508	0.512	0.516	0.5199	0.5239	0.5279	0.5319	0.5359
0.1	0.5398	0.5438	0.5478	0.5517	0.5557	0.5596	0.5636	0.5675	0.5714	0.5753
0.2	0.5793	0.5832	0.5871	0.591	0.5948	0.5987	0.6026	0.6064	0.6103	0.6141
0.3	0.6179	0.6217	0.6255	0.6293	0.6331	0.6368	0.6406	0.6443	0.648	0.6517
0.4	0.6554	0.6591	0.6628	0.6664	0.67	0.6736	0.6772	0.6808	0.6844	0.6879
0.5	0.6915	0.695	0.6985	0.7019	0.7054	0.7088	0.7123	0.7157	0.719	0.7224
0.6	0.7257	0.7291	0.7324	0.7357	0.7389	0.7422	0.7454	0.7486	0.7517	0.7549
0.7	0.758	0.7611	0.7642	0.7673	0.7704	0.7734	0.7764	0.7794	0.7823	0.7852
0.8	0.7881	0.791	0.7939	0.7967	0.7995	0.8023	0.8051	0.8078	0.8106	0.8133
0.9	0.8159	0.8186	0.8212	0.8238	0.8264	0.8289	0.8315	0.834	0.8365	0.8389
1.0	0.8413	0.8438	0.8461	0.8485	0.8508	0.8531	0.8554	0.8577	0.8599	0.8621
1.1	0.8643	0.8665	0.8686	0.8708	0.8729	0.8749	0.877	0.879	0.881	0.883
1.2	0.8849	0.8869	0.8888	0.8907	0.8925	0.8944	0.8962	0.898	0.8997	0.9015
1.3	0.9032	0.9049	0.9066	0.9082	0.9099	0.9115	0.9131	0.9147	0.9162	0.9177
1.4	0.9192	0.9207	0.9222	0.9236	0.9251	0.9265	0.9279	0.9292	0.9306	0.9319
1.5	0.9332	0.9345	0.9357	0.937	0.9382	0.9394	0.9406	0.9418	0.9429	0.9441
1.6	0.9452	0.9463	0.9474	0.9484	0.9495	0.9505	0.9515	0.9525	0.9535	0.9545
1.7	0.9554	0.9564	0.9573	0.9582	0.9591	0.9599	0.9608	0.9616	0.9625	0.9633
1.8	0.9641	0.9649	0.9656	0.9664	0.9671	0.9678	0.9686	0.9693	0.9699	0.9706
1.9	0.9713	0.9719	0.9726	0.9732	0.9738	0.9744	0.975	0.9756	0.9761	0.9767
2.0	0.9772	0.9778	0.9783	0.9788	0.9793	0.9798	0.9803	0.9808	0.9812	0.9817
2.1	0.9821	0.9826	0.983	0.9834	0.9838	0.9842	0.9846	0.985	0.9854	0.9857
2.2	0.9861	0.9864	0.9868	0.9871	0.9875	0.9878	0.9881	0.9884	0.9887	0.989
2.3	0.9893	0.9896	0.9898	0.9901	0.9904	0.9906	0.9909	0.9911	0.9913	0.9916
2.4	0.9918	0.992	0.9922	0.9925	0.9927	0.9929	0.9931	0.9932	0.9934	0.9936
2.5	0.9938	0.994	0.9941	0.9943	0.9945	0.9946	0.9948	0.9949	0.9951	0.9952
2.6	0.9953	0.9955	0.9956	0.9957	0.9959	0.996	0.9961	0.9962	0.9963	0.9964
2.7	0.9965	0.9966	0.9967	0.9968	0.9969	0.997	0.9971	0.9972	0.9973	0.9974
2.8	0.9974	0.9975	0.9976	0.9977	0.9977	0.9978	0.9979	0.9979	0.998	0.9981
2.9	0.9981	0.9982	0.9982	0.9983	0.9984	0.9984	0.9985	0.9985	0.9986	0.9986

E.3 Table of Some Quantiles of t-Distribution

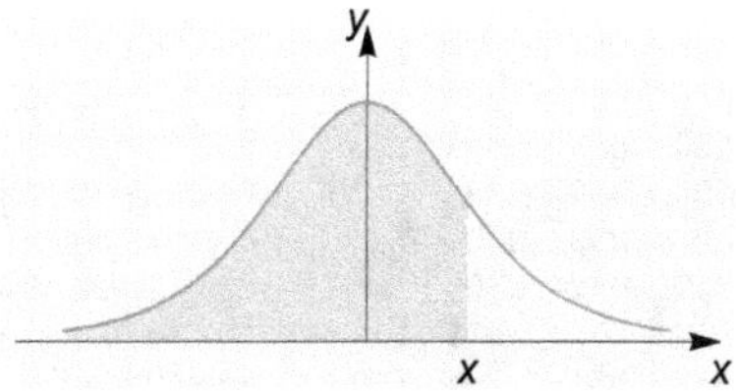

PDF for t-distribution with $n = 5$ degrees of freedom and with the quantile $t_{0.025} = 2.571$ is drawn.

The numbers in the interior of the table are quantiles t_α for $n = 3, 4, \ldots, 99$. For higher n, one uses the $N(0,1)$–table.

$F(x) = 1 - \alpha$ / n	0.750	0.800	0.850	0.900	0.925	0.950	0.975	0.990	0.995	0.999	0.9995
3	0.7649	0.9785	1.25	1.638	1.924	2.353	3.182	4.541	5.841	10.21	12.92
4	0.7407	0.941	1.19	1.533	1.778	2.132	2.776	3.747	4.604	7.173	8.61
5	0.7267	0.9195	1.156	1.476	1.699	2.015	2.571	3.365	4.032	5.893	6.869
6	0.7176	0.9057	1.134	1.44	1.65	1.943	2.447	3.143	3.707	5.208	5.959
7	0.7111	0.896	1.119	1.415	1.617	1.895	2.365	2.998	3.499	4.785	5.408
8	0.7064	0.8889	1.108	1.397	1.592	1.86	2.306	2.896	3.355	4.501	5.041
9	0.7027	0.8834	1.1	1.383	1.574	1.833	2.262	2.821	3.25	4.297	4.781
10	0.6998	0.8791	1.093	1.372	1.559	1.812	2.228	2.764	3.169	4.144	4.587
11	0.6974	0.8755	1.088	1.363	1.548	1.796	2.201	2.718	3.106	4.025	4.437
12	0.6955	0.8726	1.083	1.356	1.538	1.782	2.179	2.681	3.055	3.93	4.318
13	0.6938	0.8702	1.079	1.35	1.53	1.771	2.16	2.65	3.012	3.852	4.221
14	0.6924	0.8681	1.076	1.345	1.523	1.761	2.145	2.624	2.977	3.787	4.14
15	0.6912	0.8662	1.074	1.341	1.517	1.753	2.131	2.602	2.947	3.733	4.073
16	0.6901	0.8647	1.071	1.337	1.512	1.746	2.12	2.583	2.921	3.686	4.015
17	0.6892	0.8633	1.069	1.333	1.508	1.74	2.11	2.567	2.898	3.646	3.965
18	0.6884	0.862	1.067	1.33	1.504	1.734	2.101	2.552	2.878	3.61	3.922
19	0.6876	0.861	1.066	1.328	1.5	1.729	2.093	2.539	2.861	3.579	3.883
20	0.687	0.86	1.064	1.325	1.497	1.725	2.086	2.528	2.845	3.552	3.850
21	0.6864	0.8591	1.063	1.323	1.494	1.721	2.08	2.518	2.831	3.527	3.819
22	0.6858	0.8583	1.061	1.321	1.492	1.717	2.074	2.508	2.819	3.505	3.792
23	0.6853	0.8575	1.06	1.319	1.489	1.714	2.069	2.5	2.807	3.485	3.768
24	0.6848	0.8569	1.059	1.318	1.487	1.711	2.064	2.492	2.797	3.467	3.745
25	0.6844	0.8562	1.058	1.316	1.485	1.708	2.06	2.485	2.787	3.45	3.725
26	0.684	0.8557	1.058	1.315	1.483	1.706	2.056	2.479	2.779	3.435	3.707
27	0.6837	0.8551	1.057	1.314	1.482	1.703	2.052	2.473	2.771	3.421	3.69
28	0.6834	0.8546	1.056	1.313	1.48	1.701	2.048	2.467	2.763	3.408	3.674
29	0.683	0.8542	1.055	1.311	1.479	1.699	2.045	2.462	2.756	3.396	3.659
30	0.6828	0.8538	1.055	1.31	1.477	1.697	2.042	2.457	2.75	3.385	3.646

$F(x)=1-\alpha$ n	0.750	0.800	0.850	0.900	0.925	0.950	0.975	0.990	0.995	0.999	0.9995
31	0.6825	0.8534	1.054	1.309	1.476	1.696	2.04	2.453	2.744	3.375	3.633
32	0.6822	0.853	1.054	1.309	1.475	1.694	2.037	2.449	2.738	3.365	3.622
33	0.682	0.8526	1.053	1.308	1.474	1.692	2.035	2.445	2.733	3.356	3.611
34	0.6818	0.8523	1.052	1.307	1.473	1.691	2.032	2.441	2.728	3.348	3.601
35	0.6816	0.852	1.052	1.306	1.472	1.69	2.03	2.438	2.724	3.34	3.591
36	0.6814	0.8517	1.052	1.306	1.471	1.688	2.028	2.434	2.719	3.333	3.582
37	0.6812	0.8514	1.051	1.305	1.47	1.687	2.026	2.431	2.715	3.326	3.574
38	0.681	0.8512	1.051	1.304	1.469	1.686	2.024	2.429	2.712	3.319	3.566
39	0.6808	0.8509	1.05	1.304	1.468	1.685	2.023	2.426	2.708	3.313	3.558
40	0.6807	0.8507	1.05	1.303	1.468	1.684	2.021	2.423	2.704	3.307	3.551
44	0.6801	0.8499	1.049	1.301	1.465	1.68	2.015	2.414	2.692	3.286	3.526
49	0.6795	0.849	1.048	1.299	1.462	1.677	2.01	2.405	2.68	3.265	3.5
59	0.6787	0.8478	1.046	1.296	1.459	1.671	2.001	2.391	2.662	3.234	3.463
69	0.6781	0.8469	1.044	1.294	1.456	1.667	1.995	2.382	2.649	3.213	3.437
79	0.6776	0.8462	1.043	1.292	1.454	1.664	1.99	2.374	2.64	3.197	3.418
89	0.6773	0.8457	1.043	1.291	1.452	1.662	1.987	2.369	2.632	3.184	3.403
99	0.677	0.8453	1.042	1.29	1.451	1.66	1.984	2.365	2.626	3.175	3.392

E.4 Table of the χ^2-Distribution

The area to the left of $x = \chi^2_\alpha(n)$ is $F(x) = 1 - \alpha$, where $F(x)$ is the CDF of $\chi^2(n)$–distribution. Table for $F(x) = P(\xi \leq x) = 1 - \alpha$ where $F(x)$ is the CDF of $\chi^2(1)$–, that is $n = 1$, the number of degrees of freedom.

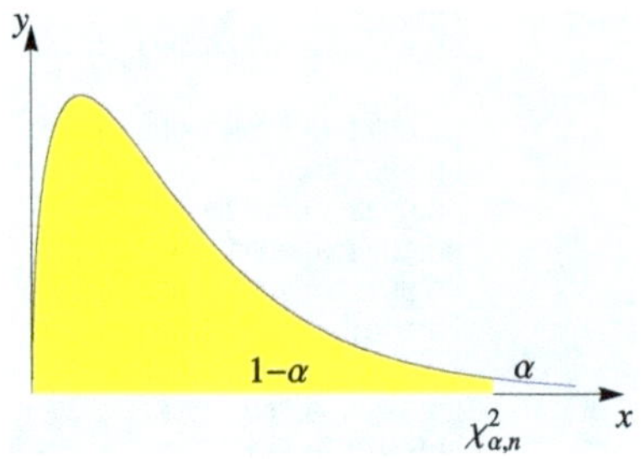

$F(x) = 1 - \alpha n = 1$ ↓	0.0005	0.0010	0.005	0.010	0.025
x	$3.93 \cdot 10^{-7}$	$1.57 \cdot 10^{-6}$	$3.93 \cdot 10^{-7}$	$1.57 \cdot 10^{-4}$	$9.82 \cdot 10^{-4}$
$F(x) = 1 - \alpha$	0.05	0.10	0.20	0.25	0.50
x	0.00393	0.0158	0.064	0.102	0.46

The following two tables contain $x = \chi^2_{1-\alpha}(n)$ for positive integers $n = 2, 3, \ldots$. The numbers inside the table are x-values; $x = \chi^2_\alpha(n)$ for corresponding n.

$F(x) = 1 - \alpha$ / n	0.0005	0.0010	0.005	0.010	0.025	0.05	0.10	0.20	0.25
2	0.00100	0.00200	0.0100	0.0201	0.0506	0.103	0.211	0.446	0.575
3	0.0153	0.0243	0.0717	0.115	0.216	0.352	0.584	1.01	1.21
4	0.0639	0.0908	0.207	0.297	0.484	0.711	1.06	1.65	1.92
5	0.158	0.210	0.412	0.554	0.831	1.15	1.61	2.34	2.67
6	0.299	0.381	0.676	0.872	1.24	1.64	2.20	3.07	3.45
7	0.485	0.598	0.989	1.24	1.69	2.17	2.83	3.82	4.25
8	0.710	0.857	1.34	1.65	2.18	2.73	3.49	4.59	5.07
9	0.972	1.15	1.73	2.09	2.70	3.33	4.17	5.38	5.90
10	1.26	1.48	2.16	2.56	3.25	3.94	4.87	6.18	6.74
11	1.59	1.83	2.60	3.05	3.82	4.57	5.58	6.99	7.58
12	1.93	2.21	3.07	3.57	4.40	5.23	6.30	7.81	8.44
13	2.31	2.62	3.57	4.11	5.01	5.89	7.04	8.63	9.30
14	2.70	3.04	4.07	4.66	5.63	6.57	7.79	9.47	10.2
15	3.11	3.48	4.60	5.23	6.26	7.26	8.55	10.3	11.0
16	3.54	3.94	5.14	5.81	6.91	7.96	9.31	11.2	11.9
17	3.98	4.42	5.70	6.41	7.56	8.67	10.1	12.0	12.8
18	4.44	4.90	6.26	7.01	8.23	9.39	10.9	12.9	13.7
19	4.91	5.41	6.84	7.63	8.91	10.1	11.7	13.7	14.6
20	5.40	5.92	7.43	8.26	9.59	10.9	12.4	14.6	15.5
21	5.90	6.45	8.03	8.90	10.3	11.6	13.2	15.4	16.3
22	6.40	6.98	8.64	9.54	11.0	12.3	14.0	16.3	17.2
23	6.92	7.53	9.26	10.2	11.7	13.1	14.8	17.2	18.1
24	7.45	8.08	9.89	10.9	12.4	13.8	15.7	18.1	19.0
25	7.99	8.65	10.5	11.5	13.1	14.6	16.5	18.9	19.9
26	8.54	9.22	11.2	12.2	13.8	15.4	17.3	19.8	20.8
27	9.09	9.80	11.8	12.9	14.6	16.2	18.1	20.7	21.7
28	9.66	10.4	12.5	13.6	15.3	16.9	18.9	21.6	22.7
29	10.2	11.0	13.1	14.3	16.0	17.7	19.8	22.5	23.6
30	10.8	11.6	13.8	15.0	16.8	18.5	20.6	23.4	24.5
31	11.4	12.2	14.5	15.7	17.5	19.3	21.4	24.3	25.4
32	12.0	12.8	15.1	16.4	18.3	20.1	22.3	25.1	26.3
33	12.6	13.4	15.8	17.1	19.0	20.9	23.1	26.0	27.2
34	13.2	14.1	16.5	17.8	19.8	21.7	24.0	26.9	28.1
35	13.8	14.7	17.2	18.5	20.6	22.5	24.8	27.8	29.1
36	14.4	15.3	17.9	19.2	21.3	23.3	25.6	28.7	30.0
37	15.0	16.0	18.6	20.0	22.1	24.1	26.5	29.6	30.9
38	15.6	16.6	19.3	20.7	22.9	24.9	27.3	30.5	31.8
39	16.3	17.3	20.0	21.4	23.7	25.7	28.2	31.4	32.7
40	16.9	17.9	20.7	22.2	24.4	26.5	29.1	32.3	33.7
45	20.1	21.3	24.3	25.9	28.4	30.6	33.4	36.9	38.3
50	23.5	24.7	28.0	29.7	32.4	34.8	37.7	41.4	42.9
60	30.3	31.7	35.5	37.5	40.5	43.2	46.5	50.6	52.3
70	37.5	39.0	43.3	45.4	48.8	51.7	55.3	59.9	61.7
80	44.8	46.5	51.2	53.5	57.2	60.4	64.3	69.2	71.1
90	52.3	54.2	59.2	61.8	65.6	69.1	73.3	78.6	80.6
100	59.9	61.9	67.3	70.1	74.2	77.9	82.4	87.9	90.1

$F(x) = 1 - \alpha$ / n	0.500	0.750	0.800	0.900	0.925	0.950	0.975	0.995	0.999	0.9995
1	0.455	1.32	1.64	2.71	3.17	3.84	5.02	7.88	10.8	12.1
2	1.39	2.77	3.22	4.61	5.18	5.99	7.38	10.6	13.8	15.2
3	2.37	4.11	4.64	6.25	6.90	7.81	9.35	12.8	16.3	17.7
4	3.36	5.39	5.99	7.78	8.50	9.49	11.1	14.9	18.5	20.0
5	4.35	6.63	7.29	9.24	10.0	11.1	12.8	16.7	20.5	22.1
6	5.35	7.84	8.56	10.6	11.5	12.6	14.4	18.5	22.5	24.1
7	6.35	9.04	9.80	12.0	12.9	14.1	16.0	20.3	24.3	26.0
8	7.34	10.2	11.0	13.4	14.3	15.5	17.5	22.0	26.1	27.9
9	8.34	11.4	12.2	14.7	15.6	16.9	19.0	23.6	27.9	29.7
10	9.34	12.5	13.4	16.0	17.0	18.3	20.5	25.2	29.6	31.4
11	10.3	13.7	14.6	17.3	18.3	19.7	21.9	26.8	31.3	33.1
12	11.3	14.8	15.8	18.5	19.6	21.0	23.3	28.3	32.9	34.8
13	12.3	16.0	17.0	19.8	20.9	22.4	24.7	29.8	34.5	36.5
14	13.3	17.1	18.2	21.1	22.2	23.7	26.1	31.3	36.1	38.1
15	14.3	18.2	19.3	22.3	23.5	25.0	27.5	32.8	37.7	39.7
16	15.3	19.4	20.5	23.5	24.7	26.3	28.8	34.3	39.3	41.3
17	16.3	20.5	21.6	24.8	26.0	27.6	30.2	35.7	40.8	42.9
18	17.3	21.6	22.8	26.0	27.2	28.9	31.5	37.2	42.3	44.4
19	18.3	22.7	23.9	27.2	28.5	30.1	32.9	38.6	43.8	46.0
20	19.3	23.8	25.0	28.4	29.7	31.4	34.2	40.0	45.3	47.5
21	20.3	24.9	26.2	29.6	30.9	32.7	35.5	41.4	46.8	49.0
22	21.3	26.0	27.3	30.8	32.1	33.9	36.8	42.8	48.3	50.5
23	22.3	27.1	28.4	32.0	33.4	35.2	38.1	44.2	49.7	52.0
24	23.3	28.2	29.6	33.2	34.6	36.4	39.4	45.6	51.2	53.5
25	24.3	29.3	30.7	34.4	35.8	37.7	40.6	46.9	52.6	54.9
26	25.3	30.4	31.8	35.6	37.0	38.9	41.9	48.3	54.1	56.4
27	26.3	31.5	32.9	36.7	38.2	40.1	43.2	49.6	55.5	57.9
28	27.3	32.6	34.0	37.9	39.4	41.3	44.5	51.0	56.9	59.3
29	28.3	33.7	35.1	39.1	40.6	42.6	45.7	52.3	58.3	60.7
30	29.3	34.8	36.3	40.3	41.8	43.8	47.0	53.7	59.7	62.2
31	30.3	35.9	37.4	41.4	42.9	45.0	48.2	55.0	61.1	63.6
32	31.3	37.0	38.5	42.6	44.1	46.2	49.5	56.3	62.5	65.0
33	32.3	38.1	39.6	43.7	45.3	47.4	50.7	57.6	63.9	66.4
34	33.3	39.1	40.7	44.9	46.5	48.6	52.0	59.0	65.2	67.8
35	34.3	40.2	41.8	46.1	47.7	49.8	53.2	60.3	66.6	69.2
36	35.3	41.3	42.9	47.2	48.8	51.0	54.4	61.6	68.0	70.6
37	36.3	42.4	44.0	48.4	50.0	52.2	55.7	62.9	69.3	72.0
38	37.3	43.5	45.1	49.5	51.2	53.4	56.9	64.2	70.7	73.4
39	38.3	44.5	46.2	50.7	52.3	54.6	58.1	65.5	72.1	74.7
40	39.3	45.6	47.3	51.8	53.5	55.8	59.3	66.8	73.4	76.1
45	44.34	50.98	52.73	57.51	59.29	61.66	65.41	73.17	80.08	82.88
50	49.33	56.33	58.16	63.17	65.03	67.50	71.42	79.49	86.66	89.56
60	59.33	66.98	68.97	74.40	76.41	79.08	83.30	91.95	99.61	102.7
70	69.33	77.58	79.71	85.53	87.68	90.53	95.02	104.2	112.3	115.6
80	79.33	88.13	90.41	96.58	98.86	101.9	106.6	116.3	124.8	128.3
90	89.33	98.65	101.1	107.6	110.0	113.1	118.1	128.3	137.2	140.8
100	99.33	109.1	111.7	118.5	121.0	124.3	129.6	140.2	149.4	153.2

E.5 F-Table

The quantiles $F_{n_1,n_2;0.05} = x$ for which the CDF, $F(n_1, n_2; x) = 0.95$, and $n_1 = 1, 2, \ldots, 10$ and, $n_2 = 1, 2, \ldots, 20, 29, 39, 49$.

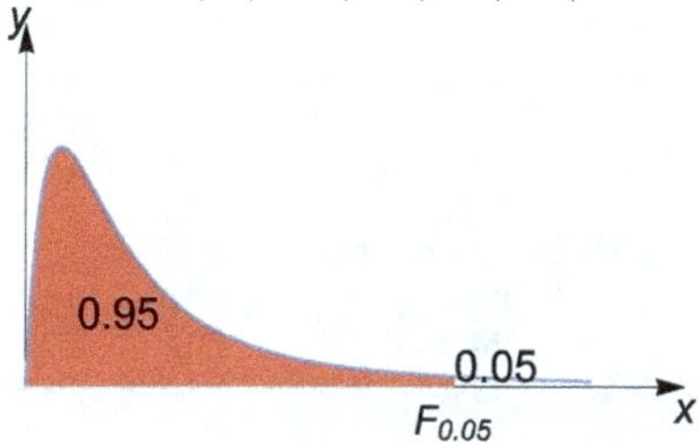

A PDF $f(n_1, n_2; x)$ for a F–ratio distribution and its 95%–quantile
$x = F_{0.05} = F_{n_1,n_2;0.05}$.

For instance, $x = F_{1,2;0.05} = 18.51$, due to the following table.

$n_1 \rightarrow$ $n_2 \downarrow$	1	2	3	4	5	6	7	8	9	10
1	161.45	199.50	215.71	224.58	230.16	233.99	236.77	238.88	240.54	241.88
2	18.51	19.00	19.16	19.25	19.30	19.33	19.35	19.37	19.38	19.40
3	10.13	9.55	9.28	9.12	9.01	8.94	8.89	8.85	8.81	8.79
4	7.71	6.94	6.59	6.39	6.26	6.16	6.09	6.04	6.00	5.96
5	6.61	5.79	5.41	5.19	5.05	4.95	4.88	4.82	4.77	4.74
6	5.99	5.14	4.76	4.53	4.39	4.28	4.21	4.15	4.10	4.06
7	5.59	4.74	4.35	4.12	3.97	3.87	3.79	3.73	3.68	3.64
8	5.32	4.46	4.07	3.84	3.69	3.58	3.50	3.44	3.39	3.35
9	5.12	4.26	3.86	3.63	3.48	3.37	3.29	3.23	3.18	3.14
10	4.96	4.10	3.71	3.48	3.33	3.22	3.14	3.07	3.02	2.98
11	4.84	3.98	3.59	3.36	3.20	3.09	3.01	2.95	2.90	2.85
12	4.75	3.89	3.49	3.26	3.11	3.00	2.91	2.85	2.80	2.75
13	4.67	3.81	3.41	3.18	3.03	2.92	2.83	2.77	2.71	2.67
14	4.60	3.74	3.34	3.11	2.96	2.85	2.76	2.70	2.65	2.60
15	4.54	3.68	3.29	3.06	2.90	2.79	2.71	2.64	2.59	2.54
16	4.49	3.63	3.24	3.01	2.85	2.74	2.66	2.59	2.54	2.49
17	4.45	3.59	3.20	2.96	2.81	2.70	2.61	2.55	2.49	2.45
18	4.41	3.55	3.16	2.93	2.77	2.66	2.58	2.51	2.46	2.41
19	4.38	3.52	3.13	2.90	2.74	2.63	2.54	2.48	2.42	2.38
20	4.35	3.49	3.10	2.87	2.71	2.60	2.51	2.45	2.39	2.35
29	4.18	3.33	2.93	2.7	2.55	2.43	2.35	2.28	2.22	2.18
39	4.09	3.24	2.85	2.61	2.46	2.34	2.26	2.19	2.13	2.08
49	4.04	3.19	2.79	2.56	2.29	2.2	2.13	2.08	2.08	2.03

$n_1 \rightarrow$ n_2 $\downarrow$	11	12	13	14	15	16	17	18	19	20
1	242.98	243.91	244.69	245.36	245.95	246.46	246.92	247.32	247.69	248.01
2	19.4	19.41	19.42	19.42	19.43	19.43	19.44	19.44	19.44	19.45
3	8.76	8.74	8.73	8.71	8.7	8.69	8.68	8.67	8.67	8.66
4	5.94	5.91	5.89	5.87	5.86	5.84	5.83	5.82	5.81	5.8
5	4.7	4.68	4.66	4.64	4.62	4.6	4.59	4.58	4.57	4.56
6	4.03	4.0	3.98	3.96	3.94	3.92	3.91	3.9	3.88	3.87
7	3.6	3.57	3.55	3.53	3.51	3.49	3.48	3.47	3.46	3.44
8	3.31	3.28	3.26	3.24	3.22	3.2	3.19	3.17	3.16	3.15
9	3.1	3.07	3.05	3.03	3.01	2.99	2.97	2.96	2.95	2.94
10	2.94	2.91	2.89	2.86	2.85	2.83	2.81	2.8	2.79	2.77
11	2.82	2.79	2.76	2.74	2.72	2.7	2.69	2.67	2.66	2.65
12	2.72	2.69	2.66	2.64	2.62	2.6	2.58	2.57	2.56	2.54
13	2.63	2.6	2.58	2.55	2.53	2.51	2.5	2.48	2.47	2.46
14	2.57	2.53	2.51	2.48	2.46	2.44	2.43	2.41	2.4	2.39
15	2.51	2.48	2.45	2.42	2.4	2.38	2.37	2.35	2.34	2.33
16	2.46	2.42	2.4	2.37	2.35	2.33	2.32	2.3	2.29	2.28
17	2.41	2.38	2.35	2.33	2.31	2.29	2.27	2.26	2.24	2.23
18	2.37	2.34	2.31	2.29	2.27	2.25	2.23	2.22	2.2	2.19
19	2.34	2.31	2.28	2.26	2.23	2.21	2.2	2.18	2.17	2.16
20	2.31	2.28	2.25	2.22	2.20	2.18	2.17	2.15	2.14	2.12
29	2.14	2.1	2.08	2.05	2.03	2.01	1.99	1.97	1.96	1.94
39	2.04	2.01	1.98	1.95	1.93	1.91	1.89	1.88	1.86	1.85
49	1.99	1.96	1.93	1.9	1.88	1.85	1.84	1.82	1.80	1.79

Appendix F

Key Concepts

F.1 Symbols

The most common mathematical symbols

(i) Binary operators

$$+,\ -,\ \cdot,\ \div,\ \oplus,\ \otimes,\ \times$$

(ii) Bounds

$$\max,\ \min,\ \sup,\ \inf$$

(iii) Cardinalities

$$0,\ 1,\ 2\ \ldots,\ \aleph_0,\ c,\ 2^c$$

(iv) Differentiation symbols

$$\frac{d}{dx},\ D,\ f',\ \frac{\partial}{\partial x},\ \nabla,\ \Delta$$

(v) Equalities and similarities

$$=,\ \equiv,\ \approx,\ \sim,\ \simeq$$

(vi) Function symbols

$\blacksquare^a$, $|\blacksquare|$, $e^{\blacksquare}$, exp, $a^{\blacksquare}$, ln, lg
$D_{\blacksquare}$, $R_{\blacksquare}$, (Domain and range)
sin, cos, tan, cot, sec, csc

$$\begin{cases}\text{arcsin}\\ \text{arccos}\end{cases} \quad \begin{cases}\text{arcsec}\\ \text{arccsc}\end{cases} \quad \begin{cases}\text{arctan}\\ \text{arccot}\end{cases}$$

sinh, cosh, tanh, coth, sech, arccsc

$$\begin{cases}\text{arcsinh}\\ \text{arccosh}\end{cases} \quad \begin{cases}\text{arcsech}\\ \text{arccsch}\end{cases} \quad \begin{cases}\text{arctanh}\\ \text{arccoth}\end{cases}$$

(vii) Geometric symbols

$$\perp,\ \|,\ \angle$$

(viii) Inequalities

$$\neq,\ /\ ,\ \leq,\ \geq,\ <,\ >$$

(ix) Integrals and sums

$$\int,\ \int_a^b,\ \iint_D,\ \iiint,\ \oint,\ \sum,\ \sum_{k=m}^{n}$$

(x) Limits

$$\lim_{x\to a} y = b, \quad x \to a \Rightarrow y \to b,$$

$$\limsup_{x\to a}, \quad \liminf_{x\to a}$$

(xi) Logical symbols (Boolean algebra)

$$\forall,\ \exists,\ \wedge,\ \vee,\ \Leftrightarrow,\ \Longleftarrow,\ \Rightarrow$$

(xii) Number spaces

$$\mathbb{N},\ \mathbb{Z},\ \mathbb{Q},\ \mathbb{R},\ \mathbb{C},\ \blacksquare_+,\ \blacksquare^n : n \in \mathbb{Z}_+$$

(xiii) Set theoretic symbols

$$\{\cdot\},\ \subset,\ \subseteq,\ \supset,\ \supseteq,\ ^c,\ \complement,\ \setminus,\ \Delta$$

F.2 General Notation

(i) $v(\cdot)$, $v(\cdot,\cdot)$, etc...: function v of one variable, two variables, etc...

(ii) $v(\cdot, b)$: partial mapping $x \to v(x, b)$.

(iii) $\text{supp } v = \overline{\{x \in X; v(x) \neq 0\}}$: support of a function v.

(iv) $\text{osc}(v; A) = \sup_{x,y\in A} |v(x) - v(y)|$.

(v) v_A or $v_{|A}$: restriction of a function v to a set A.

(vi) $P(A) = \{v_{|A}; \forall v \in P\}$, where P is an arbitrary function space defined over a region containing the set A.

(vii) $\text{tr } v$, or simply v: trace of the function v.

(viii) $R(v) = \frac{a(u,v)}{(u,v)}$: Rayleigh quotient.

(ix) $C(a)$, $C(a,b)$, etc...: Arbitrary constants depending on only a, only a, b, etc...

(x) $\overset{\circ}{A}$: Interior of the set A.

(xi) ∂A: Boundary of a set A.

(xii) $\bar{A}$: Closure of a set A.

(xiii) card A: number (cardinality) of elements in a set A.

(xiv) diam A: diameter of a set A.

(xv) $\mathcal{C}_A$, or $\mathcal{C}_X A$, or $X \setminus A$: complement of the subset A of a set X.

(xvi) $\Longrightarrow$: implies.

F.3 Derivatives and Differential Calculus

$Dv(a)$, or $v'(a)$: first (Frechet) derivative of a function v, at a point a.

$D^2v(a)$, or $v''(a)$: second (Frechet) derivative of a function v, at a point a.

$D^kv(a)$: kth (Frechet) derivative of a function v, at a point a.

$$D^kv(a)h^k = D^kv(a)(h_1, h_2, \cdots, h_k),$$

if $h_1 = h_2 = \cdots = h_k = h$.
$\mathcal{R}_k(v; b, a) = v(b) - \{v(a) + Dv(a)(b - a) + \cdots + 2\frac{1}{k!}D^kv(a)(b-a)^k\}$

$$\left.\begin{aligned} \partial_i v(A) &= Dv(a)\boldsymbol{e}_i, \\ \partial_{ij} v(a) &= D^2v(a)(\boldsymbol{e}_i, \boldsymbol{e}_j) \\ \partial_{ijk} v(a) &= D^3v(a)(\boldsymbol{e}_i, \boldsymbol{e}_j, \boldsymbol{e}_k,) \end{aligned}\right\},$$

used also for vector-valued functions.

$J_F(\hat{x}) = \det(\partial_j F_i(\hat{x}))$ = Jacobian of a mapping.

$$F : \hat{x} \in \mathbb{R}^n \to F(\hat{x}) = \Big(F_i(x)\Big)_{i=1}^n \in \mathbb{R}^n.$$

$\text{div } \boldsymbol{v} = \sum_{i=1}^n \partial_i v$.
$\nabla v(a) = (\partial_i v)_{i=1}^n$, denoted also as grad $v(a)$.
$\Delta v = \sum_{i=1}^n \partial_{ii} v$
$\Delta \boldsymbol{v} = (\Delta v_i)_{i=1}^n$.
$|\alpha| = \sum_{i=1}^n \alpha_i$, for multi-index $\alpha = (\alpha_1, \ldots, \alpha_n) \in \mathbb{N}^n$.

$$D^\alpha v(a) = D^{|\alpha|}v(a)(e_1, \ldots, e_1, e_2, \ldots, e_2, \ldots, e_n, \ldots, e_n),$$

where in each chain of $e_k, \ldots, e_k$:s, $k = 1, \ldots, n$, i.e. each e_k is repeated α_k-times.
$\boldsymbol{\nu} = (\nu_1, \nu_2, \ldots, \nu_n)$: outward unit normal vector.
$\partial_\nu = \sum_{i=1}^n \nu_i \partial_i$: (outward) normal derivative operator.

$\boldsymbol{\tau} = (\tau_1, \tau_2)$: unit tangent vector along boundary of a plane region.
$\partial_{\boldsymbol{\tau}} v(a) = Dv(a)\boldsymbol{\tau} = \sum_{i=1}^2 \tau_i \partial_i v(a)$.
$\partial_{\boldsymbol{\nu},\boldsymbol{\tau}} v(a) = D^2v(a)(\boldsymbol{\nu}, \boldsymbol{\tau}) = \sum_{i,j=1}^2 \boldsymbol{\nu}_i \tau_j \partial_{ij} v(a)$.
$\partial_{\tau,\boldsymbol{\tau}} v(a) = D^2v(a)(\tau, \boldsymbol{\tau}) = \sum_{i,j=1}^2 \tau_i \tau_j \partial_{ij} v(a)$.

F.4 Differential Geometry

(i) $(a_{\alpha\beta})$: First fundamental form of a surface.
(ii) $a = \det(a_{\alpha\beta})$.
(iii) $(b_{\alpha\beta})$: Second fundamental form of a surface.
(iv) $(c_{\alpha\beta})$: Third fundamental form of a surface.
(v) $(\Gamma^{\alpha}_{\beta\gamma})$: Christoffel symbols.
(vi) $v_{|\beta}$, $v_{|\alpha\beta}, \ldots$: the covariant derivatives along a surface.
(vii) $ds = \sqrt{a}\, d\xi$: surface element.
(viii) $\frac{1}{R}$: Curvature of a plane surface.

F.4.1 *General notations for a vector space*

$B(a; r) = \{x \in X; ||x - a|| < r\}$.
$\mathcal{L}(X; Y)$: Space of the continuous linear mappings from X to Y.
X': dual of the space X.
$\langle \cdot\, , \cdot \rangle$: duality pairing between a space and its dual.
$x + Y = \{x + y; y \in Y\}$.
$X + Y = \{x + y; x \in X,\, y \in Y\}$.
$X \oplus Y = \{x + y; x \in X,\, y \in Y\}$, if
$X \cap Y = \{0\}$.
X/Y: quotient of X w.r.t. Y.
$\boldsymbol{V}_{e_\lambda, \lambda \in \Lambda}$: vector space spanned by the vectors e_λ, $\lambda \in \Lambda$.

I: identity operator.
$\hookrightarrow$: inclusion by continuous injection.
$\subset_c$: inclusion by compact injection.
$\dim X$: dimension of a space X.
$\ker A = \{x \in X; Ax = 0\}$.

F.4.2 *Notation of special vector spaces*

Below Ω denotes an open connected subset of $\mathbb{R}^n$.

Inner product in $L_2(\Omega)$: $(u, v) = \int_\Omega uv\, dx$.
Inner product in $(L_2(\Omega))^n$): $(\boldsymbol{u}, \boldsymbol{v}) = \int_\Omega \boldsymbol{u} \cdot \boldsymbol{v}\, dx$.

$\mathcal{C}^m(\Omega)$: m-times continuously differentiable functions in Ω.

$\mathcal{C}^\infty(\Omega)$, the space of infinitely differentiable functions $f : \Omega \to \mathbb{R}$. This space can be expressed as $\mathcal{C}^\infty(\Omega) = \bigcap_{m=0}^{\infty} \mathcal{C}^m(\Omega)$.

$$\mathcal{C}^{m,\alpha}(\Omega) = \{v \in C^m(\bar{\Omega}); \forall\, \beta,\, |\beta| = m,\, \exists\, \Gamma_\beta, \forall x, y \in \Omega : |\partial^\beta v(x) - \partial^\beta v(y)| \leq \Gamma_\beta ||x - y||^\alpha\},$$

with norm $||v||_{\mathcal{C}^{m,\alpha}(\Omega)} = \max_{|\beta|=m} \sup_{x,y\in\Omega,\,(x\neq y)} ||x - y||$.

$\mathcal{D}(\Omega) = \{v \in \mathcal{C}^\infty(\Omega); \text{supp } v \text{ compact.}\}$, $\mathcal{D}'(\Omega)$: space of distributions over Ω.

$$H^m(\Omega) = \{v \in L^2(\Omega); \forall\alpha,\ |\alpha| \le m;\ \partial^\alpha v \in L_2(\Omega)\}.$$

$$H_0^m(\Omega) = \text{ closure of } \mathcal{D}(\Omega) \text{ in } H^m(\Omega).$$

$$||v||_{m,\Omega} = \left(\sum_{|\alpha|\le m} \int_\Omega |\partial^\alpha v|^2\,dx\right)^{1/2}, \quad ||v||_{m,\Omega} = \left(\sum_{|\alpha|=m} \int_\Omega |\partial^\alpha v|^2\,dx\right)^{1/2}.$$

$||\boldsymbol{v}||_{m,\Omega} = \left(\sum_{i=1}^n ||N||(v_i)^2_{m,\Omega}\right)^{1/2}$, (for functions $\boldsymbol{v} = (v_i)_{i=1}^n$, in $(H^m(\Omega))^n$).

$|\boldsymbol{v}|_{m,\Omega} = \left(\sum_{i=1}^n |v_i|^2_{m,\Omega}\right)^{1/2}$, (for functions $\boldsymbol{v} = (v_i)_{i=1}^n$, in $(H^m(\Omega))^n$).

$$\boldsymbol{W}^{m,p}(\Omega) = \{v \in L^p(\Omega); \forall\alpha, |\alpha| \le m, \partial^\alpha v \in L^p(\Omega)\}.$$

$\boldsymbol{W}_0^{m,p}(\Omega)$= closure of $\mathcal{D}(\Omega)$ i $\boldsymbol{W}^{m,p}(\Omega)$.

$$||v||_{m,p,\Omega} = \left(\sum_{|\alpha|\le m} \int_\Omega |\partial^\alpha v|^p\,dx\right)^{1/p}, \quad 1 \le p < \infty.$$

$$||v||_{m,\infty,\Omega} = \max_{|\alpha|\le m}\left\{\text{ ess } \cdot \sup_{x\in\Omega} |\partial^\alpha v(x)|\right\}$$

(denotes also the norm in $C^m(\bar{\Omega})$).

$$||v||^\star_{m,\infty,\Omega} = \text{ norm in the dual space of } \boldsymbol{W}^{m,p}.$$

$$||v||_{m,p,\Omega} = \left(\sum_{|\alpha|=m} \int_\Omega |\partial^\alpha v|^p\,dx\right)^{1/p}, \quad 1 \le p < \infty.$$

$$||v||_{m,\infty,\Omega} = \max_{|\alpha|=m}\left\{\text{ ess } \cdot \sup_{x\in\Omega} |\partial^\alpha v(x)|\right\}.$$

$$||v||_{m,\Omega} = \left(\sum_{i=1}^{n} \int_{\Omega} |D^m v(x)(e_i^m)|^2 \, dx \right)^{1/2}$$

$$||v||_{m,p,\Omega} = \left(\sum_{i=1}^{n} \int_{\Omega} |D^m v(x)(e_i^m)|^p \, dx \right)^{1/p}.$$

$$||v||_{\varphi;m,\Omega} = \left\{ \int_{\Omega} \varphi \sum_{|\beta|=m} |\partial^\beta v|^2 \, dx \right\}^{1/2},$$

$m = 0, 1, \ldots$ (weighted semi-norms).

$||v||_{m,\infty,K} = \sup_{x \in K} ||D^m v(x)||_{\mathcal{L}_m(\mathbb{R}^n;\mathbb{R})}$, for $v : K \subset \mathbb{R}^n \to \mathbb{R}$.

$||F||_{m,\infty,\hat{K}} = \sup_{\hat{x} \in \hat{K}} ||D^m F(\hat{x})||_{\mathcal{L}_m(\mathbb{R}^n;\mathbb{R})}$, for $F : \hat{K} \subset \mathbb{R}^n \to \mathbb{R}^n$.

$H^{1/2}(\Gamma) = \{r \in L^2(\Gamma); \exists v \in H^1(\Omega);\ \mathrm{tr}\ v = r \text{ on } \Gamma\}$ with norm

$||r||_{H^{1/2}(\Gamma)} = \inf ||v||_{1,\Omega};\ v \in H^1(\Omega),\ \mathrm{tr}\ v = r \text{ on } \Gamma\}$,

and with dual space $H^{-1/2}(\Gamma)$.

$\langle \cdot, \cdot \rangle_\Gamma$: duality pairing between the spaces , $H^{-1/2}(\Gamma)$ and $H^{1/2}(\Gamma)$.

$W_0^p(\mathbb{R}^n) =$ completion of $\mathcal{D}(\mathbb{R}^n)$ with respect to the norm $|\cdot|_{p,\mathbb{R}^n}$.

$H(\mathrm{div};\Omega) = \{\boldsymbol{q} \in (L^2(\Omega))^n;\ \mathrm{div}\ \boldsymbol{q} \in L^2(\Omega)\}$ with norm

$||\boldsymbol{q}||_{H(\mathrm{div};\Omega)} = |\boldsymbol{q}|_{0,\Omega}^2 + |\ \mathrm{div}\ \boldsymbol{q}|_{0,\Omega}^2$.

F.5 Generalized Functions

Notations

$\int_K \varphi(x)\,dx \sim \sum_{l=1}^{L} \omega_l \varphi(b_l)$: quadrature rule with weights ω_l and nodes b_l.

$\hat{E}(\hat{\varphi}) = \int_{\hat{K}} \hat{\varphi}(\hat{x})\,d\hat{x} - \sum_{l=1}^{L} \hat{\omega}_l \hat{\varphi}(\hat{b}_l)$: quadrature function on $\hat{K}$.

$E_K(\varphi) = \int_K \varphi(x)\,dx - \sum_{l=1}^{L} \omega_{l,K} \varphi(b_{l,K})$: quadrature error functional on
$K = F_K(\hat{K})$, with $\omega_{l,K} = \hat{\omega}_l J_{F_K}((\hat{b}_l)$, $b_{l,K} = F_K(\hat{b}_l)$.

F.5.1 *Finite element related concepts*

Notations

$\mathcal{T}_h$: *triangulation* of a set $\bar{\Omega}$.
X_h: finite element space with no boundary data.
$X_{0,h} = \{v \in X_h; v_h = 0 \text{ on } \Gamma := \partial\Omega\}$.
$X_{00,h} = \{v \in X_h; v_h = \partial_\nu v = 0 \text{ på } \Gamma := \partial\Omega\}$.
V_h: finite element space with boundary data.
Σ_h= the set of degrees of freedom of the finite element space X_h.
φ_h or φ_{kh}, $1 \le k \le M$: degrees of freedom of X_h.
$(w_k)_{k=1}^M$: basis functions in a finite element space X_h or V_h.
$\mathcal{N}_h$: the set of nodes in a finite element space X_h.
$\Pi_h v$: X_h-interpolant of a function v.
dom $\Pi_h = \mathcal{C}^s(\bar{\Omega})$, $s = \max_{K\in\mathcal{T}_h} s_K$.
$H(\text{div}\,,\Omega) := \{v \in L_2(\Omega)^d; \text{div}\, v \in L_2(\Omega)\}$, $\Omega \in \mathbb{R}^d$.
$\kappa(A)$ spectral condition number for the matrix A.

$\sigma(A)$ spectrum of the matrix A.
$\rho(A)$ spectral radius of the matrix A.
$x'y$ Euclidean scalar product of vectors x och y.
$||x||_A = \sqrt{x'Ax}$ (the energy norm).
$||x||_\infty = \max_i |x_i|$ (the maximum norm).
$H^s(\Omega)^d := [H^s(\Omega)]^d$
$H^1_\Gamma(\Omega) := \{v \in H^1(\Omega), v(x) = 0,\ x \in \Gamma := \partial\Omega\}$.
$H(\text{div}\,,\Omega) := = \{\tau \in L_2(\Omega);\ \text{div}\,\tau \in L_2(\Omega)\}$.
$H(\text{ rot }\Omega) := \{\eta \in L_2(\Omega)^2; \text{rot }(\eta) \in L_2(\Omega)\}$, $\Omega \subset \mathbb{R}^2$.
$H^{-1}(\text{div}\,,\Omega) := \{\tau \in H^{-1}(\Omega)^d; \text{div}\,\tau \in H^{-1}(\Omega)\}$, $\Omega \subset \mathbb{R}^d$.

F.6 Filter

Notations

A *discrete signal* is a double-sequence $X := \{x_k\}_{k=-\infty}^{\infty}$ (or $X := \{x(k)\}_{k=-\infty}^{\infty}$):

$$X = (\ldots, x_{-2}, x_{-1}, x_0, x_1, x_2 \ldots), \quad x_k \in \mathbb{R}, \text{ (OR } \mathbb{C}).$$

X has *bounded energy* if $x \in \ell^2$ (i.e., $\sum_{k=-\infty}^{\infty} |x_k|^2 < \infty$).
A *Filter* is an operator $H : X \mapsto Y$ ($Y = HX$ is a signal).
H is linear if $H(\alpha X + \beta Y) = \alpha HX + \beta HY \quad \alpha, \beta$ scalars.
H is *time invariant* if

$$H(SX) = SH(X), \quad (Sx)_k = x_{k-1}, \quad \forall X$$

$$\delta = \{\delta_k\}_{k=-\infty}^{\infty} \qquad \delta_k = \begin{cases} 1, & k = 0, \\ 0, & \text{else.} \end{cases}$$

$$X = (\ldots, x_{-2}, x_{-1}, x_0, x_1, x_2 \ldots) = \sum_{n=-\infty}^{\infty} x_n S^n \delta$$

$h = H\delta$ is called *impulse response* of the filter H.

$$H(S^n\delta) = S^n(H\delta) = S^n h$$

$$Y = HX = H\left(\sum_{n=-\infty}^{\infty} x_n S^n \delta\right) = \sum_{n=-\infty}^{\infty} x_n S^n h = \sum_{n=-\infty}^{\infty} x_n h_{\cdot - x}$$

Discrete convolution: $Y = \{y_k\}_{k=-\infty}^{\infty} \Longrightarrow y_k = \sum_{n=-\infty}^{\infty} x_n h_{k-x} = h * x.$

A *bounded-impulse filter* (FIR) has only finitely many $h_k \neq 0$.

Definition F.1. A linear time-invariant (LTI) filter is *causal*, if

$$h_k = 0 \qquad \text{for} \quad k < 0.$$

Auto-correlation: $X \star Y = \sum_{n=-\infty}^{\infty} x_{n+\cdot} y_n \Longrightarrow (x \star y)_k = \sum_{n=-\infty}^{\infty} x_{n+k} y_n.$

H: A LTI filter with impulse response h, and

$$H(\omega) = |H(\omega)|e^{i\Phi(\omega)}.$$

Then, $|H(\omega)|$ is called the amplitude of $H(\omega)$ and $\Phi(\omega)$ its phase function.

H has linear phase if $\omega \mapsto \Phi(\omega)$ is linear.

H symmetric if $h_k = h_{-k}$.

H is anti-symmetric if $h_k = -h_{-k}$.

The *group delay* of H is $\tau(\omega) = -\dfrac{d\Phi(\omega)}{d\omega}$.

Haar base consists of two family of functions:

$$\varphi_k = \begin{cases} \frac{1}{\sqrt{2}}, & k = 0, 1, \\ 0, & \text{else}, \end{cases} \qquad \psi_k = \begin{cases} \frac{1}{\sqrt{2}}, & k = 0, \\ -\frac{1}{\sqrt{2}}, & k = 1, \\ 0, & \text{else}. \end{cases}$$

$$\left(\varphi^{(2n)}\right)_k := \varphi_{k-2n}, \qquad \left(\varphi^{(2n+1)}\right)_k := \psi_{k-2n}.$$

Coordinates for a sequence $X = (x_k)_{k=-\infty}^{\infty}$ are given by

$$\begin{cases} C_{2n} : & = \langle x, \varphi^{(2n)} \rangle = \frac{1}{\sqrt{2}}(x_{2n} + x_{2n+1}) \quad \text{(mean-value)} \\ C_{2n+1} : & = \langle x, \varphi^{(2n+1)} \rangle = \frac{1}{\sqrt{2}}(x_{2n} - x_{2n+1}) \quad \text{(difference)} \end{cases}$$

$(\varphi^{(2n)})_k$ and $(\varphi^{(2n+1)})_k$ are basis functions in $\ell^2(\frown)$ and therefore

$$x_k = \sum_n C_n \varphi_k^{(n)} = \sum_n C_{2n} \left(\varphi^{(2n)}\right)_k + \sum_n C_{2n+1} \left(\varphi^{(2n+1)}\right)_k.$$

Bibliography

Abramowitz, M. and Stegun, I. *Handbook of Mathematical Functions with Formulas, Graphs and Mathematical Tables.* New York: De vore Publications, Inc. (1965).

Adams, R. A. and Essex, C. *Calculus. A complete Course*, 10th ed. Pearson (2021).

Adams, R. A. *Sobolev Spaces.* New York: Academic Press (1975).

Alonso, M. and Finn, E. J. *University Physics.* Vol. III. Boston: Addison-Wesley (1968).

Apostol, T. M. *Calculus.* Vol. I & II. Second Edition. New York: John Wiley & Sons, Inc. (1967).

Arnold, V. I. *Ordinary Differential Equations*, 2nd ed. (translated from Russian by R. A. Silverma). Cambridge MA and London: MIT Press (1980).

Asadzadeh, M. *Lecture Notes in Fourier Analysis.* (Available through author's web-site). Gothenburg: Chalmers University (2004).

Asadzadeh, M. *Analys och Linjär Algebra*, 2nd ed. Lund: Studentlitteratur (2007).

Asadzadeh, M. *An Introduction to the Finite Element Method for Differential Equations.* New York: Wiley (2020).

Asadzadeh, M. and Holmåker, K. *An Introduction to Fourier Analysis and Applications.* To appear (2004).

Asadzadeh, M. and Emanuelsson, R. *Flervariabelanalys (available upon request).*

Atkinson, K. *An Introduction to Numerical Analysis.* 2nd ed. New York: Wiley (1989).

Aubin, J. K. *Approximation of Elliptic Boundary-Value Problems.* New York: Wiley (1972).

Axler, J. *Linear Algebra Done Right.* 3rd ed. Heidelberg and New York: Springer Cham (2015).

Babuska, I. and Aziz, A. K. Survey lectures on the mathematical foundation of the finite element method. In: *The Mathematical Foundation of the Finite Element Method with Applications to Partial Differential Equations* (ed. A.K. Aziz). New York: Academic Press (1972).

Baker, A. *A concise Introduction to the Theory of Numbers*. London and New York: Cambridge University Press (1984).

Bank, J. and Newman, D. J. *Complex Analysis*. Third Edition. New York and London: Springer (2017).

Beckman, O. *Grundläggande Termodynamik för högskolestudier*. Stockholm: Almqvist-Wiksell (1970).

Bengzon, F. and Larson, M. *The Finite Element Method: Theory, Implementation and Practice*. Berlin, Heidelberg: Springer (2013).

Bergh, J. and Löfström, J. *Interpolation Spaces: An Introduction*. Berlin: Springer-Verlag (1976).

Birkhoff, G. and Rota, G-C. *Ordinary Differential Equations*. 4th Edition. New York, Hoboken NJ: John Wiley & Sons, Inc. (1991).

Brink, I. and Persson, A. *Analytiska Functioner*. Lund: Studentlittratur (1976).

Buffa, A. and Ciarlet, P. Jr. On trace for functional spaces related to Maxwell's equations. I & II. *Math Methods Appl. Sci.* **24** (2001).

Burden, l. R. and Faires, J. D. *Numerical Analysis*, 5th ed. Pacific Grove, CA: Brook/Cole (1998).

Butcher, J. C. *Numerical Methods for Ordinary Differential Equations*, 2nd ed. New York: Wiley (2008).

Cheney, E. W. *Introduction to Approximation Theory*, 2nd ed. Providance, RI: American Mathematical Society (2000).

Choguet, G. *Topology*. New York, London: Academic Press (1996).

Churchill, V. and Brown, J. *Fourier Series*. New York: McGraw-Hill (1985).

Cohn, P. M. *Algebra*, Vol. 1 & 2. New York: John Wiley & Sons (1977).

Davis, H. F. and Snider, A. D. *Introduction to Vector Analysis*. Boston: Allyn & Bacon. Inc. (1975).

Domar, T. *Analys II*. Gleerups. Lund (1971).

Eriksson, F. *Flerdimensionell Analys*. Lund: Studentlitteratur (1976).

Eriksson, T. and Lagerwall, T. *Klassisk Mekanik*. Stockholm: Almqvist-Wiksell (1970).

Evans, L. C. *Partial Differential Equations, Graduate Studies in Mathematics*, Vol. 19. Providance, RI: American Mathematical Society (1998).

Folland, G. B. *Introduction to Partial Differential Equations*. Princeton, New Jersey: Princeton University Press (1976).

Folland, G. B. *Fourier Analysis and its Applications*. Pacific Grove, California: Wadsworth & Cole (1992).

Golub, G. and Loan, C. V. *Matrix Computations*. Baltimore, Maryland: John Hopkins University Press (1983).

Grimmett, G. R. and Strizaker D. R. *Probability and Random Processes*. Oxford: Oxford University Press (1983).

Gustafson, K. E. *Partial Differential Equations and Hilbert Space Methods*. New York: Wiley (1980).

Hein, I. N. *Discrete Structures, Logic*. Sudbury, MA: Jones and Bartlett Publishers International (1994).

Herstein, J. L. *Topics in Algebra*. MIT, Cambridge, MA: Blaisdell Publishing Co. (1964).

Hörmander, L. *Linear Partial Differential Equations*. Fourth Printing. Berlin, Heidelberg, New York: Springer-Verlag (1976).

Hurd, A. E. and Loeb, P. A. *An Introduction to Nonstandard Real Analysis*.

Jänich, K. *Topology*. Berlin, Heidelberg, New York: Springer-Verlag (1980).

John, F. *Partial Differential Equations. Applied Mathematical Sciences*, Vol. 1. New York: Springer (1982).

Johnson, C. *Numerical Solutions of Partial Differential Equations by the Finite Element Method*. Lund: Studentlitteratur (1991).

Krylov, V. I. *Approximate Calculus of Integrals*. New York: Macmillan Press (1962).

Ladyzhenskaya, O. A. *The Boundary Value Problem of Mathematical Physics*. New York: Springer (1985).

Larson, R. and Edwards, B. *Calculus. International Metric Edition*. Boston: Cengage Learning Inc. (2022).

Larsson-Leander, G. *Astronomi och Astrofysik*. Lund: Gleerups (1971).

Larsson, S. and Thomee, V. *Partial Differential Equations with Numerical Methods*. Texts in Applied Mathematics, Vol. 45. Berlin: Springer-Verlag (2003).

Lebovitz, N. *Ordinary Differential Equations*. Pacific Grove, CA: Brooks/Cole (2002).

Lennerstad, H. *Serier och Transformer*. Lund: Studentlittratur (1999).

Mikhlin, G. S. *Variational Methods in Mathematical Physics*. Moscow: MIR (1957).

Moore, H. *MATLAB for Engineers*, 2nd ed. London: Pearson International Edition (2009).

Nagle, R. K. and Saff, E. B. *Differential Equations and Boundary Value Problems*. Boston: Addison Wesley (1993).

Nakos, G. and Joyner, D. *Linear Algebra*. Washington, DC: Thomson Publishing Inc. (1998).

Oden, J. T. and Demkowicz, L. F. *Applied Functional Analysis*. Boca Raton, London, New York: CRC Press (1996).

Ostrowski, A. M. *Solution of Equations ans System of Equations.* Cambridge, MA: Academic Press (1966).

Phillips, E. R. *Introduction to Analysis and Integration Theory.* New York: Dover Publishing, Inc. (1984).

Råde, L. and Westergren, B. *Mathematics Handbook*, 5th ed. Lund: Sdudentlitteratur (2003).

Rice, J. R. *The Approximation of Functions*, Vol. 1 & 2. Boston: Addison-Wesley (1969).

Ringström, U. *Elektronik och Kretslŕa.* Stockholm: Almqvist-Wiksell (1970).

Rudin, W. *Real and Complex Analysis*, 3rd ed. New York: McGraw-Hill (1974).

Rudin, W. *Principles of Mathematical Analysis*, 3rd ed. New York: McGraw-Hill (1976).

Scott, L. R. *Numerical Analysis.* NJ: Princeton University Press (2011).

Sharipo, L. *Introduction to Abstract Algebras.* New York: McGraw-Hill (1975).

Shearer, M. and Levy, R. *Partial Differential Equations: An Introduction to Theory and Applications.* NJ: Princeton University Press (2015).

Simmons, G. F. *Introduction to Topology and Modern Analysis.* International Student Edition, New York: McGraw-Hill (1963).

Spiegel, M. R. *Laplace Transforms.* New York: McGraw-Hill (1965).

Stewart, G. W. *Matrix Algorithms: Basic Decompositions*, Vol. I. Philadelphia, PA: Society of Industrial and Applied Mathematics (1998).

Stewart, G. W. *Matrix Algorithms: Eigenvalue Problems*, Vol. II. Philadelphia, PA: Society of Industrial and Applied Mathematics (2001).

Stewart, I. *Galois Theory.* London: Chapman & Hall (1973).

Strang, G. *Introduction to Applied Mathematics.* Cambridge, MA: Wellesely-Cambridge Press (1986).

Strang, G. *Introduction to Linear Algebra*, 5th ed. Wellesley, MA: Wellesley-Cambridge Press (2022).

Strang, W. *Partial Differential Equations. An Introduction*, 2nd ed. New York: Wiley (2008).

Stroud, A. H. *Approximate Calculation of Multiple Integrals.* Englewood Cliffs, NJ: Prentice-Hall (1971).

Taylor, M. E. *Partial Differential Equations. Basic Theory. Applied Mathematical Sciences*, Vol. 115. New York: Springer-Verlag (1996).

Thomee, V. *Galerkin Finite Element Methods for Parabolic Problems, Lecture Notes in Mathematics,* Vol. 1054. New York: Springer-Verlag (1984).

Verga, R. S. *Matrix Iterative Analysis, Springer Series of Computational Mathematics,* Vol. 27. Berlin: Springer-Verlag (2009).

Wahlbin, L. *Superconvergence in Galerkin Finite Element Methods, Series Lecture Notes in Mathematics,* Vol. 1605. Berlin: Springer-Verlag (1995).

Wilde, I. F. *Lecture Notes on Complex Analysis.* London: Imperial College Press (2006).

Wilkinson, J. H. *The Algebraic Eigenvalue Problem.* Oxford: Oxford University Press (1995).

Wolfram, S. *Mathematica: A System for Doing Mathematics by Computer*, 2nd ed. Boston: Addison-Wesley Publishing Company, Inc. (1991).

Yosida, K. *Functional Analysis.* New York: Springer-Verlag (1996).

Zwillinger, D. *Standard Mathematical Formulae*, 31st ed. Boca Raton, FL: Chapman & Hall CRC (2003).

Supplementary Material

The supplementary material includes full proofs of the theorems within the book.

Online access is automatically assigned if you purchase the ebook online via www.worldscientific.com.

If you have purchased the print copy of this book or the ebook via other sales channels, please follow the instructions below to download the files:

1. Go to: https://www.worldscientific.com/r/q0393-supp or scan the below QR code.

2. Register an account/login.
3. Download the files from: https://www.worldscientific.com/worldscibooks/10.1142/q0393#t=suppl.

For subsequent access, simply log in with the same login details in order to access.

For enquiries, please email: sales@wspc.com.sg.

Index

C

D

E

M

W

X

Y

Z

www.ingramcontent.com/pod-product-compliance
Lightning Source LLC
LaVergne TN
LVHW010322220926
848484LV00004BB/828